Contents

II. The Field Measurements

III. The Possible Utilization of Geothermal Energy

IV. Summarized Results

List of Contributors

Alheid, H.-J., Geophysikalisches Institut der Universität, 4630 Bochum
Althaus, E., Mineralogisches Institut der Universität, 7500 Karlsruhe

Balke, K.-D., Institut und Museum für Geologie und Paläontologie der Universität, 7400 Tübingen
Bartelsen, H., Institut für Geophysik der Universität, 2300 Kiel
Bamford, D., British Petroleum Company, Geophysics Research Division, Exploration and Production, Brittanic House, Moor Lane, London, Great Britain
Behrens, J., Institut für Angewandte Geophysik der Technischen Universität, 1000 Berlin
Berktold, A., Institut für Angewandte Geophysik der Universität, 8000 München
Blohm, E.-K. Niedersächsisches Landesamt für Bodenforschung, 3000 Hannover
Buntebarth, G., Geophysikalisches Institut der Technischen Universität, 3392 Clausthal-Zellerfeld

Dietrich, H.-G., Stadt Urach, 7432 Urach

Einsele, G., Institut und Museum für Geologie und Paläontologie der Unversität, 7400 Tübingen
Emter, D., Geowissenschaftliches Gemeinschaftsobservatorium Heubach 206, 7620 Wolfach
Ernst, W., Institut und Museum für Geologie und Paläontologie der Universität, 7400 Tübingen

Friedrichsen, H., Mineralogisches Institut der Universität, Abteilung Geochemie, 7400 Tübingen
Fritz, J., Ing.-Büro Fritz, 7432 Urach
Fromm, K., Niedersächsisches Landesamt für Bodenforschung, 3000 Hannover

Giese, P., Institut für Geophysik der Freien Universität, 1000 Berlin
Gregarek, R., Institut und Museum für Geologie und Paläontologie der Universität, 7400 Tübingen

Haenel, R., Niedersächsisches Landesamt für Bodenforschung, 3000 Hannover; Priv.-Doz. am Institut f. Angew. Geophysik der Technischen Universität, 1000 Berlin
Hahn, A., Niedersächsisches Landesamt für Bodenforschung, 3000 Hannover

Jentsch, M., Gewerkschaften Brigitta und Elwerath Betriebsführungsgesellschaft mbH, 3000 Hannover

Kemmerle, K., Institut für Angewandte Geophysik der Universität, 8000 München
Koller, B., Mineralogisches Institut der Universität, Abt. Geochemie, 7400 Tübingen; present address: Volksmarsweg 3, 7080 Aalen 8
Koziorowski, G., Institut und Museum für Geologie und Paläontologie der Universität, 7400 Tübingen
Krey, Th., Firma Prakla-Seismos GmbH, 3000 Hannover

Leiber, J., Geologisches Landesamt Baden-Württemberg, 7800 Freiburg

Loeschke, J., Institut und Museum für Geologie und Paläontologie der Universität, 7400 Tübingen
Lueschen, E., Institut für Geophysik der Universität, 2300 Kiel

Maier, U., Institut und Museum für Geologie und Paläontologie der Universität, 7400 Tübingen
Makris, J., Geophysikalisches Institut der Universität, 2000 Hamburg
Mäussnest, O., Institut für Geophysik der Universität, 7000 Stuttgart 1
Meidl, J., Firma Preussag AG, 3000 Hannover
Meissner, R., Institut für Geophysik der Universität, 2300 Kiel
Metzker, S., Institut und Museum für Geologie und Paläontologie der Universität, 7400 Tübingen
Müller, K., Institut für Angewandte Geophysik der Technischen Universität, 1000 Berlin

Plaumann, S., Niedersächsisches Landesamt für Bodenforschung, 3000 Hannover
Pucher, R., Niedersächsisches Landesamt für Bodenforschung 3000 Hannover
Prodehl, C., Geophysikalisches Institut der Universität, 7500 Karlsruhe

Richards, M.-L., Institut für Geophysik der Universität, 3400 Göttingen
Riepe, L., Geophysikalisches Institut der Technischen Universität, 3392 Clausthal-Zellerfeld
Rodemann, H., Niedersächsisches Landesamt für Bodenforschung 3000 Hannover
Roters, B., Institut für Angewandte Geophysik der Technischen Universität, 1000 Berlin
Rummel, F., Geophysikalisches Institut der Universität, 4630 Bochum

Schädel, K., Geologisches Landesamt Baden-Württemberg, 7800 Freiburg
Schmoll, H., Firma Prakla-Seismos GmbH, 3000 Hannover
Schmucker, U., Institut für Geophysik der Universität, 3400 Göttingen
Schneider, G., Institut für Geophysik der Universität, 7000 Stuttgart 1
Schopper, J.-R., Geophysikalisches Institut der Technischen Universität, 3392 Clausthal-Zellerfeld
Schweizer, R., Geologisches Landesamt Baden-Württemberg, 7800 Freiburg
Staffens, H.-J., Institut und Museum für Geologie und Paläontologie der Universität, 7400 Tübingen
Steveling, E., Institut für Geophysik der Universität, 3400 Göttingen
Steinwachs, M., Niedersächsisches Landesamt für Bodenforschung, 3000 Hannover
Stenger, R., Mineralogisches Institut der Universität, 7800 Freiburg
Strayle, G., Geologisches Landesamt Baden-Württemberg, 7800 Freiburg

Teichmüller, M., Geologisches Landesamt Nordrhein-Westfalen, 4150 Krefeld
Tödt, K.-H., Geophysikalisches Institut der Universität, 2000 Hamburg

Villinger, E., Geologisches Landesamt Baden-Württemberg, 7800 Freiburg
Villinger, H., Institut für Angewandte Geophysik der Technischen Universität, 1000 Berlin

Walther, Ch., Institut für Geophysik der Universität, 2300 Kiel
Werner, D., Institut für Geophysik der Universität, ETH, 8093 Zürich, Switzerland
Werner, J., Geologisches Landesamt Baden-Württemberg, 7800 Freiburg
Wimmenauer, W., Mineralogisches Institut der Universität, 7800 Freiburg
Winter, R.-B., Geophysikalisches Institut der Universität, 4630 Bochum
Wohlenberg, J., RWTH Aachen, 5100 Aachen
Wöhrl, Th., Geophysikalisches Institut der Universität, 4630 Bochum

Zoth, G., Niedersächsisches Landesamt für Bodenforschung, 3000 Hannover.

Foreword

World demand for energy is currently increasing by about 5 % every year. Though this rate may diminish somewhat in the future, reserves of fossil fuels are continuously decreasing. The research for new and alternative sources of energy during the last decade has created great interest in geothermal energy. The huge reservoir of geothermal energy is practically inexhaustible, but in most cases its economic extraction presents formidable difficulties. The estimated contribution of geothermal energy to the world energy production is less than 1 %. Every opportunity should be taken to exploit geothermal energy and thus to reduce man's dependence on hydrocarbons and coal.

Certain areas in the Federal Republic of Germany are of interest for exploiting geothermal energy. Naturally, areas of positive geothermal anomaly are the most favourable. It has been known for a long time that subsurface temperatures are higher than normal in the region around Urach in the Swabian Alb. The Urach geothermal anomaly has been the subject of systematic investigations in a research programme supported by a number of institutions.

In addition to extensive geological and geophysical investigations, a borehole was drilled through the sedimentary cover and about 1700 m into the basement. This borehole provided an opportunity for studying hot water extraction from the sedimentary cover and for carrying out experiments on the use of the hot-dry-rock technique in the basement rocks.

The present volume summarizes the results of these investigations and contains a considerable amount of new information. It should be mentioned that the Urach project has encouraged other communities to look into the possibility of using local geothermal energy for heating purposes. The practical experience obtained from this, the first large-scale research project on geothermal energy to be carried out in the Federal Republic of Germany, will provide other communities, and the government too, with valuable basic data for future geothermal studies.

P. GIESE
Chairman of the
Forschungskollegium Physik des Erdkörpers

Preface

The sudden rise of oil prices in 1973 generated increased interest in alternative sources of energy. This interest naturally extended into the field of geothermal energy. A worldwide search for geothermal energy began and research into possibilities of its utilization was started.

In the Federal Republic of Germany, the Forschungs-Kollegium Physik des Erdkörpers (FKPE) with its chairman Prof. Dr. W. Kertz, an association of the heads of the Institutes of Geophysics in the German universities and the geophysical departments of governmental geoscientific research institutes, set up a geothermal working group to conduct a geothermal survey using geophysical methods and to prepare basic data relevant to the utilization of geothermal energy.

Two areas in the Federal Republic of Germany with relatively high subsurface temperatures are suitable as potential test areas: the Upper Rhine Graben and the Urach area in the Swabian Alb. After examination of existing data, the Urach area was chosen as the more suitable test area since it is geologically more homogenous than the Upper Rhine Graben. The figures shows the temperature distribution in the Urach area at the beginning of the project.

In the town of Urach, a small group had been set up, quite independently, to sink a borehole for geothermal water for heating purposes. The Urach group and the FKPE group, having very similar objectives, decided to merge.

The aims of this geothermal working group were as follows:

1. to examine whether surface geophysical methods can yield information about the existence and size of geothermal anomalies, which could otherwise be obtained only by direct measurement in costly deep boreholes; and
2. to determine the horizontal and vertical extent of the Urach geothermal anomaly and, if possible, the nature and location of the heat soure.

Altogether 13 specialist groups were involved in the Urach Geothermal Project: geology, hydrogeology, mineralogy, geochemistry, geothermics, magnetotellurics, magnetics, gravimetry, seismology, refraction seismics, reflection seismics, microseismics and rock mechanics.

A research borehole (9°22'45" E, 48°30'48" N, 426 m above sea level) was sunk to augment our knowledge of the geology and structure of the area and to enable in-situ measurement of rock parameters to be made. Further determinations of rock parameters were made on core samples in the laboratory.

Two other important objectives were:

3. examination of the aquifers and the possibility of utilizing geothermal groundwater for heating and/or medicinal purposes in the existing spa; and
4. examination of the basement with a view of future extraction of geothermal energy utilizing the hot-dry-rock concept.

The project was started in the autumn of 1977 and completed at the end of 1980. The work was coordinated by annual meetings of all the participating groups. The individual groups' results are presented in this volume.

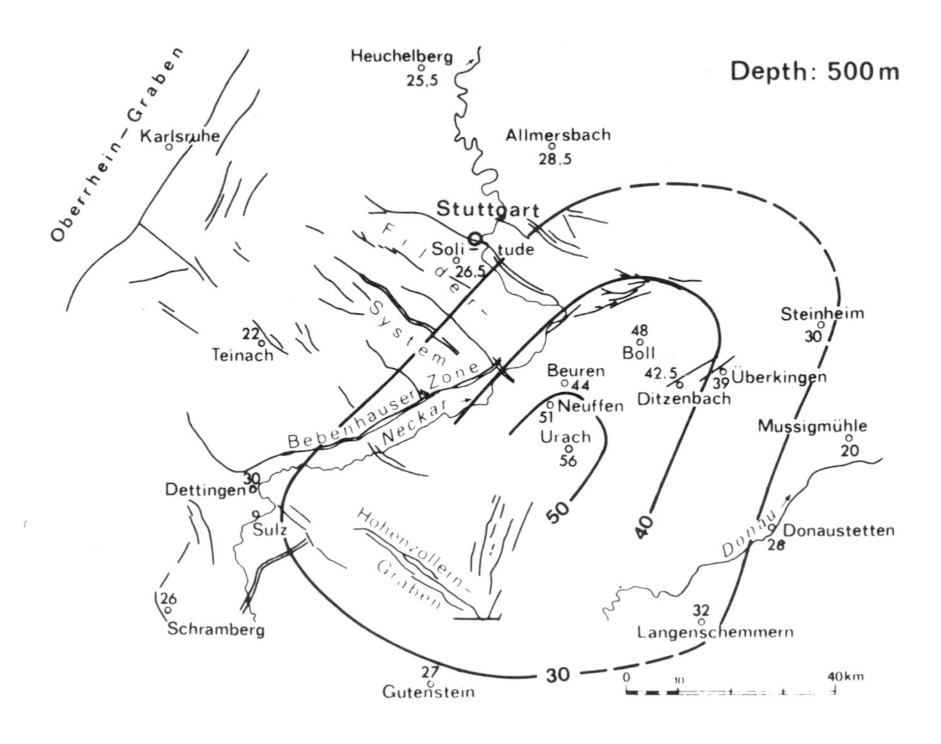
Depth: 500 m
Oberrhein – Graben
Karlsruhe
Heuchelberg
25,5
Allmersbach
28,5
Stuttgart
Soli- tude
26,5
Filder-System
Teinach
22
Bebenhauser Zone
Neckar
Beuren
44
Neuffen
51
Urach
56
48
Boll
42,5
Ditzenbach
39
Überkingen
Steinheim
30
Mussigmühle
20
Dettingen
30
Sulz
Hohenzollern-Graben
Schramberg
26
Donau
Donaustetten
28
Langenschemmern
32
Gutenstein
27
30
40
50
0
10
40 km

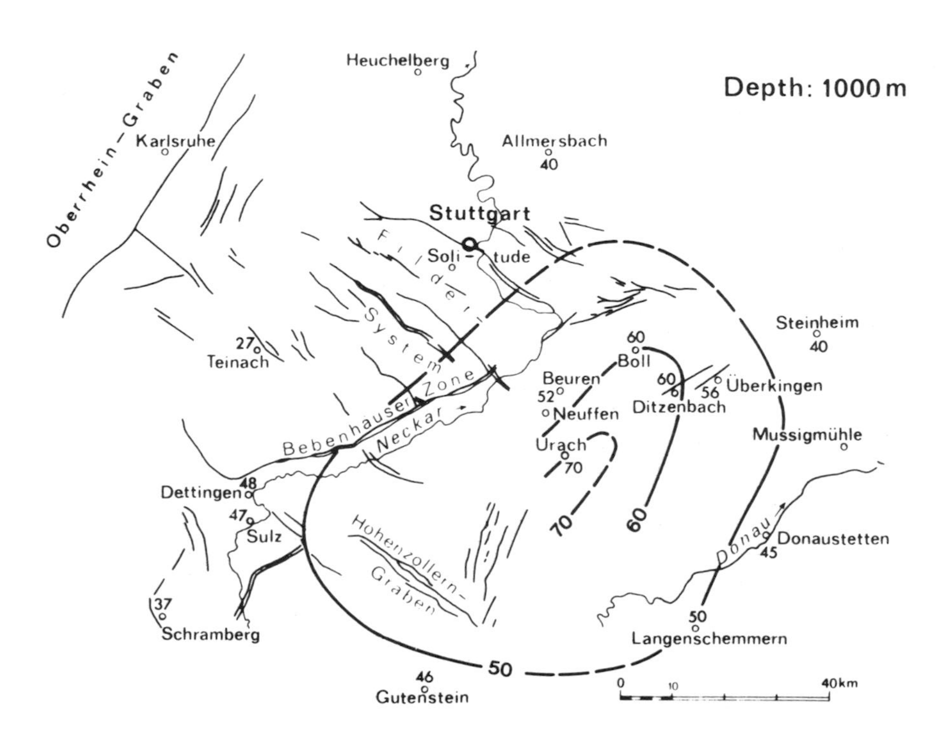
Depth: 1000 m
Oberrhein – Graben
Karlsruhe
Heuchelberg
Allmersbach
40
Stuttgart
Soli- tude
Filder-System
Teinach
27
Bebenhauser Zone
Neckar
Beuren
52
Neuffen
Urach
70
60
Boll
60
Ditzenbach
56
Überkingen
Steinheim
40
Mussigmühle
Dettingen
48
Sulz
47
Hohenzollern-Graben
Schramberg
37
Donau
Donaustetten
45
Langenschemmern
50
Gutenstein
46
50
60
70
0
10
40 km

The Urach Geothermal Project aroused the interest of other local communities and groups and, as a result, feasibility survey on the utilization of geothermal energy have been planned for other areas. The progress of these activities will also be discussed in this volume.

It should be mentioned that the idea of utilizing geothermal energy in this area is not altogether new; it was suggested during the 1950s by MAUZ (1950) and KIDERLEN (1955).

Four research groups on geothermics have been set up by the Institutes of Geophysics at the Universities of Berlin, Braunschweig, Clausthal-Zellerfeld and Karlsruhe.

The extensive research work within the Urach Geothermal Project was made possible by generous financial support from the Ministry of Research and Technology (BMFT), the German Research Foundation (DFG) and the Commission of the European Community (EG). A considerably financial load was borne by the specialist groups and, especially for the research borehole, by the local community in Urach. The geothermal working group received valuable guidance from Prof. MAYER-GÜRR (formerly BEB), Dr. NEUMANN (KFA/BMFT), Dr. MARCHANDISE (EG), Mrs. STAROSTE (EG), and Mr. HEINLE (formerly BEB), to whom we would like to express our thanks. We also owe our gratitude to Mr. PASCHER, the Mayor of Urach, and to the people of Urach for their appreciative interest and helpful support. We are indebted to Dr. ALLARD and Dr. NEWCOMB, who kindly checked the English version of the papers.

R. HAENEL

References

Mauz, J. (1950): Warum unterbleibt die Erdwärme-Nutzung in Württemberg. – Bohrtechnik, Brunnenbau, 1. Jg., H. 8: 243–246.

Kiderlen, H. (1955): Die technische Ausnützung der Erdwärme in Mittelwürttemberg. – Archiv des Geol. Landesamt Baden-Württemberg, Freiburg.

BEB = Gewerkschaften Brigitta und Elwerath Betriebsführungsgesellschaft mbH
KFA = Kernforschungsanlage Jülich

Legend for figures.

Temperature distribution at depths of 500 m and 1000 m below the surface in the Urach area, the so-called Urach anomaly. The figures at the circles and on the isolines are temperatures in °C.

The Urach Geothermal Project, p. 5–6;
Schweizerbart'sche Verlagsbuchhandlung, Stuttgart, 1982

I. THE RESEARCH BOREHOLE

Magnetic Investigation of the Proposed Urach Drilling Site

R. PUCHER

with 1 figure

Abstract: A magnetic survey was made for the area of the proposed drilling site. The existence of a volcanic pipe was thus excluded.

One of many preparatory operations for the drilling of the Urach borehole was to make sure that there was no volcanic pipe or dike at the site of the proposed research borehole. The magnetic method is best suited for detecting basaltic material.

Proton magnetometer measurements of the total intensity of the magnetic field were made and corrected by using the diurnal variations taken from recordings of the earth's magnetic field made with an interval of one minute at the base station.

Measurements were made at the proposed drilling site in a grid of 20 m between measuring points. Nearby, measurements were made along profiles of 200–2000 m length. To estimate the magnetic noise in an area known to contain no volcanic material, a profile was measured near the Urach waterfall. In addition, several profiles were measured over volcanic pipes with Quaternary overburden (south of Wittlingen and southwest of Hengen).

Profiles from areas known to contain volcanic rock and a profile from an area known to contain no volcanic rock were compared with the profiles from the proposed drilling site. Figure 1 gives an example. As a result of the magnetic survey the existence of a volcanic pipe within the area of the proposed drilling site was excluded.

Reference

Pucher, R. (1977): Bodenmagnetische Untersuchung in Urach. – Report, NLfB Hannover, Archive No. 76933.

Fig. 1. Profiles of the connected values of the total intensity Δ T of the earth's magnetic field; the proposed drilling site is marked by a cross-hatched section (see next page).

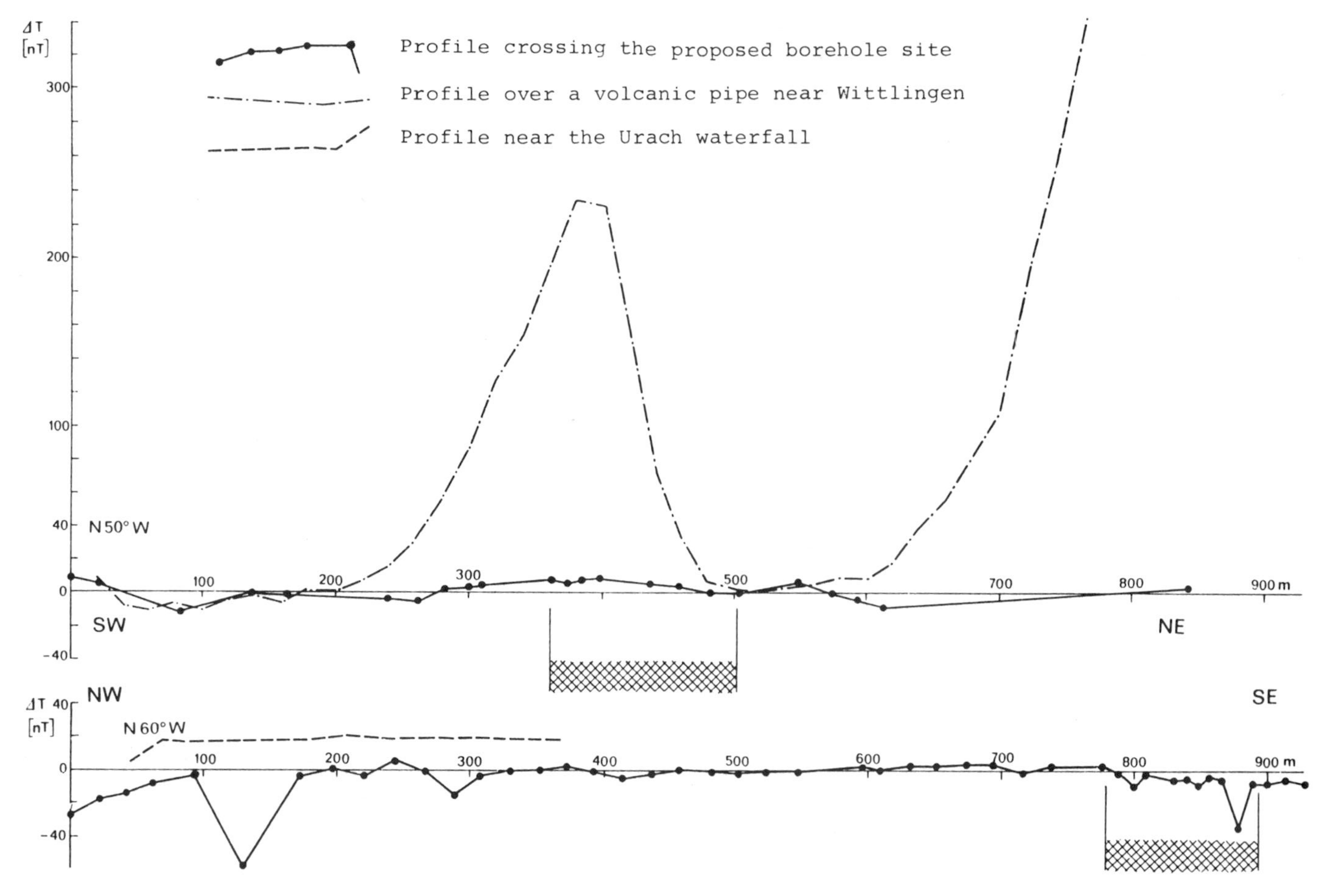

Fig. 1. Profiles of the connected values of the total intensity $\triangle$ T of the earth's magnetic field; the proposed drilling site is marked by a cross-hatched section.

The Urach Geothermal Project, p. 7–35;
Schweizerbart'sche Verlagsbuchhandlung, Stuttgart, 1982

Technical Details of the Geothermal Well Urach 3 Planning and Realization

H.-G. DIETRICH

with 6 figures and 10 tables

Abstract: Until recently the exploration of the geothermal anomaly of Urach (Swabian Alb, SW-Germany, south of Stuttgart) was limited to water-bearing sedimentary horizons, mostly in the "Upper Muschelkalk" of the Middle Triassic. By means of the research borehole Urach (deep well Urach 3) at the centre of this anomaly, exploration was extended to the basement for the first time. Originally the well was planned to a depth of 2100 m or 2500 m but the final depth is 3334 m below surface (2908 m below mean sea level) with a 1732 m penetration of basement. At a drilling diameter of 8 1/2" in this research-well, 7" casing was landed and cemented at a depth of 3320 m. The well head end-flange is suitable for a well head pressure up to 700 bar which is expected for the later frac operations.
The following mean results are: There are no severe problems when drilling through basement for the hot dry rock concept. Average drilling progress is expected to be 1.5–2.5 m/h. Best tools when drilling in the basement were medium-hard insert bits (type 3 JS–5 JS, J 44 and M 88) for drilling and very hard diamond bits (type C 24) for coring. In spite of rock tensions the uncased hole (up to 1500 m!) offers no casing problems. A final casing is necessary because of decomposed zones.
The Urach 3 well has achieved all its planned targets, at least from a technical and geological point of view. It is believed, therefore, that this is an important contribution to a future large-scale technical utilization of geothermal energy stored in the crystalline basement.

1. Introduction

Exploration drilling on the geothermal anomaly of Urach, South Germany, had previously stopped when the warm water-bearing Triassic horizons were reached. CARLE (1975) summarized that within this anomaly (which has been known for 150 years) thermal and mineral water was found in Jurassic, Keuper, Muschelkalk and, partly, also in Bunter. Only with the research project (Urach 3 well) which was planned in 1975 with the target "of investigating the geothermal anomaly of Urach, including a possible economic utilization", was the basement underlying the sedimentary series to be explored as well. A sequence of at least 500 m of crystalline rock down to a final depth of at least 2500 m was planned to be drilled in order to get a satisfactory insight into the petrography of the basement.

With a view toward the possible economic production of hot water two problems had to be taken into consideration:

– Investigation and testing in the sedimentary series of thermal water which could perhaps be used for heating or other purposes.

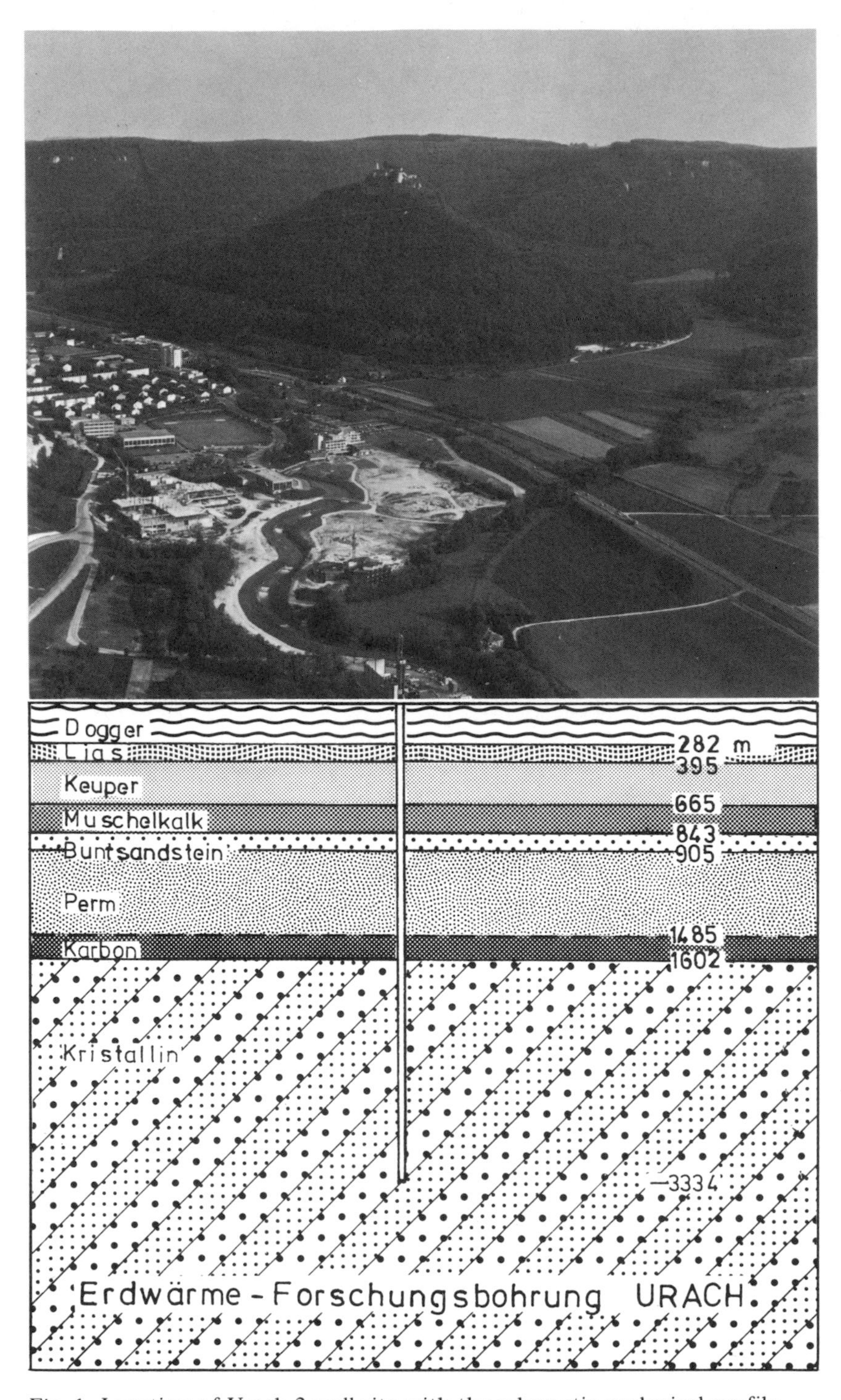

Fig. 1. Location of Urach 3 well site with the schematic geological profile.

– Investigation of the underlying basement in view of a possible application of the hot-dry-rock method.

2. Location

The central part of the anomaly was selected for the drilling location (see HAENEL and SCHÄDEL this volume). This selection took into consideration the proximity to a possible consumer, which of course should be situated within a distance of a few kilometres because of the well-known difficulties in transporting heat. Hence, the well site had to be located at the edge of the city of Urach.

In order to avoid the need to drill through the well-known Upper Jurassic cavernous strata, the well was located at the bottom of the "Swabian Alb Escarpment". As shown in Fig. 1 (with the geological section), the well site, which is 426 m above sea level, is located some 450 m NW of the Urach Thermal Bath and around 50 m east of the River Erms. This little river served as a water supply to the well.

Since many Tertiary volcanic pipes occur in this part of the country (MAUSSNEST 1974, this volume), magnetic surveys (PUCHER 1977, this volume) were carried out in order to avoid drilling through an old volcano hidden by fluviatile gravels.

3. Technical Aspects of the Well

When the money became available at the end of 1976, Preussag, a German oil producing company, was chosen as a drilling contractor. On the basis of the experience of this company and of the available experience and results of the American HDR research well at Los Alamos (RUMMEL 1977) a slightly changed and improved drilling schedule was adopted (Tab. 1). At this time we believed that the final depth (dotted line in Fig. 2) would be 2100 to 2500 m, depending on the depth of the top of the basement which (according to SCHÄDEL, DIETRICH this volume) was expected at 1600 m. Since no correlation with neighbouring wells was available, the possible occurence of a Rotliegend series down to 2000 m was tentatively included into a maximum final depth of 2500

Table 1. Planned casing schemes (LH = Liner Head, P = Predrilling).

Original plan (Dec. 1975)				Modified Plan (Point of Time July 1977)		
Depth (m) about	Diameter drilling	Diameter casing	Casing	Diameter casing	Diameter drilling	depth (m) about
30	24"	24"	Conductor	about 28"	about 31 1/2"	26–30
–	–	–		–	P: 12 1/4"	640
470	17 1/2"	13 3/8"	Anchor Pipe	18 5/8"	23"	640
–	–	–		–	P: 12 1/4"	750
850	12 1/4"	9 5/8"	Casing	13 3/8"	17 1/2"	1075
1900	8 5/8"	7" LH 800 m	Liner	9 5/8" LH 975 m	12 1/4"	1065 (2050)
2500	6" 1/4	–	Liner	7" LH 1550 m	8 1/2"	2100 (2500)

m. A detailed drilling and casing programme was worked out and presented by the drilling contractor Preussag together with service companies, Hoechst (drilling mud), Halliburton (cementation), Weatherford (casing) and Schlumberger (well logging). It is believed that all the technical problems involved in both drilling and completing the well, which could be foreseen at that time, were included in the drilling programme (see ERNST & HIEBLINGER 1979). Some features of this programme were:

A clay-fresh-water drilling mud should normally be used. It was to be replaced by a saturated salt-water mud when salty layers were encountered.

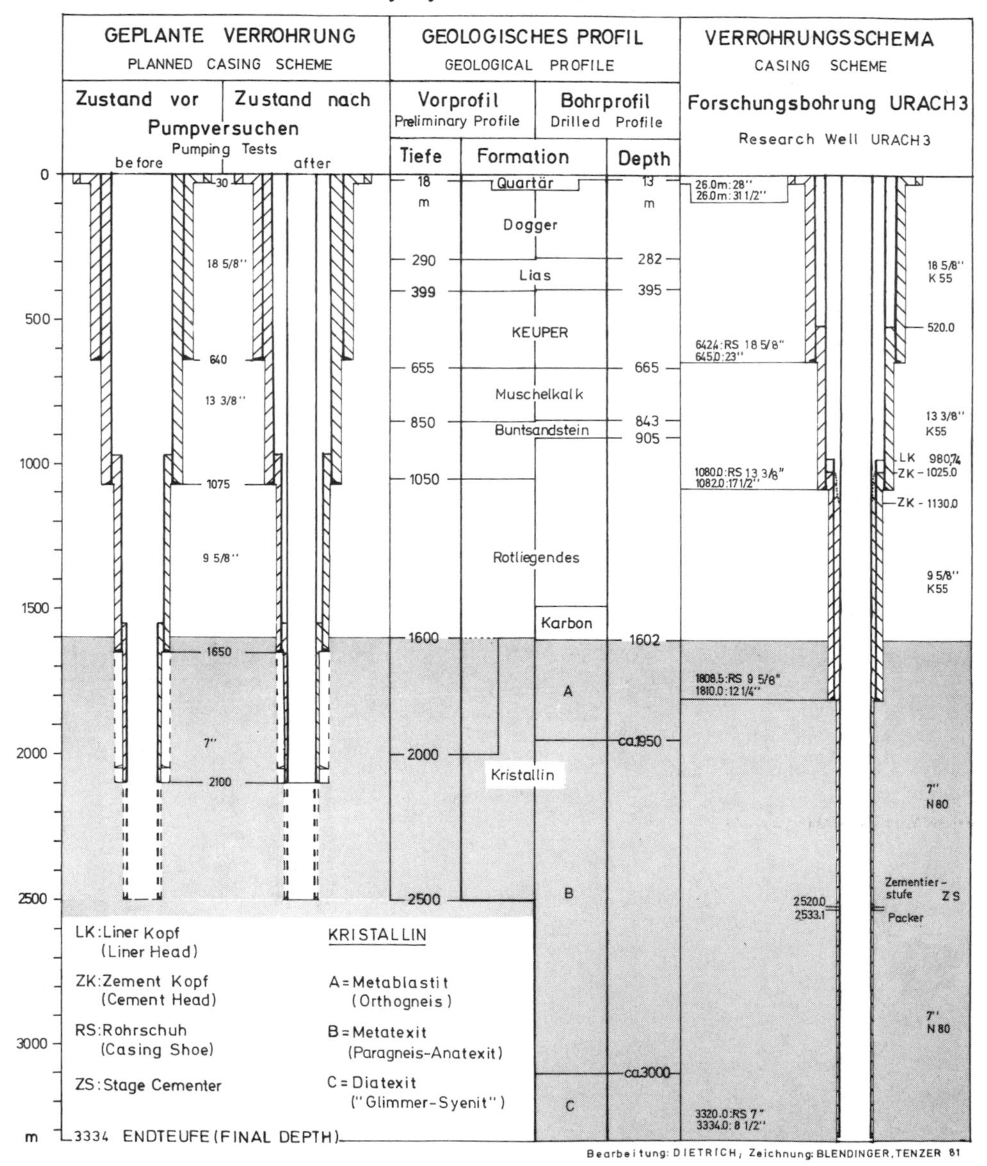

Fig. 2. Planning and realization of Urach 3 geothermal well.

Since no quantitative interpretation of Schlumberger logs was possible in boreholes with extremely large diameters, logging was to be done in pre-drilled holes with a smaller diameter (for instance a 12 1/4" hole for the planned 23" borehole section).

The same is true for the 17 1/2" borehole section to 750 m, which was predrilled with a 12 1/4" diameter in order to carry out pump tests or open flow tests (in the uncased hole) for thermal water expected in the Lower Keuper and Upper Muschelkalk. Finally, open-hole tests were intended in several parts of the uncased crystalline basement after the final depth had been reached. All other expected water horizons were to be tested by perforating the casing after the well had reached the final depth and had been cased. Perforations and tests are run hole upwards of course.

Cement bridges had to be set below each perforation in order to separate the various horizons. The bridges had to be drilled out and the 7" liner had to be prolonged up to the well head. According to Fig. 2, all later frac operations could therefore be carried out within a sound casing scheme.

For the final 7" casing, N-80 quality was chosen in view of the outside pressure and the expected high pressures and stress during the later frac operations. For the upper casing, K-55 quality, which is corrosion-proof, was chosen.

For all cementations the Class G cement was recommended. Even for the 18 5/8" casing, a PZ 350 cement was not used, since this cement and its hardening time are not standardized. A light cement with a lower hardness generally used for plugging mud loss horizons cannot be recommended, in view of the expected high pressures during frac operations. Potential loss zones therefore had to be cemented-off before casing was run.

Because of high costs and technical risks during perforations, setting and drilling out of cement plugs and long stand-by hours during pumping tests within the 18 5/8" casing section, it was decided (shortly before the well was spudded in the autumn of 1977) to carry out the planned tests within a smaller-diameter auxiliary well (Urach 4) which was to be drilled with a smaller rig.

The Urach 3 well was drilled by a mobile drilling outfit: Cabot-Franks-Explorer 900, the main technical details of which are compiled in Tab. 2. Its capacity is around 3000 m depth at 5" DP, and so it offered sufficient clearance for possible fishing jobs, deepening etc. The casing scheme for a 23"–17 1/2"–12 1/4" and 8 1/2" hole was in accordance with the capacity of the outfit and the desired drilling pressures of 10–20, 20, 20–25 and 15 tons, respectively. For the casing run, the drill pipes, drill collars and other tools see Fig. 3. The 4" drill pipe combination shown in the same figure was selected after it had been decided to deepen the well to maximally 3350 m.

For cost reasons the well was not deviated. According to the drilling contract the well had to be bottomed within a circle the radius of which should not exceed 5 % of the final depth. A limitation was set by the so-called DOG-LEG-stipulation (RISCHMÜLLER & WINKLER 1978).

4. Drilling of the well

First results and summarizing descriptions regarding the drilling procedure and the technical completion of Urach 3 well were given by DIETRICH & SCHÄDEL (1978), SAUER (1978), DIETRICH (1979), ERNST & HIEBLINGER (1979) and DIETRICH et al. (1980). More detailed reports on the drilling phase and technique are found in reports and in publications of the drilling contractor by BLOCKSDORF (1980) as well as by MEIER & ERNST (1981).

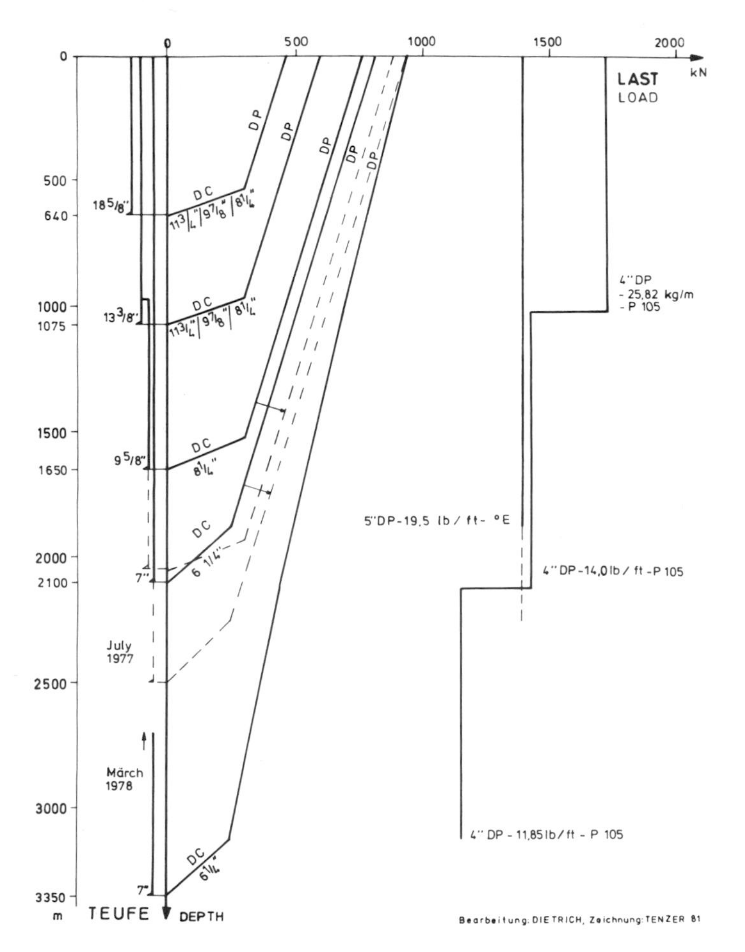

Fig. 3. Planned combinations of drill string, with drill pipe tensile data used (compiled and amended from data and drawings from the Preussag Company).

4.1 Working Procedures

The preparation of the drilling location was done, and all further protective measures were taken in accordance with legal specifications and general technical standards. By order of the local Mining Authority, the well site was asphalted in order to protect the nearby stream, the River Erms; furthermore, this asphalted area had to be surrounded by a cemented ditch with oil separators. In order to muffle the noise of the engines a wall 10 m high and 38 m long was erected; in the course of time it had to be extended to 51 m. Running of the 700 mm (about 28") 26 m long conductor was done by a subcontractor (Menning) with a light hoist; all other work has been done with the Cabot 900 drilling rig described in Tab. 2.

Until the final depth had been reached a bentonite (Tixoton)-CMC- or Bentotite-CMC-Polymer-freshwater drilling mud was used. The respective additions of CMC (Na-carboxy-methylcellulose and polymer as various Tylose-makes (VHR, B 77, VSV, SVH) as well as

Table 2. Technical specification for drilling rig of research well Urach 3.

Self-Propelled Drilling Rig Cabot-Franks-Explorer 900	
Derrick/Mast:	Telescoping Type
Height:	34, 14 m (112')
Hook load rating:	1.42 MN (319 700 lb)
Gross nominal capacity:	1.63 MN (366 100 lb)
Substructure	
Type:	Cabot
Height:	4,58 m
Max. rotary-table sub. capacity:	1.56 MN
Rotary Table	
Type:	Wirth RTS 25 1/2"
Max. input:	331/250 kW (450/340 PS)
Max. drive table:	250/185 min^{-1}
Travelling Block and Hook	
Type:	Mc Kissich WEB 667–250
Max. Load:	2.21 MN
Sheaves:	5
Reeving:	10-times
Drawwork	
Type:	Cabot-Franks-Model 2346
Max. Input:	670 kW (900 HP)
Max. Line pull:	0,23 MN (50 700 lb)
Max. Line Speed:	13,7 m
Line diameter:	1 1/8"
Drive Group (Engines)	
Type:	2 Caterpillar D 343 TAJ 276/313 kw (370/420 HP) each for drawworks, rotary table and pump 7 1/4" x 12"
Type:	1 Mercedes Benz-MB 820 Bb 552 kW (740 HP) for mud pump 7 1/4" x 12" H
Mud Pumps	
Type:	1 Wirth Duplex 7 1/4" x 12" H
Max. Input:	360 kW (485 HP)
Max. Pressure:	5"–192 bar/985 l/min (2785 lb/in^2/260 gal/min)
Max. Output:	7 1/4"–92 bar/2 150 l/min (1334 lb/in^2/568 gal/min)
Type:	1 Wirth Duplex 7 1/4" x 16" H
Max. Input:	493 kW (661 HP)
Max. Pressure:	5"–210 bar/1 200 l/min (3045 lb/in^2/317 gal/min)
Max. Output:	7 1/4"–100 bar/2 640 l/min (1450 lb/in^2/697 gal/min)
Swivel	
Type:	Gardner Denver SW 200
Static Load:	1,96 MN
Dynamic Load:	1,03 MN
Max. Pressure:	350 bar
Mud Tank System	
Mud Tanks:	6
Mud Mixing:	Mission 6" x 8"
Shale Shaker:	2 Itag, Double Vibrator
Desander/Desilter:	Pioneer 12 x 4"
Desander-Pump:	Mission 6" x 8"
Blow-out Preventer	
Type:	20"–200 lb/in^2 each for Hydrill/Shaffer 1 x 12"–500 lb/in^2 Hydrill 1 x 12"–500 lb/in^2 Shaffer-Double
BOP Closing Unit:	Koomey BOP T 20-160-35
Accumulator Volume:	629 l
Procharge Pressure:	207 bar

other additions (for example soda, defoamers and others) were used in accordance with the advice of mud service companies. Mud losses in the sediments as well as in the basement (see DIETRICH & SCHÄDEL 1978) could be successfully treated either by usual lost circulation materials (mica and cellophane flakes "KZR") and highly viscous pills of bentonite and/or by reducing the weight of the drilling mud. In spite of such measures total mud losses occurred in the Upper Muschelkalk at a depth of 692 to 700 m.

Whilst these losses could be fought by using cement PZ 350 F, a zone of severe mud losses in the basement (about 1777 m to 1810 m deep) had to be overcome by cement class G mixed with silica flour. Sealing of these two zones was successful only after 2 and 6 trials (pressure gradient tests) after having pumped in some 24 tons of cement and 49 tons of deep well cement, respectively, with plugs of bentonite. Additional pressure tests were carried out as formation-gradient-test and for checking the casing strings.

When these plugging-cementation operations in the Upper Muschelkalk had been successfully completed and the well had been enlarged to 17 1/2" to a depth of 705 m it was decided (contrary to the original drilling schedule of Tab. 1) to pre-drill with 12 1/4" the total 17 1/2" section down to the casing depth of 1082 m, in order to avoid similar unforeseen complications.

Furthermore, it was decided during the drilling to deepen the well immediately to a final depth of no more than about 3350 m: it was believed that fracturing at greater depth and higher temperature would deliver better results. Addittional information regarding the structure and drillability of the basement was expected in view of possible future geothermal projects, e.g. in the basement of South Germany.

It goes without saying that a sufficient number of cores and cuttings (every 2 metres) were taken. Furthermore, the usual well logs including caliper logs (for a better interpretation) were run (SP, IES, DLL, MLL, HDT, BHC, FDC/LDT, CNL, GRN, GRL). Electrical Sonic logs (BHC-profiles) have been further supported by two seismic geophone measurements after a depths of 1082 m and 3331 m had been reached. By a vibro-seismic survey (air pulser) at a depth of 1082 m it was attempted to obtain hints on the exact depth of the top of the basement; unfortunately this trial failed. For this reason we failed to core the transition zone sediment/basement.

Hydrogeological tests (planned with the assistance of Preussag) were carried out in open (Upper Muschelkalk, Crystalline) as well as in cased (after perforations) sections of the borehole. Tab. 3 gives a summary and more details on the perforations of the 9 5/8" and 13 3/8" casing.

Drill stem tests were conducted by Halliburton with the use of RTTS-packers and Bourdon tube pressure recording device. Pumping tests performed in the Upper Muschelkalk with the use of a Pleuger Underwater Pump, type F 87/6; all other tests were carried out by means of a Byron Jackson Centrilift type I-43 B pump (supplied by Fragro); its cable withstands temperatures up to 149 °C.

Continuous temperature and production control surveys were run during the pumping tests by the Preussag and Schlumberger services (PCT, HRT) for the in situ investigation of the sections tested (location of the inflow, rate, temperature etc). During the pumping tests on the Muschelkalk (influx area at about 235 m) an eccentrically mounted 2" PVC pipe was used for running the measuring devices. For the centrilift operations, however, (at 786 m and 958 m) a dual-completion outfit consisting of 2 7/8" tubing and 7" casing (dual hanger 7" XCSG-2 3/8" in casing spool 13 5/8" x 5000 type WF) was necessary. A mechanically recording Hügel pressure gauge and electronic measuring devices (recording at the surface) of the Preussag Wire Line Service, as well as a Maihak pressure gauge (mounted at the bottom of the pump) and a Gearhart-Owen-Device

Table 3. Casing perforations in Urach 3 borehole by Schlumberger Service and Preussag Service (av. 13 shots/m and 4 shots/ft. respectively).

Date	Depth (m)	Casing (in.)	Nos. of shots	Type of perforation	Phasing	Remarks
6/ 4/78	1784 –1775	9 5/8"	120	3 1/2" Strip-Jet	0°	
	1768 –1764	9 5/8"	54	3 1/2" Strip-Jet	0°	
	1781.2–1776	9 5/8"	68	3 1/2" Strip-Jet	0°	Post-Perforation
	1561 –1568	9 5/8"	39	3 1/2" Strip-Jet	0°	
	1537.5–1532.5	9 5/8"	65	3 1/2" Strip-Jet	0°	
	1529 –1526.5	9 5/8"	35	3 1/2" Strip-Jet	0°	
	1522 –1509	9 5/8"	169	2 1/8" Strap-Jet	0°/180°	
	1506 –1505	9 5/8"	13	2 1/8" Strap-Jet	0°/180°	
	1500 –1498	9 5/8"	26	2 1/8" Strap-Jet	0°/180°	
	1491 –1485.5	9 5/8"	72	2 1/8" Strap-Jet	0°/180°	
6/ 9/78	1547 –1545	9 5/8"	26	4"-CGSL-Hyperjet	90°	Post-Perforation
	1537.5–1535	9 5/8"	33	4"-CGEL-Hyperjet	90°	Post-Perforation
	1528 –1525.7	9 5/8"	30	4"-CGSL-Hyperjet	90°	Post-Perforation
	1522 –1520	9 5/8"	26	4"-CGSL-Hyperjet	90°	Post-Perforation
	1431 –1510	9 5/8"	40	4"-CGEL-Hyperjet	90°	Post-Perforation
	1510 –1509.3	9 5/8"	8	4"-CGEL-Hyperjet	90°	Post-Perforation
	1506 –1505	9 5/8"	13	4"-CGSL-Hyperjet	90°	Post-Perforation
6/11/78	1368.5–1362.5	9 5/8"	80	4"-CGEL-Hyperjet	90°	
	1047 –1042.5	9 5/8" + 13 3/8"	60	4"-CGEL-Hyperjet	90°	
	933 – 929	13 3/8"	53	4"-CGEL-Hyperjet	90°	
	917 – 911.5	13 3/8"	73	4"-CGEL-Hyperjet	90°	
6/13/78	871.3– 863.3	13 3/8"	40	4"-CGEL-Hyperjet	90°	
	862 – 856.5	13 3/8"	73	4"-CGEL-Hyperjet	90°	
	855 – 849	13 3/8"	80	4"-CGEL-Hyperjet	90°	
	846 – 845	13 3/8"	13	4"-CGEL-Hyperjet	90°	
	843.5– 840.5	13 3/8"	40	4"-CGEL-Hyperjet	90°	
	839 – 834	13 3/8"	66	4"-CGEL-Hyperjet	90°	
6/23/78	1780 –1775	13 3/8"	64	HK 4"	120°	Post-Perforation
	1768 –1764	13 3/8"	52	HK 4"	120°	Post-Perforation

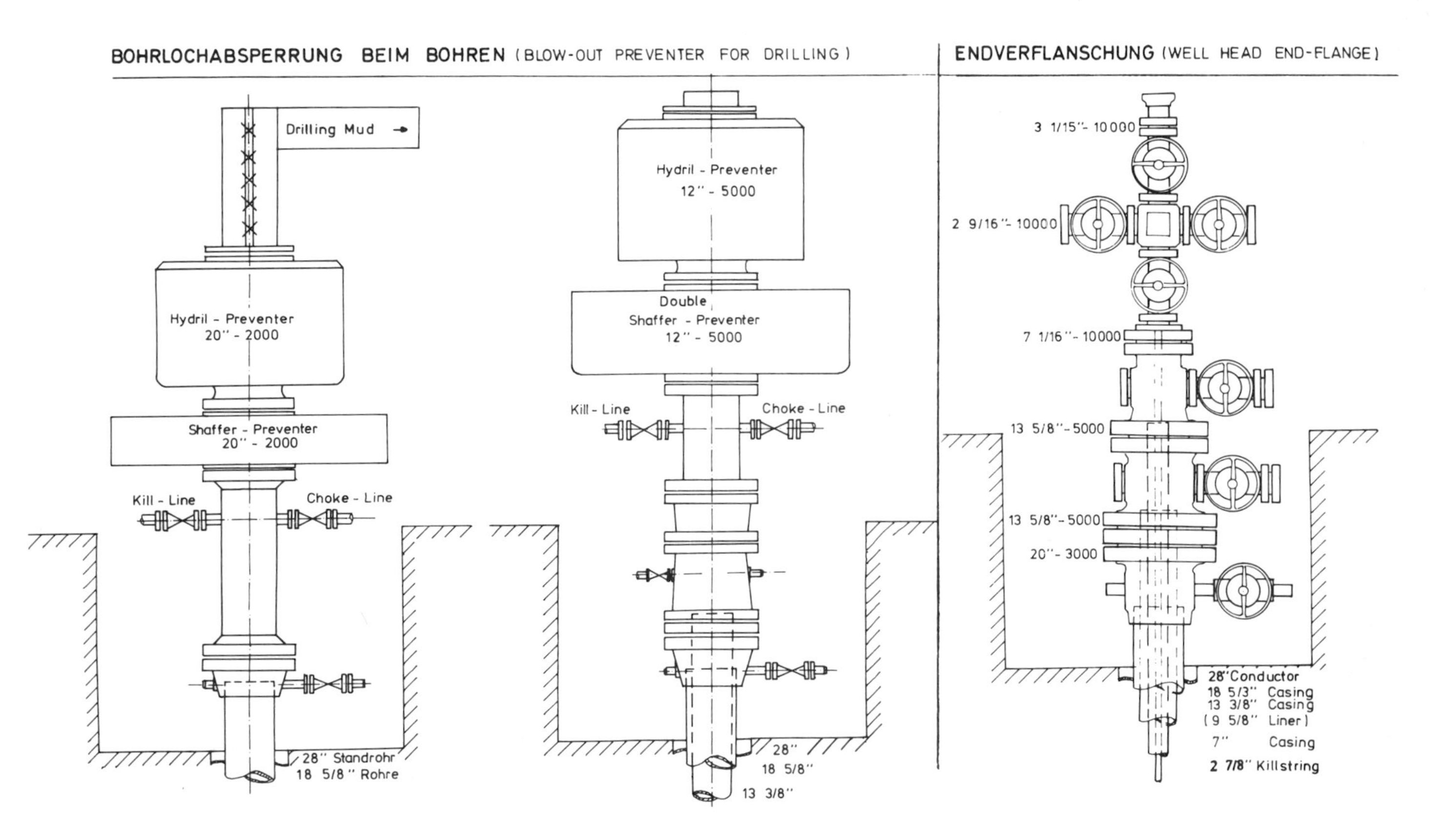

Fig. 4. Borehole blow-out prevention at Urach 3.

(for temperature) in the drill pipe were used for checking the water level in the borehole during the tests.

The temperature survey during the pumping tests was a part of the geothermal investigations. A special temperature survey during the drilling operations and after the completion of the well was carried out by the Niedersächsisches Landesamt für Bodenforschung (NLfB); electrically as well as mechanically recording apparatus (for instance Kuster-gauge) were thereby employed. Generally these were only punctual measurements and, therefore, additional continuously recording surveys were necessary; they were carried out by Schlumberger and Preussag. Other temperature data obtained by maximum thermometers and temperature strips were used for interpretation too, although these data were less exact.

A continuous measurement of the circulating drilling mud (entrance and exit) with hourly plotted data served also for the technical safety of the well (for instance composition of the mud).

For the some purpose, gas logging (hydrocarbons and CO_2) of the mud was carried out. Logging was done by means of a chromatograph and an infrared device of the Frommholz Measuring Service. Greater gas accumulations were not expected in the Urach 3 well; however, it should be remembered that the nearby exploration well Buttenhausen 1 had a severe CO_2-blowout (WIRTH 1958). Preussag, therefore, installed a blowout preventer (shown in Fig. 4), including a Koomey control system described in Tab. 2.

As already mentioned, pressure tests for checking the cementation of casing and pressure-gradient surveys were carried out as a routine. The same is true for extra caliper runs before open-hole tests and running of casing as well as after redrilling operations. The top of the cement behind casing was checked by temperature measurements. In view of later fracturing operations, additional special sonic logs were run by Schlumberger (CBL-VDL) in order to test the quality of the cement behind the 9 5/8" liner and of the final 7" casing.

Degree and direction of the deviation of the borehole was measured by Teledrift and Eastman instruments and other methods (see chapter 5) during drilling and after completion.

More data and results of the drilling procedure, the technical completation of the well and its deviation are given below. Interpretations and descriptions of geoscientific samples and measurements are found in other chapters of this volume.

4.2 Time schedule

After some preliminary work (including running of the conductor pipe) the research well Urach 3 spudded in on 3rd October, 1977 and reached its final depth of 3334 m (final diameter 8 1/2") on 2nd May, 1978. After completion of an extensive testing and measuring programme the final 7" casing was run and cemented on 14th July, 1978. The well was completed on 21st July, 1978. A summary of the course of all operations including a drilling progress curve, a drilling and casing scheme and a geological section, is given in Tab. 4 and shown in Fig. 5.

4.3 Results of drilling and coring

According to plan (see chapter 3) the Urach 3 well was drilled in four sections with diameters of 23", 17 1/2", 12 1/4" and 8 1/2". The well was deepened to the capacity of the Cabot 900 drilling rig and the final depth of 3334 m marks a record for this outfit

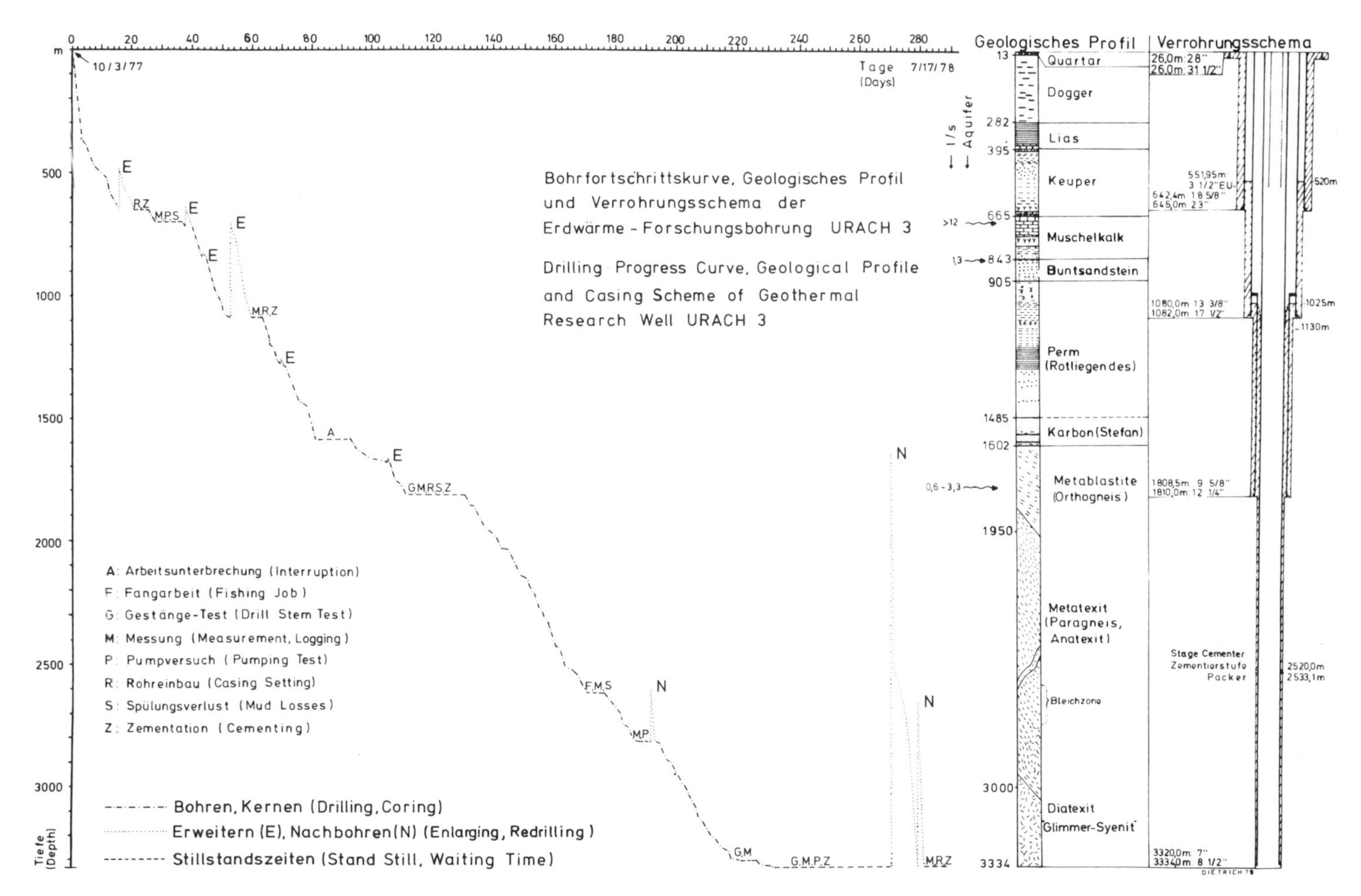

Fig. 5. Drilling progress curve, geological profile and casing scheme of the Urach 3 geothermal research well.

Table 4. Chronological summary of the geothermal research borehole Urach 3 in 1977–78.

Date executed 1977/78	Borehole section/ depths in metres	Current core no.	Principal operations (CS: casing shoe, P: predrilling, PB: predrilled borehole, RB: roller bit, TM: temperature measurement WT: waiting time
8. 8.– 9.26 1977	0 – 26	–	Preparation of the well site, placement of the conductor pipe and fixation poles for the noise barrier
9.27–10.2	–	–	Delivery of drilling rig, rig up; noise barrier mounted
10. 3–12	26 – 488.5	–	Drilling with 23" RB thru Dogger and Lias into Keuper; TM with Kuster-gauge (NLfB)
10.12–18	488.5– 645	1– 3	P with 12 1/4" RB in Keuper, TM by NLfB
10.18–19	26 – 645	–	Schlumberger measurements: IES, FDC, BHC-GR
10.19–22	488.5– 645	–	12 1/4" PB enlarged to 23"
10.22–26	0 – 642.4	–	BGT measurement; installation of 18 5/8" casing, cementation with API-cement CLASS G; WT as cement hardens, drilling on CS
10.26–29	645 – 700	4	P with 12 1/4" RB; from 687–700 m difficulties with slight and total mud losses overcome
10.29–11. 7	645 – 700	–	Pumping test incl. misc. jobs in the open hole section; Schlumberger measurements: GRN, PCT; TM with Kuster-gauge (NLfB)
11. 7.– 9	645 – 716	–	Test with uranin (sodium-flourescein) tracer to determine an hydraulic connection between Urach 3 and Urach 1 + 2: uranin is injected into Muschelkalk aquifer; rock is sealed with cement PZ 350 F; cement drilled on with 12 1/4" RB and borehole deepened
11. 9–10	645 – 705	–	12 1/4" PB enlarged to 17 1/2"
11.10–23	716 –1082	5–12	P with 12 1/4" RB thru Muschelkalk, Bunter and into Upper Rotliegendes; TM by NLfB
11.23–24	645 –1082	–	Schlumberger measurements: IES, FDC, BHC/GR
11.24–12. 1	716 –1082	–	12 1/4" PB enlarged to 17 1/2"
12. 1– 2 1977	642.4–1082	–	Schlumberger measurements: BGT, GRL and seismic geophone measurements by PRAKLA SEISMOS
12. 2– 5	0 –1080	–	Running 13 3/8" casing, cementation with API cement CLASS G WT for cement hardening, determination of cement head; drilling on CS with 12 1/4" RB
12. 5–23	1082 –1586	13–20	Drilling with 12 1/4" RB thru Upper and Lower Rotliegendes into Carboniferous (Stefanian); TM (NLfB)
12.23– 1. 2 1978	–	–	Interruption of work, TM (NLfB)
1. 2– 3	1080 –1586	–	Borehole check trip, redrilling with 12 1/4" RB
1. 3–10	1586 –1664.7	21–24	Drilling with 12 1/4" RB through Car-

Date executed 1977/78	Borehole section/ depths in metres	Current core no.	Principal operations (CS: casing shoe, P: predrilling, PB: predrilled borehole, RB: roller bit, TM: temperature measurement WT: waiting time
			boniferous into crystalline basement (metablastite): crystalline at 1602 m
1.11–13	1664.7–1668.4	25	Total core loss: core, core catcher, orienting shoe, steel ball size and 1" steel ball remain in the borehole: core drilled up with 8 1/2" RB, fishing jobs for the iron pieces with 7" fishing magnet.
1.13–21	1668.4–1810	26–27	Drilling with 12 1/4" RB; P with 8 1/2" RB for core 27; overcoming mud losses at 1777 m; TM (NLfB)
1.21–22	1082 –1810	–	Schlumberger measurements: DLL, MLL, BGT, HDT, FDC/GR, BHC; TM (NLfB)
1.22–24	1759.6–1779.9	–	Open Hole Straddle Test with anchor shoe on joint aquifer; TM (NLfB)
1.24– 2. 2	1679 –1810	–	6 sealing operations with modified API cement CLASS G on joint aquifer; drilling on cement, borehole prepared for running liner
2. 2	0 – 815	–	Running 9 5/8" liner to 815 m
2. 3– 4	–	–	WT for a suitable setting tool joint for the liner setting equipment
2. 4– 9	980.4–1808.5	–	Run with 9 5/8" liner finished, liner cemented with API cement CLASS G; WT for cement hardening, determination of cement head; drilling on cement, plugs and CS with 8 1/2" RB
2. 9–20 1978	1810 –2017.3	28–30	Drilling with 8 1/2" RB, coring attempts partly without success (run 30 b); TM (NLfB)
2.20	980 –1808	–	Schlumberger measurements: CBL, VDL; TM (NLfB)
2.21– 3.12	2117.3–2503	31–34	Drilling with 8 1/2" RB, successful coring with Navi-Drill in runs 33 + 34; TM (NLfB); the bearings of two rollers of the RB remain in borehole
3.12–14	2479 –2517.5	–	Drilling with 8 1/2" RB with fishing pipe in 2 runs coarse rock fragments, 32 bearings and 850 g of iron parts brought to the surface
3.14–19	2517.5–2597	35	Drilling with 8 1/2" RB, redrilling from 2560 to 2588 m; during coring, Navi-Drill employed with success
3.19	2597	36 a	Coring attempt with Navi-Drill: stop at 2585 m, borehole bottom not reached
3.19–21	2550 –2598	–	Borehole repeatedly redrilled: high torgue values and excessive overload when pulling (up to 25 · 10^3 daN), hydraulic jar taken to drill string
3.21–22	1809 –2582	–	Schlumberger measurements: DLL, BGT, FDC-CNL, BHC; TM (NLfB)
3.22–24	2587 –2595.7	–	3 drill- and -fishing jobs for rock pieces with junk basket: stops at run depths of 2587, 2588, and 2594.6 m resp.; 30 kg of rock pieces extracted

Date executed 1977/78	Borehole section/ depths in metres	Current core no.	Principal operations (CS: casing shoe, P: predrilling, PB: predrilled borehole, RB: roller bit, TM: temperature measurement WT: waiting time
3.24– 4. 3	2594.6–2802	36 b	Drilling with 8 1/2" RB, slight mud losses at depth 2606 m
4. 4– 8	1810 –2802	–	Preparation and execution of pumping test in the non-cased borehole section in crystalline: practically no backflow out of the rock; TM by Schlumberger (HRT) and NLfB (Kuster-gauge): following pumping test, measurement possible only to 2603 m
4. 8–14	2568 –2803.5	–	Redrilling and drilling with 8 1/2" RB, slight mud losses overcome
4.14–15	2872.3–2891	–	Redrilling and drilling with 8 1/2" Diamond bit in combination with 6 3/4" Navi-Drill; rapid wear of diamond bit
4.15–16	2872.3–2884	37 a	Bottom not reached during coring with Navi-Drill: Navi-Drill damaged and core tube plugged with rubber parts
4.16– 5.10	2891 –3301.5	37b–44	Drilling with 8 1/2" RB with slight mud losses, problems in coring due to core jamming and heat effects; TM (NLfB)
5.10–11	3294.9–3301.5	–	Open-Hole test with anchor shoe incl. misc. jobs and Schlumberger measurement (BGT): flow rate determined about 0.02 l/sec.
5.11–20	3301.5–3331	45–48	Drilling with 8 1/2" RB with slight mud losses; problems in coring
5.20	800 –3331	–	Seismic geophone measurement by PRAKLA-SEISMOS
5.20–22	3331 –3334	49	Drilling with 8 1/2" RB: final depth is reached
5.22–24	800 –3334	–	Schlumberger measurements: DLL, HDT, FDC-CNL/LDT-CNL, BHC-GR; problems in transitory jamming during the 2nd measurement with FDC-CNL (stop at 3272 m, fixed at 3243 m); redrilling with 8 1/2" RB
5.24–25	–	–	Drill collars (6 1/4", 5") broken and deposed
5.25	3334	–	TM by NLfB with Kuster gauge
5.25–30	1808.5–3334	–	Pumping test incl. misc. jobs in non-cased section of crystalline: a dual completion outfit is used as test equipment (the employed centrilift submersible pump with a high-temperature cable installed on 2 7/8" tubing in a 7" protective liner next to a 2 3/8" tubing string intended for measurements).
5.29– 6.2	1808.5–2640	–	Schlumberger measurements: PCT, HRT: it is not possible to reach borehole bottom or to determine influx areas: instruments stop at depths of 2567, 2583.4, 2685 and 2606.3 m.
6. 2– 4	1750 –2640	–	Pumping in of bentonite freshwater mud from 2640 to 1875 m and setting a

Date executed 1977/78	Borehole section/ depths in metres	Current core no.	Principal operations (CS: casing shoe, P: predrilling, PB: predrilled borehole, RB: roller bit, TM: temperature measurement WT: waiting time
			cement bridge with API cement CLASS G and bentonite in area of 9 5/8" CS Open-Hole; cement head at 1796.5 m; WT for hardening of cement
6. 4.	0 –1796.5	–	Borehole circulated with 105 cm^3 freshwater
6. 4– 7	1485.5–1784	–	Perforation of 9 5/8" casing in front of joint aquifer in crystalline as well as the expected water bearing horizons of the Carboniferous and the Lowest Rotliegendes: pumping test incl. misc. jobs with dual completion
6. 7– 9	1597.7–1650	–	Setting a cement bridge with modified API cement CLASS G: cement head at 1597.7 m; WT for cement hardening
6. 9–19	834 –1561	–	Perforation of 9 5/8" and 13 3/8" casings in front of expected water bearing horizons of Rotliegendes, Bunter and Muschelkalk and 3 pumping tests with dual completion: low flow rates; TM by PREUSSAG Measuring Service, Schlumberger and NLfB
6.19	0 –1595	–	Casing run with 9 5/8" scraper
6.19–22	911.5–1561	–	4 straddle tests at expected water bearing horizons in Carboniferous and Rotliegendes
6.22–23	834 –1561	–	Check trip with 8 1/2" RB; 9000 l mud pressed into perforated sections, sealing successful
6.23	1595.5–1670	–	Drilling on the upper cement bridge
6.23–24	1765 –1780	–	9 5/8" casing postperforated with 4" casing guns; run with 9 5/8" scraper to 1796.5 m
6.24–28	1764 –1784	–	Pumping test incl. misc. jobs at joint aquifer in crystalline
6.28–29	834 – 871.4	–	3 m^3 bentonite plugs pressed into the rock, perforations sealed
6.29–30	1671 –1796.6	–	Joint aquifer with 3 m^3 bentonite plug and 5 m^3 modified API cement CLASS G sealed, cement head at 1650 m; WT for cement hardening; drilling on cement with 8 1/2" RB
7. 1	1796.6–2510	–	Drilling on the lowest cement bridge (1796.7 to 1838.4 m) with 8 1/2" RB; borehole redrilled to 2510 m
7. 1– 8	2510 –3334	–	Redrilling with 8 1/2" RB with slight mud losses; high torgue values and excessive overload when pulling
7. 8– 9	1808.5–2049	–	Schlumberger measurement (BGT): final depth not reached, gauge stops at 2674 and 2749 m
7. 9–10	2645 –3334	–	Redrilling with 8 1/2" RB terminated
7.11	2640 –3332	–	BGT-measurement
7.11–12	0 –3334	–	6 1/2" drill collars broken during extraction and put in storage

Date executed 1977/78	Borehole section/ depths in metres	Current core no.	Principal operations (CS: casing shoe, P: predrilling, PB: pre-drilled borehole, RB: roller bit, TM: temperature measurement WT: waiting time
7.12–14	0 –3320	–	Running 7" terminal casing with Open-Hole packer and 2-stage cementer; cementation with modified API cement CLASS G and Pozmix-Special cement; WT for hardening of cement
7.15–17	0 –3334	–	Drilling on stage cementer, plugs and CS with 5 7/8" RB; borehole circulated with freshwater; drill string broken during extraction and put in storage
7.18–21	0 – 872.6	–	Running 2 7/8" tubing as kill string, completion of final flanges and Christmas tree with stuffing box; rig down; rig transport; noise barrier demounted and transported, misc. jobs
7.21	–	–	Transfer of the drilling location on the city of Urach, the commissioner of Urach 3.

(ERNST & HIEBLINGER 1981). Down to 2802 m a 5" drill pipe, and subsequently a 4" drill pipe was used (see Fig. 2). In general three pipe, and subsequently a 4" drill pipe was used (see Fig. 2). In general three stabilizers (Christensen), one at the bit and two on the drill pipe, were used. A shock sub (Christensen) had to be used when drilling through siliceous sandstone of the Stubensandstein in order to avoid an extreme axial strain on the drill pipe. As a result of rock tensions in the crystalline there was a tendency toward chipping of this rock (torque and overload when pulling, reaming); therefore a hydraulic SERVCO jar 6 1/4" OD was added to the drill pipe below 2600 m.

A summary of some important drilling data (without detailed figures) is given in Tab. 5. As shown in this table the pressures applied, the rotational speed and the pumping rates changed, since they were dependent on the technical and/or geological conditions. The same is true for the weight of the drilling mud. To give an example: light-weight muds indicate that a zone of mud loss had to be fought by lightening the weight of the mud. The lowest density of the mud was 1.03 when drilling through the Upper Muschelkalk, where a total mud loss occurred between 692 and 700 m (the mud level was about 100 m below surface at that time).

As shown in Tab. 5 and 6 a lot of pre-drilling with smaller diameters (12 1/4", 8 1/2", or 5 7/8") had to be done in various sections of the hole in order to get a better base for the interpretation of electrical logs and of pumping tests, as well as for more efficient coring (Core no. 26, 27, 48). These sections were later reamed with 23", 17 1/2", 12 1/4" or 8 1/2". Larger sections of drilling are marked by "E" on the drilling progress curve of Fig. 5.

Table 5. Summary of bit record (C = Coring, E = Enlarging, P = Predrilling).

Borehole section (m)	Formation name	Drilling-diameter (in.)	Drilling thrust (10^3 daN)	Rotary speed (min^{-1})	Pumping rate (l/min)	Pressure drop (bar	Mud density (g/cm^3)	av. Penetration rate (m/h)
26 – 488.5	Dogger	23"	6–15	60–110	3500	40–45	1.18–120	2.33
	- Keuper	C: 8 15/32"	6–9	60–100	900–1000	45–55	1.12–1.16	1.56
488.5– 645	Keuper	P: 12 1/4"	2–10	90–120	2600–3000	35–90	1.10–1.16	2.60
		E: 23"	12–15	90–110	3500	40–45	1.15–1.18	2.09
		C: 8 15/32"	5–9	90–120	900	20–50	1.07–1.15	1.58
645 –1082	Keuper	P: 12 1/4"	6–12	60–120	1800–2960	40–80	1.07–1.16	2.69
	- Rotliegendes	E: 17 1/2"	10–15	80–110	3000–3400	45–60	1.05–1.12	2.96
		C: 8 15/32"	5–7	100	900	55–60	1.15–1.16	0.95
1082 –1602	Rotliegendes	12 1/4"	10–16	100–110	2000–2200	80–95	1.14–1.16	2.29
	- Carboniferous		18–20	55				1.83
		C: 8 15/32"	6–8	70–100	900–1150	50–65	1.13–1.14	0.40
1602 –1810	Crystalline	P: 8 1/2"	7–10	50–60	1500	70–75	1.13–1.19	1.50
	Basement	12 1/4"	8–20	55–80	1800–2300	45–100	1.14	1.62
		C: 5 27/32"	4–6	90	700	100	1.11	0.23
		P: 5 7/8"	5–6	60	800	100	1.11	2.00
1810 –3334	Crystalline	C: 8 15/32"	5–8	80–90	900–1000	45–110	1.11–1.16	0.61
	Basement		5–7	av. 45 (250)	1100	90–100		1.72
		8 1/2"	7/10–15	45–90	1200–1300	65–115	1.11–1.16	1.30
			7–11	av. 50 (350)	1170–1470	135–160	1.12	1.56

Table 6. Summary of all types of bits used in the four sections of overburden (0–1602 m) and crystalline basement (1602–3334 m) of Urach 3 (E = Enlarging, O = Outdrilling, P = Predrilling, R = Redrilling; CS = Casing Shoe, FD = Final Depth).

Borehole section (m)	Formation (m)	Bit size (in.)	Bit type	Bit no. for each type	Remarks
26 – 488.5	– 283 Dogger – 395 Lias – 488.5 Keuper	23"	TS 25	3	

488.5– 645	–	645	Keuper	E: 23"	TS 25	2	
	–	668	Keuper	P: 12 1/4"	DGH	12	incl. 0 of cement seal and E of core sections
	–	843	Muschelkalk		TS 2	2	incl. 0 of cement and 18 5/8"-CS
645 –1082	–	905	Buntsandstein	E: 17 1/2"	TS 3, DGH	2	
					TS 5	1	
	–	1082	Rotliegendes				
	–	1485	Rotliegendes		DGH	8	incl. 0 of cement and 13 3/8"-CS and E of core sections
	–	1602	Carboniferous	12 1/4"	3 JS	1	
1082 –1810				P: 8 1/2"	V2H	1	Out-Drilling of core No. 25, core catcher, orienting shoe and steel ball size
					9 JS	(1)	P for a more efficient coring for core nos. 26 and 27
	–	1810	Crystalline Basement	12 1/4"	V2H	2	incl. E of core sections and 0 of cement seals
					RG2B, 5JS, 7JS	1	
					V2H	2	incl. 0 of cement and 9 5/8"-CS
					3 JS	10	
					4 JS, 9 JS, M 88	3	
					F5, J44, J55	2	
					5 JS, S 88	1	
				8 1/2"	MD210 Special	1	Run with diamond bit in combination with 6 3/4" Navi Drill
1810 –3334	– FD	3334	Crystalline Basement		L4HJ	3	Core catcher of core run 36 a drilled, 0 of cement bridges, R
					V2HJ	3	R before running of 7" final casing
				5 7/8"	F 5	1	P for core no. 46 (5 27/32"-crown)
					M4NJ	1	0 of cement, stage cementer, cement and 7"CS
							R open hole (7"-CS to FD)

In general, Smith-Tool bits were used besides SMF, Security and Hughes bits. All types of bits which have been used for drilling, pre-drilling and enlarging are listed in Tab. 6. For drilling through sediments medium-hard tooth bits were preferably used; drilling in the basement was done by medium-hard to very hard insert bits.

Average drilling penetration rates are shown in Tab. 5 as an arithmetical average without considering stand-by times and various bit performances. Thus the penetration rate averaged 2 to 3 m/h in the sediments and 1.3 to 2 m/h in the basement. The metreage drilled by each bit, (a specially interesting figure), was between 50 and 65 m; it was more or less independent of the type of bit. Taking into consideration various factors (bit stands still, drilling times, drilling penetration rates etc.) we believe that the crystalline basement is best drilled by medium-hard insert bits of type 3 JS to 4 JS, J 44 and M 88: the maximum performance was 100.6 m drilled by one bit with an average progress of 1.72 m/h in paragneiss.

Table 7. Performance data during coring in crystalline basement (1602 m–3334 m) of Urach 3 with 8 15/32" diamond core bits from Christensen Company

Coring run (No.)	Depth/ core section (m)	Core recovery (m/%)	Type of core bit	Serial no.	Drilling process	Down hole motor OD (in.)	Coring time (h)
21	1634.0–1635.7	1.0 / 59.8	C 20	P 2745 G	Rotary	–	2.2
22	1635.7–1643.7	7.1 / 88.8	C 20	P 2745 G	Rotary	–	27
23	1660.0–1661.8	0.15/ 8.3	C 23	P 2883 G	Rotary	–	6.2
24	1661.8–1664.7	2.0 / 69.0	C 20	P 2770 G	Rotary	–	9.5
25	1664.7–1668.4	0 / 0	C 23	P 2871 G	Rotary	–	12
26	1675.0–1677.9	2.2 / 75.9	C 23	P 2885 G	Rotary	–	9.8
27	1750.5–1756.8	6.3 /100	C 24	P 2886 G	Rotary	–	12
28	1853.0–1862.0	7.8 / 86.7	C 24	P 2886 G	Rotary	–	13.5
29	1945.0–1953.0	7.4 / 92.5	C 24	P 2913 G	Rotary	–	11.5
30 a	2013.5–2017.3	3.8 /100	C 24	P 2913 G	Rotary	–	8
30 b	2017.3	0 / 0	C 24	P 2886 G	Rotary	–	1.7
31	2124.2–2133.2	8.2 / 91.1	C 24	P 2971 G	Rotary	–	19
32	2189.5–2195.7	5.8 / 93.5	C 24	P 2983 G	Rotary	–	8.4
33	2302.4–2311.4	8.7 / 96.7	C 24	P 2991 G	Navi-Drill	6 3/4"	3.5
34	2412.0–2421.0	8.7 / 96.7	C 24	P 2991 G	Navi-Drill	6 3/4"	4.7
35	2517.5–2521.3	3.7 / 97.4	C 24	P 2991 G	Navi-Drill	6 3/4"	5
36 a	(2585-attempt)	–/–	C 24	P 3024 G	Navi-Drill	6 3/4"	3
36 b	2795.7–2802.0	5.6 / 88.9	C 24	P 3024 G	Rotary	–	10.1
37 a	(2884-attempt)	–/–	C 24	P 3024 G	Navi-Drill	6 3/4"	2
37 b	2930.0–2939.0	8.5 / 94.4	C 24	P 3024 G	Rotary	–	10.2
38	3053.0–3060.0	5.5 / 78.6	C 24	P 3028 G	Rotary	–	9
39	3125.0–3132.0	3.1 / 44.3	C 24	P 3028 G	Rotary	–	10
40	3205.0–3211.0	5.8 / 96.7	C 24	P 3152 G	Rotary	–	9.1
41	3250.0–3258.2	7.9 / 96.3	C 24	P 3153 G	Rotary	–	11.9
42	3290.0–3294.7	4.3 / 91.5	C 24	P 3153 G	Rotary	–	7.4
43	3294.7–3298.4	3.7 /100	C 24	P 3153 G	Rotary	–	5.8
44	3298.4–3301.0	2.1 / 80.8	C 24	P 2971 G	Rotary	–	6.2
45	3302.0–3306.2	3.6 / 85.7	C 24	P 3162 G	Rotary	–	7.8
46	3321.0–3321.9	0.6 / 66.7	C 24)+	P 3140 G	Rotary	–	4
47	3323.0–3325.3	1.2 / 52.2	C 24	P 3162 G	Rotary	–	8
48	3226.0–3330.6	3.1 / 67.4	C 24	P 3204 G	Rotary	–	10.1
49	3331.0–3334.0	2.0 / 66.7	C 24	P 3204 G	Rotary	–	5.8

The good performance of the 3 JS-bit should be mentioned: it was the only insert bit which was used in the Permo-Carboniferous. It drilled about 185 m at an average penetration rate of 1.83 m/h.

About one tenth of the total depth has been cored (see DIETRICH this volume). For the total of 49 coring operations, Christensen double tube core barrels were used (6 3/4" x 4" x 60 ft and 4 3/4" x 2 5/8"x 30 ft, the latter for core no. 46 only) in connection with 8 15/32" and 5 7/8" core bits, respectively. Diamond core bits were supplied by Christensen, too (types C 20, 23 and 24). Oriented coring operations were also supervised by the Christensen Service.

When coring in the sediments the penetration rates were 1 to 2 m/h, as compared with 0.3 to 0.7 m/h in the basement; in Tab. 5 the arithmetical averages are also shown.

For the 20 coring runs in the sediments we needed only 2 core bits of type C 20, whereas coring in the hard and abrasive crystalline basement proved to be extremely

Penetration rate	Rotary speed	Pumping rate	Drilling thrust	Wear of diamond core bit	Diamond core bit run current	Remarks
(m/h)	(min^{-1})	(l/min)	(10^3 daN)		(no.)	
0.77	100	900	7–8	0. k.	11. run	
0.30	80	900	7–8	O-Ring	12. run	
0.29	70	900	7–8	O-Ring	1. run	
0.31	70	900	6–7	smooth	1. run	
0.31	60–80	1050	6	O-Ring	1. run	core, core catcher, orienting shoe, 1" steel ball and steel ball size lost in the borehole
0.30	70	1000	6	(o. k.)	1. run	
0.53	70	1150	6	o.k.	1. run	
0.67	80	1000	6–8	(o. k.)	2. run	
0.70	80	1000	6–7	o. k.	1. run	
0.48	80	1000	6–7	O-Ring	2. run	
–	80	1000	6–7	O-Ring	3. run	run wihtout coring progress; no core cut
0.47	80	1000	6	o. k.	1. run	
0.74	90	1000	7	O-Ring	1. run	
2.57	45/250	1100	6–7	25 %	1. run	
1.90	45/250	1100	6–7	50 %	2. run	
0.76	45/250	1100	5–6	O-Ring	3. run	
–	–	–	–	o. k.	1. run	hole bottom (2597 m) not reached; core catcher lost in the borehole
0.62	80	1000	6–8	o. k.	2. run	
–	–	–	–	o. k.	3. run	hole bottom (2891 m) not reached; Navi Drill damaged; core barrel plugged with rubber parts of the Navi Drill
0.88	80	1000	7	O-Ring	4. run	
0.78	90	1000	7	(o. k.)	1. run	
0.70	90	1000	7–8	O-Ring	2. run	
0.66	90	1000	7–8	O-Ring	1. run	
0.69	90	1000	7	o. k.	1. run	
0.63	90	1000	7–8	o. k.	2. run	
0.64	90	1000	7–8	O-Ring	3. run	
0.42	90	1000	7–8	O-Ring	2. run	
0.54	100	1000	7–8	(o. k.)	1. run	
0.23	90	700	4–6	smooth	1. run)+	5 27/32" diamond core bit
0.29	90	1000	5–8	80% smooth	2. run	
0.46	90	1000	5–8	o. k.	1. run	
0.52	90	900	5–8	60% smooth	2. run	

Table 8. Coring in crystalline basement at Urach 3. Average drilling progress under consideration of all coring runs for each 8 15/32" diamond core bit (exception:+ = 5 27/32" core bit) without or with use of down-hole motors (Navi-Drill) respectively.

Type of core bit		Serial no.	Grand Total: Number of coring runs	Grand Total: Edge life/ Drilling time (h)	Grand Total: Distance cored (m)	Penetration rate (av. m/h)	Remarks
C 20		P 2745 G	2	29.2	9.7	0.33	
C 20		P 2770 G	1	9.5	2.9	0.31	
C 23		P 2871 G	1	12.0	3.7	0.31	
C 23		P 2883 G	1	6.2	1.8	0.29	
C 23		P 2885 G	1	9.8	2.9	0.30	
C 24	131818	P 2886 G	3	25.5/27.2	15.3	0.60/0.56	1 run without progress; no core cut (= 1.7 h)
C 24	131818	P 2913 G	2	19.5	11.8	0.61	
C 24	131818	P 2971 G	2	25.2	11.6	0.46	
C 24	131891	P 2983 G	1	8.4	6.2	0.74	
C 24	131897	P 2991 G	3	13.2	21.8	1.65	3 runs with Navi-Drill
C 24	131897	P 3024 G	4	20.3/25.3	15.3	0.75/0.60	incl. 2 runs with Navi-Drill: in both cases hole bottom not reached (3 h + 2 h = 5 h); loose pieces of rock rolling around under bit, no core cut
C 24	131897	P 3028 G	2	19.0	14.0	0.74	
C 24	131897	P 3140 G	1	4.0	0.9	0.23	
C 24	131897	P 3152 G	1	9.1	6.0	0.68	
C 24	131897	P 3153 G	3	25.1	16.6	0.66	
C 24	131897	P 3162 G	2	15.8	6.5	0.41	
C 24	131897	P 3204 G	2	15.9	7.6	0.48	

difficult. The C 20 and 23 core bits initially used showed achievements of only 0.3 m/h (with one exception of 0.8 m/h). This achievement could be doubled (0.6 m/h on the average) by using a C 24 type (which actually was the hardest core bit used in Europe at that time!).

Whilst coring in the basement, severe wear of core bits and core barrels as well as a repeatedly poor core recovery (see Tab. 7) resulted from the above mentioned difficulties.

Since the coring progress in the basement was partly unsatisfactory we used a down-hole motor besides the conventional rotary drilling process; the quieter operation of this motor was believed to deliver better core results (see JÜRGENS & MARX 1979). Indeed, the coring penetration rate with this Navi-Drill was extremely high, namely up to 2.6 m/h. As shown in Tab. 8 the Navi-Drill improved the drilling progress by a factor of 3 and also provided a remarkably better core recovery.

However, it was feared that loosened rubber parts of the Navi-Drill could cause some trouble and lead to severe drilling problems by plugging the core barrel and the fluid channels for the mud. The same problems occurred in the Vorderriss 1 well (BURGER & RISCHMÜLLER 1979). Improvements of these parts as well as new conceptions are under way according to JÜRGENS & MARX (1979).

Similar satisfactory penetration rates were achieved with a diamond core ejector bit for ultrahard formations (MD 210 special) in combination with a 6 3/4" Navi-Drill; this combination showed a penetration rate of 1.56 m/h, that is, a value similar to those achieved by conventional rotary drilling (see Tab. 5). Because of the low service life of this diamond bit (a total of only 18.7 m) the rotary method was more economical (compare PANHORST 1978). Because of the above-mentioned disadvantages we were forced to discontinue drilling by the down-hole motor method. Rock bits proved to be the best tool for both drilling and reaming, not only because of splitting and caving of the rocks.

Technical problems could also arise while taking oriented cores. At temperatures above 90 to 100 °C a heat shield had to be mounted in order to prevent melting of the film of the multi-shot camera. It was advisable to limit the measurements to two to three within the first 1.5 metres cored; otherwise the wear of the knives in the orienting shoe was too severe and furthermore the core easily became stuck during times of standstill.

Fortunately we had no severe fishing job. Only some small iron parts had to be retrieved from the bottom of the hole by means of a fishing magnet and fishing pipe. Junk basket runs removed some larger rocks lying at the bottom.

4.4 Completion of the well

According to plan (Fig. 2) the well was cased with 18 5/8", 13 3/8", 9 5/8" and 7" casing (Mannesmann). The required cement float shoes and cement float collars were supplied by Baker, casing centralizers by Weatherford. All casing strings were run to the surface, with the exception of the 9 5/8" liner. For running this liner hydraulic liner hanger equipment of Texas Iron Works (TIW) was used. According to Tab. 4, the final 7" casing string was run and cemented from 12 to 14 July. (Originally the 7" casing had been planned as a liner, and a prolongation to the surface after testing of the various water-bearing horizons was foreseen).

More detailed data on casing and cementation are given in Tab. 9. Additional data have been compiled by MEIER & ERNST (1981).

Because of the mud losses below 2600 m the 7" casing had to be cemented in two steps in order to obtain a satisfactory cementation within the sections which were to be fractured afterwards. Before the cementation, the weight of the drilling mud was reduced (to 1.02 g/cm^3) by circulating CMC-water in order to avoid fracturing of the rocks by the drilling fluid.

The device for the two-stage cementation, a DV-Cementer of Halliburton and a 7"-X-Line landing collar, are located at a depth of 2520 m and 3296 m respectively. The required 7"-Casing Open-Hole Packer (type Brown, C 75, 32 lb/ft, API-LTC of Edeco) is situated at 2533.1. m.

As checked by a CBL-VDL-Log the cementation has succeeded well even in caved large-diameter areas. Water and mud remained behind the casing only between 2252.7 and 2566.0 m (that is, just below the stage cementer).

In general an unblended (DYKERHOFF) or dryly blended deep-drilling cement Class G of Halliburton was used; a slight exception is the Pozmix Light Cement (1.60 to 1.61 g/cm^3). The 13 3/8" is the only casing which was cemented on a saltwater base.

Only the 18 5/8" anchor casing was cemented to surface in order to test and to produce potential water-bearing horizons. The top of the cement (Tab. 9) of the 9 5/8" liner and of the 7" casing was met a little deeper than calculated: 980 m and 900 m respectively.

Table 9. Casing and cementation data of well Urach 3 below 28" conductor pipe (0–26.0 m) (W-4 = silica flour; NF-1 and D-Air = antifoamer and defoamer;

Casing					
OD (in.)	Depth (m)	Nominal weight (lb/ft)	Grade	Joint type	Guiding and floating of caising
18 5/8"	0– 642.4	96.5	K 55	STC	Cement Float Collar and Cement Float Shoe
13 1/8"	0– 598.16 –1080.0	61.0 69.0	K 55 K 55	STC STC	Cement Float Collar and Cement Float Shoe
9 5/8"	(liner) 980.47 –1808.5	36.0	K 55	STC	TIW LG Setting Collar, drillable Backoff Bushing, Hydro Hanger Single Cone, HS-Landing Collar and LS Set Shoe
7"	0–2445.77	26.0	N 80	LTC	Complete Twostage Cementer (D.V. Tool) in combination with an external Casing Packer
	2445.77 –2533.14	29.0	N 80	LTC	
	2533.14 –3320.00	29.0	N 80	E X L	Cement Float Collar with Bypass Baffle and Cement Float Shoe

In areas of overlapping, the different cementations proved to be tight, as was shown by pressure tests (up to 200 bar) in the annuli between 13 3/8" – 18 5/8" and 7" – 9"/8" – 13 3/8" casing (see Tab. 2 and 6); these tests were carried out before the fracturing operations in 1979. It was only in the course of these fracturing operations that slight leakages were observed between the two last-named casing strings and in the 7" casing section of the stage cementer.

Furthermore a very slight casing collapse of no more than 4 to 5 mm (lead-cast) was recorded in the lowermost part of the 7" casing between 3276 and 3284 m.

At the well head the well has been completed by a final flange and a christmas tree for pressures up to 700 bar (Cameron) and a stuffing box for running wire line instruments (Type 3" x 5000 lb/in^2 of Celler Maschinenfabrik).

Table 9. HR-12 and Saturated Salt Water = cement retarder; Salt = friction reducer; SSR = Sub-Surface Release).

Cementary plug method	Cement type	Cement slurry density kg/l	Weight of cement mixtures kg	Top of cement
TPCJ with Bottom Plug and Top Plug	Class G	1.92	144 180	up to surface
Typical primary Cementing (TPCJ) Job with Bottom Plug and Top Plug	Glass G + W – 4 + Salt + NF – 1 based on salt water	1.95 1.96	43 800	cement head about 520 m
SSR-Cementing Plug Method with TIW Liner Wiper Plug and Pump down Plug	Class G + W – 4 + Salt + HR – 4 + D – Air	1.94 1.96	31 500	cement head 1025 m
Two-stage Cementing with Free Fall Plug Set (Shut-off Plug, Opening Plug, Closing Plug	Special Pozmix + HR – 4	1.60 1.62	10 110	cement head about 1130 m
	Class G + W – 4 + Salt + HR – 4 + D – Air	1.90	29 207	
First stage Bottom Bypass Plug	Class G + W – 4 + HR – 12 + D – Air	1.98	64 147	

5. Deviation logging

For several reasons an exact knowledge of the deviation of the well was necessary. This is true for both degree and direction.

Single shot measurements were continously undertaken by non-directional survey instruments (Teledrift and Eastman). A continous Schlumberger deviation log was run before each casing run; it is compiled in Fig. 6, curve 1.

For comparison purposes another deviation log was run within the final 7" casing;

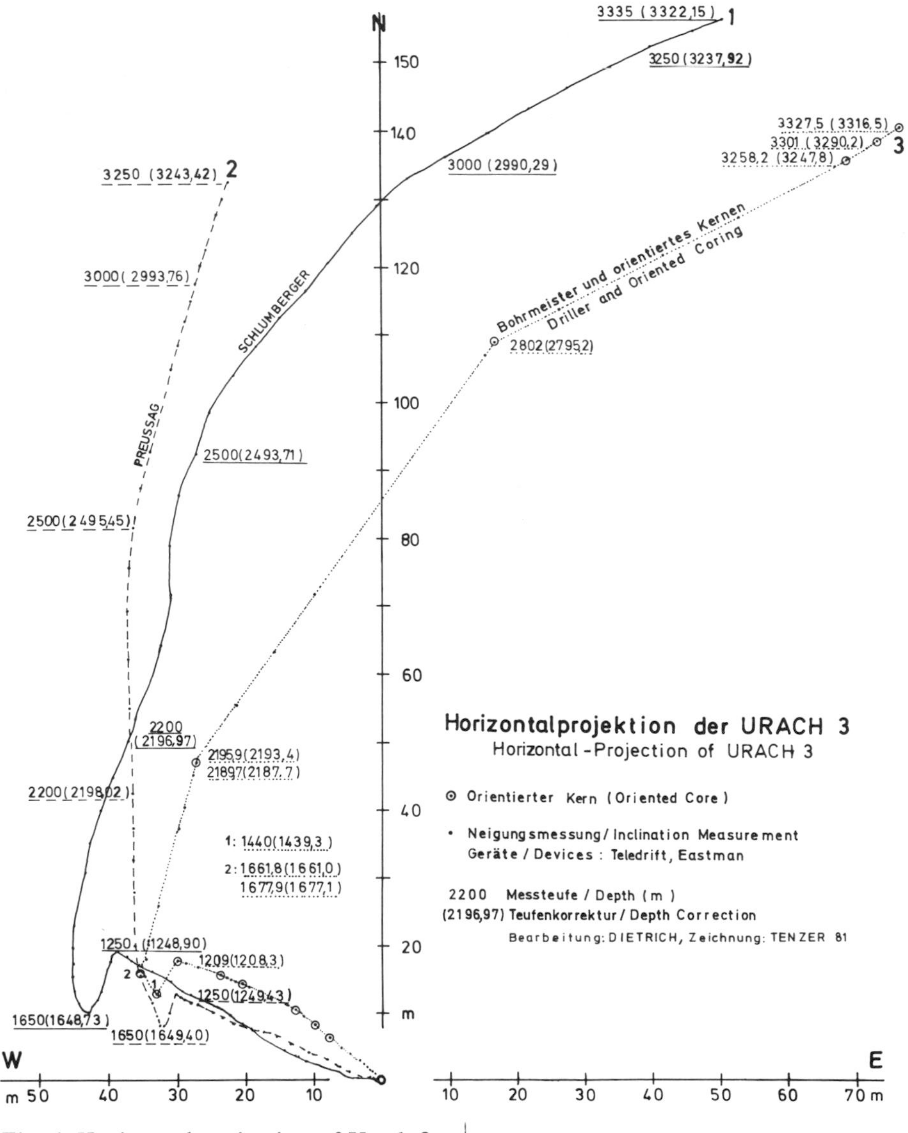

Fig. 6. Horizontal-projection of Urach 3.

it was done with the 2.5" Humphrey Gyro-compass and the result is shown in Fig. 6, curve 2. These gyroscopic directional surveying instruments are said to be extremely accurate with tolerance of no more than 2 % of the total deviation.

The two curves differ considerably, especially below 2500 m; above 2500 m differences are about 23 m in the horizontal and 1.7 m in the vertical direction. At 3250 m – the deepest point of curve 2 – the differences are 65 m and 5.5 m, respectively. Maybe there was an instrumental technical defect between 2500 and 3250 m, since the angles appear to be too low. Therefore driller measurements, for instance with Teledrift, in connection with oriented coring, have been compiled in a third curve. The general NE trend of the deviation between 2500 m and final depth has been verified by this measurement, although the number of oriented curves was limited. The Humphrey-Gyro measurement is to be repeated at a later date. According to the Schlumberger deviation log the landing point is about 165 m 20° NE of the surface point; the "loss" in depth is about 12 m.

The general deviation represented by all three curves is in accordance with the general dipping of the sediments (s. MAYER-GÜRR 1968). The sudden change of the deviation from NW to SW at 1250 m is probably caused by tectonics (see DIETRICH this volume)

Table 10. Well data bank geothermal research projekt Urach.

Well name : Urach 3
Drilling rig. : Cabot 900/2

On loc mooring : 27.09.77
Released unmooring : 21.07.78
Total metres drilled : 3334 m

Kind of work	h	%
1. Rig Up	65.0	0.91
2. Conductor	-.-	-.-
3. Drilling	1730.5	24.33
4. Redrilling	264.0	3.71
5. Enlarging	275.0	3.87
6. Underreaming	-.-	-.-
7. Cond. Hole/Circulation/Check Trips	48.5	0.68
8. Roundtrips	1486.5	20.90
9. Coring (incl. Circulation)	519.5	7.30
10. Running Casing	64.0	0.90
11. Running Liner	32.5	0.46
12. Cementing incl. Waiting Time and Drilling Casing Shoe	139.0	1.95
13. Flanging Up	61.5	0.87
14. Lost Circulation	235.5	3.31
15. Well Control	-.-	-.-
16. Case/Dev Control	46.5	0.66
17. Logging Evaluation	288.5	4.06
18. Formation Testing, Pumping Test	1143.0	16.07
19. Well Completion	15.0	0.21
20. Abandoning	-.-	-.-
21. Fishing Stuck Pipe	87.0	1.22
22. Rig Down	53.0	0.75
23. Repairs Drilling Equipment	97.5	1.37
24. Waiting for Weather	-.-	-.-
25. Waiting for Service	58.5	0.82
26. Waiting for Orders	16.0	0.23
27. Extra Downtime	385.0	5.42
Grand Total	7112.0	100.00

whilst the second changing tendency in the crystalline basement (beyond 1602 m) is believed to occur in connection with the overturn of the gneiss and the rock tensions in the crystalline basement.

6. Summary

The Urach 3 well, located in the central part of the Urach geothermal anomaly, spudded in on 3 October, 1977 and reached its final depth of 3334 m after 231 days on 22 May, 1978; a total of 1732 m of crystalline basement was therby penetrated

The average daily penetration rate of 14.4 m includes all drilling, coring, measuring, casing, and cementation operations, as well as tests including pumping of water-bearing horizons and stand-by times (Fig. 5). Fifty-nine days after reaching the final depth, after an extensive measuring and testing programme, after intensive reaming operations in the basement and after running and cementing of the 7" casing, the well was completed on 21 July, 1978.

Special attention is called to the scientific programme by drilling particulars compiled in Tab. 10. A share of 20 % of the total time was spent for measurements, tests and experiments. 32 % of the time was needed for real drilling operations and 21 % was needed for pulling and running the drill pipes including changing bits. In this connection it should be mentioned that the drilling outfit reached its technical capacity; this fact certainly contributed to the log running and pulling times.

With the completion of the well in a technically unobjectionable way, as described in this report, a first important target of the research project has been reached: We have here a well cased borehole, in which we can measure and observe the temperature of the earth's crust deep in the crystalline basement over a prolonged time; we can carry out fracturing operations at greater depths according to the hot-dry-rock concept and we can carefully investigate a possible economic utilization of the geothermal anomaly.

Acknowledgements

We want to thank all those persons who lent their help to the technically successful completion of this well. Special thank must be given to Mr. ADOLF HEINLE, retired drilling superintendent of BEB, whose knowledge and experience contributed so much to the technical achievement during the project.

References

American Petroleum Institute (1976): API Recommended Practice for Drill Stem Design and Operating Limit. – API RP G 7, 7th Edition, Dallas.
Blocksdorf, B. (1980): Erdwärme-Forschungsbohrung Urach. – Unveröffentlichte technische Abschlußarbeit an der Bohrmeisterschule Celle.
Burger, E. & Rischmüller, W. (1979): Die Bohrung Vorderriß 1 aus bohrtechnischer Sicht. – Erdoel-Erdgas, **95**: 336–344.
Carle, W. (1975): Die Mineral- und Thermalwässer von Mitteleuropa. Geologie, Chemismus, Genese. – Wissenschaftl. Verlagsges. (Stuttgart).
Dietrich, H.-G. (1979): Ergebnisse der Forschungsbohrung Urach. – 39. Jahrestagung, DGG, C 4–2, 179, Kiel 8.12. April 1979.

Dietrich, H.-G., Haenel, R., Neth, G., Schädel, K. & Zoht, G. (1980): Deep Investigation of the Geothermal Anomaly of Urach. – In: Strub, A.S. & Ungemach, P. (Eds.): Advances in European Geothermal Research International Seminar on the Results of EC Geothermal Energy Research, 2nd, p. 253–266 Strasbourg, D. Reidel Publ. Comp., Dordrecht/Boston/London.

Dietrich, H.-G. & Schädel, K. (1978): Untersuchung der geothermischen Anomalie in Urach auf eine mögliche Nutzung durch eine Untersuchungsbohrung bis tief ins Kristallin (ET 4023 B). – Progr. Energieforsch. u. Energietechn. 1977–1980, Statusrep. 1978 – Geotechnik und Lagerstätten, **1**: 79–85, Jülich (Projektleitung Energieforsch. (PLE), KFA Jülich).

Ernst, P. & Hieblinger, J. (1979): Aspekte der Nutzung geothermischer Energie. – Erdoel-Erdgas, **95**: 412–420.

Jürgens, R. & Marx, C. (1979): Neue Bohrmethoden für die Erdölindustrie. – Erdoel-Erdgas **95**: 132–140.

Mäussnest, O. (1974): Die Eruptionspunkte des Schwäbischen Vulkans. – Teil I. Z. Deutsch. Geol. Ges., **125**: 23–54.

Mayer-Gürr, A. (1968): Erschließung und Ausbeutung von Erdöl- und Erdgasfeldern. – In: Bentz, A. & Martini, H.J.: Lehrbuch der Angewandten Geologie, Bd. 2, Teil 1: 672–917, Stuttgart, F. Enke Verlag.

Meier, U. & Ernst, P.L. (1981): Drilling and Completion of the Urach 3 HDR Test Well. – International Conference on Geothermal Drilling and Completion Technology, **6**: 1–24, Albuquerque, New Mexico.

Panhorst, H.-J. (1978): Tiefbohren mit Diamant-Werkzeugen – Stand und Entwicklung. – Bergbau, **29**: 13–16.

Pucher, R. (1977): Bericht über Bodenmagnetische Untersuchungen in Urach. – Report NLfB Hannover, Archive No. 76 933.

Rischmüller, W. & Winkler, K. (1978): Erfahrungen mit neuen Geräten für die Bohrlochverlaufsüberwachung in gerichteten und nichtgerichteten Bohrungen. – Erdoel-Erdgas **94**: 318–324.

Rummel, F. (1977): Das Hot Dry Rock Projekt des Los Alamos Scientific Laboratory zur Gewinnung geothermischer Energie. – Berichte der KFA Jülich, Nr. 1410, Projektleitung Nichtnukleare Energieforschung.

Sauer, K. (1978): Geologie und Technik der Forschungsbohrung geothermisches Feld Urach. – Vortrag, 26. Haupttagung, DGMK, Berlin, Okt. 1978.

With, E. (1958): Die Schichtenfolge der Erdölaufschlußbohrung Buttenhausen 1, Schwäbische Alb. – Jber. u. Mitt. oberrh. geol. Ver., N.F. **40**: 107–128, Stuttgart.

The Urach Geothermal Project, p. 37–39;
Schweizerbart'sche Verlagsbuchhandlung, Stuttgart, 1982

The Buntsandstein (Lower Triassic) of the Urach 3 Borehole

J. LEIBER

Abstract: The Buntsandstein of borehole Urach 3 amounts to about 65 m of strongly reduced thickness. According to the used classification the Upper Buntsandstein comes up to the used classification the Upper Buntsandstein comes up to 23 m. The Middle Buntsandstein about 32 m, divided in Hauptkonglomerat with Kristallsandstein (12 m) and Eckscher Horizont (20 m). The Lower Buntsandstein can be estimated to be about 10 m. Bausandstein is not developed and so the Ecksche Horizont follows immediately below the Hauptkonglomerat. A petrographical description of this Buntsandstein is given.

The treatment of the Buntsandstein from Urach 3 was rendered difficult by taking drill cuttings only at intervals of two metres as well as by the impurity of rock samples. Only the exact examination especially of the shape of quartz grains permitted a classification of the strongly reduced thickness of Buntsandstein. It is assumed that a red mudstone layer in the core at 843.3 m depth is the upper border of the Buntsandstein. In another core at 935–936 m depth, parts of the Upper Permian Karneol-Dolomit Horizont were found. According to detailed analysis, it can be assumed that the Permian-Triassic boundary lies at a depth of about 908 m. Thus, the total thickness of the Buntsandstein amounts to about 65 metres.

1. **Upper Buntsandstein** (Röt-sequence 8)

A layer of red and also green mudstone, at least 30 cm thick, borders the basal sandstone of the Lower Muschelkalk. Below this in the core follow about 80 cm of pale red, coarse-grained, slanted sandstone beds with green clay gall. Down to about 860 m depth thereafter come brightly coloured, quartizitic, predominantly fine-grained but also to a minor degree mediumgrained sandstone rocks which are interbedded with red mudstone. Below this down to 866 m, whitish, light-pink and light-violet, fine- and seldom medium-grained sandstone rocks are encountered. White, sandy dolomite is also found in the lowest two metres, as are single carnelian fragments in this basal region. This is most likely the Karneol-Dolomit Horizont (VH 2) of the Buntsandstein. Thus, the thickness of the Upper Buntsandstein (including VH 2) amounts to about 22.5 metres.

2. Middle Buntsandstein

The only formations of the Middle Buntsandstein in the southern German classification are the Hauptkonglomerat including Kristallsandstein and the Ecksche Horizont.

2.1 **Hauptkonglomerat and Kristallsandstein** (Hardegsen-sequence 6)

Below the Upper Buntsandstein, sandstone rocks become clearly more coarsely grained. Their colour is an intense pink and below 870 m also red, and the sandstone becomes conglomeratic. The roundness of the quartz grains and the content of pebbles (predominantly milky quartz) persists down to about 878 m. The uppermost 2–2.5 m can be compared to Kristallsandstein because of the poorer roundness of the quartz grains. The remainder down to 878 m is to be considered as Hauptkonglomerat. Thus, the thickness of the Hauptkonglomerat together with the Kristallsandstein amounts to about 12 m.

2.2 **Eckscher Horizont** (Gelnhausen-sequence 2)

In the absence of the entire Bausandstein, the Ecksche Horizont follows immediately below the Hauptkonglomerat. This sandstone is red, coarse-grained and fine-gravelly, presumably slightly consolidated. Besides pebbles of milky quartz, also subrounded idiomorphous quartz has been found. This sequence, which also differs from the Hauptkonglomerat in the poorer roundness of the quartz grains, extends down to about 898 m. The thickness of the Ecksche Horizont amounts to about 20 m.

3. **Lower Buntsandstein** (Bröckelschiefer-sequence 1)

The Lower Buntsandstein in this borhole consists of red, medium-grained, presumably clayish bond sandstone. Dolomitic binding is also present. The quartz grains are only subrounded, and idiomorphous quartz grains sometimes occur too. Since the roundness of the quartz grains clearly diminishes once again at 908 m depth, and white dolomite and carnelian fragments then become frequent below this, it can be assumed that the Permian-Triassic boundary is located at this depth. Thus, the thickness of the Lower Buntsandstein can be estimated to be about 10 m.

4. **Paleogeographical Considerations**

This strongly reduced Buntsandstein is surprising and indicates the marginal region of the Buntsandstein basin. Further the Buntsandstein here exists in a facies and thickness which cannot be correlated to the deposits in the surrounding boreholes. The formation and thickness of the Hauptkonglomerat is comparable with the area of Königsfeld (Black Forest) (LEIBER & MÜNZING 1979, 1980). It was also surprising that the Hauptkonglomerat lies directly over the Ecksche Horizont in the absence of Bausandstein. Indeed, these conditions are known from the Schramberg-Königsfeld area (Black Forest); there the thickened Lower Buntsandstein is similar with regard to formation and thickness.

The location of the Urach 3 borehole at the edge of the Buntsandstein basin is, therefore, as far as the Middle and Lower Buntsandstein are concerned, directly comparable with the area of Schramberg-Königsfeld. It should be mentioned that the results of Urach 3 have shown that the thickness of the Buntsandstein in the boreholes of Albershausen and Überkingen presented by (CARLÉ 1971) now requires a revision.

References

Carlé, W. (1971): Die Tiefbohrungen auf mineralisiertes Thermalwasser in Bad Überkingen, Landkreis Göppingen, Baden-Württemberg. – Jh. Ges. Naturkde. Württemberg, **126**, 36–87, Stuttgart.

Leiber, J. & Münzing, K. (1979): Perm und Buntsandstein zwischen Schramberg und Königsfeld (Mittlerer Schwarzwald). – Jh. geol. Landesamt Baden-Württemberg, **21**, 107–136, Freiburg i. Br.

– (1980): Ergänzende Beiträge zur 4. Auflage von Blatt 7817 Rottweil. – In: SCHMIDT, M.: Erläuterungen zu Blatt 7817 Rottweil. – Geol. Karte Baden-Württ., 4. Aufl., 1980 [im Druck].

The Urach Geothermal Project, p. 41–47;
Schweizerbart'sche Verlagsbuchhandlung, Stuttgart, 1982

Petrology and Geochemistry of the Basement Rocks of the Research Drilling Project Urach 3

R. STENGER

with 3 figures and 2 tables

Abstract: The drilled crystalline rocks of the research borehole Urach 3 belong to the ENE-striking Moldanubian basement complex, thus representing a connexion between the pre-Variscan basement of the Black Forest and the Bavarian Massif. The Urach complex consists mainly of three rock units: 1. Coarse-grained metablastic plagioclase-biotite-(hornblende-)paragneiss. 2. Medium – to coarse-grained metablastic or metatectic plagioclase-biotite-(cordierite-)paragneiss. 3. Fine – to medium-grained diatectic rocks of plutonic character. All rock types are occasionally affected by retrograde processes of hydrothermal origin leading to distinctive alterations of the mineral components. The paragenesis of the alteration minerals is white mica (muscovite) + chlorite + calcite. According to chemical analysis, the Urach basement rocks may correspond to pelitic and Ca-rich greywacke compositions, with relatively low SiO_2 and high FeO, MgO and CaO contents. The hydrothermally altered rock portions show slightly lower contents of TiO_2, FeO, Na_2O, Sr and Ba. Lower values of the electrical resistivity and the sonic wave velocity in the depth ranges between 2050 and 2200 m and between 2500 and 3000 m may also be caused by strong hydrothermal activity.

Introduction

During the drilling of Urach 3 27 cores with a total length of 137 m were collected from the basement rock section which was penetrated from 1604 to 3334 m (final depth). Rock lithologies were determined by microscopic description, modal analysis and geochemistry, as well as by macroscopic examination.

The drilled rocks are part of the ENE-striking Moldanubian basement complex which crops out in the Black Forest and in the Bavarian Massif and which is covered by a thick Mesozoic series in the intermediate area. The Urach crystalline complex consists largely of three main rock units (from top to bottom):

1. Coarse-grained metablastic plagioclase-biotite-(hornblende-)gneiss.
2. Medium – to coarse-grained metablastic or metatectic plagioclase-biotite-(cordierite-)gneiss.
3. Fine – to medium-grained diatectic rocks of plutonic character.

The rocks are usually in a fresh condition, but throughout the whole sequence there are alteration zones, mostly of small volume, which may be caused by hydrothermal activity. At various depths within the borehole, granitoid or diatectoid rocks up to several metres in thickness are intercalated with relatively sharp (possibly tectonic) contacts to the wall rocks; in their petrographic and chemical properties they are different from the main rock unit 3 mentioned above. More often thin aplitic dykes a few centimetres in width, which in places are slightly foliated, cut the rocks. Occasionally the metablastic plagioclase-biotite-(hornblende-)gneisses, and especially the plagioclase-biotite-(cordierite-)gneisses contain mostly rounded enclaves with diameters up to 20 cm. They comprise fragments of strongly altered paragneisses and amphibolites. The diatectic rocks of the main unit 3 lack such fragments; only one gneiss body of about 2 m thickness with sharp contact to its wall rock was drilled. As for the petrographic and chemical characteristics, the gneiss is totally identical to the gneiss unit 2 of the above classification.

Open or healed joints of varying systems are ubiquitous in all rock units; the more brittle rocks additionally contain release joints, probably caused by mechanical processes during the drilling. The open or healed joints are occasionally striated and coated or healed with carbonates, chlorite, quartz, feldspar or even ore (pyrite).

Petrography

1. Metablastic Plagioclase-Biotite-(Hornblende-)Gneiss

The plagioclase-biotite-(hornblende-)gneiss forms the upper part of the crystalline complex. It is a very coarse-grained massive rock of bluish-grey colour displaying only weak orientation of biotites. Slight inhomogeneities result from variations in grain size and mica content.

The microscopic fabric is characterized by strongly metablastic relations, the dominant feature of which is plagioclase metablastesis. Less coarse-grained portions show a granoblastic matrix of quartz and plagioclase with coarse biotite flakes. The mineral assemblage is weakly zoned plagioclase (core: mole content = 45 to 52 % An, rim: 38 to 45 % An) + reddish-brown biotite + quartz ± slightly greenish hornblende ± colourless garnet. Accessories are apatite, zircon, ore and graphite. For further details see STENGER (1979, 1980).

2. Metablastic or Metatectic Plagioclase-Biotite-(Cordierite-)Gneiss

Macroscopic examination displays an inhomogenous, often more or less altered rock varying in petrographic features, either of coarse-grained metablastic fabric or differentiated by metatexis. Comparatively weak metatexis produces segregation into leucocratic portions rich in quartz and feldspar and melanocratic portions rich in biotite. There are occasionally frequent enclaves of relictic rocks in the gneiss as well as irregularly shaped metatectoid bands and schlieren, rounded biotite concentrations and bluish or greenish clusters of cordierite altered to pinite.

The rock portions most strongly affected by metatexis reveal a microscopic structure which is differentiated into melanosomes and coarser grained leucosomes, with variable mineral contents and modal compositions, respectively. The melanosomes comprise biotite and cordierite (completely altered to greenish or pale-brown pinite) aggregated with each other as main constituents, and plagioclase and quartz in minor amounts. The leucosomes consist of unzoned plagioclase (sometimes with antiperthitic domains) and quartz, additional microperthitic potash feldspar, and sometimes cordierite in well-shaped crystals may enter the assemblage. Rock portions less or non-affected by metatexis show metablastic relations resembling the plagioclase-biotite-(hornblende-)gneisses of the main unit 1 (see above). Accessory minerals are colourless garnet, zircon, ore, rare apatite and graphite. Sillimanite is not a constituent of the mineral assemblage.

3. Diatectic Rocks of Plutonic Character

This series of rocks, which comprises the bottom of the drilled crystalline complex may be divided into two subgroups: a dark grey fine-grained type and a medium-grained type. The grain size is variable in both directions, but no contacts or transitions between the rock types are confirmed by core samples. Slight orientation of biotite and plagioclase can be observed in both rocks. A common and very characteristic feature of both types are small-scaled nebulitic melt patches consisting of quartz and feldspar. The grain size in the vicinity of such patches varies very strikingly.

Microscopic features are characterized by a perfectly homophanous hypidiomorphic-granular fabric, with more or less well developed lineation caused by alignment of biotite flakes. The mineral assemblage is given by zoned plagioclase (core = 45–55 % An, rim = 35–45 % An) + quartz + reddish-brown biotite ± microperthitic potash feldspar ± slightly greenish or brownish hornblende. The classification according to modal analysis displays granodioritic to tonalitic and quartzmonzodioritic compositions.

The average quantitative mineral compositions of the different rock types are given in Tab. 1. Additionally, the compositions of 89 rock samples are plotted in the triangle diagram quartz-feldspars-mafic minerals (see Fig. 1). There is a clear difference between the metatectic plagioclase-biotite-(cordierite-)gneisses on one side and the diatectic rocks on the other; the metatectic gneisses are richer in quartz on the average, whereas

Table 1. Average modal compositions of the basement rocks of the research borehole Urach 3.

Volume content in %	Plagioclase-biotite-(hornblende-)gneiss	Plagioclase-biotite-(cordierite-) gneiss	Fine-grained diatexites	Medium-grained diatexites
Plagioclase	44.6	29.1	47.2	46.5
Quartz	19.0	29.3	18.8	17.4
Potash feldspar	- -	5.6	4.4	9.8
Biotite	32.1	21.2	28.6	24.6
Hornblende	3.2	- -	0.4	1.1
Cordierite	- -	13.5	- -	- -
Accessories (incl. garnet)	1.1	1.3	0.6	0.6

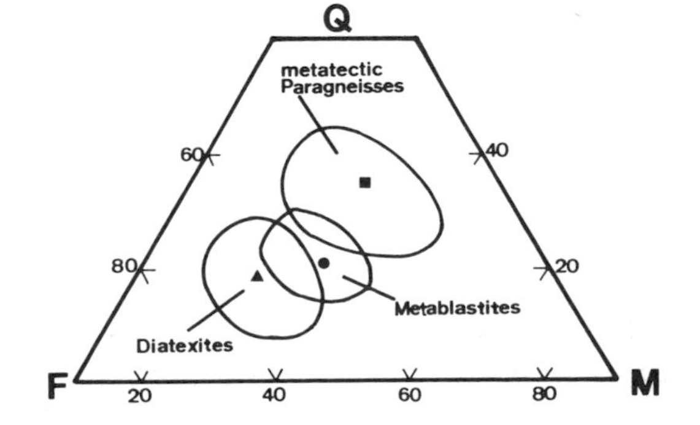

Fig. 1. Modal compositions of the basement rocks of the research borehole Urach 3, plotted in the diagram quartz-feldspars-mafic minerals.

the metablastic plagioclase-biotite-(hornblende-)gneisses occupy an intermediate position between the other groups. Compared with similar metamorphic rocks of the Black Forest, the Urach diatexites and part of the metablastic gneisses reveal a tendency to more mafic compositions.

Hydrothermal Alterations

All rocks of the crystalline complex, especially the plagioclase-biotite-(cordierite-)gneisses, are occasionally affected by retrograde processes of hydrothermal origin. These usually proceed from joints or thin veinlets and comprise zones of a few millimetres up to a metre width on both sides; in extreme cases they may occupy far larger volumes. The rocks then display pale greenish or almost white colours with additional orange or brownish touch; they are strongly interspersed with thin veinlets filled with carbonate, quartz, feldspar (adularia) and ore (pyrite). The primary fabric relationships are seriously changed; hence identification of the original mineral content may be difficult. In particular the changes are as follows:

- Plagioclase is altered into fine-flaky white mica and/or calcite; occasionally fine-crumbly, slightly greenish chlorite can be found.
- Potash feldspar is far less affected; only minor growth of white mica.
- Biotites are completely bleached, either by chloritization or by alteration into white mica; simultaneously segregation of iron- and especially titanium-bearing phases (limonite, rutile needles, leukoxene).
- Hornblende is, when present, altered early into a mixture of chlorite and calcite.
- Quartz remains largely unaffected; only strong undulating extinction.
- In strongly altered rock portions calcite may be settled as small sparry crystals between other minerals.

No significant estimation of the pT-conditions under which the retrograde processes occur can yet be deduced from the paragenesis of these alteration minerals.

Geochemistry

44 samples, which are representative for each rock unit distinguished with regard to petrographic evidence, were chosen from different core sections to obtain the bulk chemical composition of each type. Within the units of the plagioclase-biotite-(cordierite-) gneisses and the diatectic rocks, both fresh and hydrothermally altered samples were analysed. The major and minor oxides as well as some trace elements were determined by X-ray fluorescence spectrometry; FeO was determined by titration with $KMnO_4$, and the values for H_2O, CO_2 and S were obtained by using gravimetric (H_2O) and automatic titration methods (CO_2, S).

The data show that no larger variations of the element concentrations occur within the particular groups. Even the hydrothermally altered rocks do not differ decisively from the fresh ones, although some shifting in element concentration may occur, for instance Ti, Fe, Na, Sr and Ba. Certain element ratios, on the other hand, display more pronounced differences, as K/Sr, Ca/Sr, K/Ba, Rb/Sr (higher in altered rocks) and Ba/Rb (lower in altered rocks).

The chemical compositions of the rocks are given as average values in Tab. 2. Compared with those of the paragneisses of the Black Forest, the bulk compositions are relatively basic because of lower SiO_2 and higher FeO, MgO and CaO (see Fig. 2). The Urach basement rocks may correspond to pelitic and Ca-rich greywacke compositions. Metagrey-

Table 2. Average chemical composition of the basement rocks of the research borehole Urach 3.

Weight content in %	Plagioclase-biotite-(hornblende-) gneiss	Plagioclase-biotite (cordierite-) gneiss		Diatexites	
		fresh	altered	fresh	altered
SiO_2	56.31	63.29	62.22	58.58	57.71
TiO_2	1.21	0.90	0.79	1.55	1.43
Al_2O_3	17.08	16.00	15.29	16.17	15.35
FeO	5.91	5.06	4.55	5.75	4.67
Fe_2O_3	1.24	0.96	0.72	0.78	0.60
MnO	0.12	0.09	0.09	0.10	0.11
MgO	4.72	2.72	2.39	2.92	2.46
CaO	3.61	1.50	2.00	4.19	3.81
Na_2O	2.85	2.35	1.62	3.14	2.05
K_2O	2.58	3.29	3.16	3.30	3.35
P_2O_5	0.30	0.11	0.12	0.81	0.77
H_2O	2.58	2.41	2.70	1.62	2.56
CO_2	1.04	1.05	4.15	0.71	4.87
S	0.21	0.19	0.13	0.09	0.09
Rb	103	126	130	134	165
Sr	309	182	86	384	160
Ba	963	939	470	1211	514
K/Sr	90	220	447	86	224
Ca/Sr	117	95	250	109	254
K/Ba	27	41	72	27	84
Rb/Sr	0.362	0.863	1.800	0.347	1.105
Ba/Rb	9.28	7.26	3.67	9.36	3.00

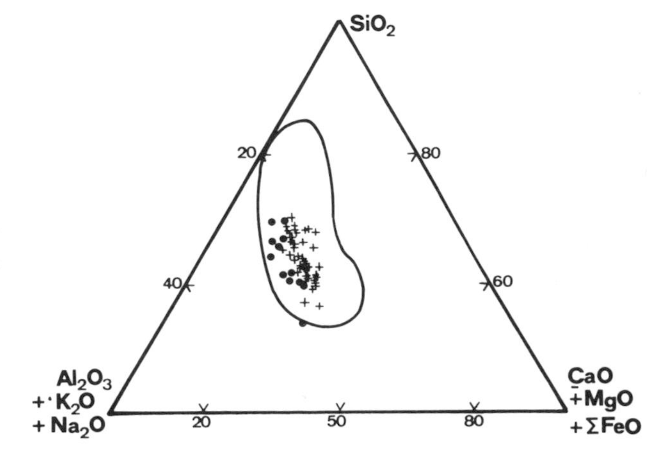

Fig. 2. Chemical compositions of the basement rocks of Urach 3 (crosses), gneisses of the drilling project Nördlingen (filled circles) and paragneisses of the Black Forest (bordered field).

wacke gneisses richer in SiO_2, which are frequent in the Black Forest, do not occur within the analysed rocks of the Urach drilling project, or within the gneisses of the research borehole Nördlingen/Ries (GRAUP 1977).

Within the Urach basement complex, the plagioclase-biotite-(hornblende-)gneisses show the highest values of FeO and MgO and a relatively high CaO content, thus reflecting the high modal biotite and plagioclase content. The diatexites show the highest values of CaO, corresponding to the considerable plagioclase content rich in An. With regard to the alkali elements, the plagioclase-biotite-(cordierite-)gneisses and the diatexites are more K-pronounced (alkali ratio $Na_2O/K_2O < 1$), whereas the plagioclase-biotite-(hornblende-)gneisses display higher Na-contents ($Na_2O/K_2O > 1$).

Some microprobe work was done to obtain the chemical composition of the main mineral constituents. The biotites of the plagioclase-biotite-(hornblende-)gneiss are comparatively rich in Mg (molar ratio Fe/Mg = 0.8–0.9), whereas the biotites of the other rock units have higher Fe contents (Fe/Mg = 1.2–1.5 in the plagioclase-biotite(cordierite-) gneisses, Fe/Mg = 1.1–1.2 in the diatexites). Plagioclase is rich in Ca in the metablastic and diatectic rocks of the units 1 and 3 (An 45–50) with normal compositional zoning, and poor in Ca in the metatectic rocks of unit 2 (An 25) without zoning. Potash feldspar is slightly richer in K in the diatexites (Or 95) than in the metatectic gneisses (Or 90). Hornblende contains considerable amounts of Ca and Al in addition to Fe and Mg in the diatexites, whereas the crystals in the metablastic rocks of unit 1 display compositions poor in Ca and Al and rich in Fe and Mg. The composition of the garnet, which is usually unzoned, in terms of end members was found to be: $Alm_{66}Pyp_{23}Gross_6Spess_5$ in the plagioclase-biotite-(hornblende-)gneiss, and $Alm_{71}Pyp_{20}Gross_4Spess_5$ in the plagioclase-biotite-(cordierite-)gneiss. Cordierite could not be analysed because of the complete alteration into pinite.

The rock composition bears some significance with regard to possible reactions of the reservoir rocks with water injected for heat extraction. Experiments executed by ALTHAUS (1980) and ROTTENBACHER (1978) show that reactions of aqueous fluids with feldspar and mafic minerals lead to considerable amounts of material dissolution and transport. According to the results presented in this paper, Ca-bearing plagioclase is the dominant feldspar phase (30–50 % of rock volume). Within the hydrothermally altered rock portions, calcite and other carbonates occur as water-soluble minerals with weight contents up to 11 %. As described above, such carbonate-bearing rocks comprise considerable volumes of the whole rock complex drilled (about 10 to 20 %). Sulphur contents in the form of possibly decomposable pyrite range within a few tenths of a percent.

Correlation of Rock Composition and Geophysical Parameters

Among the results of the various geophysical measurements executed in the borehole and at the cores, the electrical resistivity and the sonic wave velocities display significant relations to the petrographic properties of the rocks. The occasionally strong hydrothermal alterations in the depth ranges between 2050 and 2200 m as well as between 2500 and 3000 m are outstanding because of their lower resistivities and wave velocities (see Fig. 3, reprinted from HAENEL et al. 1980). To a lesser extent, the values of the thermal conductivity of the rocks and the heat flow density are variable, which again is possibly caused by hydrothermal alteration, rather than by primary differences in rock composition. The results of rock magnetism and radioactivity measurements do not yet allow any assertions about significant relations to the petrographic features.

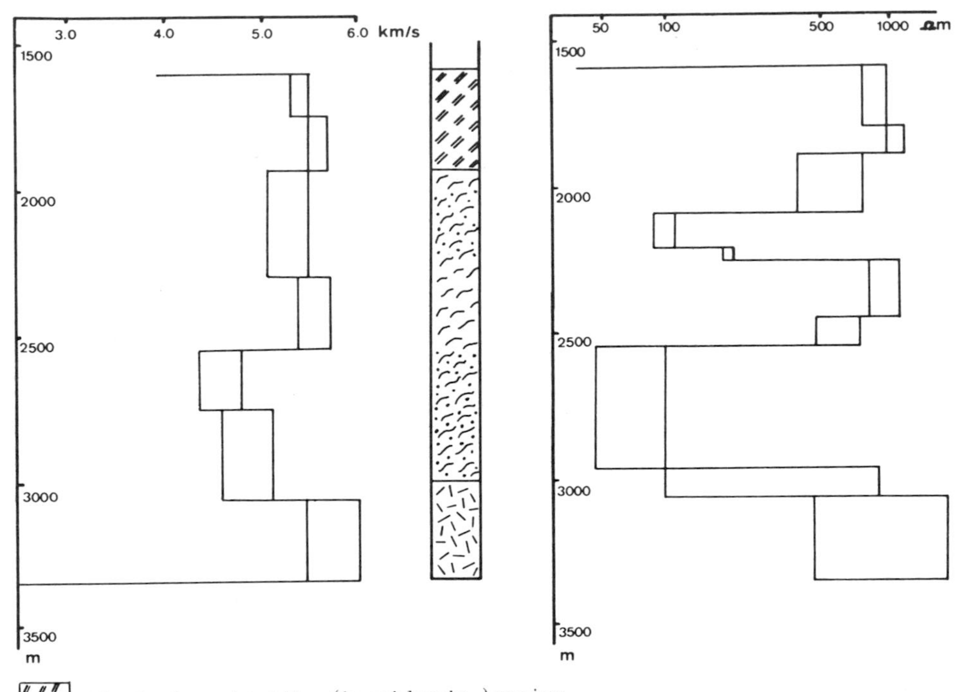

Fig. 3. Sonic wave velocities and electrical resistivities of the basement rocks of Urach 3 with additional rock profile.

References

Althaus, E. (1980): Experiments on the Interaction between Rocks and Fluids in Geothermal Systems. – 2. Int. Seminar on the Results of EC Geothermal Energy Research.

Graup, G. (1977): Die Petrographie der kristallinen Gesteine der Forschungsbohrung Nördlingen 1973. – Geol. Bav., **75**: 219–229.

Haenel, R. et al. (1980): Endbericht über Geophysikalische Untersuchungen in der Forschungsbohrung Urach. – Report, NLfB Hannover, Archive No. 85805.

Rottenbacher, K. (1978): Ermittlung von Reaktionsbeziehungen zwischen wässrigen Wärmeübertragungsmedien und den Mineralen der als Wärmespeicher wirkenden Gesteine. – Statusreport 1978, KFA Jülich, p. 27–36.

Stenger, R. (1979, 1980): Quartalsberichte zum Forschungsprojekt 3–ET 4251 A, KFA Jülich.

The Urach Geothermal Project, p. 49–58;
Schweizerbart'sche Verlagsbuchhandlung, Stuttgart, 1982

Geological Results from the Urach 3 Borehole and the Correlation with Other Boreholes

H.-G. DIETRICH

with 2 figures and 2 tables

Abstract: The sedimentary cover penetrated by the Urach 3 research borehole is 1602 m thick and comprises strata from the Jurassic to the Young Palaeozoic. The presumed existence of the Rotliegend through has been confirmed. The strata subjacent to the Rotliegend, which also form the sedimentary base, have for the first time been shown to be of Carboniferous origin. The crystalline basement, which has been drilled down to a final depth of 3334 m, can be subdivided into several units.

On the basis of comparisons between various drilling profiles, it can be assumed that the SW-NE-striking Schramberg trough extends through the Urach area all the way to Überkingen. Correlations transverse to the through axis indicate that the initially formed trough was narrower, and that areas adjacent to the north did not undergo depression until the beginning of the Upper Rotliegend.

Introduction

The stratigraphic sequence below Urach had been determined down to the Middle Muschelkalk by means of the two thermal water wells, Urach 1 and 2, drilled in 1970 and 1974, respectively. Preliminary information concerning the stratigraphic sequence thereby penetrated is given by SAUER (1971). The geological survey profile of the Urach 1 well has been published by HAHN & SCHÄDEL (in GWINNER 1974). Brief descriptions based thereon are provided by CARLE (1975) as well as GEYER & GWINNER (1979).

A general conception of the structure, thickness and stratification of the underlying, unexplored portion of the sedimentary cover is presented by SCHÄDEL (1977). A report on the geological profile actually drilled is given by SAUER (1978). The geological survey profile for the well is presented by DIETRICH & SCHÄDEL (1978), BUNTEBARTH et al. (1979) and DIETRICH et al. (1980) (see also BRENNER 1979, WOHLENBERG 1978).

New results concerning the subdivision of the lower portion of the sedimentary cover have been presented by DIETRICH & LEIBER at the meeting of the Oberrhein Geological Society (Geologischer Verein) on 9th April, 1980. These and other presentations concerning the Urach 3 borehole have been summarized by KULL (1981).

Within the scope of the present article, only a few results concerning the subdivision of the stratigraphic sequence can be discussed. The essential objective here is to compare and correlate the drilling profile of the Young Palaeozoic obtained from the Urach 3 borehole with profiles from other boreholes.

The detailed description of the stratigraphic sequence is to be presented later as a contribution to the proceedings of the Geologisches Landesamt Baden-Württemberg.

Basis for the Evaluation

For establishing the geological borehole profile, rock specimens in the form of cuttings (every 2 m) and of representative cores have been recovered according to plan. About one-tenth of the distance drilled in Urach 3 has been cored, that is, 172.8 m in the sedimentary cover and 154.6 m in the crystalline basement. Moreover, as shown in Tab. 1, some of the cores recovered were oriented, in order to elucidate the bedding conditions in the rock. The success in recovering oriented cores, as well as the core recovery in fact achieved, are likewise indicated in Tab. 1. Furthermore, the usual Schlumberger tests were conducted for the same purpose (see DIETRICH this volume). In addition, the drilling progress (reciprocal drilling speed in min/m) was plotted as a function of depth. The drilling progress log thus resulting has proven useful for the interpretation of the lithostratigraphic findings derived from cores and cuttings.

Table 1. Coring runs, distances cored, and recovery of oriented cores in the Urach 3 borehole (EMS ≙ Eastman Multishot; SSM ≙ Sperry-Sun Multishot with heat shield; K.F. ≙ no film capable of evaluation; mo ≙ Upper Muschelkalk; mu ≙ Lower Muschelkalk).

Coring run no.	Depth/ distance cored, in m	Core recovery m/%	Formation	Oriented core recovery: Oriented core	Success	Remarks
1	511.5– 520.5	8.0/ 88.9	Keuper	+	+	EMS
2	565.0– 573.5	8.0/ 94.1	Keuper	+	+	EMS
3	598.0– 607.0	9.0/100	Keuper	–		
4	659.0– 668.0	9.0/100	Keuper/mo	+	+	EMS
5	717.0– 726.0	8.0/ 88.9	Muschelkalk (mo)	–		
6	832.0– 837.8	5.5/ 94.8	Muschelkalk (mu)	+	– (+)	EMS, K.F.
7	837.8– 844.8	6.4/ 91.4	mu/Buntsandstein	+	+	EMS
8	935.0– 942.9	7.3/ 92.4	Rotliegendes	+	+	EMS
9	983.0– 992.0	8.8/ 97.8	Rotliegendes	–		
10	1017.0–1025.6	7.4/ 86.0	Rotliegendes	–		
11	1064.0–1073.0	9.0/100	Rotliegendes	+	–	EMS, K.F.
12	1073.0–1082.0	9.0/100	Rotliegendes	–		
13	1200.0–1209.0	9.0/100	Rotliegendes	+	+	EMS
14	1260.0–1269.0	9.0/100	Rotliegendes	–		
15	1269.0–1278.0	9.0/100	Rotliegendes	–		
16	1278.0–1287.0	9.0/100	Rotliegendes	–		
17	1287.0–1296.0	9.0/100	Rotliegendes	–		
18	1353.0–1362.0	9.0/100	Rotliegendes	–		
19	1422.0–1431.0	9.0/100	Rotliegendes	+	+	EMS
20	1431.0–1440.0	9.0/100	Rotliegendes	+	+	EMS
21	1634.0–1635.7	1.0/ 59.8	crystalline	–		
22	1635.7–1643.7	7.1/ 88.8	crystalline	–		
23	1660.0–1661.8	0.15/ 8.3	crystalline	+	–	EMS, no recovery
24	1661.8–1664.7	2.0/ 69.0	crystalline	–		
25	1664.7–1668.4	0 / 0	crystalline	–		
26	1675.0–1677.9	2.2/ 75.9	crystalline	+	+	EMS
27	1750.5–1756.8	6.3/100	crystalline	–		
28	1853.0–1862.0	7.8/ 86.7	crystalline	–		
29	1945.0–1953.0	7.4/ 92.5	crystalline	–		
30 a	2013.5–2017.3	3.8/100	crystalline	–		
30 b	2017.3	0 / 0	crystalline	–		
31	2124.2–2133.2	8.2/ 91.1	crystalline	–		
32	2189.5–2195.7	5.8/ 93.5	crystalline	+	+	EMS

Coring run no.	Depth/ distance cored, in m	Core recovery m/%	Formation	Oriented core recovery: Oriented core	Success	Remarks
33	2302.4–2311.4	8.7/ 96.7	crystalline	–		
34	2412.0–2421.0	8.7/ 96.7	crystalline	–		
35	2517.5–2521.3	3.7/ 97.4	crystalline	–		
36 a	(2585 test)	–/–	crystalline	–		
36 b	2795.7–2802.0	5.6/ 88.9	crystalline	+	+	EMS
37 a	(2884 test)	–/–	crystalline	–		
37 b	2930.0–2939.0	8.5/ 94.4	crystalline	–		
38	3053.0–3060.0	5.5/ 78.6	crystalline	–		
39	3125.0–3132.0	3.1/ 44.3	crystalline	–		
40	3205.0–3211.0	5.8/ 96.7	crystalline	+	–	EMS, K.F.
41	3250.0–3258.2	7.9/ 96.3	crystalline	+	+	SSM
42	3290.0–3294.7	4.3/ 91.5	crystalline	+	– (+)	SSM, K F.
43	3294.7–3298.4	3.7/100	crystalline	+	+	SSM
44	3298.4–3301.0	2.1/ 80.8	crystalline	+	– (+)	SSM, K.F. break-down
45	3302.0–3306.2	3.6/ 85.7	crystalline	+	– (+)	EMS, K.F.
46	3321.0–3321.9	0.6/ 66.7	crystalline	–		
47	3323.0–3325.3	1.2/ 52.2	crystalline	–		
48	3226.0–3330.6	3.1/ 67.4	crystalline	+	+	SSM
49	3331.0–3334.0	2.0/ 66.7	crystalline	+	– (+)	SSM, K.F. break-down

(+): Supplementary core orientation performed later by means of palaeomagnetic evaluation or juxtaposition with subsequent oriented cores.

The Geological Profile

With the Urach 3 research borehole, drilled in 1977/78, the entire thickness of the sedimentary cover was penetrated and the upper limit of the crystalline was reached for the first time in the region of the Urach thermal anomaly. It is situated at a depth of 1602 m (≙ 1176 m below mean sea level) below the drilling site. A further distance of 1732 m was drilled into the crystalline basement. Thus the bottom of the well is located at a depth of 3334 m below the surface.

The subdivision deduced from the cores and cuttings from the borehole is presented in Tab. 2. Depth corrections resulting from the dependence on the time of ascent and on the drilling progress for the drilling fluid samples have already been taken into consideration for these data. The deviation of the well from the vertical is not taken into consideration (see DIETRICH this volume). The evaluation of the deviation measurements of the Schlumberger Company yields a depth loss of 1.25 m upon reaching the upper limit of the crystalline, and of 11.85 m upon reaching the final depth.

The facies and thickness of the stratigraphic sequence penetrated agree well with the preliminary profile as far down as the Buntsandstein. Its upper boundary with the predominantly sandy to coarsely clastic Lower Muschelkalk is characterized by cored Roethian clay. After the investigations of LEIBER (this volume), who has analysed the Buntsandstein profile, the lower boundary with the Permian had to be redetermined, in contradiction to previous views. The Mesozoic stratigraphic sequence is presented by SCHÄDEL (this volume) and is partially reproduced in the simplified column profiles of Fig. 1 and 2.

Table 2. Geological survey profile of the Urach 3 borehole (drilling site: R 35 27 652, H 53 74 430; altitude above mean sea level: about 426 m; final depth: 3334 m).

Depth below surface in m	Depth referred to mean sea level in m	Formation in m		Thickness in m
– 13	+ 413	Quaternary		13
– 283	+ 143	Dogger		270
		– 76	Bajocian	63
		– 283	Aalenian	207
– 395	+ 31	Lias		112
		– 299	Toarcian	16
		– 332	Pliensbachian	33
		– 368	Sinemurian	36
		– 395	Hettangian	27
– 668	– 242	Keuper		273
		– 401	Upper Keuper	6
		– 649	Middle Keuper	248
		– 668	Lower Keuper	19
– 843	– 417	Muschelkalk		175
		– 744	Upper Muschelkalk	76
		– 796	Middle Muschelkalk	52
		– 843	Lower Muschelkalk	47
– 905		Buntsandstein		62
		– 861	Upper Buntsandstein	18
		– 895	Middle Buntsandstein	34
		– 905	Lower Buntsandstein	10
– 1485	– 1059	Permian	Rotliegendes	580
		≅ – 1247	Saxonian	342
		≅ – 1485	Autunian	238
– 1602	– 1176	Carboniferous (Stephanian)		117
– 3334	– 2908	Crystalline basement		1732
		≅ – 1950	metablastite	≅ 348
		≅ – 3000	paragneiss/metatexite	≅ 1050
		– 3334	diatexite	≅ 334

The subdivision of the Young Palaeozoic stratigraphic sequence is rendered difficult by the scarcity or lack of fossils and by the pronounced lithofacial alterations. From the upper limit of the crystalline basement to the Permian/Triassic boundary delineated by LEIBER (this volume), the sequence with a total thickness of about 700 m can be subdivided into three stratigraphic units.

The lowest section of the sedimentary cover consists of a thin basal conglomerate stratum and an overlying series of red-brown siltstones, which are in turn overlain by arcoses as well as sandstones and siltstones, in part with pronounced cross bedding and thin coal seams. A multicyclic sequence is evident in the sedimentation. Shades of grey predominate besides red-brown and greenish colorations. Palynological investigations by GREBE (in BUNTEBARTH et al. 1979) result in a classification into the Lower to Middle Stephanian.

In contrast, other palaeobotanical investigations conducted on probably equivalent stratigraphic sequences in the Schramberg region could not demonstrate the presence of Upper Carboniferous with any certainty (cf. for example FALKE 1971, LEIBER &

MÜNZING 1979). In view of this fact, it is plausible that the coal-bearing layers in the Urach profile also belong to the Lower Rotliegend.

In agreement with GEYER & GWINNER (1968), the upper limit of this section was drawn in the conventional manner where the grey rocks are succeeded by predominantly red or other coloured rocks. Accordingly, the boundary with the Autunian is drawn at 1485 m. Minor alterations in the position of the boundary may result from further analysis of the comprehensive data.

The Lower Rotliegend (Autunian) is characterized by frequent products of volcanic ascent, besides the siltstones and arcoses often with a high content of mica. The volcanic rocks include tuffs, ignimbrites and porphyrites. Large, conspicuous xenocrysts of orthoclase in fine-grained cinerite and/or tuffites could perhaps be viewed as diagenetic products, just as alkali feldspars, probably of the same age, which occur in clays with volcanic material in the Saar-Nahe area (HEIM 1960, v. ENGELHARDT 1973). Intensive silification, especially in zones of tuffs and of a few clastic sediments, occurs also in the Black Forest and in the Saar-Nahe area. It is attributed to hydrothermal processes by GEYER & GWINNER (1968), and to solutions due to thermal processes and weathering by FALKE (1965). Silification is known to occur also in products of younger volcanic ascent. Thus, WEISKIRCHNER (1980) has found that in all rocks of the "Swabian volcano" silica can be liberated by weathering and can be observed as fillings or in finely divided form.

The upper limit of the Autunian has been tentatively drawn at about 1247 m, above a layer of tuff. According to the present state of knowledge, a weak angular unconformity occurs in this section of the profile. The overlying sediments generally dip weakly toward the southeast (cf. SCHÄDEL this volume), whereas a weak dip toward the northeast is to be expected below this limit. Indications of such a dip are provided by an oriented core with a mean dip of the stratification by about 3.5°, 35° to the northeast. Various values from a dipmeter measurement performed by Schlumberger are likewise indicative of a dip of the strata in the northeastward direction. It is notable that a change in the dip of the stratification is also evident from the directional course of the borehole (cf. DIETRICH this volume). Banks with brecciated texture (cores no. 15 and 16) located just below this limit are indicative of tectonic movements, to which the disconformity can be ascribed.

It is possible that this weak angular unconformity does not in fact concur with the Autunian/Saxonian boundary, but rather represents pre-Saalian tectonic movements (e.g. KATZUNG 1972). In the region of the Black Forest too, tectonic processes which occurred in the upper part of the Lower Rotliegend and which led to a shift of the distributive province are described (e.g. GEYER & GWINNER 1968, as well as FALKE 1972, 1976).

Hence it is possible that the boundary with the Saxonian is located at a higher level. Thus it might concur with the lower dotted line in Fig. 1 and 2 (at about 1128 m) in the zone of the Saxonian. The predominantly red-brown fanglomerate and siltstone sequences, which will not be discussed here, lie above this line.

Dolomite- and carneole-bearing sediments in the upper region of the Saxonian delineated here may possibly be ascribable to the Zechstein (Upper Permian). According to various authors (e.g. FALKE 1972, 1976; BACKHAUS 1976; LEIBER & MÜNZING 1979), the sediments in question would then occur as Zechstein in Rotliegend facies.

The attempt made by KÄDING (1978) to correlate the available drilling profiles between North and South Germany is an effort in the same direction. According to his investigations, it is possible to correlate the boundary between Buntsandstein and Zechstein continously all the way to the Bad Cannstatt profile (CARLE 1975). In that area, sediments which have hitherto been ascribed to the Buntsandstein by CARLE are thereby

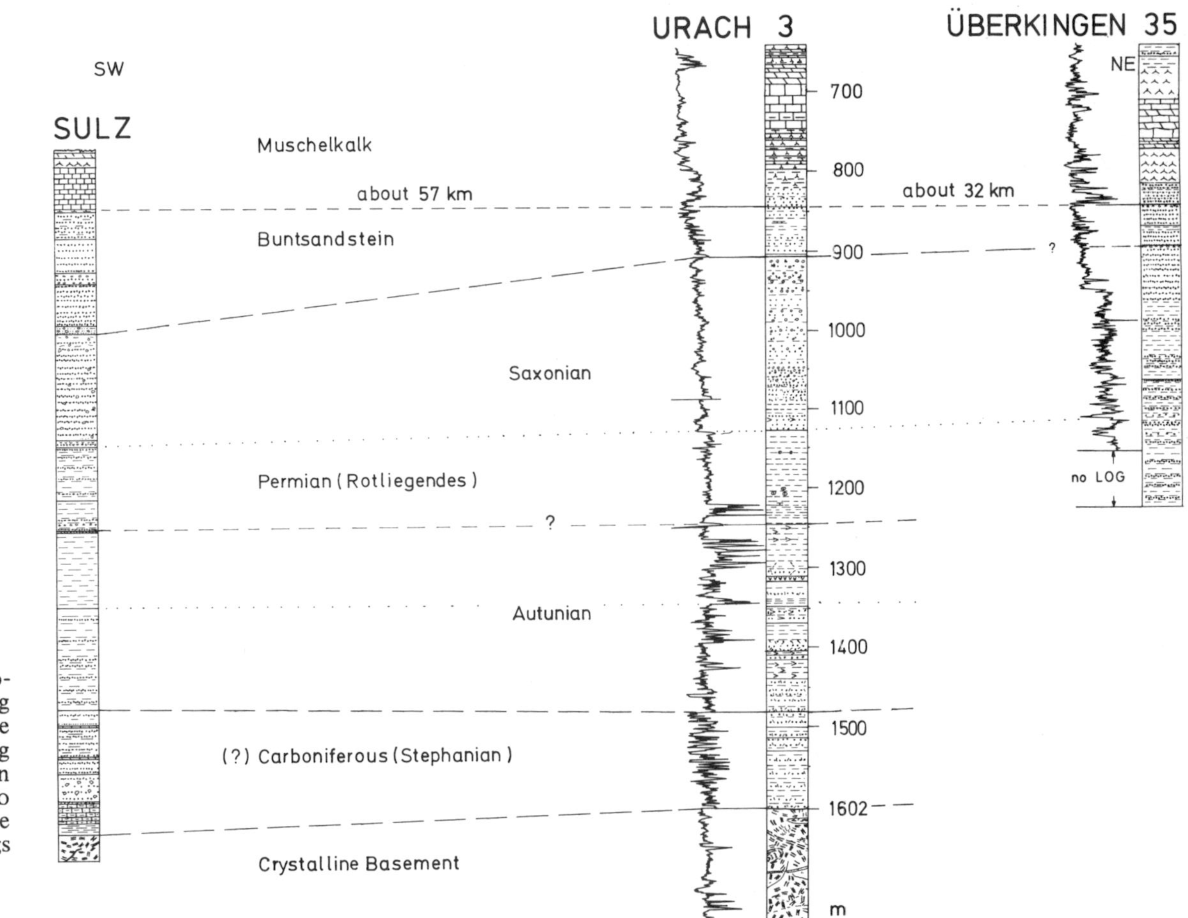

Fig. 1. Simplified litho-stratigraphic drilling profiles along the strike of the Schramberg trough and correlation endeavour, taking into account the available Gamma Ray Logs (Legend see fig. 2).

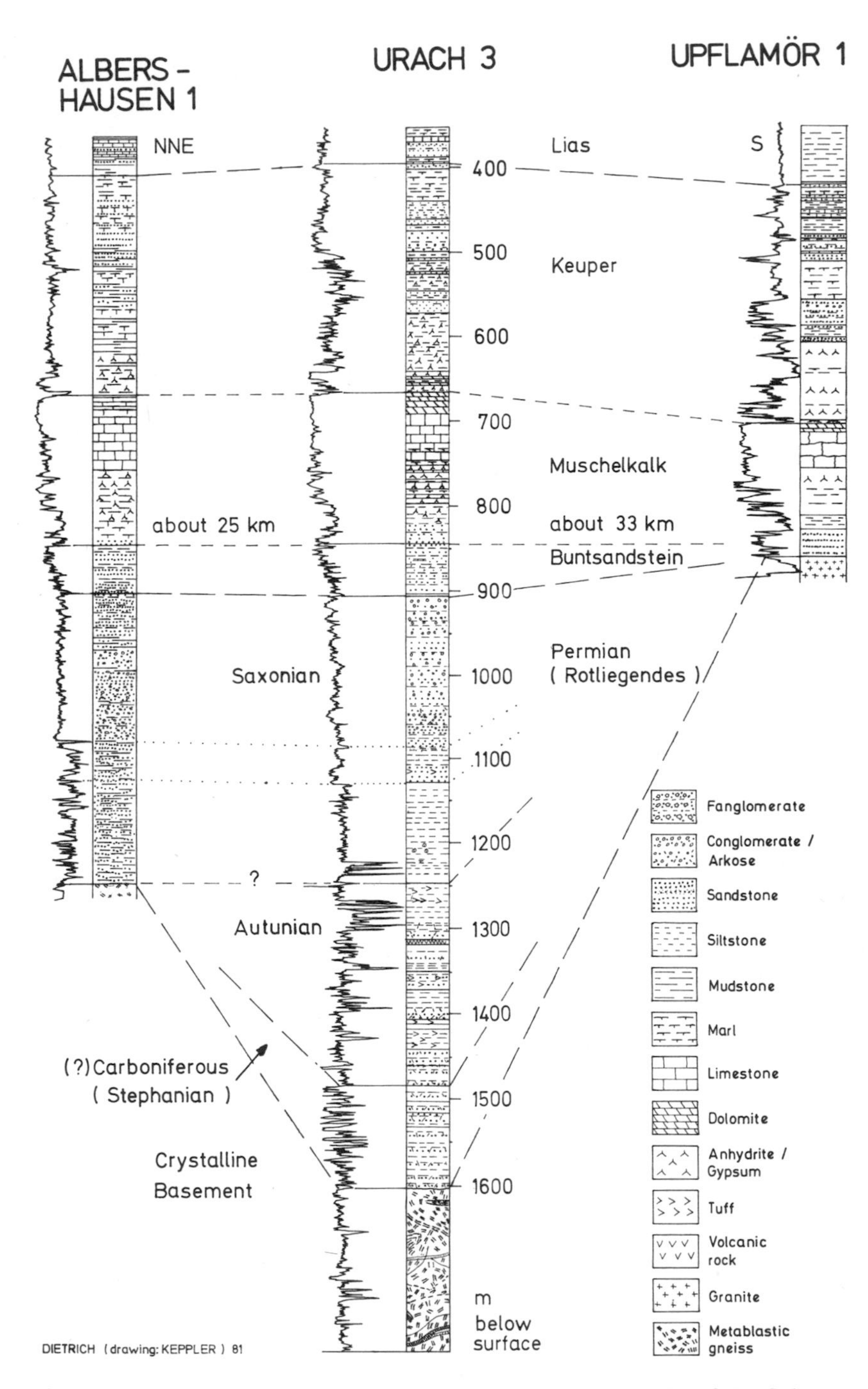

Fig. 2. Simplified lithostratigraphic drilling profiles transverse to the strike of the Schramberg trough and correlation endevour, taking into account the available Gamma Ray Logs.

classified as Zechstein. An endeavour is being made to extend this boundary to Urach, in order to permit separating similar equivalents of the Zechstein.

According to the macroscopic and the microscopic investigation, the penetrated crystalline can be subdivided into three large units: In downward succession, coarse grained metablastites ("orthogneisses") were first encountered; these are underlain by medium- to coarse-grained metatexites (paragneisses); these in turn are underlain by fine- to medium-grained diatexites ("mica-syenites"). According to STENGER & WIMMENAUER (1980), all of these rocks can be viewed as paragneisses and their metablastic and diatectic derivatives. A detailed description of the rocks is presented by STENGER (this volume) and by SCHÄDEL (this volume).

The bedding conditions in the sedimentary cover and crystalline basement, as deduced from the oriented cores, are described by DIETRICH & SCHÄDEL (1978), BUNTEBARTH et al. (1979), and SCHÄDEL (this volume). From these presentations it can be seen that the sediments usually dip slightly toward the southeast, whereas a steep northward vergency is evident in the crystalline basement. The weak angular unconformity at the Autunian/Saxonian boundary, or in the upper part of the Lower Rotliegend, which was just discussed, has not been considered in the publications mentioned.

Comparisons of Profiles

In correspondence with the presumed extension of the Schramberg trough (cf. for example FALKE 1972, 1976 and LEIBER & MÜNZING 1979), an endeavour has been made to mutually compare and correlate profiles from various boreholes, both along the strike of the trough and in the direction perpendicular to the trough axis. For a comparison of profiles along the trough axis, the profiles of the Sulz well to the southwest and Überkingen to the northeast of Urach can be employed. For a comparison perpendicular to the trough axis, the profiles of the Albershausen 1 and Upflamör 1 wells to the north and south of Urach, respectively, can be used. (Profiles are given by CARLE 1971, 1972, 1975 a, 1975 b; FRAAS 1890; SCHMIDT 1912, 1976, and WIRTH 1960, 1968, among others.)

The lithological profiles, presented in simplified form, and the Gamma Ray Logs presented simultaneously, allow a correlation both along the strike of the trough and perpendicular thereto. Profile comparisons between Sulz, Urach, and Überkingen appear to confirm the hypothesis that the intramontane trough of Upper Carboniferous origin extends continuously from Schramberg on the eastern side of the Black Forest all the way to Überkingen.

The existence of the Rotliegend trough is confirmed by the comparison of profiles between Albershausen, Urach, and Upflamör. Moreover, it can be assumed that the trough was narrower when it was formed in the Upper Carboniferous, and that regions adjacent to the north underwent depression later, in the course of the Rotliegend at the Autunian/Saxonian boundary, or in the course of the Upper Autunian. To the south, no extension of the distributive province can be detected. The Trochtelfingen borehole, for which the results have hitherto not been published, also gives no indication of a Rotliegend trough. According to thickness data of MÜNZING & LEIBER (1979), no Permian strata have been encountered by the Trochtelfingen 1 borehole. In contrast, GEYER & GWINNER (1979) indicate a thin Permian sequence.

In view of the correlations performed (Fig. 1 and 2), furthermore, LEIBER (this volume) and KÄDING (1978) conclude that the Permian/Triassic boundary should be redefined for the Albershausen, Bad Canstatt and Überkingen wells; the thickness hitherto given for the Buntsandstein should thereby be decreased.

References

Backhaus, E. (1976): The Permian/Triassic boundary in the continental area of middle Europe. – In Falke, H. (Ed.): The continental Permian in Central, West, and South Europe. – NATO advanced Study Institute Series. Series C: Mathematical and Physical Studies, C 22, 110–120, D. Reidel Publ. Comp., Dordrecht/Boston.

Brenner, K. (1979): Die natürliche Erdwärme – eine alternative Energieform. – Mineralien Magazin, 3. J, H. 8; 411–416, Stuttgart.

Buntebarth, G., Grebe, H., Teichmüller, M. & Teichmüller, R. (1979): Inkohlungsuntersuchungen in der Forschungsbohrung Urach 3 und ihre geothermische Interpretation. – Fortschr. Geol. Rheinld. u. Westf., **27**: 183–199, Krefeld.

Carle, W. (1971): Die Tiefbohrungen auf mineralisiertes Thermalwasser in Bad Überkingen, Landkreis Göppingen, Baden-Württemberg (Geologie, Hydrogeologie, Technik). – Jh. Ges. Naturkd. Württ., **126**: 36–87, Stuttgart.

– (1972): Geologie und Hydrogeologie der Mineral- und Thermalwässer von Bad Überkingen, Landkreis Göppingen, Baden-Württemberg. – Jh. geol. Landesamt Baden-Württ., **14**: 69–143, Freiburg i. Br.

– (1975 a): Die Mineral- und Thermalwässer von Mitteleuropa. Geologie, Chemismus, Genese. – Wissenschaftl. Verlagsges., Stuttgart.

– (1975 b): Die Thermalwasser-Bohrung von Stuttgart-Bad Canstatt. – Jh. Ges. f. Naturkde. Württ. **130**: 87–155, Stuttgart.

Dietrich, H.-G. & Schädel, K. (1978): Untersuchung der geothermischen Anomalie in Urach auf eine mögliche Nutzung durch eine Untersuchungsbohrung bis tief ins Kristallin (ET 4023 B). – Progr. Energieforsch. u. Energietechn. 1977–1980, Statusrep. 1978 – Geotechnik und Lagerstätten, **1**: 79–85, Jülich (Projektleitung Energieforsch. (PLE), KFA Jülich) 1978.

Dietrich, H.-G., Haenel, R., Neth, G., Schädel, K. & Zoth, G. (1980): Deep Investigation of the Geothermal Anomaly of Urach. – In Strub, A.S. & Ungemach, P. (Eds.): Advances in European Geothermal Research, 2nd, p. 253–266, Strasbourg 1980, D. Reidel Publ. Comp., Dordrecht/Boston/London.

Engelhardt, W.V. (1973): Die Bildung von Sedimenten und Sedimentgesteinen. – Sediment-Petrologie (Hsg. Engelhardt, W. v., Füchtbauer, H. & Müller, G.) Teil III. – E. Schweizerbart. Stuttgart.

Falke, H. (1965): Die Zusammenhänge zwischen Sedimentation, Regionalrelief und Regionalklima im Rotliegenden des Saar-Nahe-Gebietes. – Geol. Rdsch., **54**: 208–224, Stuttgart.

– (1971): 2. Die paläogeographische Entwicklung des Oberkarbons in Süddeutschland. – Fortschr. Geol. Rheinld. u. Westf., **19**: 167–172, Krefeld.

– (1972): The Continental Permian in North- and South Germany. – In Falke, H. (Ed.): Rotliegend. Essays on European Lower Permian. – Inter. Sedim. Petrogr. Ser., **15**: 43–113, Leiden.

– (1976): Problem of the continental Permian in the Federal Republic of Germany. – In Falke, H. (Ed.): The Continental Permian in Central, West, and South Europe. – NATO advanced Study Institute Series. Series C: Mathematical and Physical Studies, C 22, 38–52, D. Reidel Publ. Comp., Dordrecht/Boston.

Fraas, E. (1890): 7. Das Bohrloch von Sulz am Neckar. – Bericht über die 23. Versammlung des Oberrh. Geol. Vereins am 10. April 1890, p. 35–40, Stuttgart.

Geyer, O.F. & Gwinner, M.P. (1968): Einführung in die Geologie von Baden-Württemberg. – 2. Aufl., E. Schweizerbart., Stuttgart.

– (1979): Die Schwäbische Alb und ihr Vorland. – Slg. geol. Führer, **67**, 2. völlig überarb. Aufl. von B. 40: Der Schwäbische Jura, Gebr. Borntraeger, Berlin/Stuttgart.
Gwinner, M.P. (1974): Geologische Karte von Baden-Württemberg 1:25000. Erläuterungen zu Blatt 7522 Urach. – Landesvermessungsamt Stuttgart.
Heim, D. (1960): Über die Petrographie und Genese der Tonsteine aus dem Rotliegenden des Saar-Nahe-Gebietes. – Beitr. Miner. Petrogr., **7**: 281–317.
Käding, K.-C. (1978): Die Grenze Zechstein/Buntsandstein in Hessen, Nordbayern und Baden-Württemberg. – Jber. Mitt. Oberrh. geol. Ver., N. F. **60**: 233–252, Stuttgart.
Katzung, G. (1972): Stratigraphie und Paläogeographie des Unterperms in Mitteleuropa. – Geologie, Jg. **21**, 4/5: 570–584, Berlin.
Kull, U. (1981): Die Wärmeanomalie von Urach. – Naturwissensch. Rdsch. 34. Jg., H. 1: 21–23.
Leiber, J. & Münzing, K. (1979): Perm und Buntsandstein zwischen Schramberg und Königsfels (Mittlerer Schwarzwald). – Jh. geol. Landesamt Baden-Württemberg, **21**: 107–136, Freiburg i. Br.
Sauer, K. (1971): Bericht über die Thermalwassertiefbohrung Urach. – Manuskript zum Vortrag an der Volkshochschule Urach am 15. März 1971.
– (1978): Geologie und Technik der Forschungsbohrung geothermisches Feld Urach. – Vortrag, 26. DGMK-Haupttagung Berlin, 5.10.1978.
Schädel, K. (1977): Die Geologie der Wärmeanomalie Neuffen-Urach am Nordrand der Schwäbischen Alb. – Seminar on Geothermal Energy, EUR 5920, **1**: 53–60, Bruxelles-Luxembourg.
Schmidt, A. (1912): Drei Tiefbohrungen auf Steinkohle am oberen Neckar. – Württ. Jahrb. f. Statistik u. Landeskunde, Jg. 1912, H. 1: 162–173, Stuttgart.
– (1976): Geologische Karte von Baden-Württemberg 1:25000. Erläuterung zu Blatt 7617 Sulz. – Unveränderte Ausgabe der II. Auflage von 1931. Landesvermessungsamt Baden-Württemberg, Stuttgart.
Stenger, R. & Wimmenauer, W. (1980): Das Kristallin der Forschungsbohrung Urach 3 im Rahmen des süddeutschen Grundgebirges (ET 4251 A). – Progr. Energieforsch. u. Energietechn. 1977–1980, Statusrep. 1980. – Geotechnik und Lagerstätten, **1**: 61–71, Jülich (Projektleitung Energieforsch. (PLE), KFA Jülich).
Weiskirchner, W. (1980): Der Obermiozäne Vulkanismus in der mittleren Schwäbischen Alb (Exkursion C am 8. April 1980). – Jber. Mitt. Oberrhein. geol. Ver., N. F. **62**: 21–29, Stuttgart.
Wirth, E. (1960): Die Schichtfolge der Erdölaufschlußbohrung Upflamör 1, Schwäbische Alb. – Jber. Mitt. Oberrh. geol. Ver., N. F. **42**: 129–162, Stuttgart.
– (1968): Das Mesozoikum im Untergrund der Schwäbischen Alb zwischen Münsingen und Meßkirch. – Z. dt. Geol. Ges., **117**: 855–894, Hannover.
Wohlenberg, J. (1978): Geophysikalische Untersuchungen in der Forschungsbohrung Urach (ET 4131 A). – Progr. Energieforsch. u. Energietechn. 1977–1980, Statusrep. 1978 – Geotechnik und Lagerstätten, **1**: 87–100, Jülich (Projektleitung Energieforsch. (PLE), KFA Jülich).

The Urach Geothermal Project, p. 59–79;
Schweizerbart'sche Verlagsbuchhandlung, Stuttgart, 1982

Hydrogeological Results from the Urach 3 Research Borehole

H.-G. DIETRICH

with 10 figures and 3 tables

Abstract: In addition to the aquifer already known in the Upper Muschelkalk, two new thermal water storeys have been shown to exist; one is located in the Lower Muschelkalk/Buntsandstein, and the other is a fissure aquifer in the crystalline basement. A production rate of more than 1 l/s is possible from these aquifers. As far as the utilization of the geothermal energy, for instance for heating purposes, is concerned, only the Upper Muschelkalk can be considered, despite the relatively low temperature of 55 to 60 °C, because a higher rate of production (> 12.6 l/s) is feasible.
By means of a tracer test, a hydraulic communication has been shown to exist between the Urach 3 well and the two Muschelkalk wells, Urach 1 and 2, of the adjacent thermal resort. Hence the choice of a suitable location for such a project must be carefully considered, in order to avoid mutual interference.
In contrast to the upper thermal water storeys down to the Upper Muschelkalk inclusively, the aquifers of the deeper thermal water storeys can be collectively regarded as NaCl thermal water aquifers. The possibility of water circulation in the overburden and in the crystalline basement is discussed.

Introduction

The occurrence of mineral and thermal water is known in various regions of the Urach geothermal anomaly. A comprehensive description of the different types of water according to regional occurrence, together with a classification according to geological formation and chemical composition, is given by CARLE (1975).

The knowledge of deeper groundwater storeys in the central region of the Urach anomaly is due to the thermal water wells Urach 1 and 2, drilled in 1970 and 1974, respectively, and Beuren 1, drilled in 1972. By means of these three wells, thermal water-bearing strata in the Lower Lias, Rät, Middle Keuper (Stubensandstein) Lower or Letten Keuper, and Upper Muschelkalk were encountered, investigated, and in part developed for balneological purposes. The final depths of these three wells lie between 750 and 800 m below the surface, in the Middle and Upper Muschelkalk, since especially the aquifer of the Upper Muschelkalk was to be utilized. The results of the water analyses performed by Dr. Käß (Geologisches Landesamt Baden-Württemberg, Freiburg) have been reproduced by CARLÉ (1975). The various types of water include Na-HCO_3 (Lias/Rät), Na-HCO_3-Cl (Middle Keuper: Stubensandstein and Kieselsandstein), and Ca-Na-Cl-SO_4-HCO_3 (Upper Muschelkalk) acidulous water of differing composition and mineralization.

The water levels corresponding to the pressure in the three aquifers of Lias and Rät, Stubensandstein, and Upper Muschelkalk have been measured, for example in the area of the Urach 1 well, at depths of 144 m, 176 m, and 160 m below the surface, respectively.

This corresponds to altitudes of about 288 m, 256 m, and 272 m above mean sea level, respectively, (SAUER 1971, SCHÄDEL 1977), since the altitude at the well site is 432,20 m above mean sea level.

In other regions of this anomaly, such as Boll and Bad Ditzenbach, thermal water wells have also been drilled down to the Upper Muschelkalk (cf. CARLÉ 1974, CARLÉ & GROSCHOPF 1972). Only a few thermal water wells have been drilled for exploring water-bearing horizons in deeper-lying stratigraphic sequences of the Buntsandstein and Rotliegend. These include, for instance, the Bonlanden 1 well (CARLÉ 1975 a, 1975 b) at the northern periphery of the anomaly, and the Überkingen 35, 52 and 71 wells (CARLÉ 1971, 1972, DIETRICH, this volume) about 32 km northeast of Urach.

By means of Urach 3 research borehole, the hydrogeological exploration has been extended for the first time all the way to the base of the sedimentary cover; a further distance of 1732 m was drilled into the subjacent crystalline basement.

Supplementary hydrogeological investigations in the upper 500 m of the sedimentary cover, which had been planned for the Urach research project, were to be conducted by means of an auxiliary borehole (Urach 4) (DIETRICH, this volume). Because of increased expenses for the Urach 3 borehole, and for the frac and circulation experiments subsequently performed in 1979 (SCHÄDEL & DIETRICH), as well as the altered financial situation in the field of research, however, it has hitherto not been possible to drill this ancillary well.

Execution of the Influx Tests

In the Urach 3 well, all horizons and sections of the sedimentary cover below the Lower Keuper capable of yielding water, as well as those of the crystalline basement penetrated, were tested for their water-bearing capacity by means of pumping tests and/or drill stem tests. These hydrogeological tests were conducted according to plan (DIETRICH, this volume), both in the open, uncased borehole and in the cased borehole after perforation of the casing in the vicinity of horizons worthy of investigation (Fig. 1).

From Fig. 1 it can be seen that pumping tests were performed in the open borehole, both in the Upper Muschelkalk (P 1) and in the crystalline basement (P 2, P 3). Drill stem tests (DST) were performed in the open borehole at only two positions within the crystalline basement during the drilling operations (S 1, S 2).

All further flow tests were conducted after the casing strings had already been installed and perforated. These also include a pumping test (P 8) carried out in the crystalline on a fissure aquifer which had already been tested during the drilling phase (S 1).

The potential water-bearing horizons in the stratigraphic sequence of the sedimentary cover between the upper limit of the crystalline basement and the Middle Muschelkalk were investigated only from the cased borehole. This was accomplished by means of pumping tests (P 4 to P 7), as well as DST straddle tests (S 3 to S 6).

The determination of the sections to be perforated (DIETRICH, this volume) was carried out in cooperation with the service company (Schlumberger) on the basis of the geological findings and the electrical well logs.

Prior to the beginning of perforation, a preliminary cement bridge was installed in the transition zone between the 8 1/2" borehole and the 9 5/8" liner, with the cementing head at a depth of 1796.5 m, in order to mutually isolate the uncased and cased sections of the crystalline. After P 4, a second cement bridge was installed in the 9 5/8" casing (cementing head at 1597.7 m), in order to obviate any possibility of forming a hydraulic

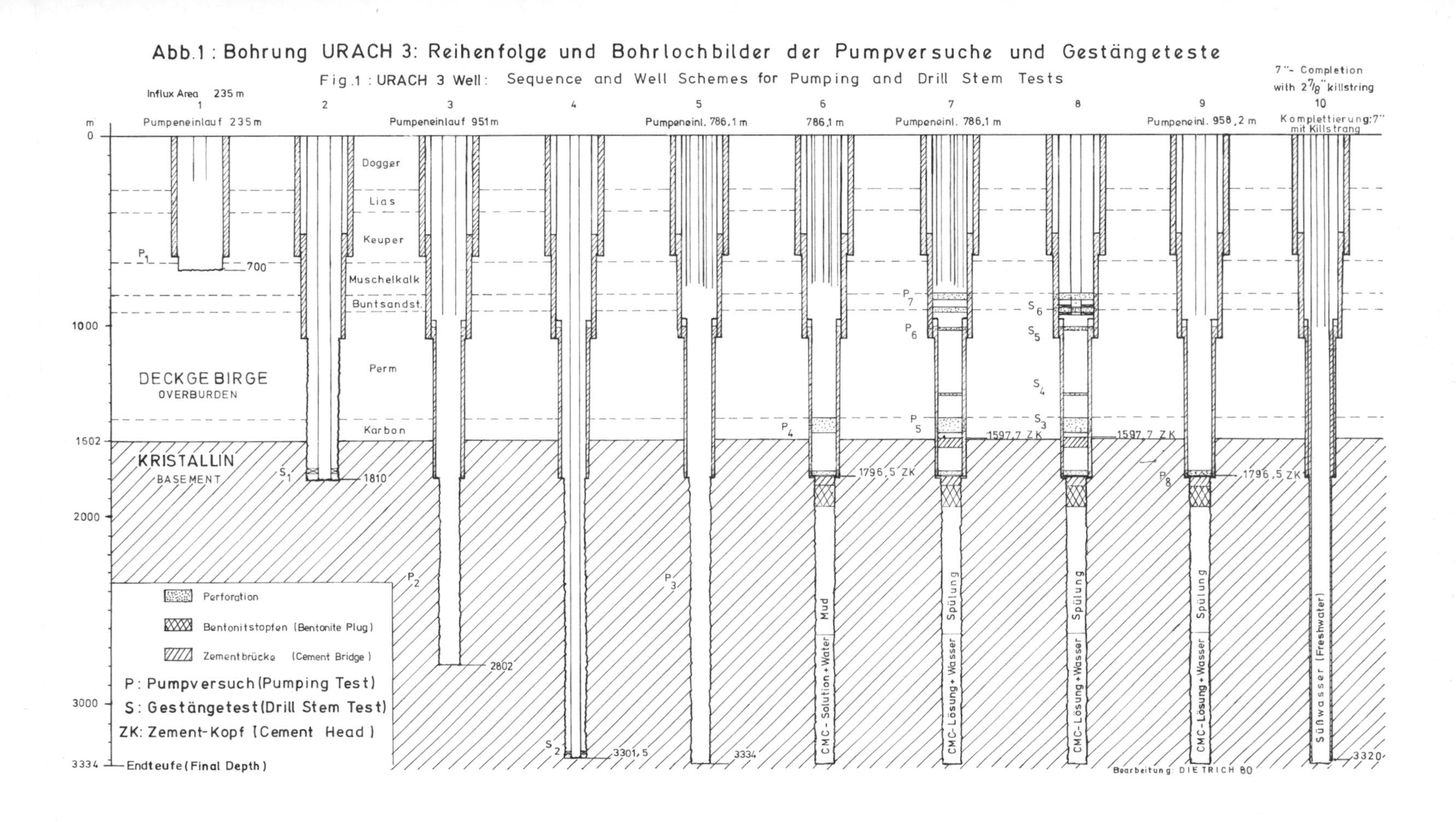

Fig. 1. Urach 3 well: Sequence and well schemes for pumping and drill stem tests.

connection through the perforation between the fissure aquifer in the crystalline basement and the aquifers to be tested in the sedimentary cover.

During the further sequence of pumping tests in upward succession, the installation of further cement bridges between the perforated sections was omitted (Fig. 2), in contrast to the original plans. Because of the low rates of influx from the entire Young Palaeozoic (cf. Tab. 1), any interference with the respective, subsequent flow tests was either unlikely or negligible.

Because of the low influx rates, DST straddle tests were carried out in a few depth intervals (Fig. 1 and 3), as a supplement to these pumping tests (P 4 to P 7). For financial reasons, this was possible only in a few cases, unfortunately. Moreover, during this test series (S 3 to S 6), several perforated sections were simultaneously included between the packers in each case, in order to reduce expenses.

By means of these straddle tests, it was feasible to obtain a few samples of thermal water under in situ conditions, in spite of the low influx rates. A so-called DCIP (Dual Closed-In-Pressure) sampler from the Halliburton Company had been installed as a sampling chamber in the test string a short distance above the upper packer. The remainder of the test equipment used in combination with the 4" drill pipe was also supplied by this company.

After the test series in the sedimentary cover had been completed, the interesting fissure zone in the upper region of the crystalline was more closely investigated by means of a further pumping test (P 8). In order to establish optimal test conditions, the fissure

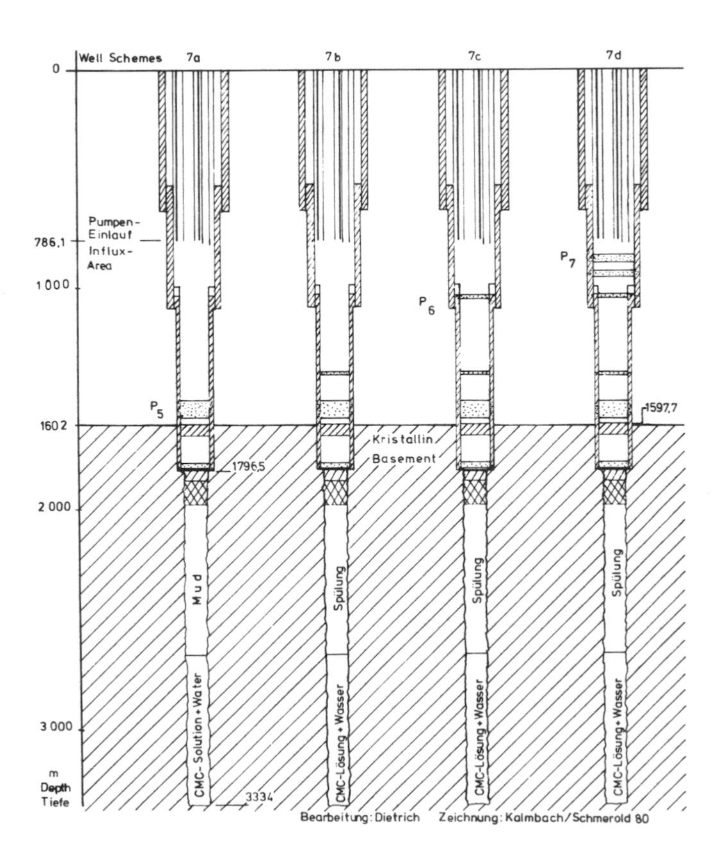

Fig. 2. Pumping tests in Urach 3 well; schemes 7 a to 7 d.

aquifer, which had been sealed off with Class G deep drilling cement during the drilling operation, had to be first reperforated and subsequently fractured hydraulically. The 9 5/8" RTTS packer installed for the purpose was set at a depth of 1749 m. With the use of the water present in the borehole (density = 1.02 g/cm^3) and an average pumping rate of 62.5 l/min, a fracturing pressure of about 200 bar thereby resulted at the well head. This corresponds to a pressure gradient of 0.215 bar/m at the preforation depth of 1764 m. Of the total volume of 250 l thereby injected, only 10 l flowed back. In two immediately following tests for determining the possible rates of injection into the fractured rock, the following values have been measured: At a constant pumping rate of 40 l/min and an injection volume of 1 m^3, the well head pressure varied between 100 and 150 bar; at a depth of 1764 m this corresponds to a pressure gradient between 0.157 and 0.187 bar/m. At a constant well head pressure of 200 bar and an injection volume of 2.35 m^3, an average pumping rate of 117.5 l/min resulted. The backflow for the two tests amounted to 9 and 150 l, respectively; this indicates the presence of a large, extensive system of fissures.

For conducting the pumping tests, a Pleuger pump was first employed; a Byron Jackson pump was later used. More detailed technical information on both types of pumps and on the required dual completion (borehole illustrations 5, 6 and 7 of Fig. 1 and 2), as well as the instruments also required for measuring pressure and temperature, is provided by DIETRICH (this volume). For completeness it must be mentioned that the drilling fluid remained in the borehole only during the straddle tests S 1 and S 2. In all

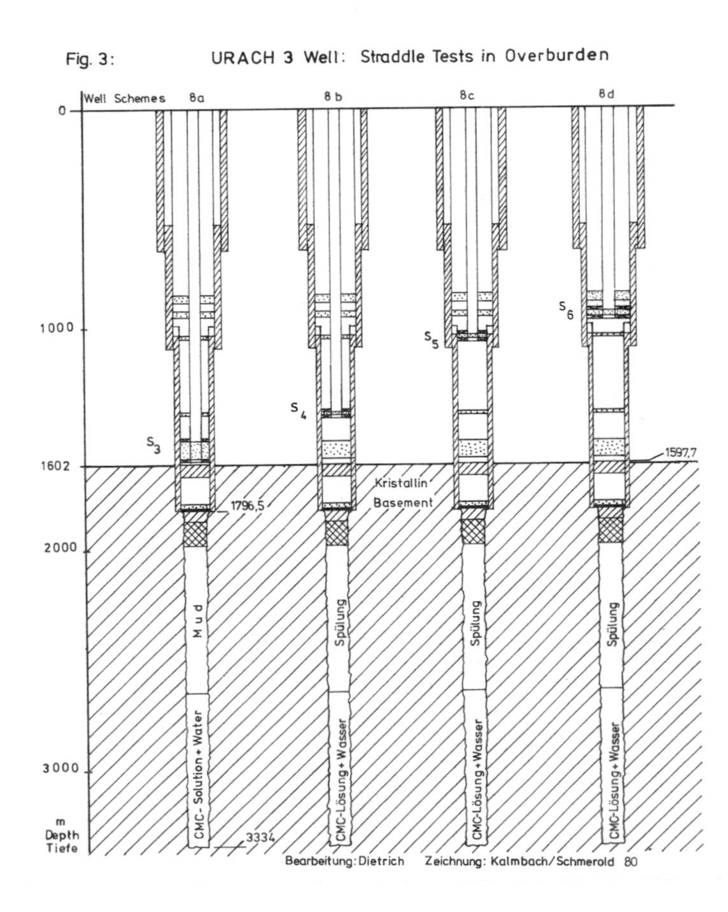

Fig. 3. Urach 3 well: straddle tests in overburden.

other cases the borehole had been flushed clean prior to the beginning of the flow tests, with the use of water from the Erms, a stream some 50 m distant from the drilling site.

After all tests had been completed, the lowest cement bridge was drilled out too. The final 7" casing was subsequently installed, and the well was completed (Fig. 1, borehole illustration 10).

Results and the Influx Tests

On the basis of the pumping tests and straddle tests conducted in the depth range between 642.4 m and 3334 m (final depth), the following survey results for the thermal water storeys between the Lower Keuper and upper limit of the crystalline (1602 m) and in the penetrated section of the crystalline basement (cf. DIETRICH, STENGER and SCHÄDEL, this volume): Below the aquifer of the Upper Muschelkalk, which has already been made accessible by the Urach 1 and 2 wells, there are only two further zones in which production rates greater than 1 l/s are feasible (cf. Fig. 7, 8, 9). These are the test zone in the Lower Muschelkalk/Buntsandstein and a fissure aquifer in the upper section of the crystalline basement. All other zones tested yielded influx rates of 0.1 l/s at most, as indicated in Tab. 1. Of course, there exists the possibility that these poorly permeable horizons can be stimulated to give a higher yield by means of further treatment such as pressure acidizing or hydraulic fracturing.

The pumping test performed in the Upper Muschelkalk (Fig. 4) yielded a flow rate of 12.6 l/s (capacity limit of the Pleuger pump used); this is by far the highest influx rate of all thermal water aquifers below Urach. A flow meter test for ascertaining the zones of influx within the section tested between 642.4 m (basal Gipskeuper) and 700 m (Upper Muschelkalk) yielded the following result: As far as the distribution of the total flow rate of 12.6 l/s over this uncased borehole section is concerned, the entire influx occurs exclusively from the Upper Muschelkalk (Fig. 5). Almost 75 per cent of the total influx was thereby shown to occur between 687 and 700 m. During the drilling phase, this section had already been the subject of particular concern, first because of gradual, and from 692 m onward because of total loss of circulation; this is indicative of highly permeable zones.

For the influx test in the Lower Muschelkalk/Buntsandstein section, the 13 3/8" casing had already been perforated at several intervals between 834.0 and 871.3 m (cf. DIETRICH, this volume). The boundary between the two formations is thereby located at a depth of 843.0 m below the surface (LEIBER, this volume). With the effected drawdown and a nearly constant dynamic water level taken into consideration, a maximal flow rate of about 1.3 l/s resulted from this pumping test, (P 7), as can be seen from Fig. 6. The highest rate of influx observed during the rise after the pumps had been turned off was about 1 l/s.

Table 1.
Investigation of thermal water-bearing layers below Urach with pumping tests and drill stem tests at Urach 3 (n. m. ≐ not measurable/not measured);)[1] = Head of cement bridge;)[2] ≐ Temperature Strips;)[3] +)[4] ≐ Fresh water level in drill pipe above upper packer before the valve was opened: 500 m and 600 m respectively;)[5] ≐ Limitation given by pump capacity).

Date	Bottom of well below surface	Test section: uncased borehole or perforated casing	Formation	Type of Test	Hydro-static water level below surface	max. Temperature within Aquifer	lowest water level below surface	max. flow rate	Cumulative volume
	m	m			m	°C	m	l/s	m³
31.10.– 7.11.77	700	642.4– 700	Lower Keuper Upper Muschelkalk	P 1: Pumping Test	160	54	162.5	12.6)[5]	4700
23. 1.78	1810	1759.6–1779.9	Crystalline basement	S 1: Open-Hole straddle test with anchor shoe	220	93)[2]	1259)[3]	3.3	8.95
6. 4.– 7. 4.78	2802	1808.5–2802.0	Crystalline basement	P 2: Pumping Test	n.m.	101,5 (2603 m)	951	n.m.	n.m.
11. 5.78	3301.5	3294.9–3301.5	Crystalline basement	S 2: Open-Hole straddle test with anchor shoe	n.m.	127)[2]	2565)[4]	0.02	0.094
27. 5.– 30. 5.78	3334	1808.5–3334.0	Crystalline basement	P 3: Pumping Test	n.m.	113,6 (2567 m)	690	0.25	26.31
5. 6.– 7. 6.78	1796.5)[1]	1485.5–1561.0 1764.0–1784.0	Carboniferous + Crystalline b.	P 4: Pumping Test	n.m.	104 (1780 m)	788	0.2	7.65
9. 6.– 10.6.78	1597.7)[1]	1485.5–1561.0	Carboniferous	P 5: Pumping Test	n.m.	n.m.	742	0.01	0.16
11.6.– 12.6.78	1597.7)[1]	911.5–1368.5 + 1485.5–1561.0	Rotliegendes + Carboniferous	P 6: Pumping Test	n.m.	88,5 (1595 m)	786	0.07	3.15
13. 6.– 16. 6.78	1597.7)[1]	834.0– 871.3 + 911.5–1368.5 + 1485.5–1561.0	Lower Muschelkalk + Buntsandstein + Rotliegendes + Carboniferous	P 7: Pumping Test	~220	93 (1595 m)	740	1.3	125
19. 6.78	1597.7)[1]	1485.5–1561.0	Carboniferous	S 3:	n.m.	88)[2]	1480	0.008	0.41
20. 6.78	1597.7)[1]	1362.5–1368.5	Rotliegendes	S 4: Straddle	n.m.	82)[2]	1362.5	0.001	0.09
21. 6.78	1597.7)[1]	1042.5–1047.0	Rotliegendes	S 5: Test	n.m.	? 77)[2]	± 1049	0	0.02
22. 6.78	1597.7)[1]	911.5– 933.0	Rotliegendes	S 6:	n.m.	? 77)[2]	911.5	0.05	0.76
25. 6.– 28.6.78	1796.5)[1]	1764.0–1784.0	Crystalline basement	P 8: Pumping Test	n.m.	n.m.	692	0.6	135

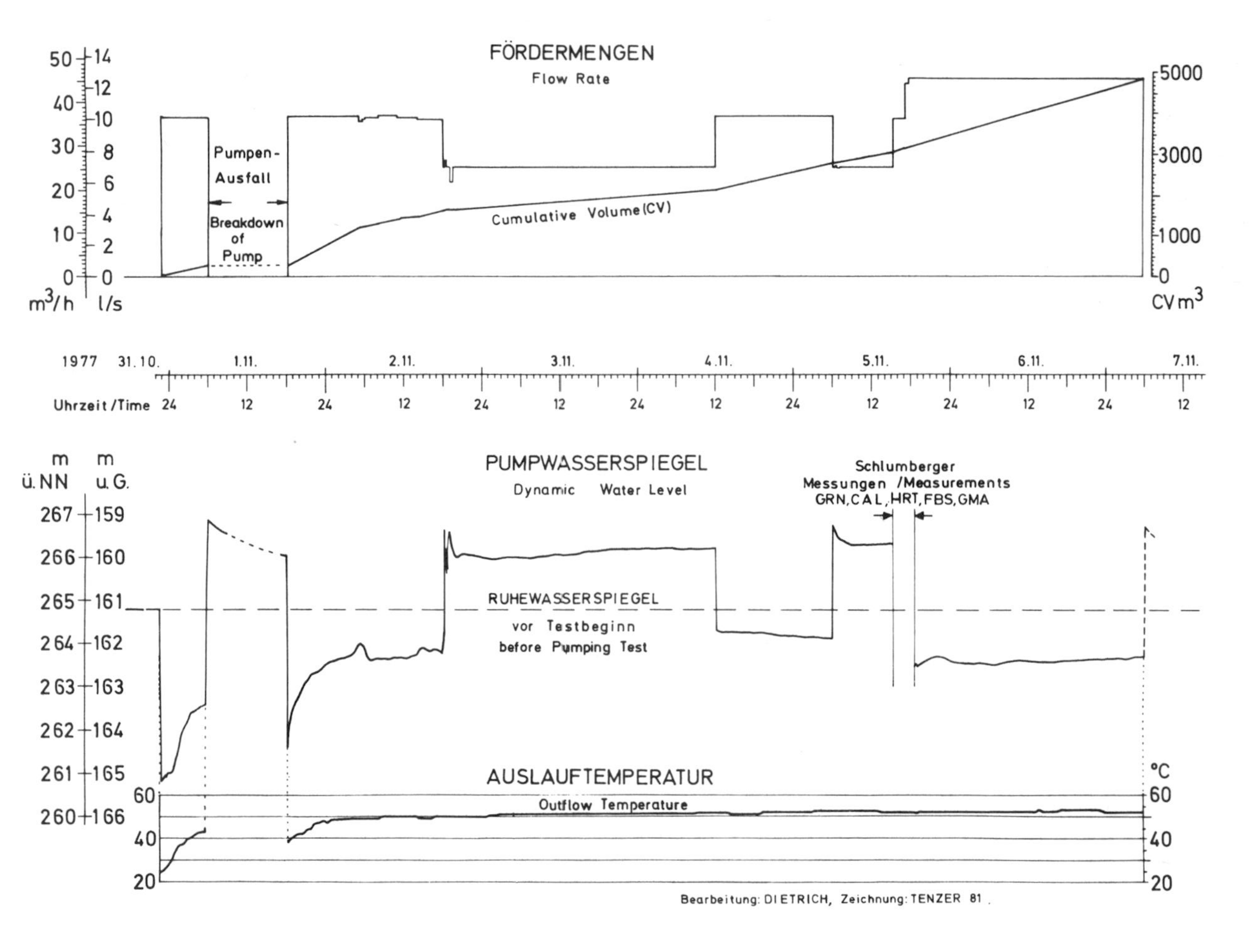

Fig. 4. Pumping test in Upper Muschelkalk at Urach 3 well.

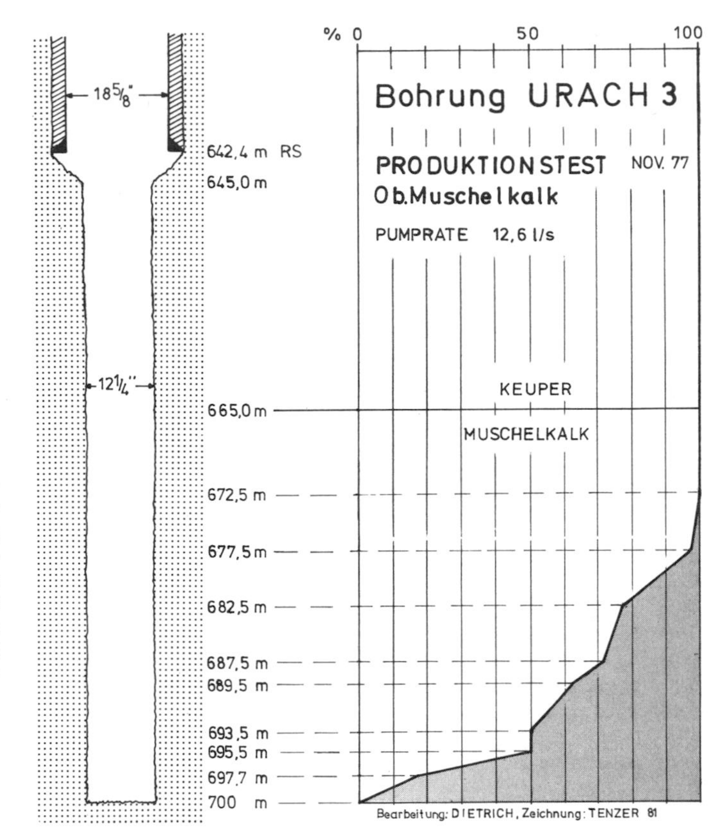

Fig. 5. Fractional distribution of the total flow race of 12.6 l/s in Lower Keuper and Upper Muschelkalk during a full-bore flowmeter test with a production combination tool (PCT) by Schlumberger (RS = casing shoe).

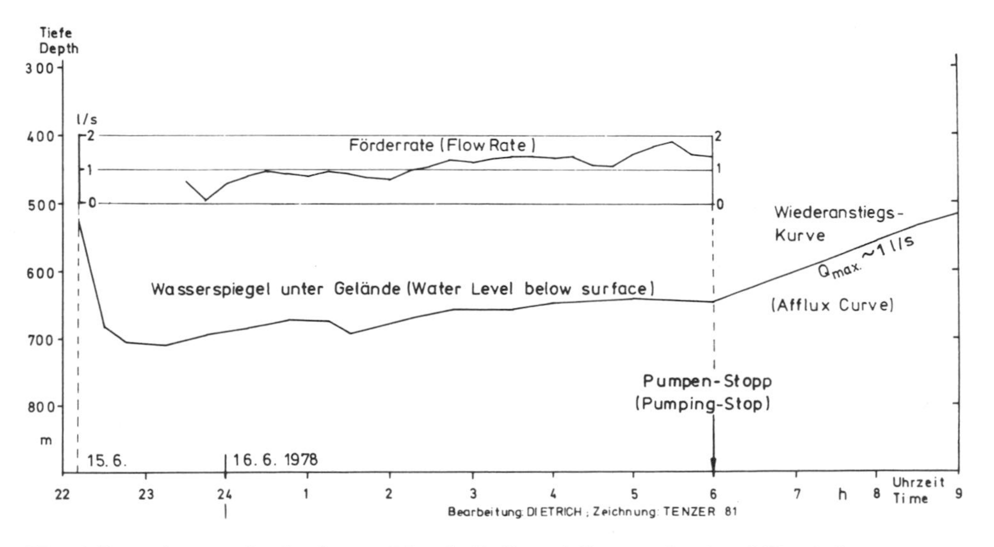

Fig. 6. Pumping test in the Lower Muschelkalk and Buntsandstein of Urach 3.

The rates of influx from the fissure aquifer at 1775 m in the upper crystalline, that is, in the region of the metablastites from 1602 to about 1950 m, amounted to maximally 3.3 l/s in a straddle test (S 1), and maximally 0.6 l/s in a pumping test (P 8). The difference between the results of the two tests can be explained by the circumstance that the maximal drawdown of the water level during the pumping test was only 692 m below the surface. During the straddle test, on the other hand, the water level was drawn down to about 1259 m below the surface, with the use of a 500 m water cushion in the 5" drill string (cf. Tab. 1).

In addition to this extensively fissured zone at 1775 m, further zones of water-bearing rock have been discovered in deeper regions of the crystalline. According to a fine-feature evaluation by Preussag, a straddle test (S 2) conducted at a depth of 3300 m in the diatexite zone (3000 m to the final depth of 3334 m) yielded an influx of 94 l, which corresponds to a flow rate of about 0.02 l/s (Tab. 1).

During the two pumping tests performed in the uncased borehole in the crystalline (P 2, P 3) between 1808.5 m (9 5/8" casing shoe) and 2802 and 3334 m, respectively, it was not possible to accurately determine the zones of influx, unfortunately. The flushing of the borehole with fresh water and the decompression resulting from the drawdown of the water column before and during the pumping tests, respectively, gave rise to unstable borehole conditions. The dislodging of rock fragments caused a constriction of the borehole at various depths below about 2600 m; consequently the temperature gauges employed could not be run into the influx zones below this level in order to locate them.

Indications of open joints below about 2600 m had already occurred during the drilling operations, as evidenced by gradual circulation losses (DIETRICH & SCHÄDEL 1978). Between 2606 m and the final depth, that is, in the region of the paragneiss and diatexite, the observed loss of circulation was between 0.1 and 0.4 m^3/h on the average, with a maximum of 1.18 m^3/h, that is, 0.03 to 0.11 l/s with a maximum of 0.33 l/s.

Open fissures, in which salt water was also observed, were encountered in several intervals of cored crystalline rock.

Numerous zones of bleaching and disintegration, especially in the lower third of the paragneiss and underlying diatexite, are indicative of hydrothermal mineralization processes (cf. STENGER, this volume). Where the alteration and disintegration are especially pronounced, the originally hard and abrasive rock has been locally transformed to highly friable, completely fragile material.

This finding too is indicative of systems of open joints and fissures in which hydrothermal convection takes place.

The deposits within the fissures include preferentially quartz, feldspar, carbonates and/or pyrite. Moreover, prominent clay gouges occur locally. Deposits of iron oxide with varying content are a characteristic feature in the paragneiss zone, whereas carbonate fillings are evidently absent and pyrite is only of subordinate importance. This circumstance indicates that carbonate and sulphur occurred only scarcely or not at all in this region of the hydrothermal convection system (e. g. PAPE 1977).

As shown in Tab. 1, the determination of the static water level was feasible only for tests at higher flow rates. For the other tests, the gradual attainment of the hydrostatic water level could not be awaited for reasons of cost, because of the very low rates of influx.

The evaluation of the first straddle test (S 1) indicates that the hydrostatic water level of the fissure aquifer in the crystalline basement lies about 206 m below the surface. A similar position is indicated for the hydrostatic water level of the Lower Muschelkalk/

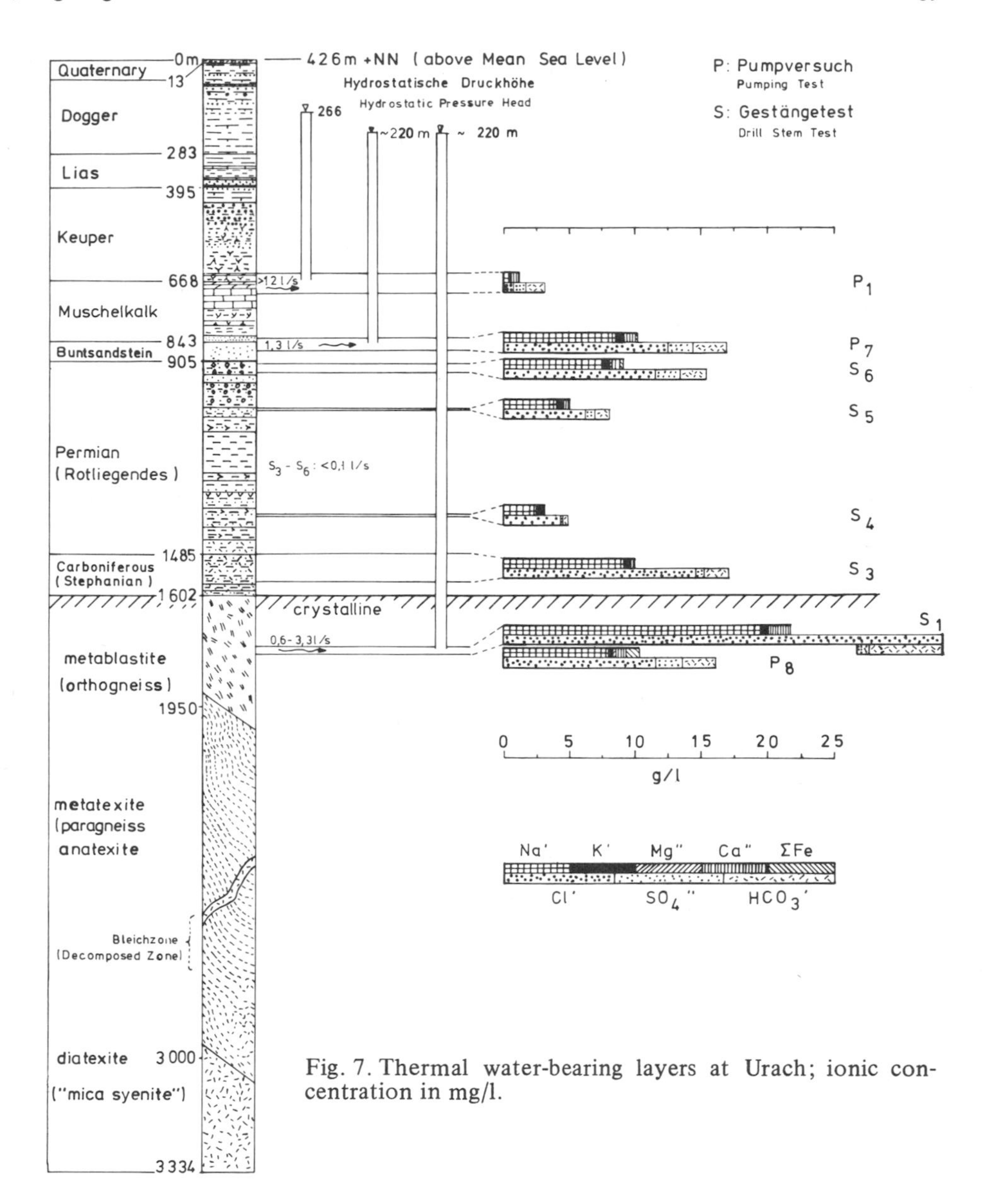

Fig. 7. Thermal water-bearing layers at Urach; ionic concentration in mg/l.

Buntsandstein section, whereas a depth of about 160 m below the surface was measured for the Upper Muschelkalk. The corresponding values, as referred to mean sea level, are presented in Fig. 7, 8 and 9.

The hydrostatic water levels thus determined reflect the respective values of the aquifer pressure only approximately. Because of the length of the water column in the borehole or in the drill string, interfering effects must be expected. This was observed especially during the sustained pumping test in the Upper Muschelkalk (Fig. 4). With increasing discharge temperature, the height of the water column in the borehole rises for comparable production rates. Thus, the hydraulic depression of the water level is superimposed with the thermal expansion of the water column. Besides the influence of temperature on the water column, further interference occurs during the pumping test, especially from the flushing of the aquifer and from the gas-lift effect due to free carbon dioxide (CO_2). According to preliminary calculations by E. VILLINGER (Geologisches Landesamt, Freiburg), the water level in the Upper Muschelkalk rises by about 5 m during this test, as a result of the temperature increase mentioned (private communication).

With this correction taken into account, a new value results for the performance-drawdown quotient thus determined. The uncorrected maximal value is about 3.5 l/s · m, whereas the correction yields a lower value of about 2.1 l/s · m.

During the pumping test in the Lower Muschelkalk and Buntsandstein (P 7), additional effects are evidently also superimposed on the hydraulically determined drawdown, as can be seen from Fig. 6.

For accurately ascertaining the values of the aquifer pressure and geohydraulic conductivity, as well as of the transmissivity and reservoir coefficients – the so-called formation constants (e.g. HOLTING 1980), – a computer programme is to be employed in the course of supplementary investigations at the Geologisches Landesamt in Freiburg. According to preliminary evaluations by E. VILLINGER (private communication), a value of $T = 31 \cdot 10^{-4}$ m^2/s results for the Upper Muschelkalk below Urach.

The values of the temperature in the aquifers between the Upper Muschelkalk and the upper crystalline basement range from 56 to 102 °C. During the pumping tests they were recorded by means of various electronically or mechanically operating temperature sensors, partly in combination with maximum thermometers. In the case of the straddle tests, temperature strip charts were usually employed. As far as the temperatures prevailing in the unperturbed rock formation are concerned, the investigations of HAENEL & ZOTH (this volume) should be considered.

Results of the Water Analyses

The water samples taken in the Urach 3 borehole were analysed at the Chemisches Institut für Umweltanalytik in Tübingen, the Chemische Landesuntersuchungsanstalt in Sigmaringen, and at the Niedersächsisches Landesamt für Bodenforschung (NLfB) in

Table 2. Ionic concentration in thermal water-bearing layers below Middle Keuper at Urach 3 research well (P = Pumping test; S = Drill stem straddle Test)

Borehole and test sections m	Formation	Cation concentration mg/l	Anion concentration mg/l	Cumulative ionic concentration mg/l
P 1 642.2– 700.0	Lower Keuper/ Upper Muschelkalk	1200.19	3103.83	4302.02
P 7 834.0– 871.3	Lower Muschelkalk/ Buntsandstein	9690.67	16942.77	26633.44
S 6 908.5– 936.0	Rotliegendes (R)	9222.13	15446.31	24668.44
S 5 1041.9–1048.6	Rotliegendes	5055.19	8187.44	12242.63
S 4 1360.4–1366.6	Rotliegendes	3150.1	5019.61	8169.62
S 3 1480.4–1565.6	R. + Carboniferous	9994.07	16351.65	26345.72
S 1 1759.6–1779.9	Crystalline Basement	21412.4	39516.75	60929.15
P 8 1769.0–1784.0	Crystalline Basement	9490.67	16132.13	25622.80

Table 3. Average values of different water analyses from production tests at Urach 3 (P: Pumping test; S: Straddle test). The concentrations of ions, SiO_2 and carbon dioxide are in mg/l; the water hardness in °dH = German hardness scale. (n.a. = not analysed).

Chemical concentration	Test section and types of test							
	P 1	P 7	S 6	S 5	S 4	S 3	S 1	P 8
	642.4 – 700.0 m	834.0 – 871.3 m	908.5 – 936.0 m	1041.9 – 1048.6 m	1360.4 – 1366.6 m	1480.4 1565.6 m	1759.6 – 1779.9 m	1769.0 – 1784.0 m
Na^+	336.0	8062.5	7466.7	4070	2545	9172.5	19533	8065
K^+	58.8	395	621.4	542.5	409.3	399	536	283.13
Mg^{++}	107.9	198.5	202	54.73	9.6	40.8	177.17	209.9
Ca^{++}	686.4	802	610	289.7	171	232.8	1146.7	824.5
$Fe^{++/+++}$	10.13	224.8	317	88.2	3.2	126.4	11.7	105.53
Mn^{++}	–	4.45	2.77	1.16	0.81	2.27	1.23	2.15
NH_4^+	0.965	3.42	2.26	8.9	11.1	20.3	6.6	2.46
Cl^-	750.55	12518.7	11509.8	6247	4316.8	14726	33096.7	11626
SO_4^{--}	992.1	1894.3	1892	762.1	214	603.6	921.7	1981
HCO_3^-	1357	2518	2040.8	1173.7	481.3	1018.7	5475.6	2522.3
NO_2^-	0.013	1.8	0.18	0.02	0.01	–	0.15	0.023
NO_3^-	2	8.16	1.55	3.6	6	1.25	21.6	1.33
PO_4^{---}	0.17	1.81	1.98	1.02	1.5	2.1	1	1.48
SiO_2	6	37	30	24	23.3	38.7	23.3	26
Free Carbon dioxide	1263	1384	3389	484	47	199	n.a.	3542
firmly bound	489.5	925	680	425	174	307	251.9	870
Total hardness	124.8	678.6	119.0	52.8	26.0	40.2	212.3	161.2
carbonate hardness of water	61.3	114.6	95.1	53.9	22.0	48.8	32.1	115.7

Hannover. The results of the various analyses have been compiled and, to the extent feasible or necessary, arithmetically averaged. The sums of the cationic and anionic concentrations, as well as the total ionic concentrations, in mg/l, are presented in Tab. 2. As can be seen from the table, the lowest value of the total ionic concentration, about 4.3 g/l, occurs in the aquifer of the Upper Muschelkalk. The ionic concentration in all deeper-lying thermal water storeys is decidedly higher than this value, whereby the highest value of 61 g/l for the total ionic concentration occurs in the fissure aquifer at a depth of 1775 m in the upper region of the crystalline (metablastite).

Since the rates of influx were often low (Tab. 1), mixing of the inflowing thermal water with the fresh water present in the borehole during the sampling cannot be excluded. Thus the values given for the ionic concentration are to be viewed, at least partially, as lower limits. This applies especially for the samples taken during the tests S 4 and S 5.

From the detailed results of these water analyses, the contents of the most important cations, anions, nonionic components (SiO_2), and carbon dioxide, as well as water hardness, are compiled in Tab. 3. In order to permit a better mutual comparison of the water-bearing horizons, the ionic mass concentrations thus determined have been converted from mg/l (or g/l) to equivalent mass concentrations and contents; they are presented as horizontal bar diagrammes in Fig. 7, 8 and 9. On the basis of the equivalent mass contents, two types of mineral water can be distinguished: The thermal water from the Upper Muschelkalk is an acidulous Ca-Na-Cl-So_4–CHO_3 mineral water with a high content of magnesium, whereas all types of water from deeper-lying aquifers can be regarded summarily as Na-Cl mineral water.

Because of the required content of free carbon dioxide, the Na-Cl water from these lower levels can be viewed as acidulous mineral water only in a few cases, that is, the water produced in the tests P 7 and S 6, as well as P 8. Because the results of the water analysis for P 8 differ markedly from those for S 1, however, uncertainties concerning the purity of the P 8 water sample arise. In view of the good agreement with the analyses of water samples from P 7 and S 6, it cannot be excluded that influx of water and/or gas from those perforated intervals, and consequently mixing, may have occurred during the P 8 test despite previous efforts to mutually isolate the horizons. This is in addition to the dilution by fresh water from the borehole.

A common feature of both types of mineral water is, however, that they represent hard to very hard brines. To the extent that reliable data are available, the relationship between the investigated or tested sections in the geological profile, the maximal influx rates, and the hydrostatic pressure head is presented together with the bar diagrammes for the chemical water analyses in Fig. 7, 8 and 9.

Age determinations have been performed on various samples by the NLfB. The analyses revealed that the absolute age of the thermal water is at least 25 000 years – samples contaminated by fresh water were not taken into consideration. The highest value obtained was more than 44 000 years. At the same time it was revealed that fossil carbon dioxide from deeper strata is evidently involved in the chemistry of the fossil water. An accurate account of these data will be presented at a later time, after additional investigations have been completed.

Water circulation in the Sedimentary Cover and Crystalline Basement

Some of the findings and results discussed above already indicate that in all sections of the penetrated crystalline (1602 to 3334 m below surface) beneath Urach zones of open

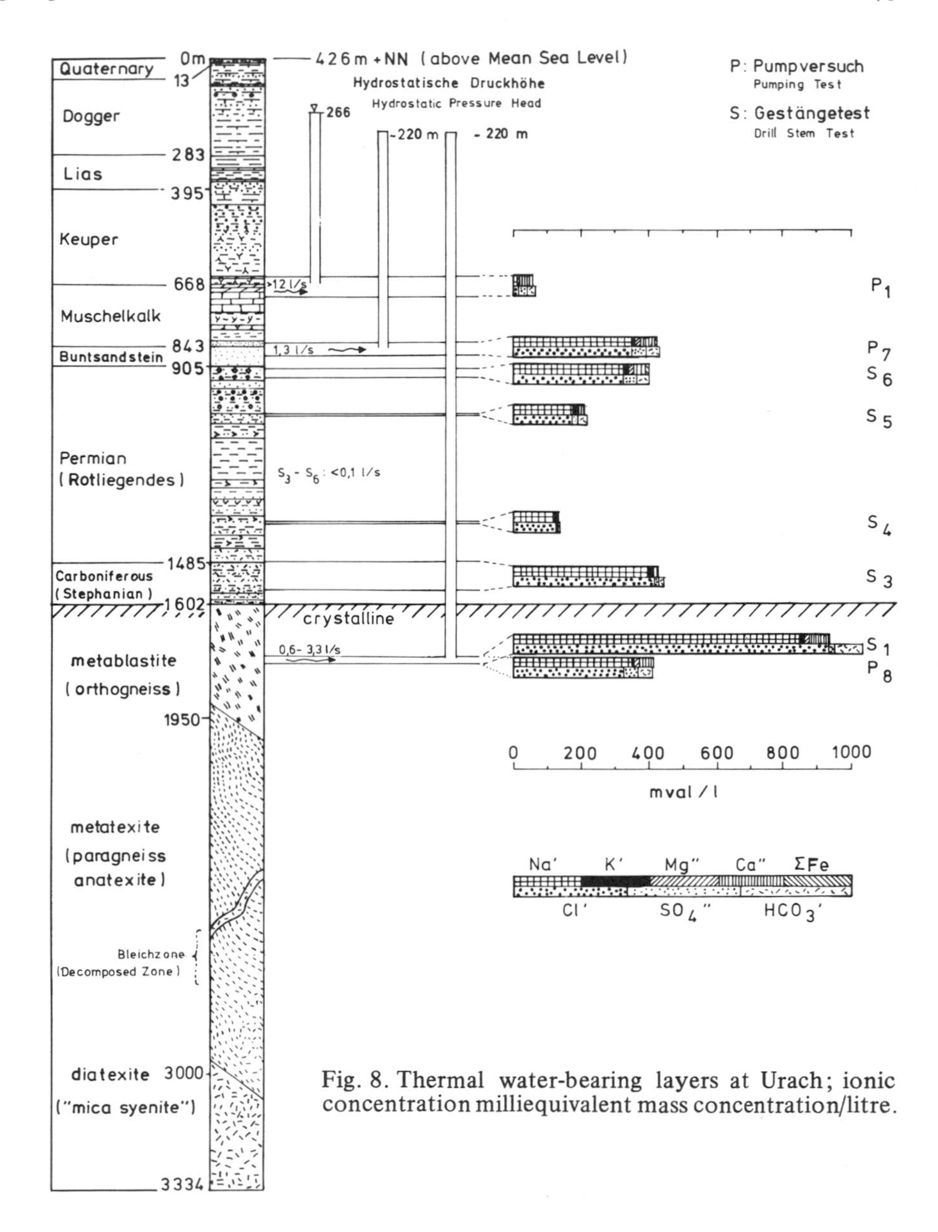

Fig. 8. Thermal water-bearing layers at Urach; ionic concentration milliequivalent mass concentration/litre.

fissures and joints in which hydrothermal convection is occurring and/or has occurred can be expected.

The decline of the water level in the completed Urach 3 well at rest also indicates that water can seep through the uncased borehole section between 3320 and 3334 m, as well as through three perforated sections located slightly higher, into systems of fissures in the diatexite encountered there (cf. SCHÄDEL & DIETRICH this volume). On the basis of the volumes injected, frac and circulation tests performed in the same depth range indicate the presence of a system of open joints capable of accommodating water in the

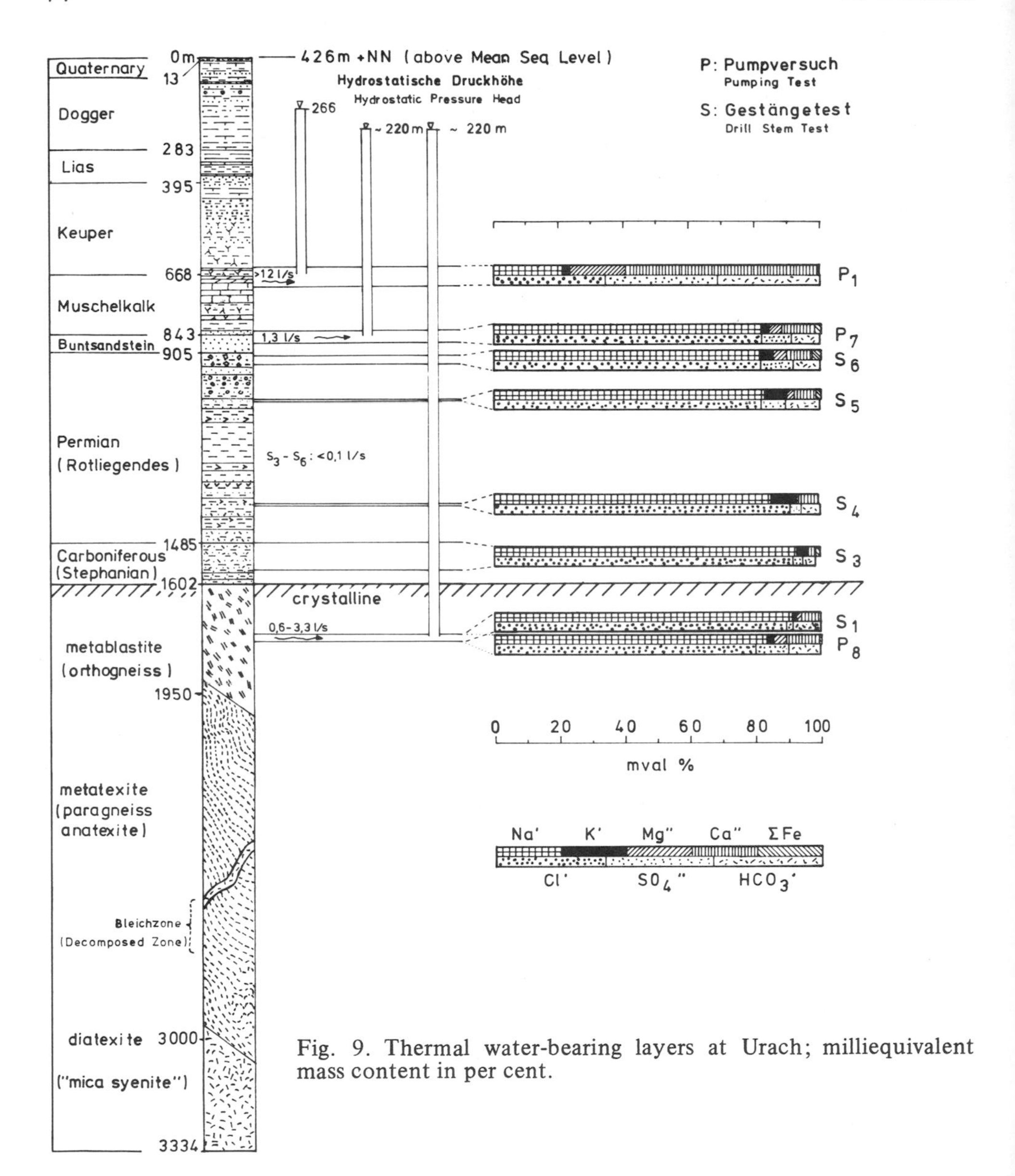

Fig. 9. Thermal water-bearing layers at Urach; milliequivalent mass content in per cent.

region surrounding even the deepest section of the borehole (SCHÄDEL & DIETRICH this volume).

Numerous zones of fissures and joints, as well as hydrothermally superimposed rock, have been revealed, both by direct investigation of core material (STENGER this volume), and by the evaluation of seismic velocity and electrical resistivity measurements performed in non-cored sections of the borehole (contributions by WOHLENBERG and STENGER this volume).

Moreover, the presence of water-bearing crystalline rock can be deduced from the results of geoelectric soundings performed in the region surrounding the Urach 3 well (BLOHM this volume).

According to the magnetotelluric investigations of BERKTHOLD & KEMMERLE (this volume), a substantial water-bearing capacity can be expected in the crystalline basement even at depths as far down as 8 km.

On the basis of geothermal investigations, HAENEL (1980) and HAENEL & ZOTH (this volume) arrive at the conclusion that heat transport by ascending water or gas has been possible for a long time, and thus that the thermal anomaly at Urach has been caused by hydrothermal convection. As an order of magnitude for the velocity of this flow, a value of about 10^{-8} cm/s or 0.3 cm/a has been estimated. As presumed by BUNTEBARTH et al. (1979, this volume), the intrinsic heat source is situated at great depths. This supposition is based on investigations of coalification which have been performed on drilling fluid samples from the Urach borehole. However, immediate evidence of water convection in the crystalline, or of the existence of ascending water, could not be secured from the Urach 3 borehole.

The good agreement between the chemical properties of the thermal water from the various aquifers and, as far as could be ascertained, between the hydrostatic pressure heads of the aquifers from the crystalline all the way to the Lower Muschelkalk, suggests that water convection is or has been possible between the crystalline basement and the ground-water storeys in the lower region of the sedimentary cover.

Flow paths for such circulation might be provided by zones of joints and faults extending to great depths in the tectonically highly distrubed basement (see BALKE et al., this volume). A further possibility for the ascent of hot water is presented by the numerous volcanic pipes, especially at the centre of the anomaly (DIETRICH et al. 1980). This becomes all the more plausible with augmenting depth, since the impermeable tuff fillings, which are dominant in the volcanic pipes near the surface, show a progressive transition to basalt with increasing depth; the latter is hydraulically more permeable to water.

It is also possible that the ascent of water may have been accelerated by gas lift. In addition to carbon dioxide, the presence of hydrocarbon gases has been ascertained, both in the sediments and in the crystalline basement.

According to the stratigraphic conditions in the bedrock (SCHÄDEL, this volume), which have been reconstructed from the evaluation of oriented cores, a further possibility for the ascent of thermal water can be derived: In the peripheral regions of the Rotliegend trough, where permeable sandstones and conglomerates of the Lower Muschelkalk and of the Buntsandstein are interlocked with those of the Young Palaeozoic and simultaneously overlie the crystalline discordantly, hydraulic connections between various zones of the sedimentary cover and crystalline basement may exist or have existed.

The Middle Muschelkalk separates the two types of mineral water determined. To the extent that the hydrostatic water levels could be ascertained in the thermal water wells, Urach 1 and 2 (SAUER 1971, SCHÄDEL 1977), and in the Urach 3 research well (Tab. 1), for the ground-water storeys above the Middle Muschelkalk, there is no evidence for the ascent of hot water from deeper levels to the surface in the region of Urach.

E. VILLINGER (this volume) arrives at the same result in the course of his hydrogeological survey of the entire region of the Urach thermal anomaly.

On the other hand, BALKE et al., as well as KOLLER & FRIEDRICHSEN (this volume) assume, with reference to the results of their hydrochemical and isotopic-geological investigations, the following: Water originating from greater depths, for example, from the transition zone between sedimentary rocks and crystalline basement, ascends to higher levels where it mixes with cooler water penetrating from above to produce thermal water of various types. The mixing models presented are only partially, or not al all, realizable with the use of the hydraulic potentials (cf. E. VILLINGER).

Utilization of the Thermal Water

Because of the low natural influx rate from the aquifers between the Lower Muschelkalk and the fissure zone in the crystalline, commercial utilization of the thermal water encountered there for extracting energy is hardly feasible, in spite of the maximal temperature of about 100 °C. However, the question of whether poorly permeable strata can be stimulated to yield a higher influx by appropriate treatment in a particular case is not considered.

On the basis of the available hydraulic data, only the Upper Muschelkalk below Urach can in principle be exploited for geothermal energy, despite the relatively low reservoir temperature of 55 to 60 °C; the feasibility of sufficiently high production rates is thereby decisive. Because of the possibility of detrimental interference with the Urach thermal wells 1 and 2 resulting from such operations in Urach 3, the wells of the thermal resort had been under observation since the beginning of drilling operations.

In the producing horizon (Upper Muschelkalk) of the Urach 1 and 2 wells, which are located at distances of about 480 and 420 m, respectively, from Urach 3, no irregularities were detected, either during the drilling of Urach 3 or during the pumping tests conducted there at various depths. Even the total circulation losses suffered during the drilling of the Upper Muschelkalk and an immediately subsequent pumping test with a duration of about a week in this formation did not cause any turbidity of the thermal springs or exert any hydraulic influence.

A hydraulic communication between the three wells could be demonstrated only with the use of the sensitive uranin (Na fluorescein) tracer.

The injection of uranin was carried out on 7th November, 1977 in the Urach 3 well. The uranin dye solution was injected through the drill pipe directly into the Upper Muschelkalk. The formation was subsequently sealed off with cement. Water samples were taken from the Urach 1 and 2 wells at regular intervals and were analysed for their uranin content by Dr. Käß at the Geologisches Landesamt in Freiburg. In Fig. 10 the quantities of water sampled daily and cumulatively from the Urach 1 and 2 wells are presented together with the uranin contents determined. With the approximately constant production of thermal water (about 9 l/s) over the entire sampling period, it can be assumed that no significant perturbing influence was thereby exerted on the tracer experiment.

The flow velocity between the wells can be calculated as the quotient of the linear distance between the wells to time of flow of the uranin. Since water was produced almost exclusively from the Urach 2 well during the tracer test, the appropriate distance is about 420 m. As referred to the first observation of uranin on 28th January, 1978, this results in a maximal value of 5.1 m/d for the velocity. For the most pronounced occurrence of uranin, observed on 7th June, 1978, this yields a value of 1.7 m/d.

These low values of the linear velocity appear to exert the strongest influence on the uranin yield. As a result of the comparatively low rate of production from the thermal wells and the low flow velocities, the uranin can remain in regions of near stagnancy for a long time. Thus, of the total injected quantity of uranin, about 3 kg, only about 12 g (about 0.4 per cent) was recovered. Of course, the flow of the water labelled with uranin to other sites of emergence cannot be excluded.

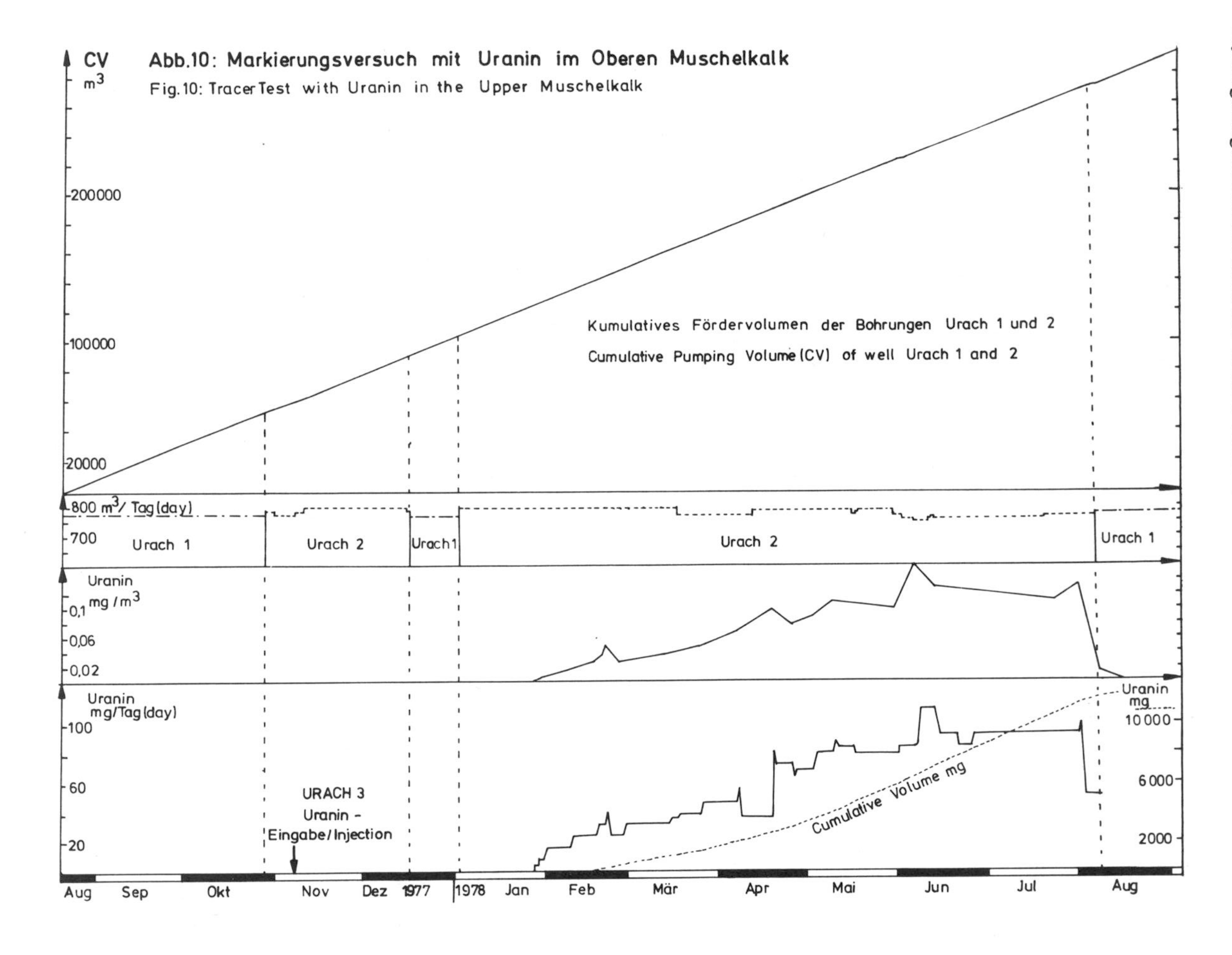

Fig. 10. Tracer test with uranin in the Upper Muschelkalk.

Because of the duration of several months for the emergence of the dye solution and of the various maxima, the throughput curve in Fig. 10 allows further conclusions: The labelled water may have flowed to the thermal wells via several separate paths with differing lengths. In the Upper Muschelkalk, the water could not have flowed through larger fissures and channels, but rather through zones only slightly karstified. Consequently, dilution and retardation occur during the main flow of the water through the aquifer.

For the effective porosity, estimated values between 0.5 and 1.0 per cent are obtained, if the total thickness of about 76 m for the Upper Muschelkalk is set equal to the aquifer thickness.

Because of the proven hydraulic communication between the Urach 1, 2 and 3 wells, long-term detrimental effects on the already existing installations of the thermal resort due to the extraction of geothermal energy from the Upper Muschelkalk by means of the Urach 3 well cannot be excluded. In order to confirm this finding, however, additional investigations should be conducted.

References

Buntebarth, G., Grebe, H., Teichmüller, M. & Teichmüller, R. (1979): Inkohlungsuntersuchungen in der Forschungsbohrung Urach 3 und ihre geothermische Interpretation. – Fortschr. Geol. Rheinld. u. Westf., **27**: 183–199, Krefeld.

Carlé, W. (1971): Die Tiefbohrungen auf mineralisiertes Thermalwasser in Bad Überkingen, Landkreis Göppingen, Baden-Württemberg (Geologie, Hydrogeologie, Technik). – Jh. Ges. Naturkd. Württ., **126**: 36–87, Stuttgart.

– (1972): Geologie und Hydrogeologie der Mineral- und Thermalwässer von Bad Über-Kingen, Landkreis Göppingen, Baden-Württemberg. – Jh. geol. Landesamt Baden-Württ., **14**: 69–143, Freiburg i. Br.

– (1974): Geologie und Hydrogeologie der Mineral- und Thermalwässer in Boll, Landkreis Göppingen, Baden-Württemberg. – Jh. geol. Landesamt Baden-Württ., **16**: 97–158, Freiburg i. Br.

– (1975 a): Die Mineral- und Thermalwässer von Mitteleuropa. Geologie, Chemismus, Genese. – Wissenschaftl. Verlagsges. (Stuttgart).

– (1975 b): Geologie und Hydrogeologie der Thermalwässer von Bonlanden, Stadt Filderstadt, Landkreis Esslingen, Baden-Württemberg. – Jber. u. Mitt. Oberrh. Geol. Ver., N.F. **57**: 21–41.

Carlé, W. & Groschopf, P. (1972): Geologie und Hydrogeologie der Säuerlinge, Mineralwässer und Thermalwässer von Bad Ditzenbach, Landkreis Göppingen, Baden-Württemberg. – Oberrhein. geol. Abh., **21**: 1–42, Karlsruhe.

Dietrich, H.-G., Haenel, R., Neth, G., Schädel, K. & Zoth, G. (1980): Deep Investigation of the Geothermal Anomaly of Urach. – In Strub, A.S. & Ungemach, P. (Eds.): Advances in European Geothermal Research, 2nd, p. 253–266 Strasbourg, D. Reidel Publ. Comp., Dordrecht/Boston/London.

Dietrich, H.-G. & Schädel, K. (1978): Untersuchung der geothermischen Anomalie in Urach auf eine mögliche Nutzung durch eine Untersuchungsbohrung bis tief ins Kristallin (ET 4023 B). – Progr. Energiefrosch. u. Energietechn. 1977–1980, Statusrep. 1978 – Geotechnik und Lagerstätten, **1**: 79–85; Jülich (Projektleitung Energieforsch. (PLE), KFA Jülich).

Haenel, R. (1978): Die Erkundung des Temperaturfeldes bis in größere Tiefen im Bereich von Urach sowie Erkundung geophysikalischer und geochemischer Methoden (ET

4027 A). – Progr. Energieforsch. u. Energietechn. 1977–1980, Statusrep. 1978 – Geotechnik und Lagerstätten, **1**: 29–37, Jülich (Projektleitung Energieforsch. (PLE), KFA Jülich).

Pape, H. (1977): Leitfaden zur Bestimmung von Erzen und mineralischen Rohstoffen: geochem. Grundlagen der Lagerstättenbildung; Bestimmungsschlüssel für Erze nach äußeren Kennzeichen; chem. Vorproben. – 1. Aufl., Ferdinand Enke Verlag, Stuttgart.

Sauer, K. (1971): Bericht über die Thermalwassertiefbohrung Urach. – Manuskript zum Vortrag an der Volkshochschule Urach am 15. März 1971.

Schädel, K. (1977): Die Geologie der Wärmeanomalie Neuffen-Urach am Nordrand der Schwäbischen Alb. – Seminar on Geothermal Energy, EUR 5920, **1**: 53–60, Bruxelles-Luxembourg.

Wohlenberg, J. (1978): Geophysikalische Untersuchungen in der Forschungsbohrung Urach (ET 4131 A). – Progr. Energieforsch. u. Energietechn. 1977–1980, Statusrep. 1978 – Geotechnik und Lagerstätten **1**: 87–100; Jülich (Projektleitung Energieforsch. (PLE), KFA Jülich).

The Urach Geothermal Project, p. 81–88;
Schweizerbart'sche Verlagsbuchhandlung, Stuttgart, 1982

Temperature Measurements and Determination of Heat Flow Denstiy

R. HAENEL and G. ZOTH

with 3 figures and 2 tables

Abstract: The highest temperature in the Urach 3 research borehole is about 143 °C at 3334 m depth, the bottom of the borehole. The mean temperature gradients at 0–300 m depth, 400–1500 m depth, and 1600–3334 m depth, are 10.5 °C/100 m, 3.9 °C/100, and 3 °C/100 m, respectively, and over the total vertical distance 4 °C/100 m.
Between the earth's surface and about 1000 m depth the temperature field is disturbed. Water movement in the rock surrounding the borehole must be assumed, probably resulting from incorrect cementation done to prevent water exchange between different strata.
The mean heat flow density is about 86 mW m^{-2}.
The radiogenetic heat production of sedimentary rock is H_S = 1.12 μW m^{-3} and for the crystalline basement H_B = 2.87 μW m^{-2} Temperatures down to 30 km depth have been estimated.
The heat flow density, as well as the temperature measurements, indicate a possible heat transport by ascending water or gas over long periods of time, thus causing the temperature anomaly. In such a case, the heat flow density would be somewhat higher than the value mentioned above, because calculated only for heat conduction.

Introduction

The objective of the geothermal investigation in the Urach research borehole is to determine the temperature distribution with depth, as well as the heat flow density. These two parameters can be expected to yield information on the location and cause of the heat. The data are also of interest for the other participants in the Urach research project, especially for the hot-dry-rock experiments.

1. Temperature Measurements

After the drilling work was finished, several measurements were carried out in 1978 and 1979 to check recovery time and to determine equilibrium temperatures. The final temperature distribution with depth is shown in Fig. 1. It can be seen that the temperature distribution is irregular between the earth's surface and about 1000 m depth.

The drilling work was interruppted several times in order that temperature measurements could be carried out. From these data and the temperature measurements in the Urach borehole 2, about 400 m away, the undisturbed or equilibrium temperature was calculated for 0 to 1600 m depth as shown in Fig. 1, curve b.

The disturbance in curve a can be interpreted in terms of water movement as discussed by KAPPELMEYER & HAENEL (1974). From this assumption and the two curves a and b in Fig. 1 it follows that at about 500 m depth water descends with a temperature of 50 °C. The measured curve is irregular, probably because of the superposition of other influences during the descent of water, for example, an influx of cooler water (between 500 and 550 m) from a higher level (400 m). Warmer water from lower levels would have the opposite effect. The irregular curve can be also effected by water movement at various distances from the borehole into the surrounding rock. Deeper than about 950 m only a small amount of water descends as indicated by the temperatures that are only slightly lower than on curve b.

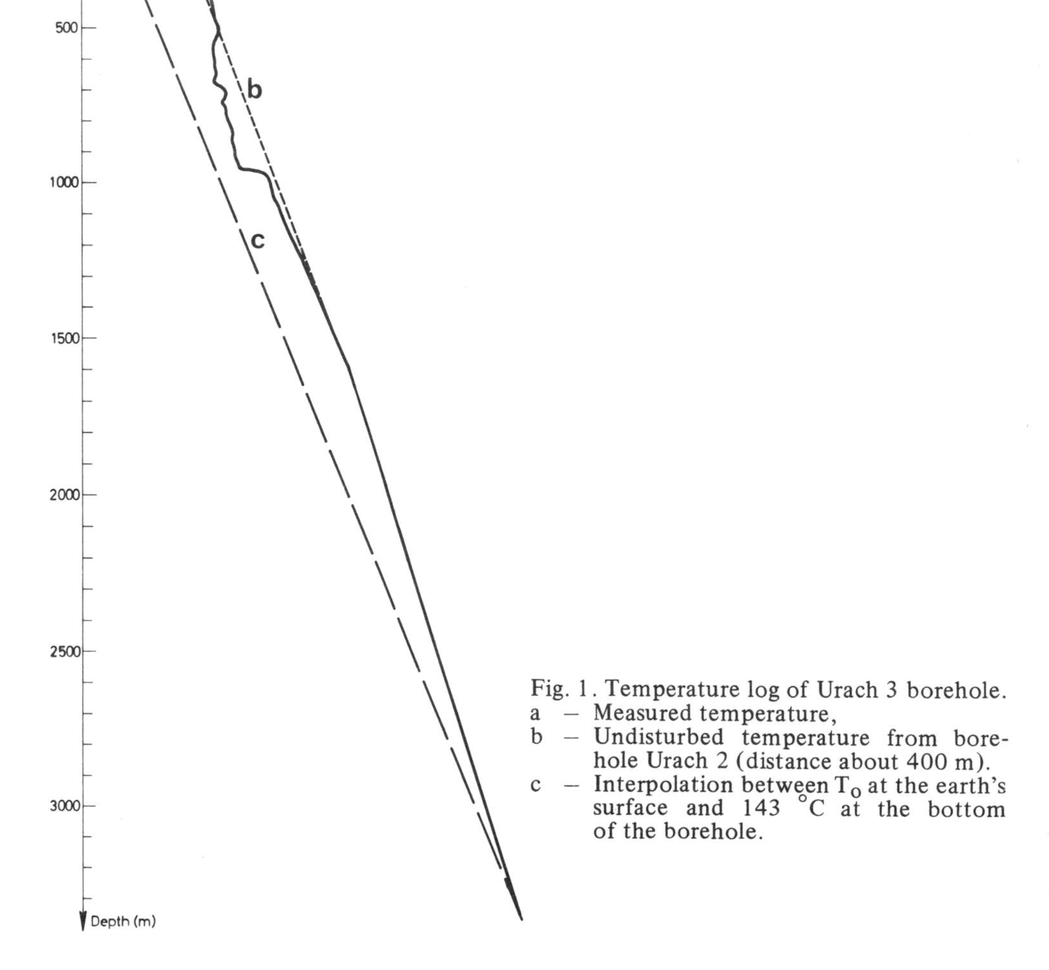

Fig. 1. Temperature log of Urach 3 borehole.
a – Measured temperature,
b – Undisturbed temperature from borehole Urach 2 (distance about 400 m).
c – Interpolation between T_0 at the earth's surface and 143 °C at the bottom of the borehole.

At about 450 m depth ascending water must influence the temperature. The temperature data allow no decision whether the water comes from one aquifer or from two closely spaced aquifers.

The reason for these water movements is probably defective cement behind the casing. See also the detailed discussion of HAENEL & ZOTH (1980). Within the crystalline basement, the measured temperature curve is very smooth and does not indicate water movement.

The temperature gradient changes with depth as follows:

0 to 300 m: 10.5 °C/100 m
400 to 1600 m: 3.9 °C/100 m
1600 to 3334 m: 3.0 °C/100 m

and the mean temperature gradient for the entire curve is:

4 °C/100 m.

The maximal temperature at the bottom of the borehole is:

143 °C.

2. Heat Flow Density Determination

The terrestrial heat flow density q is given by the Fourier equation:

$$\vec{q} = -k \operatorname{grad} T,$$

where k represents the thermal conductivity determined in the range of the measured temperature gradient grad T. In practice, only the vertical component $q_z = q$ is considered, this usually being far greater than the two horizontal components.

Unlike temperature, heat as a form of energy satisfies a conservation theorem. Heat flow density can therefore, be extrapolated to an inaccessible depth with greater certainty than can the temperature. If the thermal conducticity k and the heat production H of the rocks in question are known and if water movement and mass convection can be ruled out there, the temperatures at great depths can be given as follows with the aid of the heat flow density:

$$T(z) = T_0 + \frac{q}{k} - \frac{H \cdot z^2}{2 \cdot k} \tag{2}$$

where T_0 represents the annual mean temperature at the earth's surface.

The determination of thermal conductivity is carried out on rock samples in the laboratory by an absolute method as described by CREUTZBERG (1964), as well as by the Quick Thermal Conductivity Meter as described by ARAKAWA & SHINOHARA (1980), called QTM.

If QTM is used and the rock samples are anisotropic, a definite measuring process must be used so that the vertical component, $k_\perp$, of the thermal conductivity can be calculated. $k_\perp$ is necessary for the heat flow density determination and is calculated as follows KAPPELMEYER & HAENEL (1974):

$$k_\varphi = k_\perp \cos^2\varphi + k_\parallel \sin^2\varphi = k_\perp + (k_\parallel - k_\perp)\sin^2\varphi,$$

where k_φ (thermal conductivity in direction φ) corresponds to k_B (measured thermal conductivity for a line source B in Fig. 2) and $k_\parallel$ (thermal conductivity component parallel to the stratification) corresponds to k_A (measured thermal conductivity for a line source A in Fig. 2).

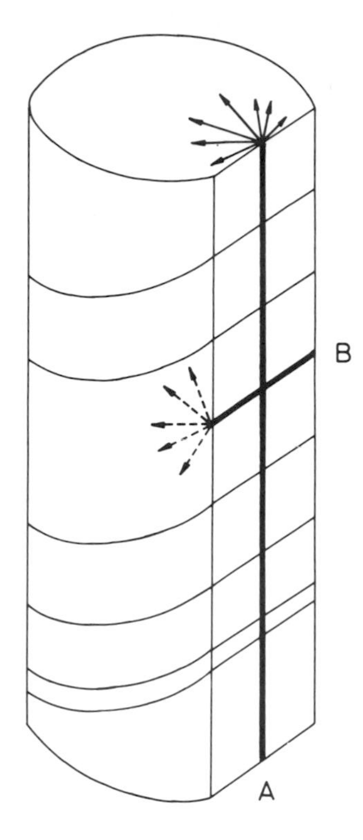

Fig. 2. Position of line source for determining the vertical component (parallel to rock cylinder axis) of thermal conductivity.

Further, the mean of the thermal resistance of a line source is given by:

$$\left[\frac{1}{k}\right]_{mean} = \frac{1}{\pi}\int_{o}^{\pi} \frac{d\varphi}{k_{\perp} + (k_{\parallel} - k_{\perp})\sin^2\varphi}$$

which leads to:

$$k_{\perp} = \frac{{k_B}^2}{k_A}$$

An improved procedure has been developed by GRUBBE (1980). About 40 rock samples have been measured by the absolute method and about 2000 by the QTM. The mean values for each 100 m of depth are given in Tab. 1 and Fig. 3. These values show that in the range of anatexite and paragneiss at about 1950 m to 3000 m depth, the thermal conductivity is significantyl higher than in the other formations.

Heat flow density is also given for each 100 m depth in Tab. 1 and Fig. 3. Heat flow density is calculated by equation (1). Corrections were made for topographic influences on the temperature gradient before the heat flow density was calculated. The mean heat flow density between 500 m and 330 m depth is about

86 mW m^{-2} (see Tab. 1 and Fig. 1).

The deviation from the mean values within the basement is possibly due to water movement within fractures and fissures.

The heat flow density between 0 and about 300 m depth can be estimated by means of the above-mentioned temperature gradient of 0.105 °C m^{-1} and the thermal conductivity of about 1.5 W m^{-1} K^{-1} (BEHRENS et al., this volume). The calculation amounts to about 160 mW m^{-2}.

If the temperature T_0 at the earth's surface is connected by a straight line with the temperature at the bottom of the borehole (line c in Fig. 1), the measured temperature curve in the basement, as well as its extrapolation by curve b in the sedimentary layers, also indicates water movements (see also KAPPELMEYER & HAENEL, page 201). Further, the temperature gradient between the earth' surface and about 300 m depth is quite high, as already above mentioned. This can be explained by ascending warm water from great depth. The possibility is given also in the crystalline basement by fissured zones as known from the Urach 3 borehole at 1759–1780 m depth.

The rate of ascent of the water can be estimated by the method of BREDEHOEFT & PAPADOPOLUS (1965) and depends on the assumed depth of the source of the ascending water:

source depth of ascending water	1600	3300	5000 m
water velocity	$3.8 \cdot 10^{-8}$	$2.1 \cdot 10^{-8}$	$1.4 \cdot 10^{-8}$ cm s^{-1}

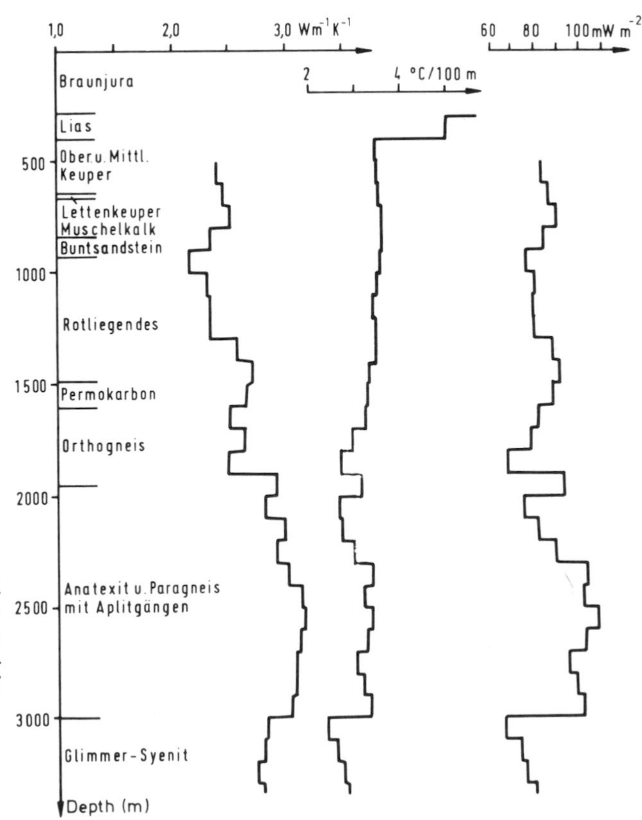

Fig. 3. Geological profile, thermal conductivity, temperature gradient, and heat flow density of Urach 3 borehole. The data are presented as a mean for steps of every 100 m depth.

Table 1. Determination of heat flow density.

Depth	Measured gradient (°C/100 m)	Topographic correction (°C/100 m)	Corrected gradient (°C/100 m)	Mean thermal conductivity ($W m^{-1} K^{-1}$)	Heat flow density ($mW m^{-2}$)
0/ 100	11.5	− 2.01	9.49		
100/ 200	10.5	− 1.79	8.71		
200/ 300	9.0	− 1.57	7.43		
300/ 400	6.0	− 0.91	5.09		
400/ 500	3.8	− 0.41	3.39		
500/ 600	3.8	− 0.345	3.46	2.4	82.9
600/ 700	3.8	− 0.289	3.51	2.46	86.4
700/ 800	3.8	− 0.241	3.60	2.52	89.7
800/ 900	3.8	− 0.202	3.60	2.34	84.2
900/1000	3.7	− 0.168	3.53	2.16	76.3
1000/1100	3.6	− 0.141	3.46	2.32	80.2
1100/1200	3.5	− 0.117	3.38	2.34	79.2
1200/1300	3.5	− 0.097	3.40	2.34	79.6
1300/1400	3.5	− 0.076	3.42	2.58	88.3
1400/1500	3.4	− 0.056	3.34	2.72	91.0
1500/1600	3.35	− 0.039	3.31	2.66	88.1
1600/1700	3.27	− 0.031	3.24	2.52	81.6
1700/1800	2.98	− 0.025	2.96	2.64	78.0
1800/1900	2.72	− 0.019	2.70	2.50	67.5
1900/2000	3.20	− 0.014	3.19	2.92	93.0
2000/2100	2.68	− 0.009	2.67	2.82	75.3
2100/2200	2.72	− 0.005	2.72	3.0	81.4
2200/2300	2.05	− 0.001	3.05	2.92	89.0
2300/2400	3.41	+ 0.002	3.41	3.02	103.0
2400/2500	3.20	+ 0.004	3.20	3.14	100.6
2500/2600	3.40	+ 0.007	3.41	3.16	107.7
2600/2700	3.27	+ 0.009	3.28	3.12	102.3
2700/2800	3.06	+ 0.011	3.07	3.08	94.6
2800/2900	3.18	+ 0.013	3.19	3.08	98.3
2900/3000	3.33	+ 0.015	3.35	3.04	101.7
3000/3100	2.33	+ 0.017	2.35	2.82	66.2
3100/3200	2.60	+ 0.017	2.62	2.80	73.3
3200/3300	2.75	+ 0.018	2.77	2.74	75.8
3300/3350	2.84	+ 0.018	2.86	2.80	80.0

That means the water velocity is on the order of 10^{-8} cm/s ≈ 0.3 cm a^{-1}. This water movement over a long period of time is then the probable reason for the temperature anomaly in the Urach region. There are also ascending gases as a postvolcanic phenomenon in the Urach area. This can also affect the temperature field. However, the two causes cannot be distinguished on the basis of geothermal measurements.

If the assumption of heat transport by ascending water from great depth is realistic, so the heat flow density calculation by equation (1) for heat transport by conduction is not allowed. The heat flow density of 86 mW m^{-2} would be too low, because the ascending water lowers the temperature gradient. On the other hand, the heat flow density of about 160 mW m^{-2} between earth's surface and about 300 m depth would be to high, because this value is strongly influenced by the transported heat from great depth to about 300 m depth.

3. Heat Production Measurements

The determination of temperature down to great depth using equation (2) also requires data for the heat production by the rock. Radiogenetic heat production has been determined for several rock samples from the Urach 3 borehole by the Institute of Geophysics, ETH Zürich, Switzerland.

Table 2. Heat production. U = uranium, Th = thorium, K = potassium, H = Heat production, HGU = Heat Generation Unit = $4.2 \cdot 10^{-7}$ W m^{-3}

No.	Depth (m)	Density (μW m^{-3})	U (ppm)	Th (ppm)	K (%)	H (HGU)	H μW m^{-3})
1	513.3	2.7	< 0.05	< 0.05	< 0.1	< 0.04	< 0.02
2	517.3	2.7	0.24	0.37	0.15	0.24	0.10
3			0.30	0.61	0.19	0.33	0.14
4	517.7	2.7	0.15	0.29	0.07	0.16	0.066
5			0.15	0.20	0.06	0.14	0.057
6	568.4	2.7	2.43	8.97	4.38	3.96	1.65
7			2.44	8.60	4.42	3.91	1.64
8	569.8	2.7	2.18	7.62	4.03	3.51	1.46
9			2.07	7.85	4.01	3.47	1.45
10	570.9	2.7	3.49	12.5	4.60	5.24	2.19
11			3.61	12.4	4.61	5.31	2.22
12	599.5	2.7	0.19	1.02	0.28	0.35	0.14
13			0.22	0.96	0.28	0.36	0.15
14	602.6	2.7	2.84	4.56	1.63	2.86	1.20
15			2.80	4.71	1.62	2.86	1.20
16	659	2.52	2.5	12.2	5.7	4.51	1.88
17	666	2.68	0.5	0.8	0.2	0.49	0.21
18	838	2.35	1.8	8.8	2.5	2.70	1.13
19	843.5	2.70	3.3	1.9	0.4	2.43	1.01
20			3.5	2.0	0.4	2.55	1.07
21	942.0	2.53	1.1	5.8	3.2	2.15	0.90
22			1.2	6.2	3.2	2.29	0.96
23	1025	2.44	2.0	12.4	4.0	3.75	1.57
24	1265	2.66	6.7	14.4	4.5	7.41	3.10
25	1440	2.70	3.6	18.5	3.6	6.09	2.54
26	1643	2.68	2.2	18.5	2.3	4.87	2.04
27	1756	2.78	2.3	16.7	2.1	4.76	1.99
28	1853	2.73	2.3	14.5	2.2	4.37	1.83
29			2.2	14.8	2.2	4.39	1.83
30	2013	2.74	1.7	13.5	2.4	3.88	1.62
31			1.5	13.6	2.4	3.78	1.58
32	2195	2.70	3.2	31.9	2.5	7.77	3.25
33	2420	2.68	2.7	17.7	2.7	5.12	2.14
34			2.9	17.5	2.7	5.23	2.18
35	2930	2.71	9.7	41.9	2.6	13.5	5.65
36			9.1	42.4	2.7	13.3	5.54
37	3128	2.78	4.3	25.7	2.3	7.62	3.18
38	3301	2.77	5.0	27.8	2.9	8.51	3.56
39	3306	2.77	4.8	27.4	3.0	8.38	3.50
40	3328	2.73	3.9	27.6	2.1	7.52	3.15

The results are shown in Tab. 2, the mean value for sediments is:

$$H_S = 1.12\ \mu W\ m^{-3}$$

and for the basement:

$$H_B = 2.87\ \mu W\ m^{-3}.$$

For depths greater than 15 km it is assumed that $H = 0.46\ \mu W\ m^{-3}$ and for depths greater than 3.5 km $k = 0.4887 \cdot (\frac{1050}{250 + T} + 3.3)\ W\ m^{-1}\ K^{-1}$ (HAENEL 1979).

From this and $q = 86\ mW\ m^{-2}$, the following temperatures have been calculated:

Depth (km)	5	10	15	20	25	30
Temperature (°C)	199	362	499	619	737	852

Because the heat flow density is probably influenced by water movements, the value of 86 mW m^{-2} would be too low. Therefore, the actual temperature must be assumed to be higher than the calculated temperatures.

References

Arakawa, Y. & Shinohara, A. (1980): Quick Thermal Conducticity Meter. – 7th EGS-Meeting, in press.

Bredehoeft, J.D. & Papadopulos, I.S. (1965): Rates of vertical groundwater movement estimated from the earth's thermal profile. – Water Resources Res., **1**: 325–328.

Creutzburg, H. (1964): Untersuchungen über den Wärmestrom der Erde in Westdeutschland. – Kali und Steinsalz, **4**: 73–108, Essen.

Grubbe, K. (1980): Vertikalbewegungen und ihre Ursachen am Beispiel des Rheinischen Schildes. – Report, NLfB Hannover, Archive No 86 025.

Haenel, R. (1979): Determination of subsurface temperatures in The Federal Republic of Germany on the basis of heat flow values. – Geol. Jb. Hannover, **E 15**: 41–49.

Haenel, R. & Zoth, G. (1980): Thermische Untersuchungen. – In: Haenel, R., Geophysikalische Untersuchungen in der Forschungsbohrung Urach. Report, NLfB Hannover, Archive No. 85 805.

Kappelmeyer, O. & Haenel, R. (1974): Geothermics with special reference to application. – Geoexplor. Mongr., S. 1, No 4, 238 pp. Gebr. Borntraeger, Berlin, Stuttgart.

The Urach Geothermal Project, p. 89–95;
Schweizerbart'sche Verlagsbuchhandlung, Stuttgart, 1982

Ancient Heat Flow Density Estimated from the Coalification of Organic Matter in the Borehole Urach 3 (SW-Germany)

G. BUNTEBARTH and M. TEICHMÜLLER

With 3 figures and 2 tables

Abstract: The degree of coalification, determined by vitrinite reflectivity, increases from 0.5 % Rm in the Dogger (at a depth of 106 m) to 1.84 % Rm in the Stephanian (at a depth of 1605 m). If the same coalification range (0.6–1.7 % Rm) is compared, the coalification gradient is much higher in the borehole Urach 3 (0.085 % Rm/100 m) than in most boreholes of North-West Germany (0.055 % Rm/100 m). These results suggest a high paleogeothermal gradient in the Urach area which, at present, represents the centre of a positive geothermal anomaly.
Depending on assumptions of different thicknesses of overburden, the temperature gradients, calculated from the coalification pattern, range from 39 °C/km to 46 °C/km. The lower temperature gradient is obtained when assuming 500 m thickness of eroded overburden, the higher one for the case that no overburden has been eroded. The mean value of 43 °C/km is higher than the mean global value. It corresponds to a heat flow density of $Q \approx 90$ mW/m^2, assuming a heat conductivity of $K = 2.2$ W/m°C. The calculated paleogeothermal gradient must have affected the organic material at the time of its deepest subsidence. Thus the high paleogeothermal gradient was active before the onset of uplifting during the Upper Tertiary, i.e. between Upper Jurassic and Lower Tertiary. The paleogeothermal regime equals more or less the present one at Urach. The rather constant thermal regime suggests an extensive heat source at great depth.

1. Introduction

The degree of coalification ("rank") of finely dispersed organic matter in sedimentary rocks is often used as a paleo-geothermometer. During subsidence of the sediments organic matter is affected by increasing rock temperature and overburden pressure. Within the uppermost few hundred metres, the rising overburden pressure causes mainly loss of porosity and moisture from the organic particles. The increasing temperature initiates chemical reactions, thus causing a rise of coal rank. Besides the degree of temperature, the duration of temperature exposure is decisive for the degree of coalification. Since the chemical reactions are irreversible, the degree of coalification can be used to determine only the maximum temperature.

One measure for the degree of coalification is the optical reflectivity of polished vitrinite particles. Microscopical reflectance measurements are a rapid and efficient method for determining the rank of finely dispersed organic matter in sedimentary rocks (STACH et al. 1975).

In this study an empirical method is applied in order to estimate an ancient thermal regime (BUNTEBARTH 1978). The fact that the degree of coalification increases con-

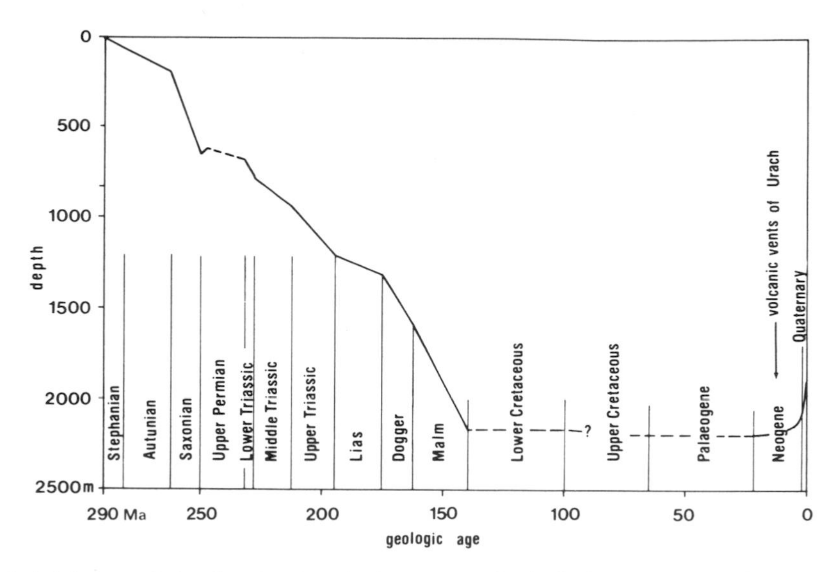

Fig. 1. Subsidence of the Stephanian during the geological history in the Urach area, on the basis of the geological section encountered in the borehole Urach 3 after R. TEICHMÜLLER (in: BUNTHEBARTH et al. 1979).

tinuously with increasing temperature and the duration of its exposure, is used to establish a relationship between the three variables: coalification, temperature and time. If the temperature gradient is constant within the investigated depth interval, the temperature is a linear function of depth. The data of both depth (z) and time (t), which are derived from observations in boreholes, are used to establish the "subsidence curve" (Fig. 1). Within vertical sections through young sedimentary basins of southern Germany, the square of vitrinite reflectivity (Rm) was found to be proportional to the integrated burial history in the following way (BUNTEBARTH 1979):

$$(1) \qquad Rm^2 = f\,(dT/dz) \int_0^t z\,(t^*)\,dt^*$$

The factor of proportionality, f, is a function of the temperature gradient (dT/dz).

To reduce the three-dimensional problem to two dimensions, eq. (1) is calculated at the plane Rm = 1.0 %. This procedure makes eq. (1) much easier to solve. Using the integrated value of eq. (1) at Rm = 1.0 %, called I, the function f is calibrated.

Actual well temperature data of the Upper Rhine Graben (M. TEICHMÜLLER 1979) and the Bavarian Molasse (M. & R. TEICHMÜLLER 1975), as well as temperature data derived from paleothermal calculations on the basis of chemical maturation parameters from wells of the Upper Rhine Graben (ESPITALIE 1979) are used. The calibration yields the inverse function (f^{-1}) of:

$$(2) \qquad dT/dz\,[^\circ C/km] = 98.7 - 14.6 \ln I\,[Ma\ km]$$

2. Coalification Pattern in the Borehole Urach 3

38 cuttings and 5 core samples were studied under the microscope for the rank of organic matter. The optical reflectivity of vitrinite particles was measured and the fluorescence properties of liptinites under blue light excitation were studied (M. TEICHMÜLLER 1974). Oil maturation of the Posidonia shale (Lias ϵ) was estimated on the basis of fluorescence measurements in ultraviolet light (method see M. TEICHMÜLLER & OTTENJANN 1977). With the microscope, cavings of Posidonia shale (which was encountered at a depth of about 290 m) could be identified down to 400–462 m where Keuper was encountered. Many samples do not contain measurable vitrinite; hence reliable reflectivity data could be obtained from 23 samples only. Tab. 1 contains the results of reflectivity measurements. The Rm values are arithmetic means of a number (n) of single measurements on a given sample. In Fig. 2 vitrinite reflectivity is plotted against depth.

Although cuttings from the Dogger and Lias are rich in organic matter, they contain mainly liptinites and rarely vitrinite. The low fluorescence intensity of the mineral-bituminous matrix in the Dogger samples suggests that the measured vitrinite reflectivity is comparable with data from coal beds.

On the other hand, the results of fluorescence measurements (BUNTEBARTH et al. 1979) on the Lias samples indicate the beginning of oil maturity. In this case the measured vitrinite reflectivity (about 0.5 % Rm) is influenced by bitumen impregnations which tend to lower the reflectivity. In a non-bituminous environment (e.g. in a coal bed) at the same depth, vitrinite would reach a reflectance of about 0.6–0.7 % Rm. This has been considered when drawing the coalification curve of Fig. 2.

Table 1. Vitrinite reflectivity of samples from the borehole Urach 3 (BUNTEBARTH et. al. 1979).

Sample no.	Kind of sample*	Depth [m]	Period	Reflectivity Rm [%]	Number of measurements n
16350	S	106	Dogger	0.5	4
16353	S	158	Dogger	0.5	4
16355	S	202	Dogger	0.7	5
16360	S	286	Lias	0.5	2
16362	S	296	Lias	0.40	10
16363	S	302	Lias	0.49	14
16364	S	308	Lias	0.59	4
16369	S	462	Keuper	0.54	4
15462	K	570	Keuper	0.86	80
15463	K	570.6	Keuper	0.83	22
15627	K	662	Keuper	0.90	41
15630	K	1267	Rotliegendes	1.33	6
16367	S	1535	Rotliegendes	1.76	50
16378	S	1550	Stephanian	1.79	50
16379	S	1574	Stephanian	1.58	33
15632	S	1582	Stephanian	1.58	32
16380	S	1582	Stephanian	1.70	15
15633/15634	S	1584	Stephanian	1.62	50
15635	S	1608	Stephanian	1.82	18
15636	S	1614	Stephanian	1.62	4
15637	S	1630	Stephanian	1.81	43
15638	S	1658	Stephanian	1.84	30

*S = cuttings, K = core

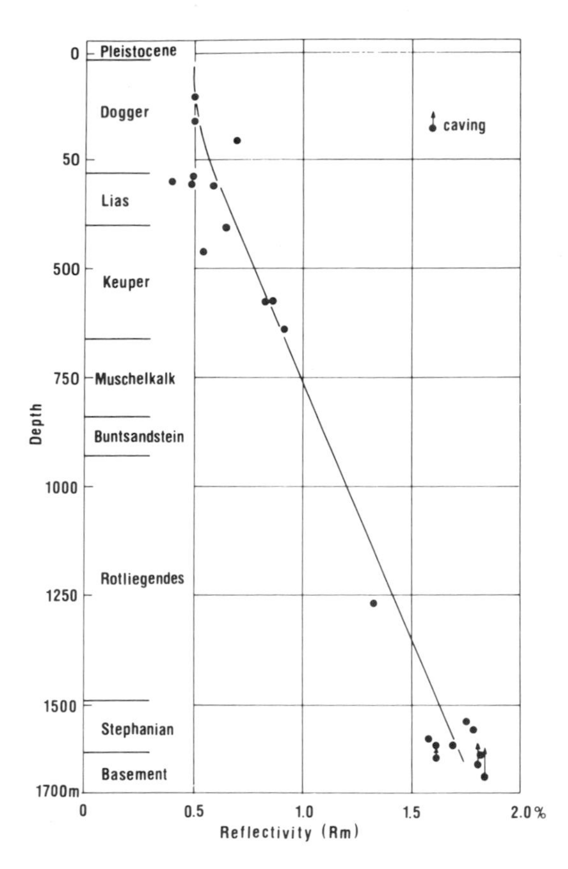

Fig. 2. Increase of coalification with depth in the borehole Urach 3, on the basis of vitrinite reflectivity.

The Keuper samples contain relatively much vitrinite proper, occurring in the form of wide bands, sometimes associated with thick cutinite. The cutinites emit an orange fluorescence with a higher intensity than cutinites from Ruhr coals of the same rank (vitrinite reflectivity). The Rotliegend samples are poor in organic matter, which is more or less oxidized, thus causing a wide scatter of reflectivity values (1.05–2.2 % Rm). A reliable value was obtained from one sample only (1.76 % Rm). No fluorescing liptinites were observed in Rotliegend samples, the probable loss of fluorescence being in agreement with the high reflectivity values.

Stephanian samples, the age of which was determined by GREBE with palynological methods (see BUNTEBARTH et al. 1979), are very rich in vitrinite, occurring in pure vitrite layers and in vitrinertites from coal beds. The coal is rich in syngenetic and epigenetic carbonates. Some vitrinite layers contain syngenetic pyrite framboids which are, however, mostly oxidized. A wide scatter of vitrinite reflectivity values (ranging between 1.5 and 1.95 % Rm) indicate that also the vitrinite is partly oxidized. Transitions from vitrinite to inertinite ("semi-vitrinite") with reflectivities between 2.0 and 2.3 % Rm are common. Altogether, the microscopic appearance of the Stephanian vitrinites indicates a fairly intense fossil oxidation which presumably was caused by percolating hot waters acting in connection with volcanic activity during the Lower Rotliegend. Tab. 1 and Fig. 2 contain only the lowest reflectivity values measured in the Stephanian.

According to the coalification curve of Fig. 2, the mean increase of reflectivity with depth is 0.085 % Rm/100 m between 300 and 1600 m depth (Lias to Stephanian).

In North-West-German boreholes the gradient is only 0.055 % Rm/100 m (M. TEICHMÜLLER in: STACH et al. 1975: 328) for the same coalification range. This comparison reflects a more rapid increase of coalification with depth in the Urach area, suggesting a relatively high paleogeothermal gradient.

3. Geothermal Interpretation

Care must be taken in reconstructing the burial history, which is necessary for applying the method in order to estimate the temperature gradient. R. TEICHMÜLLER (in: BUNTEBARTH et al. 1979) used the borehole profile of DIETRICH & SCHÄDEL (this volume) for his subsidence diagram. However, for paleogeothermal considerations, it must be taken into account that part of the Jurassic sediments probably has been eroded. Although the original thickness of the Jurassic is unknown, after LEMCKE (1974) the Alpine foredeep experienced a strong subsidence in Mesozoic times, during which a rather uniform layer of Malm was deposited with a thickness of about 500–600 m. Whether Cretaceous was deposited or not is unknown. If it was deposited, its thickness would not exceed 200 m. Taking into account an eroded overburden with a thickness of z_e the burial history z (t) can be replaced by the expression:

$$z(t) = \tilde{z}(t) + z_e$$

which, if substituted in eq. (1), yields:

$$(3) \qquad Rm^2 = f(dT/dz) \int_0^t (\tilde{z}(t^*) + z_e)\, dt^*$$

Since the integrated value is constant (c) from t = 0 until t = t' (Cretaceous) within a vertical section, the integral can be split into:

$$(4\,a) \qquad \int_0^t (\tilde{z}(t^*) + z_e)\, dt^* = \int_0^{t'(\mathrm{Cret.})} (\tilde{z}(t^*) + z_e)\, dt^* + \int_{t'(\mathrm{Malm})}^{t} (\tilde{z}(t^*) + z_e)\, dt^*$$

$$(4\,b) \qquad = c + \int_{t'}^{t} (\tilde{z}(t^*) + z_e)\, dt^*$$

The constant c, which depends on the value z_e, is found graphically after calculation of the integrated value in eq. (4 b), right hand side, for each averaged reflectivity value of Tab. 2. Fig. 3 finally allows a determination of the integrated value at the reflectivity Rm = 1.0 %, which is necessary for applying the calibration function for estimating the temperature gradient, eq. (2). If thicknesses of eroded overburden between 0 and 500 m are assumed, the temperature gradients range between 46 and 39 °C/km. According to these values, paleotemperature gradients of dT/dz = 39 to 43 °C/km are more probable than a gradient of 46 °C/km (BUNTEBARTH et al. 1979).

The temperature gradient is proportional to the heat flow density:

$$(5) \qquad Q = K\, dT/dz$$

K, as a factor of proportionality, is the heat conductivity. The mean value of 2.2 W/m °C is estimated from measurements by HAENEL & ZOTH (this volume). Applying this conductivity value with eq. (5), a heat flow density of about 90 mW/m^2 is estimated.

The question arises which period(s) was (were) governed by such a high thermal regime. The paleotemperature gradient is commonly calculated for the period of maximum temperature, i.e. the period of maximum burial. In the Urach area, the maximum burial

Table 2. Mean reflectivity of vitrinite (after Fig. 2) and integrated values of burial history depending on an additional cover of sediments, which has been eroded.

Depth z (m)	Mean reflectivity Rm (%)	Integrated values of burial history $\int z(t)\,dt$ (km Ma) with thickness of an additional cover z_e of		
		0 m	200 m	500 m
100	0.52	9.9	10.4	11.1
300	0.60	12.9	16.0	20.9
660	0.92	28.5	41.3	56.7
1250	1.42	78.3	107.0	138.6
1570	1.70	106.4	124.1	158.9
	1.00	$I_1 = 36.2$	$I_2 = 46.6$	$I_3 = 60.5$

Ma = million years

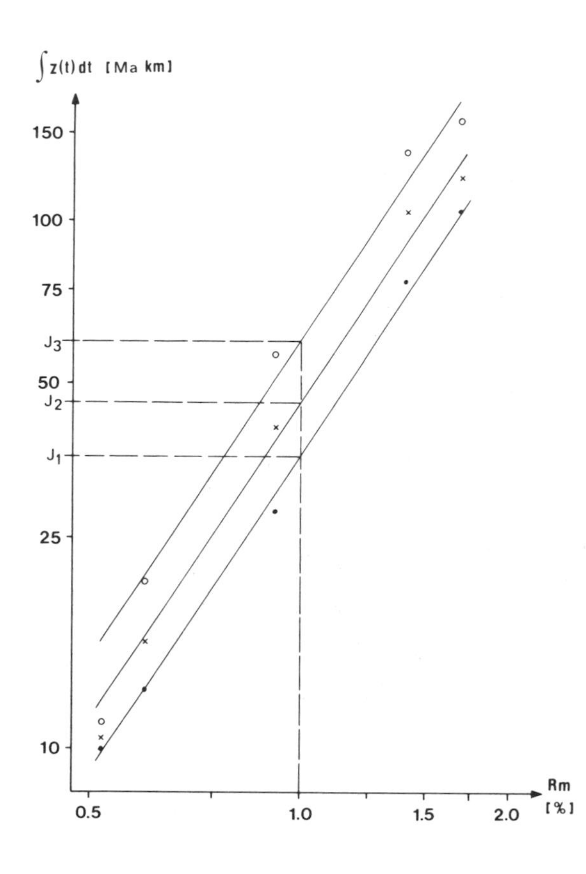

Fig. 3. Integrated burial history versus the vitrinite reflectivity in the borehole Urach 3 with thickness of eroded former overburden as parameter. : •—• 0 m, x—x 200 m, o—o 500 m.

took place during the Malm, perhaps during the Cretaceous. Because of the linear relationship shown in Fig. 3, a higher thermal regime before the Malm is improbable. Since the degree of coalification of the uppermost layers reached values of about 0.5 % Rm, the coalification reactions must have been stopped before the uplift during the Upper Tertiary. This is in agreement with the assumption presented above.

The fact that the present heat flow density (87 mW/m^2) is about the same as the calculated ancient heat flow (about 90 mW/m^2) suggests an extensive heat source at great depth which has outlasted the uplift. This may perhaps be connected with the heat source that causes the heat anomaly of the Upper Rhine Graben.

References

Buntebarth, G. (1978): The degree of metamorphism of organic matter in sedimentary rocks as a paleo-geothermometer, applied to the Upper Rhine Graben. – Pure Appl. Geophys. **117**: 83–91.

– (1979): Eine empirische Methode zur Berechnung von paläogeothermischen Gradienten aus dem Inkohlungsgrad organischer Einlagerungen in Sedimentgesteinen mit Anwendung auf den mittleren Oberrheingraben. – Fortschr. Geol. Rheinl. u. Westf. **27**: 97–108, Krefeld.

Buntebarth, G., Grebe, H., Teichmüller, M. & Teichmüller R. (1979): Inkohlungsuntersuchungen in der Forschungsbohrung Urach 3 und ihre geothermische Interpretation. – Fortschr. Geol. Rheinl. u. Westf. **27**: 183–199, Krefeld.

Espitalie, J. (1979): Charakterisierung der organischen Substanz und ihres Reifegrades in vier Bohrungen des mittleren Oberrheingrabens sowie Abschätzung der paläogeothermischen Gradienten. – Fortschr. Geol. Rheinl. u. Westf. **27**: 87–96, Krefeld.

Lemcke, K. (1974): Vertikalbewegungen des vormesozoischen Sockels im nördlichen Alpenvorland vom Perm bis zur Gegenwart. – Eclogae geol. Helv. **67**: 121–133, Basel.

Stach, E., Mackowsky, M.Th., Teichmüller, M., Chandra, D., Taylor G.H. & Teichmüller R. (1975): STACH'S textbook of coal petrology. – (Borntraeger), Berlin, Stuttgart.

Teichmüller, M. (1974): Über neue Macerale der Liptinit-Gruppe und die Entstehung des Micrinits. – Fortschr. Geol. Rheinl. u. Westf. **24**: 37–64, Krefeld.

– (1979): Die Diagenese der kohligen Substanzen in den Gesteinen des Tertiärs und Mesozoikums des mittleren Oberrheingrabens. – Fortschr. Geol. Rheinl. u. Westf. **27**: 19–49, Krefeld.

Teichmüller, M. & Ottenjann, K. (1977): Liptinite und lipoide Stoffe in einem Erdölmuttergestein. – Erdöl u. Kohle **30**: 387–398, Leinfelden.

Teichmüller, M. & Teichmüller, R. (1975): Inkohlungsuntersuchungen in der Molasse des Alpenvorlandes. – Geol. Bavaria **73**: 123–142, München.

The Urach Geothermal Project, p. 97–100;
Schweizerbart'sche Verlagsbuchhandlung, Stuttgart, 1982

Seismo-acoustic and Geoelectric Experiments within the Urach 3 Borehole

J. WOHLENBERG

with 1 figure and 1 table

Abstract: Seismo-acoustic and geoelectric experiments have been carried out in the research borehole Urach 3. Results of the BHC Sonic Log, the Induction Logs (IES), the Dual Laterologs (DLL) and the Well Shooting experiments correlate well with the geology. All depth functions show a clear discontinuity at the top of the crystalline rock, separating the lower resistivity and lower velocity material from the higher resistivity and higher velocity basement. Hard limestone layers are reflected by higher values for seismic velocities and electric resistivities between 670 m and 800 m. Lower velocities and lower resistivities between 2550 m and 3050 m depth characterise strongly altered zones in the crystalline basement.

1. Introduction

As part of the research and development programme for the exploration of the Urach geothermal anomaly, geophysical experiments were carried out within the deep drillhole Urach 3. Target of these experiments was the in situ investigation of the physical parameters of the downhole formations, in order

– to correlate these data with those obtained from available core materials, and
– to facilitate the interpretation of the geophysical exploration experiments from the surface.

The downhole experiments within the Urach 3 borehole included seismo-acoustic and geoelectric logging as listed in Tab. 1. These experiments were executed by Schlumberger. As downhole-to-surface experiments Prakla-Seismos carried out two Well Shootings. The results of these experiments are reported in the following.

Table 1.

Date	Log		Depth Interval (m)
18.10.77	Induction-Electrical Log	(IES)	26 – 645
19.10.77	BHC Sonic Log; γ-Ray	(BHC)	342 – 645
23.11.77	BHC Sonic Log; γ-Ray	(BHC)	642 –1079
23.11.77	Induction-Electrical Log	(IES)	642 –1082
21. 1.78	Simultaneous Dual Laterolog	(DLL)	1079 –1806
21. 1.78	Microlaterolog-Microlog	(MLL)	1079 –1808
22. 1.78	BHC Sonic Log	(BHC)	1080 –1807
21. 3.78	BHC Sonic Log	(BHC)	1809 –2579
22. 3.78	Dual Laterolog	(DLL)	1809 –2578.8
22. 5.78	Dual Laterolog	(DLL)	2550 –3334.5
24. 5.78	BHC Sonic Log	(BHC)	2460.0–3332.0

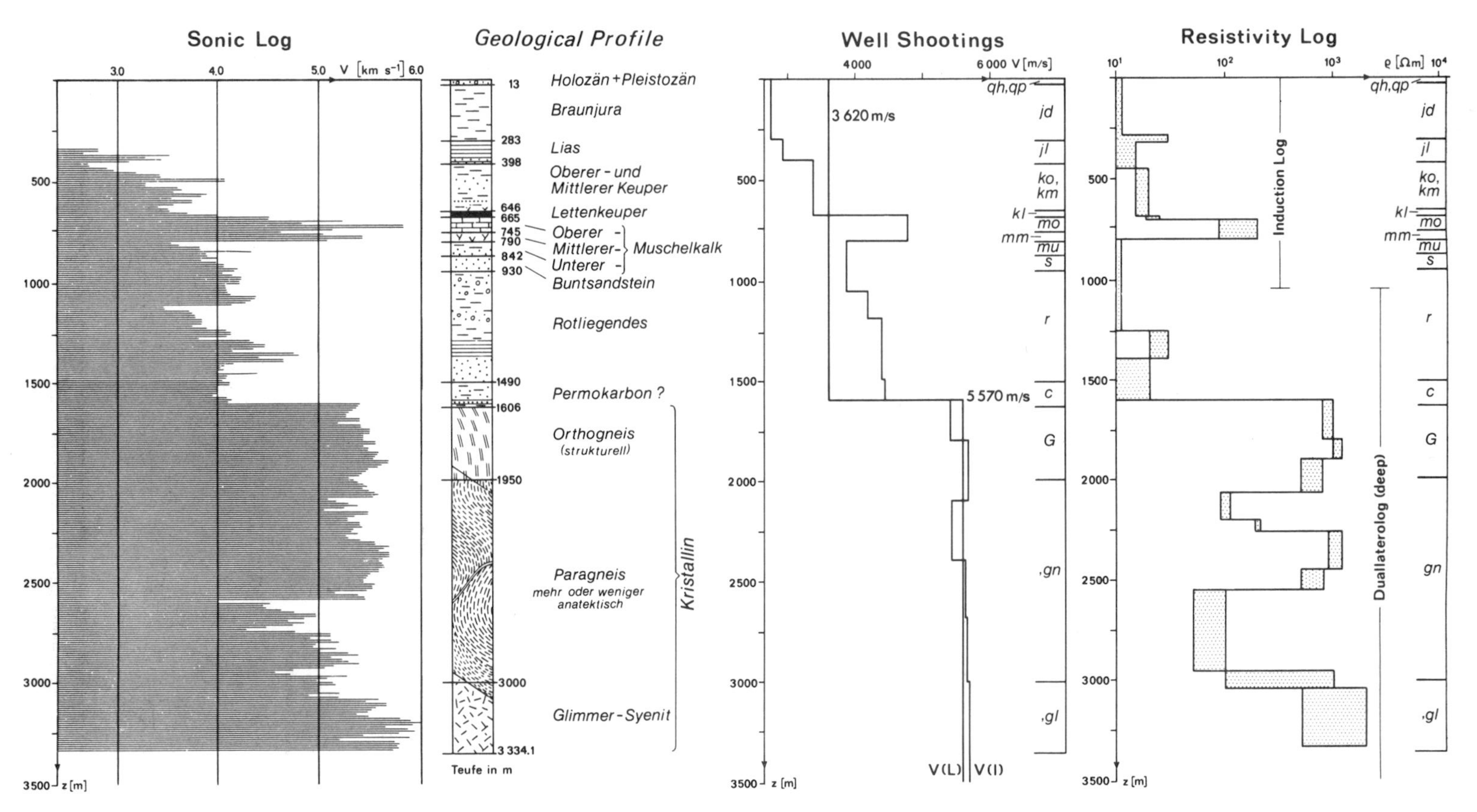

Fig. 1. Summarized results from Seismo-acoustic and Geoelectric experiments within the Urach 3. Data of the Sonic Logs, the Well Shootings and the Resistivity Logs are plotted against the geology after DIETRICH & SCHÄDEL (1978).

2. The Seismo-Acoustic Experiments

2.1 The Borehole Compensated Sonic Log (BHC)

Of all the downhole experiments, special attention was paid to the interpretation of the BHC Sonic Logs. These experiments were expected to deliver detailed information on the rigidity of the crystalline rock regarding the planned frac operations and on the seismic velocity field in the immediate environment of the borehole.

The interpretation of the BHC Sonic Logs was done by using the integrated travel-time. The seismic velocities were calculated by reading the distance travelled within a 2 ms time interval. The result is plotted as a velocity-depth function (Fig. 1).

For the depth interval from 0 to 340 m no data are available. The diameter of the drillhole was too large for any downhole equipment. Down to 660 m velocities range between 2.6 km s^{-1} and 4.1 km s^{-1} with velocities increasing with depth. Between 670 m and 800 m, velocities reach values from 4.6 km s^{-1} up to 5.8 km s^{-1}. This depth and velocity interval corresponds to the geological formation of the Muschelkalk (limestone).

Between 800 and 1600 m depth, velocities vary between 3.6 km s^{-1} and 4.8 km s^{-1} with no systematic pattern. The interpretation of the Caliper Logs suggests that the lower seismic velocities within this depth range originate from stronger variations of the borehole diameter. At 1604 m the borehole hits the crystalline basement with a net jump to higher velocities ranging from 5.2 km s^{-1} to 5.6 km s^{-1}. Below approx. 2550 m these rather constant velocities suddenly decrease to values between 4.4 km s^{-1} and 5.2 km s^{-1}. A noisy Caliper Log suggests poor borehole conditions for this part of the borehole down to a depth of 3050 m. Petrological investigations revealed strongly altered rock formations, the alterations probably being a result of chemical processes. From 3050 m to the final depth of 3334 m, velocities vary from 5.4 km s^{-1} to 6.0 km s^{-1} suggesting good borehole and uniform rock conditions.

According to the Sonic Log, the steps in the velocity-depth function at the top of the limestone formation and at the top of the crystalline basement rock represent significant discontinuities for the reflection- and refractionseismic investigations.

2.2 The Well Shooting Experiments

In addition to the BHC Sonic Log two well shooting experiments were carried through by Prakla-Seismos GmbH for further information on the seismic velocity patterns around the Urach 3 borehole.

The first experiment took place on 1st Dezember, 1977 when drilling had reached a depth of 1060 m. Seismic signals were generated by using an air-pulser at the surface approximately 180 m SE of the drillhole.

The second experiment took place on 20th May, 1978 when the borehole had reached final depth at 3334 m. The purpose of this experiment was to investigate the velocity field below 1000 m depth not only in the immediate proximity of the hole. For a better interpretation of the surface seismic experiments it seemed reasonable to have additional information on the more regional pattern of the seismic velocities around the Urach borehole 3.

In contrast to the first experiment, therefore, seismic signals were now generated by using explosives from five shotpoints around the borehole. The distances between the shot-points and the drillhole varied from 1 km to approximately 4 km. To avoid further corrections in calculating the velocities the shots were fired from only one level. This procedure required shallow drilling to depths of about 100 m. A total of 49 shots were fired by using charges between 1 kg and 10 kg. The shots were recorded with a single three-component geophone at the following levels: 3333m, 3000m, 2700m, 2400m, 2100m, 1800m, 1604m, 1500m, 1200m, 1060m, and 800m. The results of both experiments are summarized in Fig. 1. The interval-velocities V (I) for the geological formations Dogger, Lias and upper Middle-Keuper vary between 2.6 km s^{-1} and 3.0 km s^{-1}. For the Middle and Lower Keuper velocities from 3.3 km s^{-1} to 3.7 km s^{-1} are observed with intercalated layers of Gypsum-Keuper with lower velocities ranging from 3.2 km s^{-1} to 3.4 km s^{-1}. A clear velocity step to 5.2 km s^{-1} marks the top of the Limestone-formation. Towards the Bunter, velocities decrease to values around 4.0 km s^{-1} and 4.4 km s^{-1}. The top of the crystalline basement is marked by a net velocity increase to values around 5.6 km s^{-1}. This velocity-depth function is in good agreement with the results obtained from the BHC Sonic Logs.

Generalising the results of the Well Shooting experiments one can establish a twofold velocity structure with an average velocity V (L) of 3.62 km s^{-1} for the sedimentary layers and 5.57 km s^{-1} for the basement rock.

3. The Geoelectric Induction Logs (IES) and the Dual Laterologs (DLL)

The research programme around the geothermal anomaly of Urach included magnetotelluric (MT) and geoelectric experiments at the surface. The interpretation of the MT- and geoelectric sounding curves is much more reliable and easier if additional information on electric resistivities of the near surface layers is available. This information was derived from several Induction Logs (IES) and Dual Laterologs (DLL) run in the Urach 3 borehole.

The records revealed changing resistivities with depth within a wide range. Therefore, independently of a geological stratification, depth intervals with more or less comparable resistivity values were taken together (Fig. 1). The width of the resistivity variations within the depth intervals is marked by the upper and lower limits of the observed resistivity values.

The ρ versus depth function shows a clear separation of the low resistivity sedimentary layers on top of a high resistivity crystalline basement. Down to 1600 m resistivity values vary between 10 and 30 Ohm · m with intercalated layers of limestone between 700 m and 800 m depth with resistivities around 100 Ohm · m. Below 1600 m high resistivities of 1000–2000 Ohm · m are characteristic for the basement rock. Intervals of lower resistivities, 50–200 Ohm · m, within the crystalline basement are clearly correlated to altered rock types already identified on the BHC Sonic Logs (Fig. 1).

References

Schädel, K., Dietrich, H.-G. (1978): Untersuchung der geothermischen Anomalie in Urach auf eine mögliche Nutzung durch eine Untersuchungsbohrung bis tief ins Kristallin. – In: Statusreport 1978 Geotechnik und Lagerstätten, KfA Jülich 1978, pp. 79–85.

Wohlenberg, J. (1978): Zwischenbericht über die Geophysikalischen Untersuchungen in der Forschungsbohrung Urach. – Report, NLfB Hannover, Archive No. 81 388.

– (1979): Zwischenbericht über die Geophysikalischen Untersuchungen in der Forschungsbohrung Urach 3. – Report, NLfB Hannover, Archive No. 82 210.

The Urach Geothermal Project, p. 101–105;
Schweizerbart'sche Verlagsbuchhandlung, Stuttgart, 1982

Determination of Rock Densities in the Urach Borehole

S. PLAUMANN and J. WOHLENBERG

with 4 figures and 1 table

Abstract: The core material from the deep research borehole Urach 3 was used to determine rock densities from a total of 218 samples. Formation Density-Logs run in the Urach 3 provided a possibility of comparing corresponding density data. The main experience gained from this comparison is that in most cases direct density determinations cannot be replaced by Formation Density-Logs if rather accurate information is needed.

Introduction

As part of a research project for investigating methods of exploring and exploiting geothermal resources, the deep borehole Urach 3 was drilled. It was financially supported by the Federal Ministry of Research and Technology (BMFT).

Urach 3, the coordinates of which are

φ = 48 ° 30,478' N
λ = 9 ° 22,457'E
H = 426 m above sea level

(German Topographic Map 1 : 25 000 no. 7422 Lenningen)

was cored with great succes, and extensive geophysical logging was executed within the drillhole.

To provide the necessary basic information for data interpretation in geothermal, seismic and gravity studies, rock densities were determined by using the core material and the Formation Density-logs (FDC-Log).

Because of the simultaneous determination of rock densities from core samples and density logs a direct comparison of both sets of data was possible.

Densities from Core Material

Determinations of the rock density with the use of the core material from the Urach 3 borehole were carried through by the Geological Survey of Lower Saxony (NLfB). This was done by weighing the rock samples in air and in water and by using the relation

$$\rho = \frac{G_L}{G_L - G_W} \cdot \rho_W$$

with G_L = sample weight in air
G_W = sample weight in water and
ρ_W = density of water

Sufficient accuracy was achieved with $\rho_W = 1.0 \text{ g cm}^{-3}$. Only samples whose weight in air was less than 1000 g but came very near to this upper limit were taken. The error of each single density determination could thus be kept below 0.5 %.

49 cores were taken from the Urach 3 borehole (Tab. 1). Densities were determined for a total of 218 samples, 113 of which were crystalline basement rock samples. Porous rock samples were kept under water for 12 hours before being weighed in air, to assure

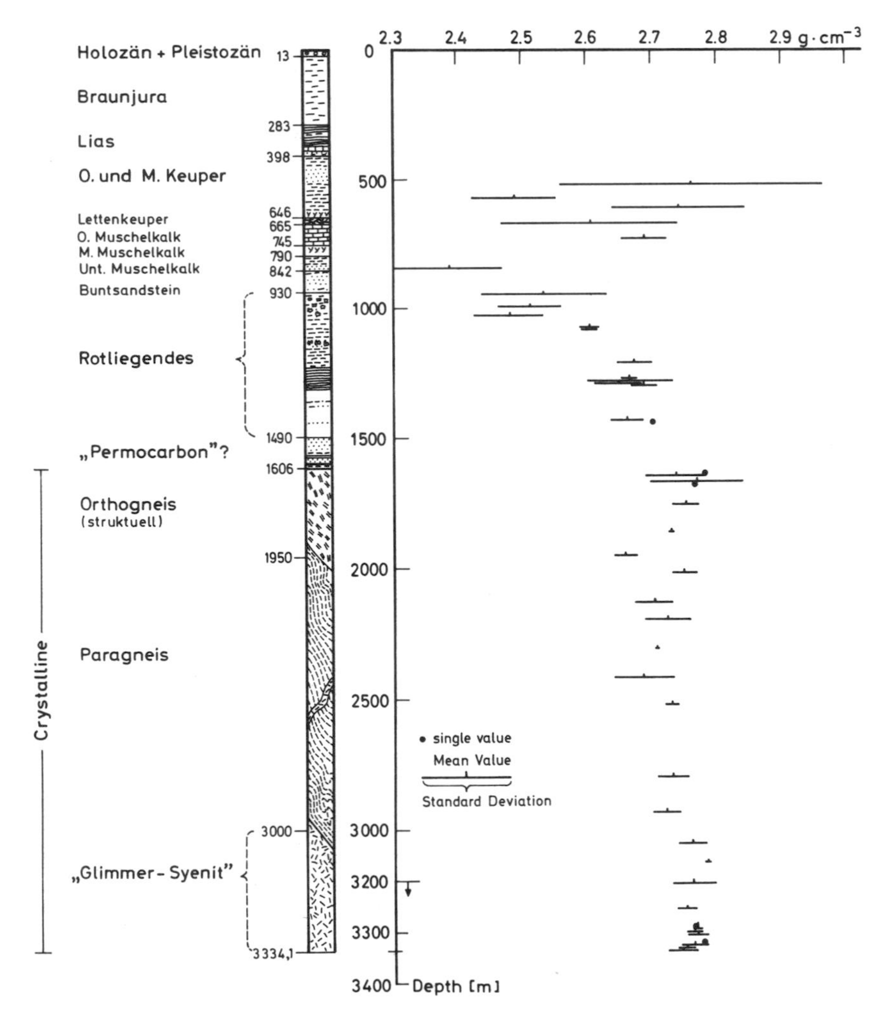

Fig. 1. Results of density determinations in the Urach 3 borehole: Mean density versus depth for the cores as listed in Tab. 1.

Table 1.

Core-No.	Depth of Core m	Length of Core	Formation	No. of Samples	ρ_{max}	ρ_{min}	ρ_m	s
1	511.5– 520.5	8 m	Keuper	4	2.952	2.562	2.762	0.202
2	565.0– 573.5	8 m	Keuper	8	2.553	2.366	2.481	0.065
3	598.0– 607.0	9 m	Keuper	5	2.862	2.594	2.743	0.103
4	659.0– 668.0	9 m	Keuper/mo	8	2.851	2.466	2.605	0.134
5	717.0– 726.0	8.5 m	Muschelkalk/mo	7	2.727	2.617	2.690	0.035
6/7	832.0– 844.8	6.5 m	Muschelkalk/mu	4	2.484	2.303	2.390	0.084
8	935.0– 942.9	7.5 m	Rotliegendes	6	2.685	2.435	2.535	0.096
9	983.0– 992.0	9 m	Rotliegendes	6	2.556	2.443	2.514	0.048
10	1017.0–1025.6	8 m	Rotliegendes	5	2.541	2.425	2.483	0.053
11	1064.0–1073.0	9 m	Rotliegendes	6	2.622	2.592	2.605	0.014
12	1073.0–1082.0	9 m	Rotliegendes	5	2.616	2.587	2.604	0.012
13	1200.0–1209.0	9 m	Rotliegendes	4	2.704	2.638	2.674	0.027
14	1260.0–1269.0	9 m	Rotliegendes	7	2.685	2.648	2.666	0.013
15	1269.0–1278.0	9 m	Rotliegendes	6	2.737	2.543	2.668	0.066
16	1278.0–1287.0	9 m	Rotliegendes	6	2.684	2.579	2.649	0.037
17	1287.0–1296.0	9 m	Rotliegendes	8	2.717	2.652	2.688	0.019
18	1353.0–1362.0	9 m	Rotliegendes	–				
19	1422.0–1431.0	9 m	Rotliegendes	9	2.694	2.627	2.662	0.026
20	1431.0–1440.0	9 m	Rotliegendes	1	–	–	2.698	
21	1634.0–1635.7	1.0 m	Crystalline	1	–	–	2.779	
22	1635.7–1643.7	8 m	Crystalline	4	2.792	2.677	2.738	0.047
23	1660.0–1661.8	0.15 m	Crystalline	–				
24	1661.8–1664.7	2.0 m	Crystalline	2	2.819	2.718	2.769	0.071
25	1664.7–1668.4	0 m	Crystalline	–				
26	1675.0–1677.9	2.2 m	Crystalline	1	–	–	2.761	–
27	1750.5–1756.8	6.3 m	Crystalline	5	2.779	2.726	2.752	0.021
28	1853.0–1862.0	7.8 m	Crystalline	2	2.734	2.728	2.731	0.004
29	1945.0–1953.0	8 m	Crystalline	6	2.675	2.626	2.659	0.018
30	2013.5–2017.3	3.8 m	Crystalline	3	2.769	2.730	2.749	0.020
31	2124.2–2133.2	9 m	Crystalline	5	2.755	2.685	2.704	0.029
32	2189.5–2195.7	6.2 m	Crystalline	3	2.763	2.703	2.723	0.035
33	2302.4–2311.4	9 m	Crystalline	3	2.709	2.704	2.707	0.003
34	2412.0–2421.0	9 m	Crystalline	8	2.751	2.631	2.686	0.046
35	2517.5–2521.3	3.8 m	Crystalline	4	2.744	2.717	2.729	0.011
36	2795.7–2802.0	5.6 m	Crystalline	7	2.779	2.703	2.730	0.023
37	2930.0–2939.0	9 m	Crystalline	8	2.748	2.688	2.721	0.022
38	3053.0–3060.0	7 m	Crystalline	6	2.781	2.721	2.760	0.022
39	3125.0–3132.0	3.2 m	Crystalline	3	2.786	2.778	2.783	0.004
40	3205.0–3211.0	6 m	Crystalline	7	2.805	2.710	2.761	0.034
41	3250.0–3258.2	8.2 m	Crystalline	5	2.768	2.731	2.751	0.015
42	3290.0–3294.7	4.7 m	Crystalline	1	–	–	2.765	–
43	3294.7–3298.4	3.7 m	Crystalline	5	2.779	2.758	2.767	0.009
44	3298.4–3301.0	2.6 m	Crystalline	4	2.774	2.748	2.763	0.012
45	3302.0–3306.2	4.6 m	Crystalline	8	2.783	2.753	2.771	0.010
46	3321.0–3321.9	0.6 m	Crystalline	1	–	–	2.773	–
47	3323.0–3325.3	1.2 m	Crystalline	3	2.778	2.740	2.763	0.020
48	3326.0–3330.6	4.6 m	Crystalline	3	2.761	2.733	2.749	0.015
49	3331.0–3334.0	2.0 m	Crystalline	5	2.782	2.718	2.745	0.024

ρ_{max} = highest density value within one sample group
ρ_{min} = lowest density value within one sample group
ρ_m = arithmetical average
s = standard deviation

All ρ values given in g/cm^3

fluid saturation of cavities and pore volumes. For the individual cores – identified by the core number –, Tab. 1 gives the number of measured samples, maximal and minimal densities, an averaged density, and the standard deviation.

A graphical presentation of results is given in Fig. 1. The averaged density values for the different cores, together with the standard deviation, are plotted versus depth. The standard deviation may be regarded as a measure for the homogeneity of the corresponding core. It has to be pointed out, nevertheless, that the selection of the samples already falsifies the results in advance, and that the samples taken characterize a core only insufficiently. This is especially true for samples taken from cores 1 to 20. Within these sedimentary rocks there is a strong variation of densities and strengths of the different layers. Thus the selection of the samples is a major factor. Therefore, the density values obtained from these cores probably are too high in general.

Density Logs and Comparison of Results

As part of the routine downhole experiments FDC-Logs were run in the Urach 3 borehole for the following depth intervals:

445.0– 645.2 m (10.10.1977)
642.7–1081.7 m (23.11.1977)
1079.5–1809.2 m (22.01.1978)
1809.5–2582 m (22.03.1978)
1808.0–3236.5 m (23.05.1978)
3100.0–3335.0 m (24.05.1978)

Thus density data for the total interval from 445 m to the final depth of 3335 m are available. The FDC-Logs were run by Schlumberger. All cored parts of the Urach 3 borehole are within the total log interval, thus offering an opportunity to compare the different density information.

The degree of conformity between both density determinations is quite different and varies from a few 0.01 g cm^{-3} to several 0.1 g cm^{-3}. In general, good correlation is observed when the caliper indicates good borehole conditions. Fig. 2 gives as an example for good conformity results of cores 5 and 32. The caliper is undisturbed in these cases. Poor borehole conditions, as given in Fig. 3 for cores 6 and 34 consequently produce poor conformity for the rock and formation densities. On the other hand, there are significant differences between the densities (e.g. cores 16 and 17, Fig. 4) although the caliper log and the compensation curve indicate smooth borehole conditions. The dashed line in the diagrams represents the density correction to compensate for the influence of sudden changes of the borehole diameter and variations of the consistency of the drilling mud.

The compensation device failed in many cases because of borehole conditions beyond the correction capacity of the density log. Errors of approximately 1 m in the depth figures influence the comparison of single density values but do not influence the general tendency.

In summarizing these results from the Urach 3 borehole investigations one can say that if very accurate density data are needed, direct data acquisition with the use of core material cannot be replaced by FDC-Logs.

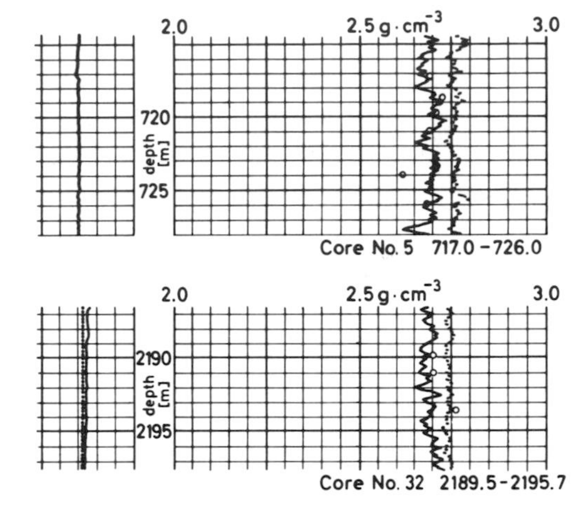

Fig. 2. Examples of rather good data correlation (cores no. 5 and no. 32).
Left: Caliper Log.
Right: Density Log (solid line) and correction curve (dashed line); Circles: Density values from core samples.

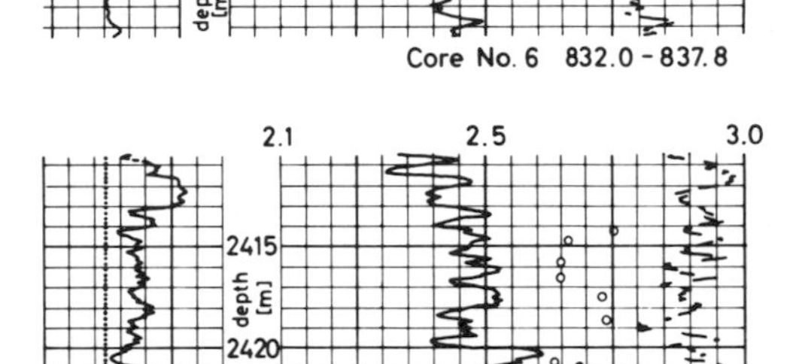

Fig. 3. Examples of a poor data correlation (cores no. 6 and no. 34).
Left: Caliper Log.
Right: Density Log (solid line) and correction curve (dashed line); Circles: Density values from core samples.

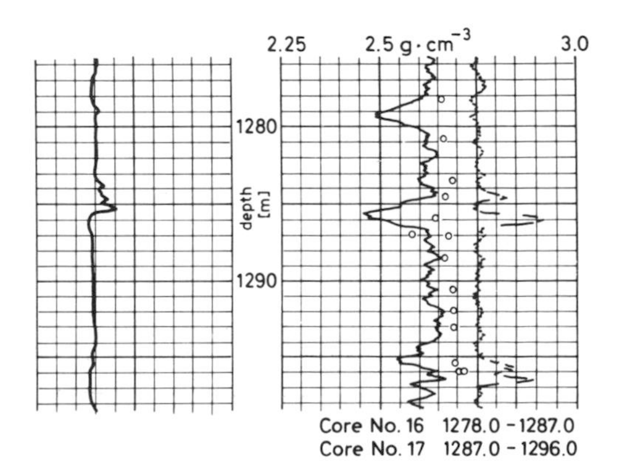

Fig. 4. Poor data correlation under good borehole conditions (cores no. 16. and no. 17).
Left: Caliper Log.
Right: Density Log (solid line) and correction curve (dashed line); Circles: Density values from core samples.

The Urach Geothermal Project, p. 107–116;
Schweizerbart'sche Verlagsbuchhandlung, Stuttgart, 1982

Magnetic Investigations on Cores of the Research Borehole Urach 3

K. FROMM

with 4 figures and 4 tables

Abstract: On oriented cores of the research borehole Urach 3, the remanent and induced magnetisations were measured. With use of partial demagnetisation, the paleoremanences were determined and compared with the orientation of the paleomagnetic field, as it is expected at Urach according to the known pole-positions for each geological formation. In most of the sediments the measured directions of remanence have been observed to approach the expected direction with increasing demagnetisation. Special attention has been devoted to the varification of the data on cores orientation.
The results are:
- Agreement with the geological ages and confirmation of the polar-wandering-path of Europe.
- Detection of a record of a polarity transition in the permean.
- Subsequent orientation of some cores.
- Application of viscous magnetisation as a tool to orient drilled cores.

In the crystalline, no basis for deriving directions of the paleoremanence has been found. The susceptibility can be correlated in several cases with grain size and degree of weathering of the rock.

1. Introduction

Of 49 cores, taken for diverse geoscientific purposes, 22 had been oriented inside the borehole especially for paleomagnetic investigations. Two justifications for a magnetic investigation of the cores existed: First it was expected to confirm geological age determinations by comparing the position of the paleomagnetic pole derived from the remanences measured with the polar-wandering-curve of Europe. The second aspect was to secure data on the magnetic susceptibilities of different materials for a quantitative evaluation of magnetic anomalies in southwestern Germany.

The magnetic measurements were performed on samples of the cores in the laboratory for rock magnetism in the Grubenhagen station of the Niedersächsisches Landesamt für Bodenforschung. Several uncertainties in the orientation had to be cleared up in the core-magazine at Urach in close contact with the geologist in charge, H.-G. DIETRICH

2. Sampling and Orientation

278 samples were taken from sediments: Keuper (core 2), Lower Muschelkalk down to Buntsandstein (cores 6 and 7), Rotliegendes (cores 8, 11, 13, 19 and 20). From the crystalline basement another 67 samples were taken (cores 26, 32, 38, 41 to 45, 47 and 48); see Fig. 1.

In order to study the homogeneity of the remanence, the samples were taken at close intervals of about 10 cm within the upper cores down to 11.

The oriented cores were marked with an Eastman Multishot-device, which cuts traces with three knives into the surface of each core and records its azimuth at every metre. The films of cores 6, 11 and 38 were defective; therefore, those marks indicate only the relative position of all sections of the cores. Nevertheless, the orientation of core 6 was made possible by geological correlation with the upper part of core 7.

In the core-magazine at Urach the samples were drilled out of the core sections perpendicular to the axis of the core, as well as through the orientation mark 1 on the core surface. The drilling position was fixed asymmetrically with respect to the core; consequently, the orientation mark appeared off centre at the top of the sample, as shown in Fig. 2. This technique enables the operator to proceed quickly to the next sample with-

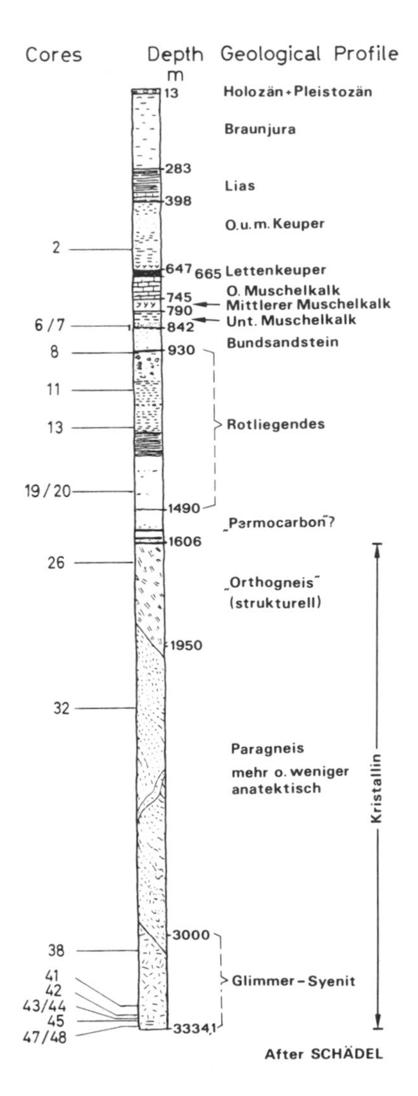

Fig. 1. Positions of the magnetically investigated cores in the borehole Urach 3.

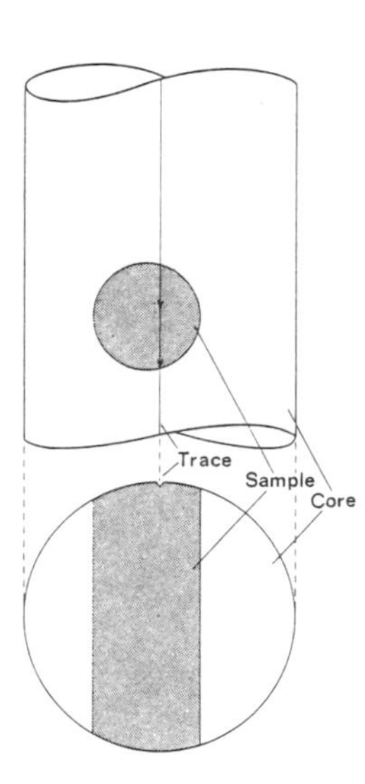

Fig. 2. Cylinderical sample drilled out of the core eccentrically and oriented by arrows along the trace of knife 1.

out losing the correct orientation of the drilled samples. The labeled samples were divided into two regular cylindres for the investigation of the magnetic susceptibility and the direction and intensity of remanence in the laboratory.

3. Paleomagnetic Investigations

3.1 General Remarks

The direction of the paleoremanence depends on the position of the magenetic pole during the forming of the rock. The pole positions for Europe are known for most of the geological formations; therefore, the direction of the expected remanence can be derived for various geological times. This was done with the use of data of MC ELHINNY (1973) and VAN DER VOO & FRENCH (1974), because it is more convenient to compare the measured directions with the expected ones than to transform all data to pole positons.

Because of various influences on the remanence of the rock during geological times, it is not possible to measure the direction of the paleoremanence directly. Especially the geomagnetic field superposes a viscous remanence, thus turning the direction of the remanence to the latest direction of the field. This viscous component of the natural remanent magnetisation (NRM) can be cancelled by partial demagnetisation. The amplitude of the demagnetising field was increased stepwise, sometimes up to an amplitude of 64 kA/m or 80 kA/m (800 Oe or 1000 Oe). After each step of the procedure the remaining partial remanence (PRM) was measured. In most cases the direction of the PRM tends toward the paleomagnetic orientation.

3.2 The Remanent Magnetisation in the Sediments

It should be mentioned that sometimes the orientation data were doubtful, because parts of the cores could have been sheared off and rotated before the orientation process in the borehole. The arising complexity of the situation becomes obvious in the discussion of the measurements of core 7 (at 843 m). Only the close spacing of samples allowed a derivation of paleomagnetic information, as well as a valid orientation. Similar doubts had to be cleared up for cores 8 and 19. No corrections were made for the dip of the sedimentary formations because of the small dip angle (up to 10 °) and for the deviation of the borehole (up to 3.5 °).

Influences of technical magnetic fields on the remanences during the sampling of the cores were not observed. A synopsis of the paleomagnetic results is given in section 3.4. A short discussion for the cores 2, 6, 7, 8, 11, 13, 19 and 20 is given in the following (see also Fig. 1, and for more details FROMM (1979)):

Core 2

33 samples of Triassic (Keuper) from 565.0 to 571.1 m. The paleomagnetic and geologic results are in agreement.

Cores 6 and 7

75 samples of Triassic (lower Muschelkalk, Upper Buntsandstein) from 832.0 to 844.3 m.

In spite of low magnetisation in the Muschelkalk, it was possible to find out that at a depth of 842.4 m core 7 had been sheared off and the upper part had been distorted in the borehole before the orientation was fixed. An inverse remanence direction (with a declination D = 242° and inclination I = – 32°) is observed at the bottom of the core, while in the upper part the remanence appears to be rotated clockwise by 200°. This cannot be caused by a paleomagnetic field reversal since the incliniation is unchanged invers across that leap of 200° in the declination (Fig. 3). DIETRICH had defined the orientation of core 6 by geological comparison with the upper part of core 7. This correlation is proved to be correct by the remanence directions in these parts, but the correction of 200° anticlockwise, as for the upper part of core 7, must be added.

The sudden change of the facies at the border Buntsandstein/Muschelkalk at 843.15 m represents a discordance with a larger time-span. From the magnetic data a time-span can neither be concluded nor excluded.

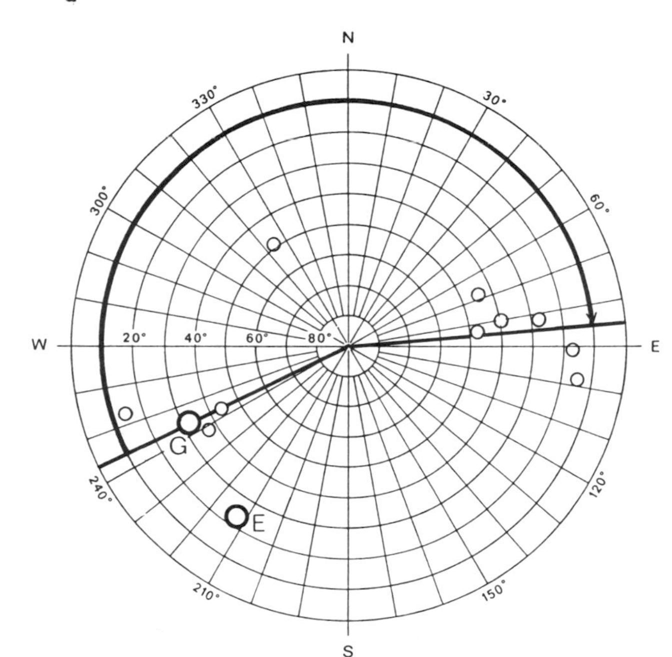

Fig. 3. Directions of remanences in the lower part of core 7. The two data sets with D = 85° and D = 242° and with negative inclinations are separated because of the rotation (see arrow) of the core by 200°. By rerotating, a common direction G is obtained; this deviates by 35° from the expected reverse direction E. The value of D = 325° is not interpreted because of a considerable spread.

Core 8

29 samples of Permian (Rotliegendes) from 935.0 to 937.7 m. The rather stable remanence showes decreasing inclination with increasing demagnetisation. The final orientation (D = 160°, I = 20°) in the upper section of the core points to a component which agrees with the expected inverse orientation (D = 200°, I = – 14°) of the Permian field. In the lower section, the orientation (D = 80°, I = 15°) tends toward the direction of the normal Permian field with a continuous transition along 1 m to the reversed direction above. Misorientation of the core can be excluded in this case.

Core 11

104 samples of Permian (Rotliegendes) from 1064.1 to 1073.0 m. During the orientation of the core the camera in the borehole did not work; the azimuth of the marks, therefore, was unknown. The declination values were calculated assuming north for the orientation mark. According to the close grouping of the orientation of NRM it can be assumed that the average value does point to the present field orientation. The mark made by the knife 1 then points toward 195°. The average orientation of PRM tends toward the normal Permian field.

Core 13

2 samples of Permian (Rotliegendes) from 1200.2 to 1204.4 m. The remanence is homogeneous and has been reduced slightly because of demagnetisation, while its orientation turns from D = 175°, I = 55° to D = 180°, I = 17° and tends towards the general direction of the inverse Permian field.

Cores 19 and 20

24 samples of Permian (Rotliegendes) from 1424.0 to 1434.8 m. The remanence can be considered more or less homogeneous; significantly higher magnetisations were found in the tuff. While the inclination of the remanence was reduced by demagnetisation from I = 60° to about I = 20° in the whole depth interval, the declination remained stable. The declination changes, however, with the depth by 160° as shown in Fig. 4.

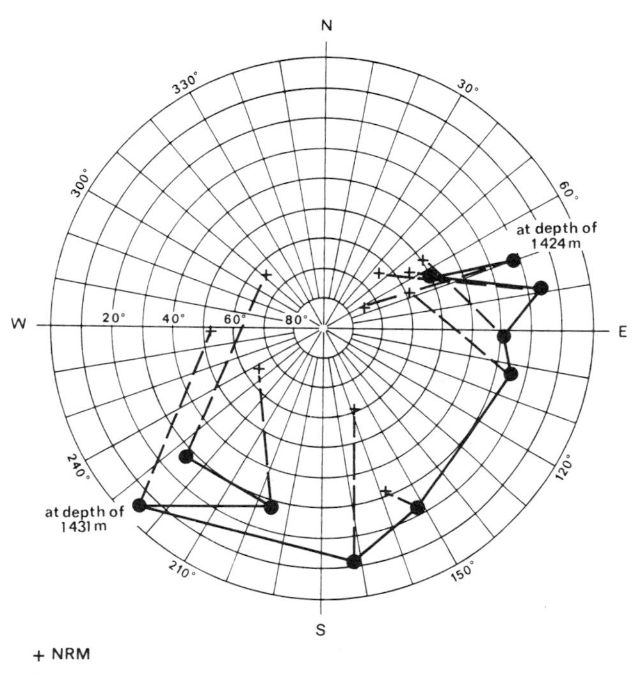

Fig. 4. Rotation of the direction of remanence from the reverse Permian field orientation (core 19/20). The average values of the NRM are connected with the corresponding values of the PRM by broken lines, which point to an approximate common centre – the present field direction (D = 0°, I = 65°).

This rotation of the remanence in core 19, which already begins in core 20, can be interpreted as a change in polarisation, which would have happened within approximately 5 000 years at a supposed rate of sedimentation of 2 mm/year in the depth-interval of 10 m.

To confirm the assumption of a change in polarisation, it must be assured that no errors have been caused by sheared off core sections. A detailed investigation of the core as well as the clockwise rotation of the drilling well excluded such orientation errors. There is also evidence from magnetic data for a change of polarisation: as we consider the decreasing inclination with each step of demagnetisation we recognize that a viscous component with orientation of the present field supperposed the paleoremanence. This is shown in Fig. 4 by broken lines from PRM to NRM, which point to a common centre – the present field direction. Hence, these directions of the PRM cannot result from rotated core sections.

Therefore, the results obtained from the cores 19 and 20 document a transition from the inverse to the normal Permian field orientation.

3.3 The Remanent Magnetisation in the Crystalline

The natural remanent magnetisation (NRM) and, after a demagnetisation in a field of 32 kA/m, the remaining partial magentisation (PRM) was measured on 67 samples from the crystalline. However, the measurements on both cylinders of nearly all the samples yielded such different orientations of their remanences that the interpreation of the resulting orientation did not appear to be useful. Even after demagnetisation, the results remained inhomogeneous in spite of considerable changes in the orientations.

The intensity of the remanences is very different. In intervals with higher magnetisations an additional demagnetisation could perhaps yield a more homogeneous paleoremanence. Prior to such experiments, however, limiting geological questions are required combined with suitable sampling of additional specimens in the corresponding intervals. Even then, however, only moderate help can be expected in answering these specific questions from the paleomagnetic investigation, as the results now presented indicate.

For a first overview, the average intensities of remanence are listed in Tab. 1.

Table 1.

Type of rock	Core	NRM mA/m	PRM mA/m	Remarks
Special composition of Orthogneiss	26	98.2	37.7	
Anatectic Paragneiss	32	1.5	0.7	
Mica-Syenite	38	4.4	4.0	Values vary as the rock is weathered
	41	5.4	11.9	Rock hardly weathered
	42	3.4	2.3	Rock hardly weathered
	43/44	21.3	7.1	Rock hardly weathered
	45	5.2	2.7	Relat. fine-grained, wathered
	47	9.2	20.0	Fine-grained to solid
	48	0.4	2.8	Slightly weathered

3.4 Results of the Paleomagnetic Investigations

Limited success was achieved in defining the paleomagnetic directions of remanences from oriented samples in order to derive estimates of their ages. The cores from sediments do show an acceptable agreement with those field orientations which are expected for the corresponding geological formations. No assessments can be derived from the remanences measured in the crystalline part, since the magnetisation is inhomogeneous or cannot be interpreted because of tectonic rotations in the strongly folded basement; corresponding data for corrections of the orientation do not exist.

The stratigraphic sequence of the sediments was well known and the measurements yielded no new data, certainly, but the results do confirm the hitherto known positions of the paleopoles for Europe. Since the polar-wandering-path for Europe is not equally well known for all the different geological formations, the results presented may help to complete the data within the time-span from Rotliegendes to Keuper.

The results for the Triassic do agree rather well – within the accuracy of the measurements – with the expected orientations. A better fit for the Permian samples is desirable (see Tab. 2). However, the deviations are always in the direction of a positive inclination due to a viscous component of the remanence, which could not be demagnetised. Therefore, the true paleomagnetic orientation probably agrees with the expected one (I = 14°, reverse I = – 14°). The true inclination may be smaller in the cores 19 and 20, which would correspond with the older age; for Carboniferous I = 0 is expected.

The first result of the paleomagnetic investigations is the confirmation of the polar-wandering-path, which will support further investigations. Especially the knowledge of the normal and reverse periods found in the sediments of the borehole Urach 3 might help to separate several sections within the formations preferably at sites in southern Germany. As a second result can be named the detection of a change of polarisation within the Rotliegendes (core 19). Such reversals are rare in the Permian and divide this formation into few epochs; therefore, they can be a useful tool for differentiating the rocks of similar age. Besides that in core 19, the change of polarisation can be studied in detail: at a supposed rate of sedimentation of 2 mm/year the polarity transition would have taken several thousand years.

The third result of the investigation of remanences lies in the subsequent orientation of some cores or sections of them. The orientation of core 11 was reconstructed, and that of the cores 6 and 7 was corrected.

There is a fourth result: It has been shown that the viscous component of the NRM, which points toward the north, can be a useful tool for finding the orientation of cores. This practice will be helpful for further investigations on cores, for that problem arises only at boreholes, in which usually the cores are not oriented.

After correcting the orientation data of some core sections the directions of remanences have been observed as shown in Tab. 2.

Table 2.

Core	Formation	Remanences Expected direction D	I	Observed direction	D	I
2	Triassic (Keuper)	21	53	normal	65	55
6	Triassic (mu)	211	– 23	reverse	245	– 32
7	Triassic (so)	211	– 23	reverse	245	– 32
8	Permian (Rotlieg.)	200	– 14	reverse	160	20
		20	14	(normal ?)	80	15
11	Permian (Rotlieg.)	20	14	normal	355	40
13	Permian (Rotlieg.)	200	– 14	reverse	180	17
19	Permian (Rotlieg.)	(transition)		normal	65	20
				reverse	151	20
20	Permian (Rotlieg.)	200	– 14	reverse	225	27

mu – Lower Muschelkalk, so = Upper Buntsandstein

4. The Magnetic Susceptibility of the Rocks

4.1 General Remarks

For measuring the susceptibility, no separate samples have been taken; therefore, values are given only for the paleomagnetically investigated depth intervals.

Table 3. Induced and remanent magnetisation in the sediments (Field in situ F = 48 000 nT).

Depth m	Induktion in situ mA/m (scale 100 – 10 – 1 – 0.1)	Q	Nat. Remanence mA/m (scale 100 – 10 – 1 – 0.1)
Core 2			
565	9.0	.2	2.2
566	9.0	.2	1.7
567	8.0	.3	2.1
568	7.6	.2	1.9
569	6.2	.3	1.6
570	7.6	.2	1.6
571	5.8	.3	1.8
Cores 6/7			
832	1.3	.1	.1
833	2.7	.1	.2
836	.8	.2	.2
837	.8	.1	.1
838	1.3	.1	.2
839	.9	.2	.2
840	.8	.2	.1
841	.4	.5	.2
842	.6	.6	.4
843	5.2	.1	.5
844	2.3	.3	.7
Core 8			
935	5.6	.5	2.9
936	5.6	.5	3.0
937	4.3	.4	1.8
Core 11			
1064	5.7	.5	2.9
1065	.0		13.7
1066	6.0	.6	3.6
1067	6.0	.6	3.5
1068	6.1	.5	3.2
1069	5.5	.5	2.7
1070	5.3	.5	2.4
1071	7.3	.4	2.6
1072	6.4	.4	2.9
Core 13			
1200	17.2	.4	6.2
1204	19.6	.4	8.7
Cores 19/20			
1424	22.5	.2	4.2
1425	11.9	.4	4.5
1426	27.0	.4	10.2
1427	46.7	.3	12.7
1428	18.9	.3	6.3
1429	30.3	.3	8.5
1430	62.5	.9	54.7
1431	63.1	.6	38.4
1432	24.4	.4	10.4
1434	69.6	.5	37.7

Table 4. Induced and remanent magnetisation in the sediments (Field in situ F = 48 000 nT).

Depth m	Induktion in situ mA/m (log scale 100 – 10 – 1 – 0.1)	Q	Nat. Remanence mA/m (log scale 100 – 10 – 1 – 0.1)
Core 26			
1675	83.7	1.2	100.0
1676	45.9	1.6	73.2
1677	52.0	1.9	101.0
Core 32			
2190	18.6	.0	.8
2191	9.9	.1	.6
2192	11.6	.1	1.7
2193	17.8	.1	1.5
2195	10.9	.1	.9
Core 38			
3053	10.7	.1	.9
3054	13.7	.1	1.2
3055	19.2	.2	4.0
3057	25.1	.3	8.0
3058	16.6	.1	1.7
Core 41			
3250	19.0	.4	6.8
3252	16.5	.7	12.1
3254	25.6	1.6	41.4
3256	100.1	.5	54.4
3257	61.1	.8	48.8
Core 42			
3290	17.0	.2	3.8
3291	15.3	.1	1.1
Cores 43/44			
3295	16.4	.1	1.7
3296	17.3	1.6	26.9
3297	21.3	1.7	36.0
3298	18.0	.1	1.1
Core 45			
3302	22.3	.5	12.2
3304	13.6	.0	.7
3305	15.3	.0	.5
Cores 47/48			
3324	34.2	2.3	78.1
3326	37.4	2.7	102.8
3327	12.0	.3	3.6
3328	13.3	.1	2.0

Since the interpretation of magnetic anomalies uses the magnetisation rather than the susceptibility, the induced in situ magnetisation (IM) and, for comparison, also the NRM and the KÖNIGSBERGER-ratio Q = NRM/IM are listed in Tab. 3 and 4. The values of IM and NRM are presented on a logarithmic scale such that their fluctuations as functions of depth are shown. Both magnetic values will be considered in the following. From the induction the abundance of magnetic material can be concluded. The remanence is also influenced by the grain size of the magnetic mineral.

As far as the remanence is concerned, it must be considered that not the arithmetic means has been given, but the according portion of the resultant vector, because for the total magnetisation of a rock body only the resultant vector is of importance. From this it follows that in cases of strong variations in the vector directions – due to inhomogeneous magnetisation or the narrow folding of the rock – even strong remanences yield small average values in the list.

4.2 Results from the Sediments

The Keuper in core 2 shows a homogeneous induction of 8 mA/m (in situ) and a homogeneous remanence four times smaller (Q = 0.25).

The Kaolinitic compacted sandstone of the Lower Muschelkalk in the cores 6 and 7 shows only low magnetisation. The induction stays around 1.1 mA/m with smaller variations; the remanence, with 0.2 mA/m, is considerably smaller. Only in the lower part of core 7, higher magnetisations indicate a different facies of the Upper Buntsandstein.

In the Rotliegend the nearly steady increase of magnetisation with depth is striking. The induction rises from core 8 through cores 11, 13, 19 and 20 from 5 mA/m to a final value of approximately 70 mA/m. In the cores 19/20 three layers with tuff show increased values (1426–27 m, 1430–31 m and 1434 m); however, the increasing trend is not simulated by them. The same tendency is observed for the remanence, but the increase is less steady.

4.3 Results from Cores of the Crystalline

In the crystalline a nearly constant induction of 15 mA/m seems to prevail; however, in several depth intervals values two to six times higher appear. Even higher contrasts are seen in the remanences that reach a ratio 1 : 100 at the same depth intervals, where high values exist for the induction.

In part, the higher magnetisations can be correlated with visible rock properties: for example the Orthogneiss in its "special content" in core 26 is much more strongly magnetised than the very homogeneous anatectic Paragneiss measured in core 32. A dependency of the magnetisation on the degree of weathering can be recognized in the Glimmer-Syenite of core 38. On the other hand, no signs which would explain the strong variation of the measured values for the only slightly weathered cores 41 and 42 can be seen. In the likewise slightly weathered cores 43 and 44 the stronger remanence appears at 3296 m–3297 m in a relatively fine-grained zone. The smaller values for core 45 can be correlated with the weathering as well as in core 48. On the other hand the fine-grained and dense material in core 47 shows a strong magnetisation.

References

Fromm, K. (1979): Die magnetischen Untersuchungen an Kernen der Forschungsbohrung Urach 3. – Report NLfB, Hannover, Archive No 84 955.

Mc Elhinny, M.W. (1973): Paleomagnetism and plate tectonics. – 206 p., Cambridge, Cambridge University Press.

Van der Voo, R. & French, R.B. (1974): Apparent polar wandering for the Atlantic-bordering continents: late Carboniferous to Eocene. – Earth Science Rev., **10**: 99–119.

The Urach Geothermal Project, p. 117–122;
Schweizerbart'sche Verlagsbuchhandlung, Stuttgart, 1982

Petrophysical Laboratory Measurements on Core Samples of Borehole Urach 3

L. RIEPE and J.R. SCHOPPER

with 2 figures and 2 tables

Abstract: Preliminary laboratory investigations on some important petrophysical parameters of core samples and borecuttings from the Urach 3 borehole had been carried out for correlations with well-logging data and for formation evaluation.
At the present state, only a tabular list of measured petrophysical data and short references to the principles of the applied measuring methods will be given.

Sample Preparation

From the delivered 14 core samples of Rotliegend (Lower Permian) and Unterer Muschelkalk (Middle Trias) Formation between 836 m and 1281 m depth about 100 plugs of 20 mm diameter had been drilled both in axial and in radial direction with respect to the borehole (marked by A and R respectively in the label of the plug). The plugs were then cut to 20 mm–40 mm length and cleaned with distilled water in a Soxhlet extraction apparatus. For well consolidated plugs the first cleaning may be achieved in an ultrasonic bath. After cleaning, the plugs were dried in a vacuum – oven at about 90 °C and a vacuum of at least 10^{-3} bar. For the measurements of the specific internal surface, the plugs were further heated and degassed, flushed with N_2 and evacuated again several times and then stored under a nitrogen atmosphere.

Porosity

The porosity, density and volume values of the plugs are measured by the "Archimedian Method" consisting of three weight determinations on the rock sample:
– in the dry state, i.e. with air filled pore space,

$$m_1 = \rho_{mtx} \cdot V_{mtx}$$

– in the wet state, i.e. with pore space completely saturated with fluid,

$$m_2 = \rho_{mtx} V_{mtx} + \rho_{por} V_{por}$$

– and in the wet state, submerged in the identical fluid, i.e. an apparent mass affected by the buoyancy of the matrix,

$$m_3 = \rho_{mtx} \cdot V_{mtx} - \rho_{por} \cdot V_{mtx}$$

From these three values, the porosity ϕ can be calculated:

$$\phi = \frac{V_{por}}{V_{por} + V_{mtx}} = \frac{m_2 - m_1}{m_2 - m_3}$$

and with the knowledge of the fluid density the different volumes and densities, too:

V_{mtx}	Matrix volume
V_{tot}	Bulk volume
V_{por}	Pore volume
ρ_{mtx}	Matrix density
ρ_{tot}	Bulk density

The matrix volume and the matrix density of the two borecuttings U/T and U/S were determined by gas porosimetry by using a Beckman® air comparison pycnometer.

Specific Internal Surface

The absolute internal surface S_{abs} of the specimen was determined by the adsorption of nitrogen at its boiling point with the use of a modified BET-method. The specific values of the internal surface are defined by the quotient of the absolute internal surface and the mass or volume respectively:

$$S_g = S_{abs}/m_{mtx}$$
$$S_{por} = S_{abs}/V_{por}$$
$$S_{mtx} = S_{abs}/V_{mtx}$$
$$S_{tot} = S_{abs}/V_{tot}$$

A large number among the selected formatized plugs showed values exceeding the measuring range of the apparatus; hence most of the measurements had to be done with small disks or cuttings of the same core material. For these cuttings the specific surfaces had to be determined by using the porosity and density values measured for "sister-plugs" from the same cores. On the assumption of sufficient homogeneity of the cores, the specific surfaces can then be derived from:

$$S_{mtx} = S_g \cdot \rho_{mtx}$$
$$S_{tot} = S_g \cdot \rho_{tot}$$
$$S_{por} = 1/\phi \cdot S_{tot}$$

These values are marked by$^{(*)}$ in Tab. 1.

Permeability

Gas permeabilities of about 30 plugs have been determined at variable mean pore pressures p $(p = p_V + p_H)/2)$. The plugs were sealed by a rubber jacket and mounted in a triaxial cell, working at a constant simulated overburden pressure of 100 bar (Fig. 1).

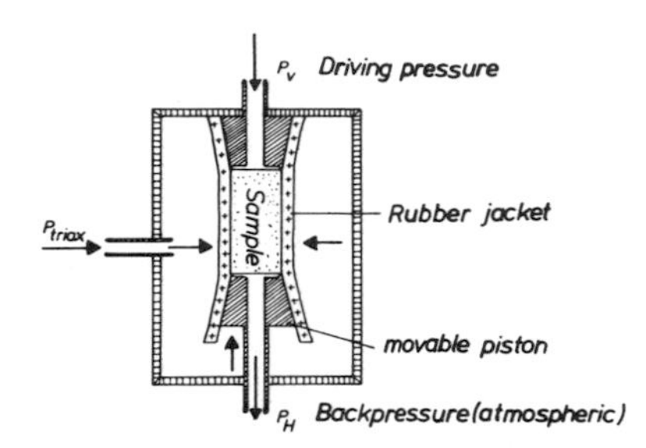

Fig. 1. Schematic illustration of the triaxial cell used for permeability measurements.

Klinkenberg-Effect

If the pore pressures were less than about 8 bar, deviations from the pure Darcy flow caused by gas slipping along the matrix walls had to be considered. In order to correct for this "Klinkenberg-Effect", the "true" permeabilities K_∞ (marked by (*) in Tab. 2) may be extrapolated for $p \to \infty$. If the permeabilities of the samples were determined

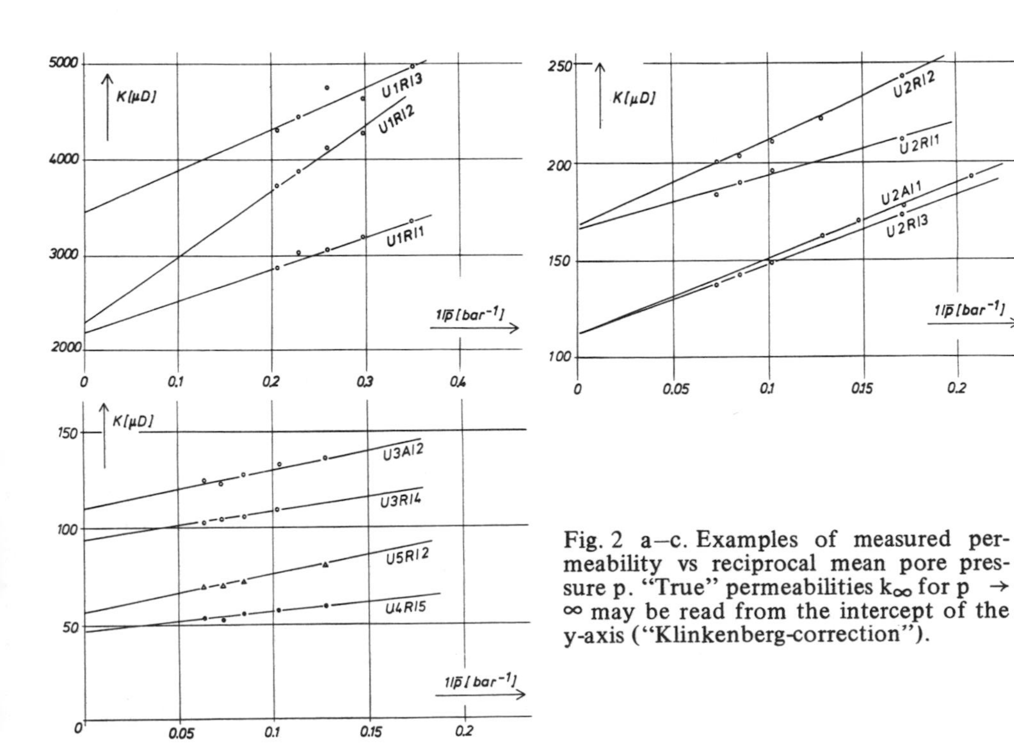

Fig. 2 a–c. Examples of measured permeability vs reciprocal mean pore pressure p. "True" permeabilities k_∞ for $p \to \infty$ may be read from the intercept of the y-axis ("Klinkenberg-correction").

at rather high pore pressures of about 30–50 bar the slip effect could be neglected. For later routine investigations, measurements at only one or two mean pore pressures p might be sufficient if the "Klinkenberg-constant" b entering the equation

$$K = K_{\infty} (1 + b/p)$$

could be generally determined. The "constant" b is a function of the mean free path of the gas molecules and the mean pore radii of the sample and thus of the true permeability K_{∞} itself. In the relationship

$$b = a \cdot K_{\infty}^{-c}$$

the empirical constants a and c have to be determined.

The preliminary data will have to be completed for final results.

"Quick-look" Gas Permeameter

In the beginning of the investigations, some permeability measurements had been performed by using a "quick-look" gas permeameter working at about atmospheric pore pressure. The plugs were sealed in a modified Hassler coreholder at a constant jacket pressure of about 6 bar. These permeabilities, listed in the last column of Tab. 2, are significantly higher than those measured in the triaxial cell. Three main reasons might explain the deviations:

1. Apparative, sealing problems occurred; additional gas flow between the mantle of the plug and the rubber jacket may occur at lower jacket pressures.
2. No "Klinkenberg-correction" could be performed, because the mean pore pressure could not be varied
3. Reduction of permeability was caused by compression under the high overburden pressure of 100 bar in the triaxial cell.

For these reasons, the values determined with the "quick-look" permeameter do not seem to be realistic.

The strong influence of the overburden pressure on the permeability could not yet be investigated; however it would certainly be of great interest for further considerations.

Results

The laboratory petrophysical data for some cores from the Urach 3 borehole are listed in Tab. 1.

Further results of permeability measurements are given in Tab. 2. Even these few cores already show a variation of permeability within about four orders of magnitude ($3000\ \mu D < K < 0.1\ \mu D$).

The coarse-grained quartz-sandstone of the Middle Triassic (U 1, U 2) reached values up to 3500 μD. The porphyry fanglomerates of the Lower Permian (U 3 – U 5) scatter within 40–100 μD.

The permeability of the shaly fanglomerates and the shaly sandstones (U 6 – U 11) decreased – depending on shaliness – down to 0.1 μD ($\doteq 10^{-19}\ m^2$). The black shales

Table 1. Results of petrophysical measurements on cores from Urach 3 drillhole.
Values marked by (*) are measured for disks or cuttings of the same core material.

Sample		Porosity	Volume, densities			Specific internal surface				Permeability
laboratory label	Depth m	ϕ %	V_{mtx} cm^3	ρ_{mtx} g/cm^3	ρ_{tot} g/cm^3	S_g m^2/g	S_{mtx} m^2/cm^3	S_{por} m^2/cm^3	S_{tot} m^2/cm^3	see table 2 μD
U 1 R / 3	836.5	12.80	5.292	2.632	2.295	1.99	5.24	35.7	4.57	3460
U 2 R / 2	838.0	10.33	4.931	2.639	2.367	2.77	7.31	63.5	6.56	167
U 3 R / 4	938.5	12.31	5.265	2.636	2.311	4.38*	11.54*	82.2 *	10.1 *	94
U 4 R / 5	939.3	11.99	5.173	2.637	2.321	7.16*	18.88*	138.6 *	16.6 *	47
U 5 R / 2	941.0	12.37	5.237	2.636	2.310	8.08*	21.30*	150.9 *	18.7 *	60
U 6 R / 4	984.5	6.14	5.443	2.673	2.509	7.54*	20.15*	308.1 *	18.9 *	3.5
U 7 R / 3	987.0	15.73	4.896	2.610	2.200	> 8.34*	> 21.77*	> 116.6 *	> 18.3 *	63
U 8 R / 4	989.8	6.63	5.292	2.689	2.511	10.54*	28.34*	399.2 *	26.5 *	0.9
U 9 / R 4	991.0	9.75	4.970	2.654	2.395	8.27*	21.95*	203.15*	19.8 *	3.6
U 10 R/1	1071.0	5.75	5.973	2.675	2.521	–	–	–	–	1.2
U 11 R/6	1073.5	4.06	5.255	2.668	2.560	5.47*	14.59*	344.9 *	14.0 *	0.13
U/T	570.62	8.42	2.483	2.628	–	> 9.70	> 3.70	–	–	–
U/S	569.66	12.36	1.412	2.662	–	3.63	1.37	–	–	–

Table 2. Results of permeability measurements on cores from Urach 3 drillhole. Values marked by (*) are extrapolated for p → ∞ ("Klinkenberg-correction").

Sample	depth [m]	Formation, rough description	Permeability [μD] p_{triax} = 100 bar	p_V [bar]	"quick-look" permeameter P_M = 6 bar
U 1 R / 1	836.5	Middle Triassic	2200*	1– 3	
U 1 R / 2	836.5		2300*	1– 3	
U 1 R / 3	836.5	Coarse grained	3460*	1– 3	8250
U 2 R / 1	838.0	quartz-sandstones	164*	4–12	
U 2 R / 2	838.0		167*	4–12	577
U 2 R / 3	838.0		114*	4–12	
U 2 A / 1	838.0		114*	3– 7	
U 3 R / 4	938.5	Lower Permian	94*	6–14	6240
U 3 A / 2	938.5		110*	6–14	
U 4 R / 5	939.3	porphyry	47*	6–14	4910
U 4 A / 2	939.3	Fanglomerates	98*	6–14	
U 5 R / 2	941.0		60*	6–14	3310
U 5 A / 1	941.0		43	38	
U 6 R / 4	984.5		3.5	38	101
U 6 A / 1	984.5	shaly	9	38	
U 7 R / 3	987.0	Fanglomerates	63*	6–14	2050
U 7 A / 3	987.0		12	38	1410
U 8 R / 4	989.8		0.87	36	78
U 8 A / 1	989.8		0.37	60	
U 9 R / 4	991.0	light	3.6	38	219
U 9 A / 4	991.0	red	12	38	
U 10 R / 1	1071.0	shaly	1.2	38	141
U 11 R / 6	1073.5	series	0.13	55	46
U 11 A / 2	1073.5		2.9	38	
U 12 A / 1	1273.4	dark shaly	< 0.1	60	
U 13 A / 5	1281.5	series	0.46	38	

from about 1280 m depth may reach values even below 0.1 μD, but they could not yet be measured under the given conditions.

On the other hand the sandstones or limestones of higher formations probably can show much higher permeabilities in the order of magnitude 1–1000 mD. For these measurements proper core material would be necessary.

Further systematic experimental and theoretical investigations for final interpretation and correlation should be recommended for future projects.

The Urach Geothermal Project, p. 123–133;
Schweizerbart'sche Verlagsbuchhandlung, Stuttgart, 1982

Geochemical Problems in Fluid-Rock Interaction

E. ALTHAUS

with 14 figures and 5 tables

Abstract: Aqueous heat extraction fluids usually are at disequilibrium with reservoir rocks and hence cause heterogeneous reactions to occur. Kind and velocity of reaction depend on temperature and fluid composition. In a streaming medium a linear relationship exists between reacting area, time, and amount of reacted material. The stronger the deviation from equilibrium compositions, the higher are reaction rates and degree of transformation. At strong deviation, stoichiometric dissolution occurs, while precipitation of secondary minerals becomes more important if equilibrium compositions are approached.
Experimental findings with mineral and rock samples have been confirmed by circulation experiments in the Urach 3 borehole. Results of fluid analyses lead to the conclusion that feldspar minerals of the basement rock are attacked by near-neutral solutions, whereas under more acid conditions calcite veins were dissolved more strongly with less marked effects on the silicates.

Introduction

Aqueous solutions are the most probable heat transfer media to be used in geothermal energy extraction from both wet and dry hot rocks. These solutions are by no means inert with respect to mineral reactions, since they usually are at chemical disequilibrium with underground rocks. This is especially true for systems in which a fluid is circulated by external action (hot dry rock systems, re-injection of exhausted fluid). Two problems must be considered: What kind of reactions will be induced, and at what rate will these proceed.

The reactions to be expected are of the same kind as those encountered in hydrothermal wall rock alteration around mineral veins. Since these are fairly well known, extrapolations to geothermal systems are possible with reasonable reliability. The question that remains open is concerned with reaction rates.

Few investigations that can be applied to the problems in question have been performed so far. Among these, the papers of LAGACHE (1965, 1976), LAGACHE & WEISBROD (1977), PETROVIC et al. (1976) and PETROVIC (1976) are important. All experiments were performed with rather fine-grained material, however. It did not seem feasible to draw conclusions from their results for the behaviour of solid wall rocks in contact with hydrothermal fluids.

The rocks most likely to be expected for geothermal reservoirs are more or less granitic in composition. Since these are composed of feldspars to 50–60 % by mass, feldspar minerals were studied as model substances. Experiments were performed in closed teflon-coated autoclaves of 10 cm^3 volume with single cleavage particles of known surface area (about 200 mm^2). The temperature varied between 120° and 200 °C; pressures were saturation pressures at given temperatures. In order to parallel the conditions in the Urach 3 borehole more exactly with experimental findings, runs were performed with small drill cores from a rock ("Glimmer-Syenit", GS) recovered from the deepest part of the borehole. A flow of solutions was simulated by replacing the reaction fluid by fresh batches every day. The compositions of the resulting solutions were analysed by atomic absorption spectroscopy, and the resulting solids examined by x-ray, microprobe, and optical methods.

Results of Experiments with Minerals

According to varying chemical conditions several kinds of reaction were observed during the experiments. The governing factor common to all of them is the ratio of metal ions to hydrogen ions. The most extreme reaction observed was total stoichiometric dissolution. For a feldspar, this can be formulated as follows:

$$\text{(1 a)} \quad NaAlSi_3O_8 + 4\,H^+ + 4\,H_2O = Na^+ + Al^{3+} + 3\,H_4SiO_4$$

$$\text{(1 b)} \quad KAlSi_3O_8 + 4\,H^+ + 4\,H_2O = K^+ + Al^{3+} + 3\,H_4SiO_4$$

with the equilibrium constants

$$K_1 = \frac{[Na^+] \cdot [Al^{3+}] \cdot [H_4SiO_4]^3}{[H^+]^4}$$

in the case of 1 b, Na^+ is replaced by K^+.

If these reactions are expected to run from left to right, the activity of H^+ must be relatively high as compared to those of the other species. Under less extreme conditions, reactions that precipitate secondary phases take place. Consequently, the dissolutions are no longer congruent, i. e. the concentrations of species in solution deviate strongly from those of the initial mineral.

The following reactions were observed to occur:

$$\text{(2)} \quad Na(K)Al\,Si_3O_8 + H^+ + 7H_2O = Na^+\,(K^+) + \underset{\text{Gibbsite}}{Al(OH)_3} + 3\,H_4SiO_4$$

$$\text{(3)} \quad 2\,Na(K)AlSi_3O_8 + 2\,H^+ + H_2O = 2\,Na^+\,(K^+) + \underset{\text{Kaolinite}}{Al_2\,[(OH)_4Si_2O_5]} + SiO_2$$

$$\text{(4)} \quad 3\,Na(K)AlSi_3O_8 + 2\,H^+ = 2\,Na^+(K^+) + Na(K)Al_2[(OH)_2AlSi_3O_{10}] + 6\,SiO_2$$

The equilibrium constants are all of the type

$$K = \frac{[Na^+]}{[H^+]}$$

At a given cation activity, the kind and direction of reaction depend mostly on the pH-value of the solution.

An example for reaction 1 is shown in Fig. 1. Starting material was a microcline pepthite from a pegmatite from Evje, Norway, whose composition is given in Tab. 1. The dependence on time is linear; thus it follows that diffusion is not an important parameter

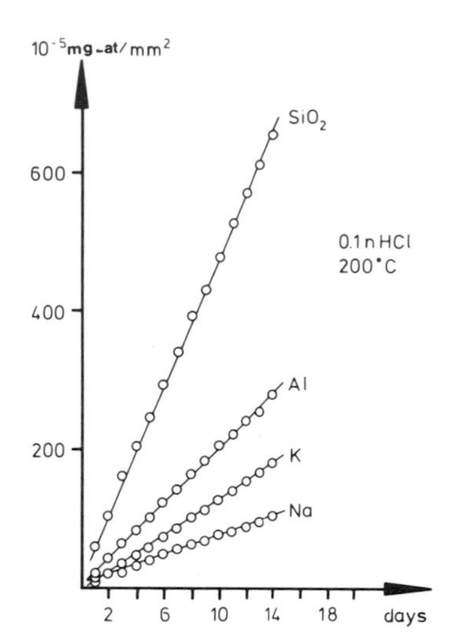

Fig. 1. Leaching of microcline perthite, acid condition. Cumulative amounts dissolved per unit of area.

Table 1. Composition of starting materials (mass content in %).

	microcline perthite	sanidine
SiO_2	67.6	68.0
Al_2O_3	16.8	17.3
Fe_2O_3 (tot. Fe)	0.18	0.08
MgO	0.32	0.02
CaO	0.09	0.07
Na_2O	2.96	2.17
K_2O	12.0	12.9
TiO_2	0.02	0.02
MnO	0.01	0.01
	99.98	100.57
K-rich component	98.6 molar content in % Or	
Na-rich component	7.1 molar content in % Or	

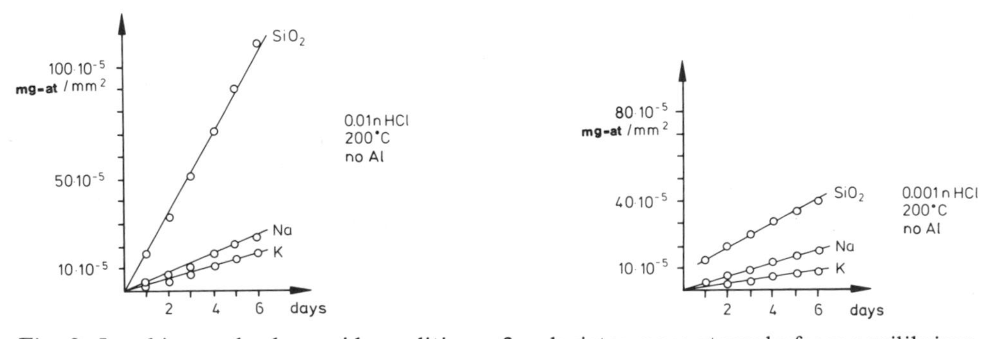

Fig. 2. Leaching under less acid conditions. 2 a deviates more strongly from equilibrium than 2 b.

in this reaction. The amounts of dissolved species (recalculated to standard surface area) are in stoichiometric feldspar proportions, namely (Na + K) : Al : SiO_2 = 1 : 1 : 3.

If the H^+-activity of the solution is lowered, reactions 2 to 4 proceed in that sequence. The ionic ratios are no longer similar to those in feldspars, but proportionality with time is still preserved in Fig. 2. The total amounts of dissolved matter decrease considerably with increasing pH-value for all species. The dependence is shown for K in Fig. 3. Roughly, it can be said that more strongly acid solutions are much more reactive than weakly acid ones. Strong dependence on flow velocity (simulated here by length of period for exchange of solutions) (Fig. 4) and on temperature also exists (Fig. 5). The most plausible measure for the poorly defined term "reactivity" is deviation from the equilibrium composition of the dissolved species.

For the K-Al-Si-H-O-system theoretical equilibrium conditions are shown in Fig. 6. This is likely to be modified by additional components, e.g. Na. For a given temperature, a section through a system at quartz saturation corresponds to Fig. 7 (data from HELGESON et al. 1969). A solution in equilibrium plots along the border line albite-microcline. All experimental solutions deviate more or less from these compositions, although solid K + Na-feldspars are present in all samples. The amount of dissolved matter decreases in the order of runs labelled F 4 to F 10, i.e. with increasing ionic ratios. The distance from equilibrium values therefore defines reactivity more quantitatively.

Volume effects are connected with the hydrothermal mineral reactions as well as with cooling. It is evident from Tab. 2 that even if a perhaps unrealistically high degree of cooling (200 K) is assumed, the effect of shrinking is smaller by more than one order of magnitude than the resulting from reaction. Hence, mineral transformation is an important tool for controlling volume changes in a geothermal rock system.

If secondary phases are formed they usually are precipitated on the rock surfaces as continuous layers. In Fig. 8 such a layer of mica is shown on top of a leached feldspar

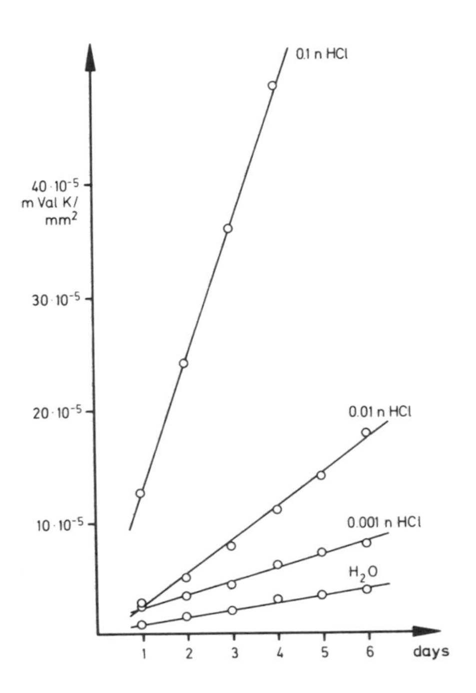

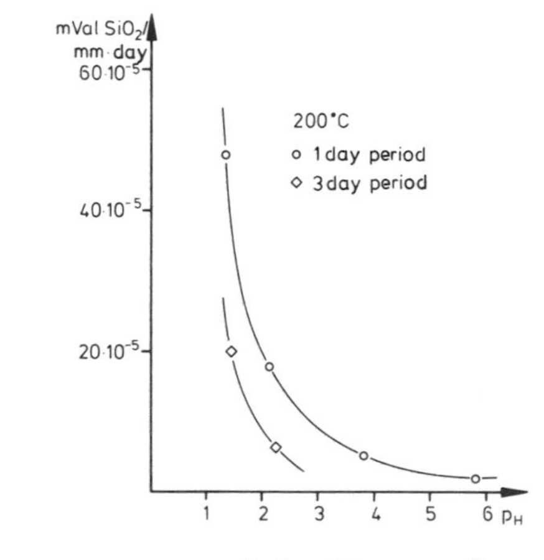

Fig. 4. Dependence of leaching on flow rate. 3 day period means slower flow compared to 1 day period.

Fig. 3. Dependence of leaching on acidity. Data for K shown as an example.

with deep etch pits. It was formed within 6 days. Prolonged treatment would increase the thickness of the layer considerably, until diffusion through the interstices becomes the velocity controlling effect retarding thus considerably heat exchange. It was shown

Table 2. Volume effects

Reaction			Δ V, solids cm^3 / mole
3 Ksp + 2 H^+	=	Ms + 6 Q + 2 K^+	− 16.4
2 Ms + 2 H^+ + 3 H_2O	=	3 KaO + 2 K^+	+ 8.6
2 Ksp + 2 H^+ + H_2O	=	Kao + 4 Q + 2 K^+	− 13.6
2 Ab + 2 H^+ + H_2O	=	Kao + 4 Q + 2 Na^+	− 4.45
An + 2 H^+ + H_2O	=	Kao + Ca^{2+}	− 1.3
100 g granite		37.43 cm^3	− 1.90
		=	− 5 %
Mineral, cooled 500 K to 298 K			**Δ V, %**
microcline (mik)			0.398
plagioclase An_{23} (plag)			0.23
quartz (q)			0.78
100 cm^3 granite		35.6 cm^3 plag 31.3 cm^3 mik 30.3 cm^3 q 2.8 cm^3 biot	0.302
Volume decrease, reaction:			5 %
Volume decrease, cooling 500 K to 298 K:			0.302 %

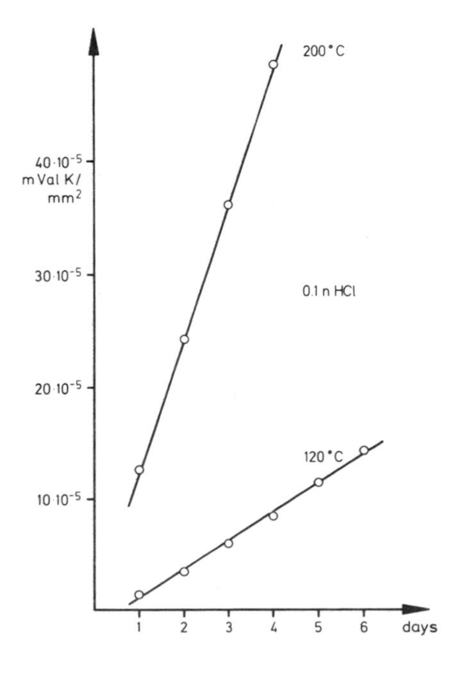

Fig. 5. Temperature dependence of leaching.

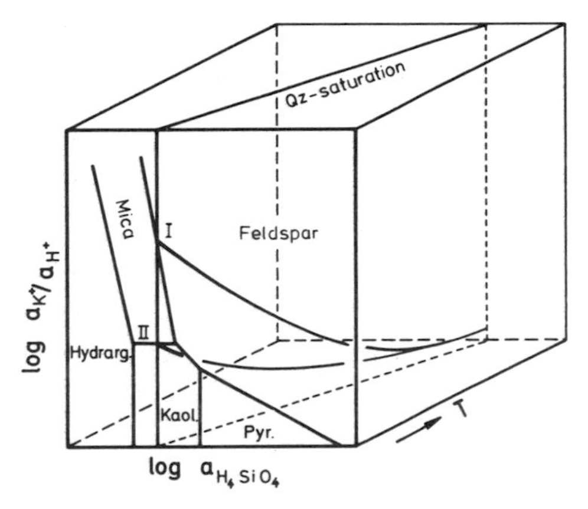

Fig. 6. Equilibrium boundaries in the system K-Al-Si-O (schematically).

by experiments in mafic systems (KRONIMUS 1979) that reactions can be directed in such a way that surfaces are sealed by an impermeable layer preventing further attack by aggressive solutions. Chemical conditions (i.e. compositions of fluids) have to be chosen in such a way as to optimise these competing effects.

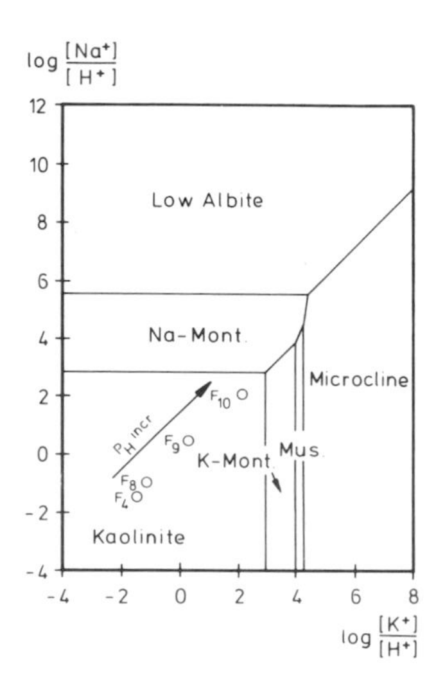

Fig. 7. Isothermal-isobaric section through system Na-K-Al-Si-H-O at quartz saturation. Equilibrium boundaries and experimental points with different degrees of reactivity.

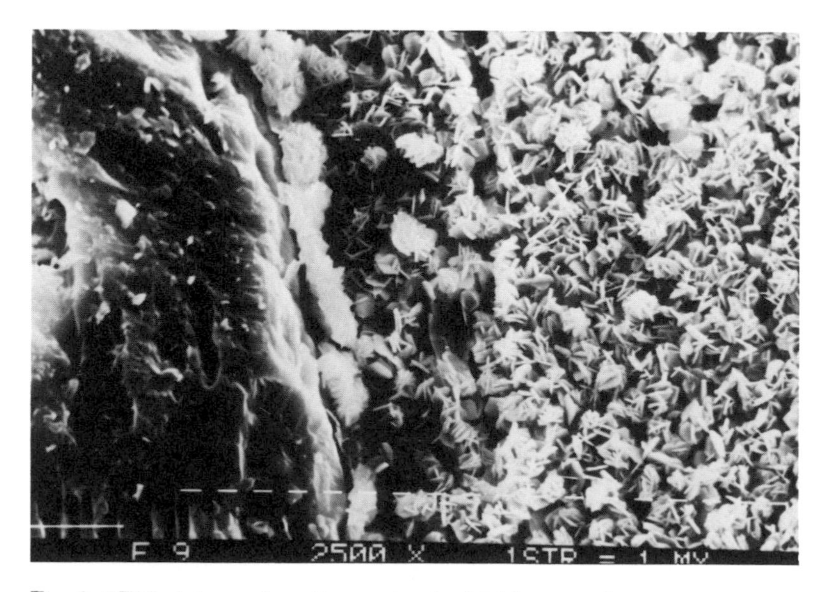

Fig. 8. SEM picture of coating on leached feldspar surface.

Results of Experiments with Rocks

In mixtures of minerals, i.e. rocks, mutual influences occur among the solid components. Hence the reactions observed in experiments with a "Glimmersyenit" from the borehole and water of the Erms stream (this water was used in the circulation experiments) differed from those performed with single minerals. The compositions of rock and water are shown in Tab. 3 and 4, respectively. The acidity of the Erms water was controlled by addition of HCl or NaOH solutions.

Table 3. Composition of starting rock for alteration experiments (Glimmersyenit).

	weight-%
SiO_2	63.3
TiO_2	2.56
Al_2O_3	15.2
Fe_2O_3 (total Fe)	6.64
MgO	2.81
MnO	0.09
CaO	2.40
Na_2O	3.43
K_2O	2.53
	98.96

Table 4. Composition of water of the Erms stream.

	ppm
SiO_2	64.2
Ca	1010
K	10
Na	113
Al	trace
Fe	0
Mg	35

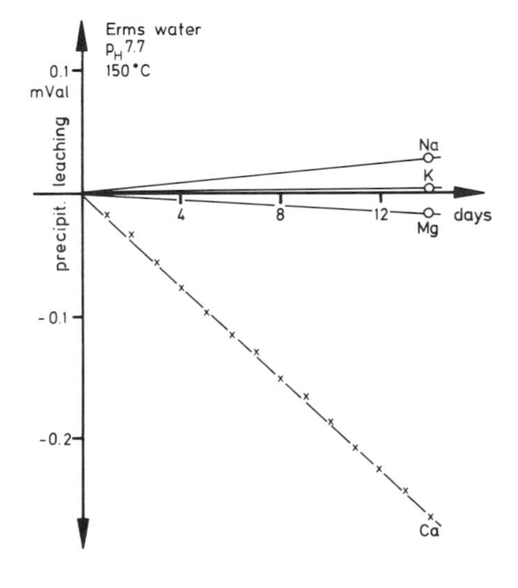

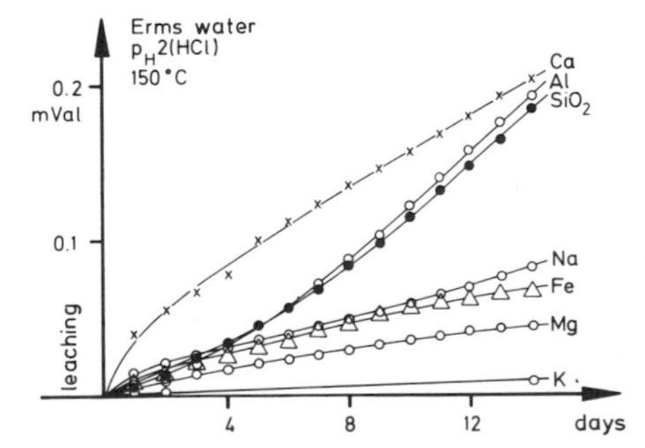

Fig. 10. Leaching experiments as in Fig. 9, but with acid solution.

Fig. 9. Leaching experiments with "Glimmersyenit" and water from the Erms stream, no change in initial composition.

The most important results of the experiments are: In approximately neutral solutions, only K and Na are leached. SiO_2 and Al are practically inert components, Ca, however, is bound in the rock by precipitation of a secondary mineral (Ca-zeolite). Mg present in very small amounts in solution is absorbed by the rock in an unidentified phase. This behaviour strongly resembles that described by CHARLES (1979) who observed precipitation of phillipsite on plagioclase in long-term runs. Under acid conditions, this behaviour changes drastically (Fig. 10). All Components of the rock are leached to a considerable extent, Ca, Al and SiO_2 becoming the most prominent constituents of the solution and K being the least concentrated. With decreasing acidity of the solution SiO_2 is mobilized to a lesser extent and is finally less concentrated than the alkalies; Ca, however, remains the most highly concentrated species.

Under alkaline conditions only SiO_2 is leached significantly. Again, Ca is precipitated from the initial solution. Neither Al nor Mg are mobilized (Fig. 11). Of course, nothing can be said about Na, since it is present in such high concentrations.

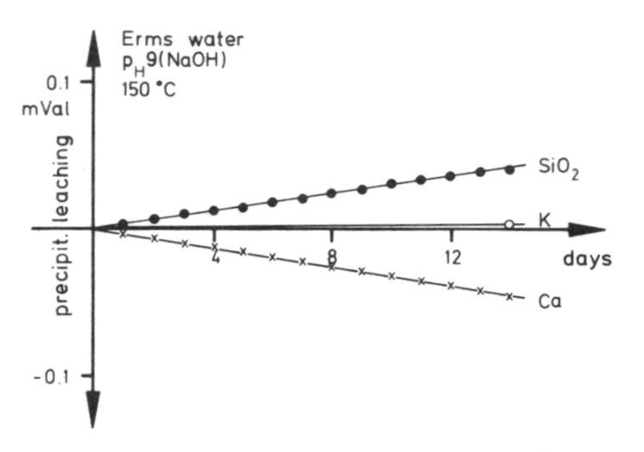

Fig. 11. Leaching experiments as in Fig. 9, but with alkaline solution.

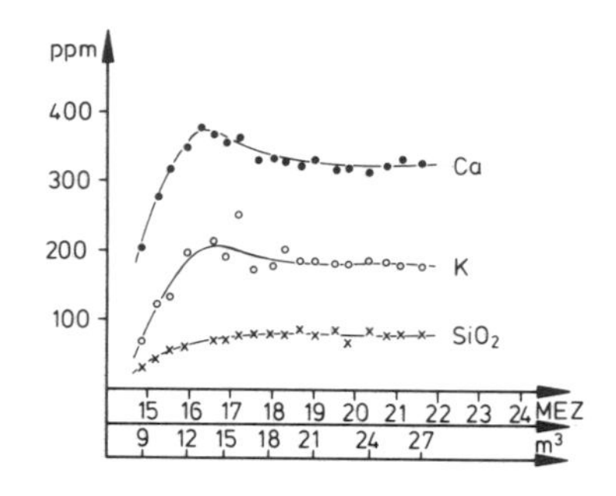

Fig. 12. Circulation experiment no. 8. Water as circulating fluid.

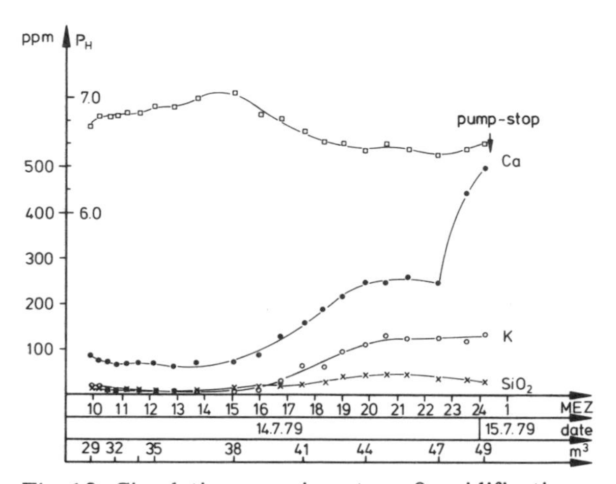

Fig. 13. Circulation experiment no. 9; acidification.

Results of Circulation Experiments in the Borehole

Several weeks after completion of the borehole, water samples were taken with an autoclave. The results of analyses are shown in Tab. 5. The solutions from the deepest part have a high natural mineralization (Na, K, Ca, SiO_2) and, most surprisingly, low pH-

Table 5. Water samples from the Urach 3 borehole (March 1979).

No	Depth below surface (m)	pH	Contents (ppm) SiO₂	Al	Fe	Ca
1	49	9.55	52.4	12.2	14.5	17.0
2	332.5	2.6	162.1	1.4	195.5	123.0
3/1	1500.8	6.45	6.3	trace	20.0	23.0
3/2	1500.8	6.45	7.1	trace	4.0	21.0
4/1	3325	3.7	174.1	0.74	269.0	148.5
4/2	3325	4.2	156.6	0.58	110.6	150.5

Mg	K	Na	Mn	Cl^-	SO_4^{2-}	Total
3.2	207.5	580.0	0.18	+	+	887.2
0.78	172.5	615.0	2.64	++	+	1272.9
0.73	4.0	30.5	0.48	+	–	85.0
0.67	4.0	30.9	0.29	+	–	67.1
0.59	150.0	530.0	3.25	++	+	1276.7
0.45	160.0	560.0	1.35	++	+	1139.6

Ti examined, but not detected
Cl^- and SO_4^{2-} semi-quantitative; + present; ++ larger amount; – not present

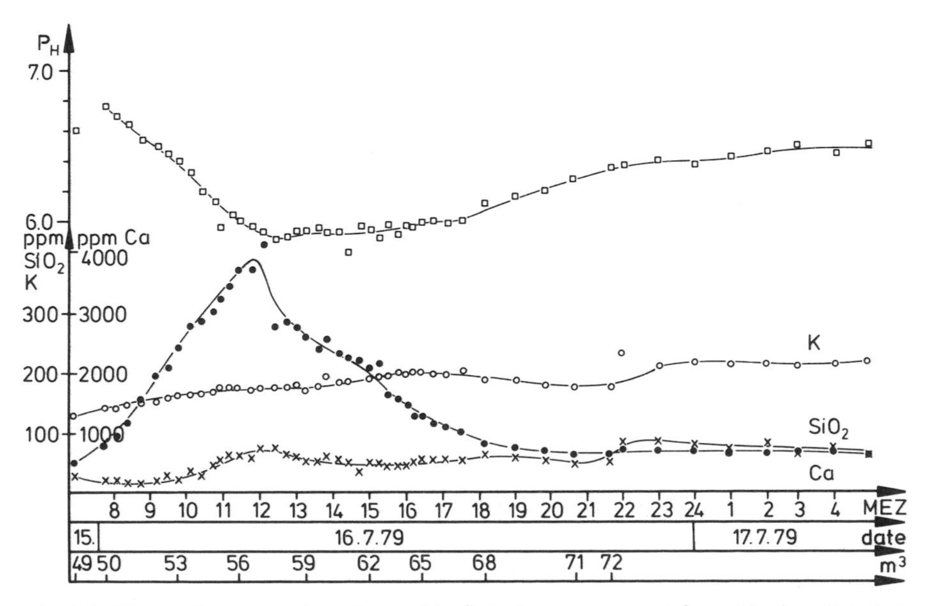

Fig. 14. Circulation experiment no. 10. Solutions recovered from the borehole later after acidification.

values. The results resemble those of the leaching experiments performed under acid conditions. It was assumed that circulation would create similar concentrations in solution. The compositions of solutions produced in a typical circulation experiment (no. 8, of 13th July 1979, between the open end of the hole and perforations no. 1, 2, and 3) are shown in Fig. 8. The result corresponds to that of experiments carried out under slightly acid conditions with the exception that the potassium content is much higher than expected. Compared to that of the water samples from the undisturbed borehole, the concentration of Ca is much higher, probably because of preferential leaching of calcite veins.

On July 14th to 16th, circulation experiments were performed with acid solutions (5000 l of 5 % HCl fed into the hole in one single charge). The results are shown in Figs. 13 and 14. It is easily seen that an "acid wave" flushed the rock system (see pH dependence on time), thus producing solutions highly concentrated in Ca, with contents of K and SiO_2 corresponding to the values measured before. It is likely that the high Ca concentration results preferentially from calcite dissolution, whereas the leaching of feldspars is less important.

The total amount of matter transported to the surface is rather high. Several hundred kilogrammes have been leached during the circulation experiments.

Conclusions

Experiments with rocks and minerals have shown that a great variety of reactions occur between fluids and solids. The factors determining kind and rate of interaction most sensitively are temperature and deviation of solution composition from equilibrium values. The most important parameter is the cation-to-hydrogenion ratio. In granitelike rocks, feldspar minerals are most likely to react, whereas quartz is more or less inert under acid and near-neutral conditions. Secondary minerals can be forced to precipitate on rock surfaces by choosing appropriate solution compositions.

For the circulation of fluids, solutions of the same type as in experiments were anticipated. Indeed it was shown by the circulation experiments that the reactions induced by the acid solutions corresponded to those derived from the previous experimental investigation. It can be concluded that most of the freight of dissolved matter resulted from the decomposition of calcite. Aluminous and silicious material, probably aluminium silicate, must have been precipitated on the rocks, because the ionic ratios in the solutions point to incongruent decomposition of alumosilicates. In the leaching experiments no increase in permeability was observed. This may possibly be due to secondary mineral precipitation on the crack surfaces.

Acknowledgement. Most of the experiments described here were performed by Dr. D. ROTTENBACHER; Mr. P. ZRINJSCAK very carefully performed many of the analyses. Their assistance is gratefully acknowledged. Financial support from the Ministry of Research and Technology as well as the Commission of the European Communities was a prerequisite for performing this study.

References

Charles, R.W. (1979): Experimental geothermal loop II. 200 °C study. – Los Alamos Scientific Report LA-7735-MS.

Helgeson, H.C.; Brown, Th.H. & Leeper, R.H. (1969): Handbook of Theoretical Activity Diagrams Depicting Chemical Equilibria in Geologic Systems. Involving an Aqueous Phase at 1 atm and 0° to 300°. – Freeman, Cooper & Company, San Francisco.

Kronimus, B. (1979): Mineralogische Experimente zur Gewinnung geothermischer Energie. – Reaktionen zwischen einer hydrothermalen, fluiden Phase und den Mineralen Olivin und Diopsid. – Diplomarbeit Karlsruhe.

Lagache, M. (1965): Contribution à l'étude de l'alteration des feldspaths, dans l'eau, entre 100 et 200 °C, sous diverses pressions de CO_2, et application à la synthése des mineraux argileux. – Bull. Soc. Fr. Mineral. Cristallogr. **88**: 223–253.

– (1976): New data on the kinetics of the dissolution of alkali feldspars at 200 °C in CO_2 charged water. – Geochim. Cosmochim. Acta **40**: 157–161.

Lagache, M. & Weisbrod, A. (1977): The system: Two alkali feldspars-KCl-NaCl-H_2O at moderate to high temperatures and low pressures. – Contrib. Mineral. Petrol. **62**: 77–101.

Petrovic, R. (1976): Rate control in feldspar dissolution – II. The protective effect of precipitates. – Geochim. Cosmochim. Acta **40**: 1509–1521.

Petrovic, R.; Berner, R.A. & Goldhaber, M.B. (1976): Rate control in dissolution of alkali feldspars – I. Study of residual feldspar grains by x-ray photoelectron spectroscopy. – Geochim. Cosmochim. – Acta **40**: 537–548.

The Urach Geothermal Project, p. 135–146;
Schweizerbart'sche Verlagsbuchhandlung, Stuttgart, 1982

Ultrasonic Velocity and Fracture Properties of the Rock Core from the Urach Borehole Crystalline Section

F. RUMMEL, H.-J. ALHEID, R.-B. WINTER and TH. WÖHRL

with 8 figures and 5 tables

Abstract: P- and S-wave velocities on cores from the crystalline rock of the Urach deep borehole, SW Germany were studied in detail by ultrasonic measurements at the drill site as well as on samples at high pressure in the laboratory. The values determined agree approximately with sonic-log data obtained in the borehole. In addition, compressive strength and frictional strength data were measured under triaxial loading. In relation to hydraulic fracturing experiments in the borhehole, laboratory hydro-frac tensile strength as well as fracture toughness and specific surface energy values were obtained from laboratory studies on selected cores. They allow an estimate of in-situ breakdown and fracture extension pressures if stress field data at depth are known.

Introduction

The Urach deep borehole offered a unique opportunity to study in detail the intrinsic physical, chemical and geological properties of the crystalline basement buried underneath the Mesozoic and Permian sedimentary formations in the Urach geothermal area of SW Germany. In the overall investigation, measurements on the core material from the basement rock may contribute to the interpretation of geophysical surface studies and borehole logging surveys. Moreover, in connection with geothermal energy extraction from hot crystalline rock formations, the fracture behaviour of crustal rock must be understood with respect to the generation of large heat exchange surfaces as well as for estimating the stability of fracture systems which could be activated by fluid injections (induced seismicity).

The crystalline section of the Urach borehole extended from a depth of about 1606 m to the bottom of the borehole at a final depth of 3334 m. The core consisted mainly of gneiss of ortho- and para-type intersected by aplitic and granitic dykes. The bottom portion below a depth of 3000 m may be described as a mica-rich syenite (DIETRICH, this volume).

For this particular study, measurements have been conducted on 26 core sections each about 9 metres in length. The measurements included the following tests:

P- and S-wave velocities were directly determined on fresh core samples at the drill-site in order to obtain immediate information on the velocity structure. High pressure P-wave velocities were measured on selected minicores to correct velocity data obtained at low pressure with respect to depth. The measurements of fracture parameters included the determination of triaxial compressive strengths, hydraulic fracturing tensile strength, fracture toughness and specific fracture surface energy.

Ultrasonic Measurements

P- and S-wave velocities were measured on the fresh core from the borehole at the drill-site. On each core section about 8 single measurements were taken at various locations on the core. In general, the core samples were unloaded during the tests.

In most cases, both P- and S-waves were generated and recorded by S-wave transducers with a resonant frequency of 2 MHz (Panametrix, V 154, 2.25/0.5). The electric pulse given to the transmitter had a main frequency between 300 and 800 kHz. For the measurements parallel to the core axis the receiver was moved along a profile on the core surface in order that velocities could be derived from travel-time-curves. The distance between the transducers varied between 10 and 80 cm. Pulse transmission normal to the core axis was performed by placing the transducers at opposite diametrical locations of the core.

A total of 95 measurements were obtained. The velocity data of the single tests are given in Tab. 1, which also shows the average velocity for each core section. P-wave velocities vary between 4.36 and 5.69 km s^{-1}, S-wave velocities between 2.57 and 3.28 km s^{-1}. The ratio v_p/v_s varies between 1.69 and 1.91 (Fig. 1).

In order to correct the velocities with respect to the lithostatic pressure at depth, P-wave velocity measurements at pressures up to 1.5 kbar were conducted on minicores with a diameter of 3 cm. The measurements were carried out in a room temperature

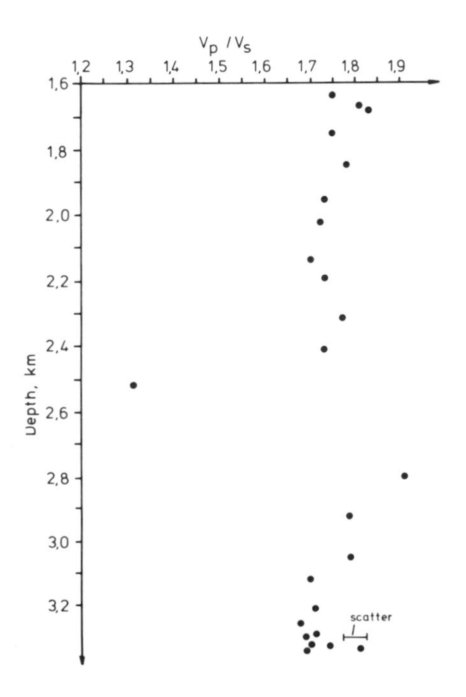

Fig. 1. v_p/v_s ratio as a function of depth for the Urach borehole crystalline section, determined by ultrasonic measurements.

3 kbar pressure vessel at hydrostatic pressure. 10 minicores of core section 37 (paragneiss) and 41 (mica syenite) were tested. The velocity-pressure dependence is presented in Figs. 2 a and 2 b. The velocity increase is about 10 per cent at a pressure of 400 bar or about 6.3 per cent per kilometre. Between pressures of 450 bar (1700 m depth) and 1.5 kbar (6000 m depth) the increase in P-wave velocity is about 4 to 5 per cent or only 1 per cent per km. In the higher pressure range the velocity increase with pressure is approximately linear.

By application of this velocity-pressure relation, the velocity values obtained at atmospheric pressure were corrected with respect to depth, assuming a lithostatic stress field and an average rock density of 2.7 gcm^{-3}. The velocity depth relation is plotted in Fig. 3. P-wave velocities vary between 4.9 and 6.4 km s^{-1}. P-wave velocities higher than 6 km s^{-1} are obtained for the upper basement rock between 1.6 and 2.4 km depth and for syenite at a depth between 3.0 and 3.3 km. Paragneiss cores are generally characterized by a P-wave velocity of about 5 km s^{-1}. This velocity depth relation is in agreement with

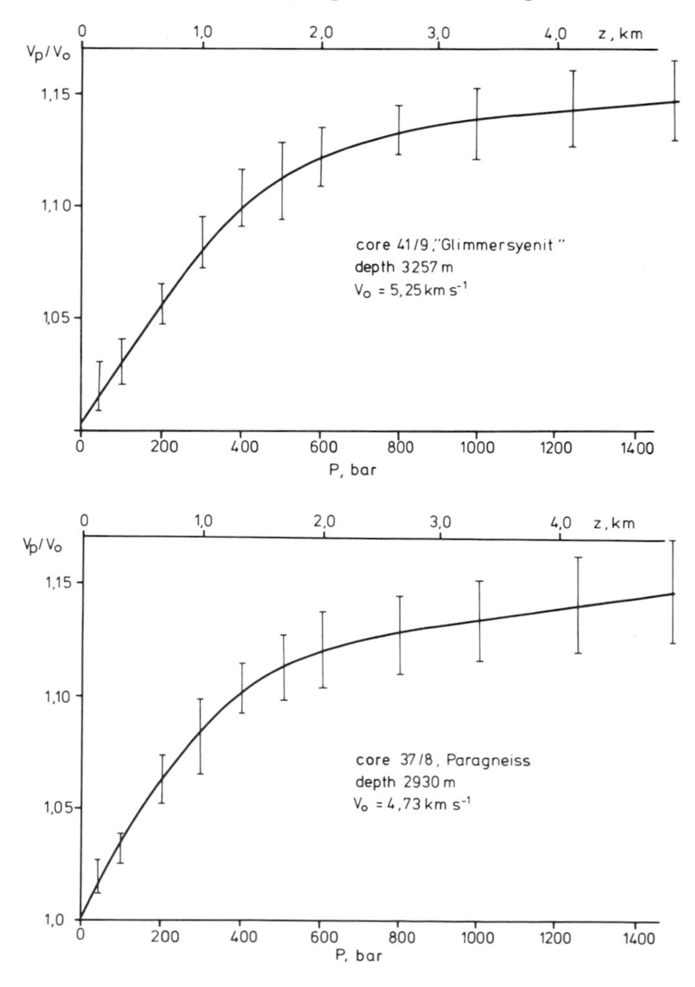

Fig. 2. Pressure dependence of P-wave velocity. V_0 velocity at atmospheric pressure: a) for cores from core section 37 at 2930 m; b) for cores from core section 41 at 3250 m.

Table 1. Ultrasonic velocity data of cores from the Urach 3 borehole.

No. of core section	Depth m	Rock type	V_p, km s^{-1}	V_S, km s^{-2}	V_p, km s^{-2}	V_S, km^{-2}
22	1635.5	coarsegrained biotite	5.65	3.28		
		gneiss with idiomorph	5.78			
		garnet (Orthogneiss)	5.5	3.13		
22	1639	dto.	5.47	3.09		
			5.47		5.39 ± 0.24	3.06 ± 0.13
			5.30	2.99		
22	1640.5	dto.	5.06	3.08		
			5.21	3.08		
22	1639.2	dto.	5.05 (1)	2.8		
24	1662.5	dto.	5.37	3.05		
			5.63	3.11	5.59 ± 0.13	3.08 ± 0.03
			5.71	3.08		
26	1675.5	dto.	5.59	2.92		
			5.41	2.88		
			5.03	2.85	5.35 ± 0.21	2.92 ± 0.07
			5.15			
26	1675.9	dto.	5.59	3.04		
			5.35			
27	1751.7	dto.	5.60	3.19		
			5.68	3.22		
			5.56	3.26	5.54 ± 0.13	3.17 ± 0.06
			5.65			
			5.31	3.10		
			5.42	3.10		
28	1854.3	orthogneiss	5.58	3.13		
			5.67	3.15		
			5.67	3.18	5.69 ± 0.10	3.19 ± 0.04
			5.52	3.20		
			5.67	3.23		
			5.8			
28	1859.2	dto.	5.79	3.25		
			5.79	3.20		
29	1945.7	orthogneiss	5.71	3.24		
			5.58	3.21	5.52 ± 0.15	3.22 ± 0.01
			5.55	3.21		
			5.61	3.21		
29	1950.7	granitic dyke	5.22	3.13		
			5.36	3.11		
			5.57	3.11		
			5.57	3.25		
			5.57	3.25		
30	2014.3	Pinite (?)	5.28	3.03	5.42 ± 0.11	3.10 ± 0.04
			5.42	3.1		
			5.42	3.1		
			5.57	3.15		
31	2129.8	paragneiss with mica	4.54	2.76	4.52 ± 0.02	2.66 ± 0.07
		bands and granoblasts	4.50			
			4.52	2.60		
			4.50	2.63		
32	2189.8	paragneiss ("Braun-	4.29	2.64	4.66 ± 0.4	2.69 ± 0.05
		glimmergneiss")	4.24	2.64		
			5.06	2.74		
			5.06	2.72		

No. of core section	Depth m	Rock type	V_p, km s^{-1}	V_s, km s^{-2}	V_p, km s^{-2}	V_s, km^{-2}
33	2310	paragneiss	5.46		5.47 ± 0.05	3.10 ± 0.02
			5.55			
			5.40	3.11		
			5.46	3.08		
34	2412.5	pegmatite	5.23	3.01	5.25 ± 0.03	3.05 ± 0.05
			5.29	3.01		
			5.23	3.12		
			5.23			
35	2519.8	paragneiss with significant mica content (Braunglimmergneiss)	4.52	3.19	4.34 ± 0.18	3.32 ± 0.12
			4.17	3.44		
36	2798.0	light and heavily altered gneiss	4.31	2.2	4.36 ± 0.05	2.28 ± 0.08
36	2801.4	dto.	4.4	2.35		
37	2930	paragneiss, fresh	4.6	2.6	4.73 ± 0.37	2.63 ± 0.17
			4.5	2.58		
37	2935.5	dto.	4.7	2.77		
37	2936.6	dto.	4.2	2.4		
37	2937.4	dto.	5.1	2.5		
37	2938.6	dto.	5.32 (1)	2.9		
38	3053.2	mica syenite	5.44	3.05	5.48 ± 0.04	2.98 ± 0.08
38	3053.6	dto.	5.51 (1)	2.9		
39	3125.6	granitic zone	5.42	3.27		
39	3126.5	dto.	5.6	3.32	5.55 ± 0.09	3.28 ± 0.03
39	3127.2	dto.	5.63	3.26		
40	3205.4	dark mica syenite	5.75	3.39	5.43 ± 0.28	3.18 ± 0.21
40	3206.5	dto.	5.07	2.8		
40	3209	dto., enclosed paragneiss relict	5.77	2.30		
40	3210.2	dto.	5.28	3.14		
			5.28	3.29		
41	3257.6	mica syenite with dark Schlieren	5.7	3.1	5.25 ± 0.45	2.93 ± 0.17
			4.8 (1)			
42	3291.5	dark syenite with blasts	5.49	3.18	5.28 ± 0.21	3.02 ± 0.16
42	3293	dto.	5.07	2.86		
43	3297	dto.	5.7	3.1	5.67 ± 0.04	3.25 ± 0.15
43	3297	dto.	5.63 (1)	3.4		
44	3299.2	dto.	5.63	3.13	5.63	3.13
44	3299.2	dto.	5.63	3.13	5.63	3.13
45	3303	dto.	5.63	3.2	5.42 ± 0.21	3.2
45	3306	dto.	5.21	3.2		
46	3321.7	mica syenite	5.44	3.2	5.44	3.2
47	3325.1	dto.	5.42	3.1	5.42	3.1
48	3326.5	dto.	5.05	2.66	5.05	2.66
48	3329.8	dto.	4.95	2.86	4.95	2.86
49	3331.3	dto.	5.2	3.08	5.2	3.08

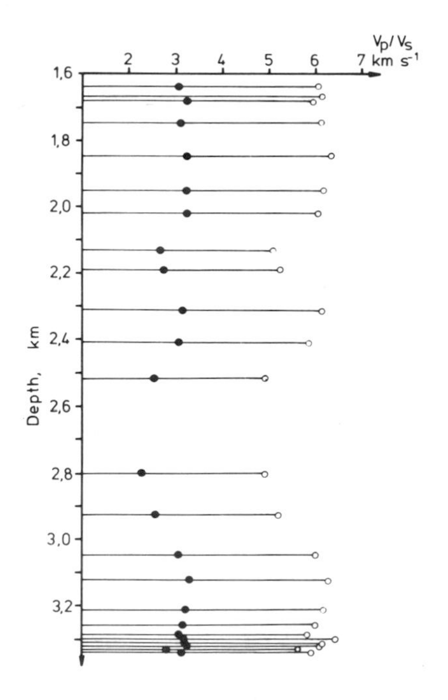

Fig. 3. Velocity-depth relation as derived from ultrasonic measurements on various core sections. ○ P velocity, ● S velocity.

the velocity structure obtained from sonic log measurements (WOHLENBERG this volume) although core velocities are generally 10 per cent higher. Seismic refraction investigations revealed a uniform P-wave velocity of 5.66 ± 0.02 km s^{-1} for the crystalline basement (JENTSCH et al., this volume).

The ultrasonic velocity data yield an average Young's modulus of 64 ± 5 Mbar and a Poisson's ratio of 0.255. Similar values are obtained from static compression tests (Fig. 4).

Stress-Strain Behaviour, Triaxial Compressive Strength and Friction

Minicores for the triaxial compression tests were drilled from core sections 37 (paragneiss at 2938 m) and 41 (syenite at 3257 m depth). The minicores were 3 cm in diameter and 6 cm long. The specimens were loaded at different constant confining pressures, S_3, in a 3-kbar triaxial pressure vessel. The axial stress, S_1, was induced by a servo-controlled, fast-acting, electrohydraulic loading system which permitted a control of the complete deformation process from purely elastic deformation to the development of the final macroscopic shear fracture..

The compressional deformation behaviour of the rock can be derived from the stress-strain curves given in Fig. 4. The curves demonstrate that the onset of inelastic deformation, which probably constitutes the beginning of microfracturing, occurs at about 75 % of the peak strength. In the post-peak region the paragneiss at higher confining pressure ($S_3 > 300$ bar) shows a rather ductile fracture behaviour which progressively leads to the development of a macroscopic shear fracture. In contrast. the syenite is characterized by extreme brittle fracture development even at confining pressures as high as 1600 bar. The brittleness leads to unstable shear fracture propagation which cannot be controlled by the loading system. During this phase the load bearing capability or strength of the rock is reduced rapidly (dashed portion of stress-strain curves in Fig. 4).

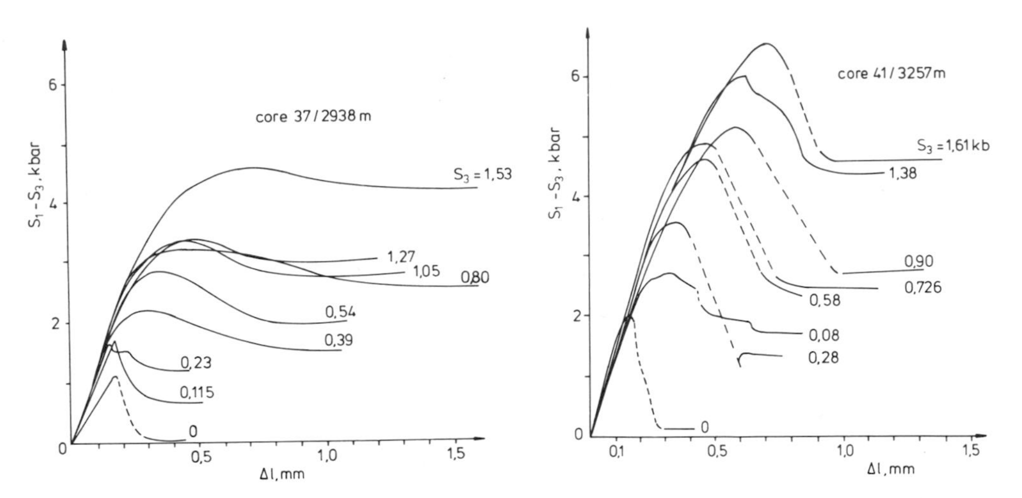

Fig. 4. Stress-strain curves for 2 rock types from the Urach borehole at various confining pressures S_3. $S_1 - S_3$ differential stress.

Table 2. Compressional strength $S_1 - S_3$ and residual strength $S_{1R} - S_3$ as a function of confining pressure S_3.

Core section	Core no.	S_3, bar	S_1, bar	$S_1 - S_3$ mbar	$S_{1R} - S_3$, bar
37	UG 8	1	1200	1200	670
	UG 9	120	1890	1770	680
	UG 1	230	1990	1760	1250
	UG 7	390	2660	2270	1520
	UG 2	540	3400	2860	1990
	UG 6	800	4040	3240	2470
	UG 3	1050	4470	3420	2890
	UG 4	1270	4650	3390	3040
	UG 5	1530	5770	4240	3790
41	U 11	1	2000	2000	210
	U 10	80	2910	2830	1780
	U 4	280	3780	3500	1240
	U 3	580	5080	4500	–
	U 9	730	5640	4910	2360
	U 8	900	6150	5250	2820
	U 5	940	6390	5450	3000
	U 6	1380	7430	6050	4360

The final deformation process in both rocks is then dominated by stable frictional sliding along the induced shear fracture surfaces at about constant stress.

Compressive strength (peak-strength in Fig. 4) and residual strength data during frictional sliding are given in Tab. 2 and plotted as a function of confining pressure S_3 in Fig. 5. The increase of both parameters with S_3 is not linear for the pressure range used. Finally, the coefficient and angle of internal friction and the coefficient and angle of sliding friction, as well as the values of "cohesion", are shown in Fig. 6 and are given in Tab. 3.

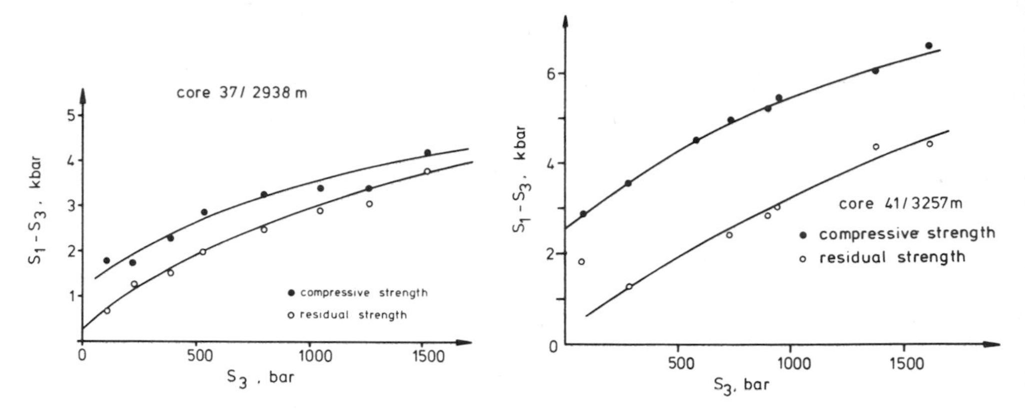

Fig. 5. Compressive strength $S_1 - S_3$ and residual strength $S_{1R} - S_3$ as a function of confining pressure S_3.

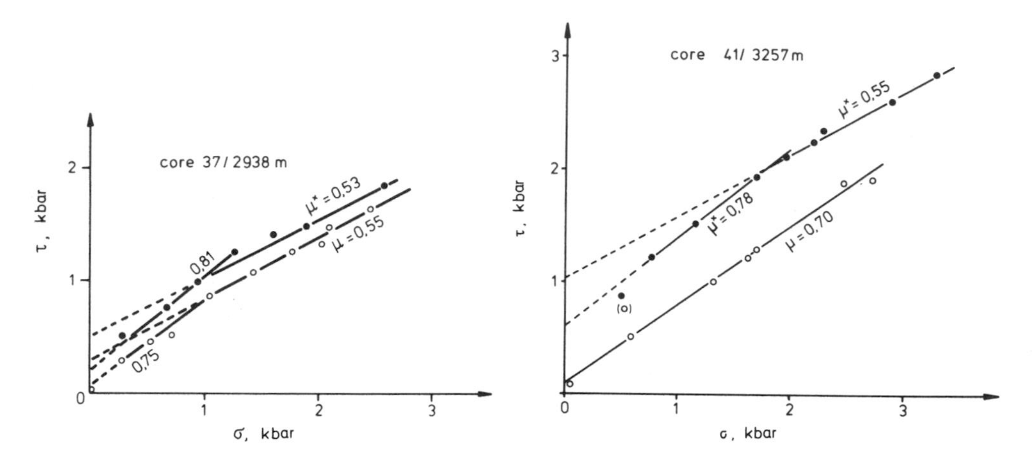

Fig. 6. Friction properties as a function of normal stress δ, τ = shear strength.

Table 3. Friction parameters.

Core section	Normal stress o, bar	Internal friction μ^*	ϕ^* in °	τ_o^*, bar	Sliding friction μ	ϕ in °	τ_o bar
37	700–1300	0.81	39	200	–	–	–
	300–1000	–	–	–	0.75	37	60
	1000–2600	0.53	28	480	0.55	29	280
41	800–1800	0.78	38	610	–	–	–
	1700–3300	0.55	29	1040	–	–	–
	500–3500	–	–	–	0.70	35	100

Hydraulic Fracturing Tensile Strength

Hydraulic fracturing tests were conducted in the laboratory at different values of confining pressure on paragneis minicores of core section 31 (2125 m depth). The crylindrical specimens contained an axial borehole of 2.5 mm diameter for fluid injection. The injection pressure rate was constant at 2 bar per second. The injection fluid was light oil. The confining pressure ranged from 150 to 600 bar.

The breakdown pressure, p_c, at which unstable tensile fracture propagation occurred is plotted versus the applied confining pressure S_3 in Fig. 7. The results may be described by the frac-formula $p_c = p_o + kS_3$. The unconfined hydraulic fracturing tensile strength p_o is 180 bar; the frac gradient k is 1.11. It should be mentioned that the theoretical frac gradient for an impermeable uncracked material should be k = 2 for the stress

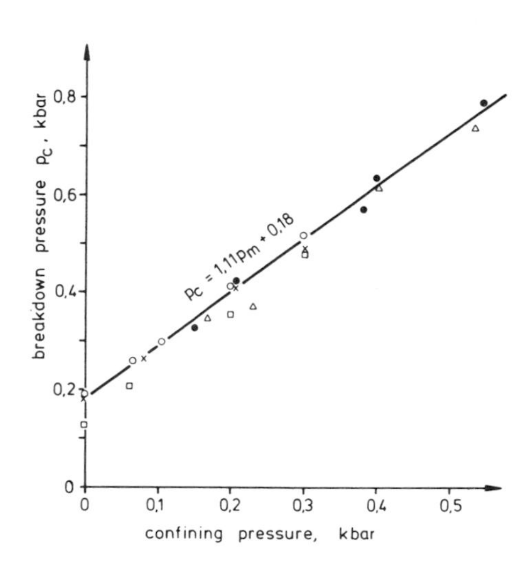

Fig. 7. Laboratory hydraulic fracturing breakdown pressure p_c as a function of confining pressure S_3 for Urach gneiss and other crystalline rocks for comparison. (Symbols: ● Urach paragneiss K 31/2, ○ Ruhr-sandstone, ◇ granite Fichtelgeb., □ marble, △ granite Falkenberg)

Table 4. Frac-results from laboratory measurements.

Rock	Density	p_o, bar	$k = \frac{dp_c}{dp_z}$	$\frac{dp_c}{d_z}$, bar m^{-1}
marble (Carrata)	2.70	128	1.16	0.307
granite (Epprechtst. Fichtelgeb.)	2.65	180	1.11	0.289
granite (Falkenbg.)	2.68	133	1.17	0.302
Urach 31 Paragneiss z = 2125 m	2.70	180	1.11	0.294

p_o breakdown pressure at $p_m = 0$
k Frac-gradient as a function of confining pressure p_m
$\frac{dp_c}{dp_z}$ Frac-gradient as a function of depth z

situation in these experiments. The discrepancy can be explained only by the assumption of preexisting cracks with an effective length of 5 mm (RUMMEL & WINTER 1980). Crack length of this size may exist in the paragneiss rock tested, if grain boundaries are considered as planes of weakness. For comparison, Tab. 4 presents values of k and p_o for some other crystalline rocks.

If the laboratory hydraulic fracturing results are applied to frac operations at great depth, the breakdown pressure may be estimated by means of the relation $p_c = 180 + 0.294 \cdot z$, where p_c is obtained in bar and z is the depth of the frac in metres. For the derivation of this formula the existence of a lithostatic stress field ($S_v = S_H = S_h$, S_v vertical principal stress, $S_{H,h}$ horizontal principal stresses) and a rock density of 2.7 gcm^{-3} were assumed. Thus, a breakdown pressure of 1150 bar (well head pressure 820 bar) is estimated for in-situ frac tests at a depth of 3300 m, if the fracture size distribution in-situ is similar to that in core samples. In the extreme case, that a joint intersects the frac interval at depth, the breakdown pressure required to re-open the joint will be 880 bar (550 bar well head pressure). If a non-lithostatic stress field is considered, the breakdown pressures may be considerably lower.

Fracture Toughness and Specific Fracture Surface Energy

Estimations on hydraulic fracture propagation in rocks may also be derived from considerations of fracture mechanics, if the intrinsic fracture-mechanical parameters are known. In this case the breakdown pressure in a vertical borehole is given by the general equation.

$$p_c = 1/h^* (K_{Ic}/R - S_H f^* - S_h g^*).$$

f*, h* and g* are numerical functions depending on the borehole radius and the length of existing microcracks and fractures. K_{Ic} is called the fracture toughness or critical stress intensity factor for the tensile fracture mode. K_{Ic} may be converted to specific surface energy by the relation $\gamma = [(1 - \nu^2) / 2 E] K_{Ic}^2$ (E = Young's modulus, ν = Poisson's ratio).

Recently, values of K_{Ic} were determined for a number of rocks (RUMMEL & WINTER 1980), with the use of the three point loading test on prismatic rock samples (WINTER

Table 5. Fracture toughness K_{Ic} and specific surface energy.

Core no.	K_{Ic}, $MNm^{-3/2}$	γ, Jm^{-2}
limestone	0.40 ± 1.00	–
Solnhofen	0.78 ± 0.06	5.1 ± 0.4
Treuchtlingen	0.99 ± 0.12	8.0 ± 2.1
sandstone		
Ruhrsandstone	1.43 ± 0.10	28 ± 4
granite		
Epprechtstein, Fichtelgeb.	1.60 ± 0.12	29.2 ± 4.5
Westerly, USA	0.90 ± 0.08	139
Carnmenellis, Cornwall	2.158	121
gneiss from Urach		
32–1	1.74	21
37–1	1.62	17
37–2	1.26	15
41–1	1.7	22

1979). K_{Ic}- and γ-values for the Urach borehole rock were determined on samples from core sections 32, 37 and 41. The results are given in Tab. 5 together with data from other crystalline rocks for comparison. K_{Ic} for the Urach rock has a value of 1.6 to 1.7 $MNm^{-3/2}$. The fracture surface energy is about 20 Jm^{-2}.

With the experimental value of the fracture toughness the critical fluid pressure p_c for propagation of existing fractures may be calculated if the principal stresses and the length of the fractures are known. To demonstrate the effect of both parameters, a calculation for the Urach borehole was conducted. The result is given in Fig. 8. It was assumed that the fluid pressure is applied at a depth of 3300 m, that the major principal horizontal stress S_H is equal to the overburden pressure $S_v = g \cdot g \cdot z$, and that the fracture plane is vertical. The borehole radius was taken as R = 6.25 cm. In addition, it must be mentioned that the fluid pressure on the fracture plane was assumed to be constant and equal to the pressure in the frac interval in the borehole.

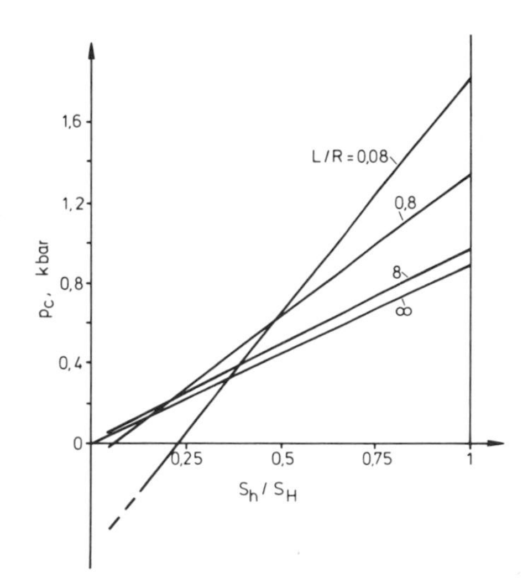

Fig. 8. Crack extension pressure p_c as a function of stress ratio S_h/S_H, S_v = 890 bar, K_{Ic} = 1.65 $MNm^{-3/2}$.

For the case of a lithostatic stress field the fluid pressure for crack extension is naturally always greater than the overburden stress of about 890 bar. For "intact rock" with an intrinsic crack length of L = 5 mm (L/R = 0.08), as used for the laboratory hydraulic fracturing tests, the critical in-situ pressure for crack extension is greater than 1.8 kbar. This is in agreement with the results from hydraulic fracturing laboratory tests if one considers the difference in the ratio between crack length and borehole radius for the two cases. For extremely small values of S_h/S_H the calculation also shows that small fractures are unstable and propagate even without fluid pressurization.

Although the exact stress field at depth in the Urach area is unknown, it can be assumed from tectonic considerations that the stress field is nearly to lithostatic. If we further agree that the in-situ rock mass in frac-intervals of several metres cannot be described as intact rock such as laboratory specimens but always contains fractures of considerable size, a breakdown pressure of about 900 bar should be expected. Since the in-situ experiments in Urach revealed much lower peak pressures (SCHÄDEL & DIETRICH, this volume), it must be concluded that existing fractures were opened by pressurization but apparently did not propagate.

Acknowledgement

The present work was carried out under the BMFT project ET 4131 A in collaboration with the project management at NLfB, Hannover.

References

Rummel, F. & R.-B. Winter (1980): Laboruntersuchungen zum Bruchvorgang in Gesteinen beim Hydraulic Fracturing Experiment. – Ber. f. BMFT-DGMK Projekt ET 3023 Al, Inst. f. Geophys., Ruhr-Universität Bochum.

Winter, R.-B. (1979): Bestimmung der Bruchzähigkeit und der spez. Oberflächenenergie von Gesteinen im Dreipunktbiegeversuch. – Dipl. Arbeit, Inst. f. Geophys., Ruhr-Univ. Bochum.

The Urach Geothermal Project, p. 147–156;
Schweizerbart'sche Verlagsbuchhandlung, Stuttgart, 1982

II. THE FIELD MEASUREMENTS

The Geology of the Heat Anomaly of Urach

K. SCHÄDEL

with 9 figures

Abstract: In the Urach area, Mesozoic sedimentary rocks (800–1200 m thick) overlie a crystalline basement. The strata dip at an angle of 1–2° towards the southeast. Palaeozoic sedimentary rocks (600 m thick) between the Mesozoic and the crystalline basement are limited to a trough crossing the centre of the anomaly. – During the Tertiary, the crystalline basement and the overlying rock were pierced by about 250 volcanic pipes. – The crystalline rocks are jointed; the fissures, even in greater depths, are filled with salt water; with increasing depth, the salt water presumably passes into hydrothermal solutions.

1. Introduction

This paper presents a geological survey of the heat anomaly of Urach, whereas stratigraphical or palaeontological problems will not be discussed. Furthermore, a statement about the crystalline basement below the heat anomaly and about the crustal structure will be ventured. The objective is to stimulate interpretation of the results of our measurements from the geological point of view rather than to convey facts alone.

According to present knowledge, the heat anomaly of Urach involves an area of oval shape which extends over a distance of approximately 60 km in length from WSW to WNW. Its centre near Urach covers an area of 300 km^2 (see Fig. 1).

This anomaly is situated in Southwest Germany's cuesta landscape, which is composed of Mesozoic strata bedded upon the crystalline basement. The uplifts of the Black Forest and the Vosges Mountains together with the Central Rhine Rift Valley (100 km further west) and the sinking of the Molasse basin in front of the Alps (100–150 km to the south) are the determining tectonic factors of the cuesta landscape and thus of the heat anomaly as well; it is not possible, however, to relate the anomaly directly to both factors.

The cuesta landscape of southwestern Germany was formed by the slight tilt of sedimentary strata (the strata dip in general at an angle of 1–2° towards the southwest) during the arching of the Rhine shield. During millions of years of geological history, erosion and denudation worked upon the hard and soft layers to fashion a great, regularly formed relief, which is seen today on the earth's surface and which covers the anomaly. The following rule illustrates this process in simplified terms: Hard rocks are cleared to form steps, whereas soft layers are eroded to form the surface area between the steps. The more resistant the layer is, the higher it towers today. Of all the hard strata which form steps, the Upper Jurassic (Malm) is the highest. Its cuesta, jagged as a result of erosion, crosses over the anomaly nearly parallel to its longitudinal axis. It reaches an elevation of 700–800 m above sea level. The Neckar Valley, the contour base line of the relief, at an elevation of about 270–300 m lies around 20–25 km to the north. The Neckar is receiving stream for all of the water-bearing, sedimentary strata. The land on both sides of the Neckar lies at an elevation of 380–450 m and forms a hilly relief, which is overlooked by the cuesta of the Upper Jurassic some 300 m higher. The plateau of the Upper Jurassic is the Swabian Alb, which is karstified. The aquiferous valleys on the northern edge are 200–300 m in height and deeply incised. The Urach 3 research borehole as well as both thermal water wells of Urach are situated in such a deeply incised valley, in the valley of the Erms. Thick neogenic sediments as well as natural hot springs are absent from the entire area.

Fig. 1. The geological situation around Urach and a profile crossing this area.

2. Mesozoic Sedimentary Rocks

2.1 Jurassic Sedimentary Rocks

Altogether, Jurassic sediment is 800 m thick; the entire thickness, however, is only roughly preserved in the southern part of the anomaly. On the northern edge it is extensively abraded, and on the eastern and southern edges the lower parts are reduced in their development.

The classification of Jurassic sediment is as follows:

Malm	= 400 m limestone and marlstone
Dogger	= 300 m mudstone with limestone and sandstone beds
Lias	= 100 m mudstone and limestone

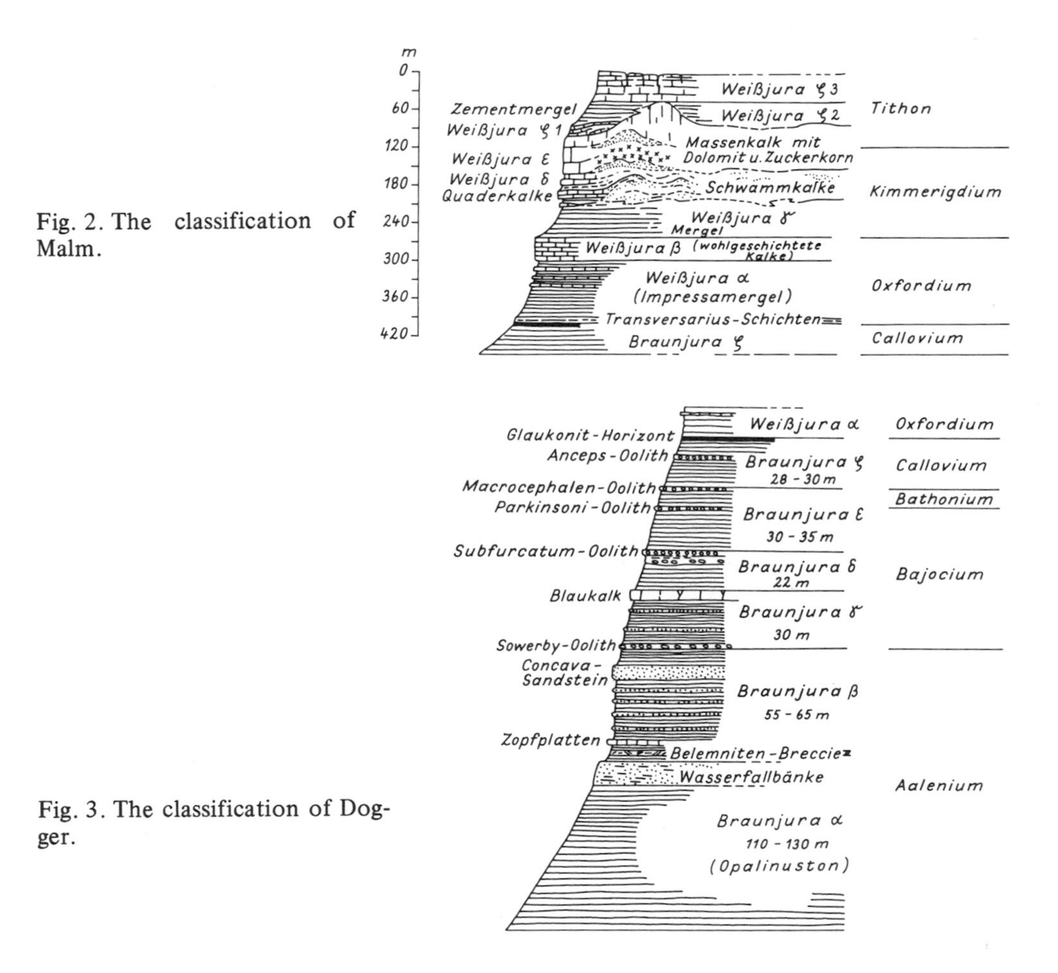

Fig. 2. The classification of Malm.

Fig. 3. The classification of Dogger.

The subclassification of the **Malm** is illustrated in Fig. 2. The 100–150 m thick compact limestone sequence (Kimmeridgian) above the middle is important. It forms the surface of the Swabian Alb and thus of large portions of the anomaly. The compact limestone contains fissures and caves (volume share 0.5–1.5 %), which are filled with water the base and permits circulation of air above, though they are sometimes sealed with loam.

Stratified limestone lies above and below the compact limestone with thick or thin marlstone intercalations.

The **Dogger** consists to 80–90 % of blue-grey mudstone, which turns brown upon weathering, of iron-bearing calcareous sandstone and oolithic limestone beds. The mudstone is cleaved by compaction in the plane of bedding, that is, after sedimentation it contracted to 1/4–1/10 of its original thickness as a result of dehydration. The degree of compaction is less in the case of limestone and sandstone. The mudstone is slightly permeable to water; the lime- and sandstone conduct some water in clefts (see Fig. 3).

In the **Lias**, blue-grey mudstone also predominates, although the share of limestone and calcareous marlstone (including bituminous oil shale) is higher than in the Dogger (see Fig. 4). The Lias alpha and the Rhaetic sandstone contain a utilizable thermal water bed.

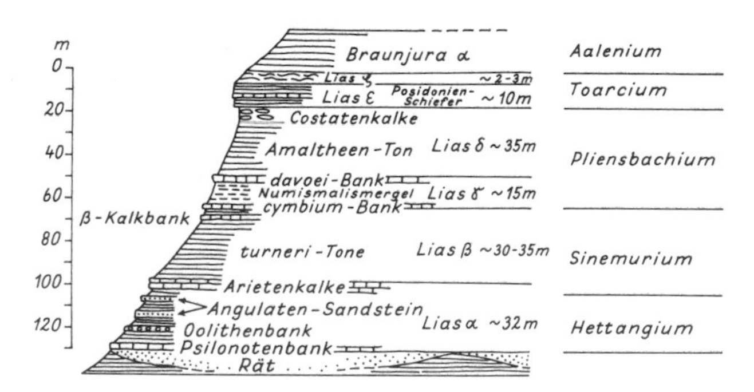

Fig. 4. The classification of Lias.

2.2 Triassic Sedimentary Rocks

Triassic sediment consists of:

Keuper = 250 m sandstone, mudstone and gypsum
Muschelkalk = 160 m marine dolomite rock, limestone, marlstone and sandstone, gypsum and salt
Buntsandstein = 80–120 m terrestrial sandstone

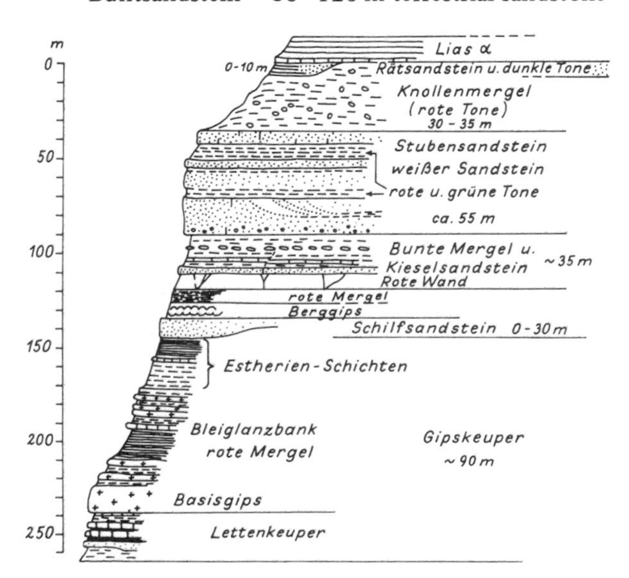

Fig. 5. The classification of Keuper.

Keuper rocks are red, green- to olive-grey in colour. The sequence is shown in Fig. 5. Water circulation is known to occur in the deeper underground layers from Stubensandstein (highly mineralized thermal water).

The upper rock sequence of the **Muschelkalk**, 60–80 m thick, the "Hauptmuschelkalk" or "Obere Muschelkalk", is significant. Its dolomite rock and limestone are brittle and acoustically impedant with transient times for longitudinal waves faster than 5000 m/s. It yields the best reflection of the sedimentary nappe. Moreover, the Hauptmuschelkalk contains the most important thermal water aquifer of the anomaly in a fissure zone in Trigonodus-Dolomite. The arenaceous facies at the basis of the Muschelkalk also conducts thermal water, and constitutes a common water-bearing layer with that of the **Buntsandstein**. The stratigraphic sequence is illustrated in Fig. 6.

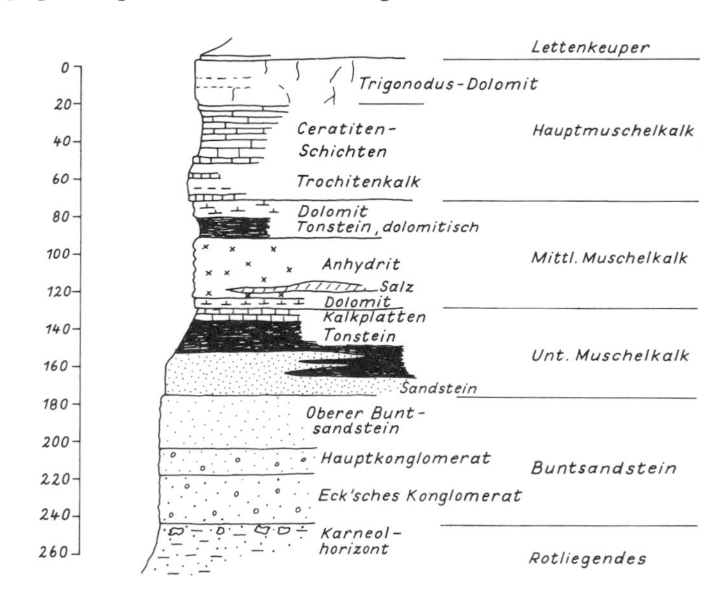

Fig. 6. The classification of Buntsandstein.

3. Palaeozoic Sedimentary Rocks

The anomaly of Neuffen-Urach, extending from WSW to ENE, coincides approximately with the shape of a basin embedded in the crystalline floor, which is several hundred metres thick and filled with Permian (Rotliegend) and Upper Carboniferous sedimentary rocks. It is exposed at the earth's surface near Schramberg in the Black Forest and is therefore called "Schramberg Trough". Seismic refraction soundings conducted during field measurements revealed its contour more accurately. According to the results, its greatest depth lies south of Urach. The thickness of the sediment there amounted to about 800 m whereas 700 m were penetrated at the Urach 3 borehole.

Buntsandstein (Lower Triassic) and Rotliegend (Lower Permian) lie one upon the other without discordance; this is also the case within the Lower Permian and Upper Carboniferous sediments. As far it was discernible in the Urach 3 borehole as well as in other boreholes in the Lower Permian trough, the sediments are bedded horizontally one over the other. Hence, it follows that the area of the heat anomaly shows a relatively uniform sedimentation since the Upper Carboniferous period.

3.1 Permian Sedimentary Rocks

To the north and to the south, that is, on both sides of the trough, Permian sediment is missing or is represented by arcoses, which are only a few tens of metres thick. In Schramberg the edge of the trough is visible and appears as a nearly vertical upslide. Therefore, one may think of the edges of the trough as upslides which were sedimentated syngenetically. Nevertheless, it is possible that the edges of the trough rise beyond the upslide in its higher.

The sedimentary deposits of the trough show a regular sequence which, amazingly enough, agrees in detail with other deposits of the Schramberg Trough as well as with those of the entire southwest of Germany. These are as follows:

red fanglomerates and mudstone (about 300 m)
dark and violet-red, coloured mudstone (about 100 m)
green (igneoaqueous) mudstone with limestone, in which dioritic arcoses are embedded (about 200 m)

Thus, Permian sediment consists chiefly of mudstone; even the matrix of the fanglomerates is clayey.

3.2 Upper Carboniferous (?) Sedimentary Rocks

In the underlying Upper Carboniferous (in question) sediment, white, then grey sandstone and greywackes predominate; they bear coal streaks near the base.

3.3 Permian Volcanism

This section cannot be concluded without considering Permian volcanism. The depression of longitudinal troughs falls in the asturic phase of Variscan orogenesis. It was accompanied by a very acidic volcanism, which produced quartz porphyrite nappes but certainly intrusive domes with ignimbrites as well. Quartz porphyrite dikes can be considered to have been the roots of this volcanism. They presumably extend down to granitic-granodioritic subvolcanoes, although their existence has not yet become evident. The conveyed products, quartz porphyritic slides and pyroclastic ignimbrites in mudstone, were deposited as sedimentary rocks in the Permian trough.

4. The Crystalline Basement

The crystalline basement in the area of the heat anomaly belongs to the central Moldanubic crystalline belt of the Variscan Mountains. Metablastite (orthogneis), Metatexite (paragneiss) and Diatexite (glimmersyenit, quartz monzonite) were found in the Urach 3 research borehole. As can be deduced from their structure, these rocks are for the most part strongly slanted, or show folds or crumplings with greatly varying fold axes. A fundamental discordance exists between the Variscan crystalline basement and the Palaeozoic sedimentary rock bedded upon it.

Metablastite, Metatexite and Diatexite, which are classified by WIMMENAUER as the so-called monotonous series of the Moldanubicum, are not the only rocks of the crystalline basement. One portion of the volcanic channels expelled other crystalline rocks: garnetiferous cordierite gneiss, kornelite gneiss and dike granite. Fig. 7 illustrates uncovered basal rocks. The entire gneiss formation has a relatively high content of mica (biotite) and is therefore more basic and denser than granite. If granite were substituted for gneiss further to the north, as in the case of the northern part of the Black Forest, the differences in gravity which were found during the gravimetric survey would be explained.

Graphite is embedded in the gneiss, which also contains carbon dioxide and hydrocarbons (methane), presumably in the form of small bubbles in the minerals.

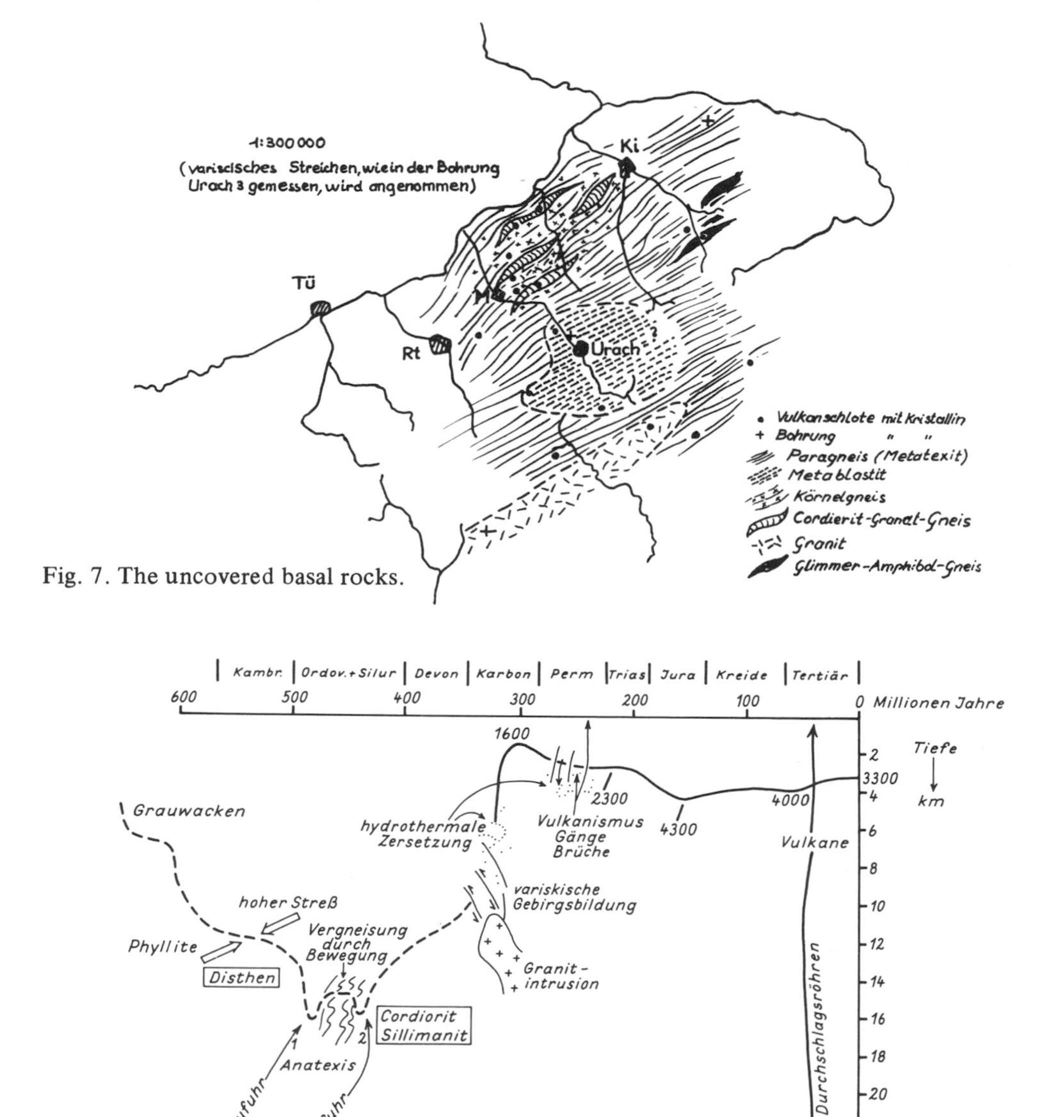

Fig. 7. The uncovered basal rocks.

Fig. 8. The geological history of the crystalline basement. The approximate depth of crystalline basal rocks in the Urach 3 borehole relative to "today = 330 m depth".

The crystalline basement is streaked with granite dikes, with aplite, with dikes and belts of a bright, brittle rock, which originated from the gneiss through diatexis. Its identification is difficult because it has been strongly alterated by hydrothermal water. It differs from the gneiss in its significantly higher gamma-ray activity.

Fig. 8 illustrates the genesis of the crystalline basement and its geological age. The idea that a primary high pressure metamorphosis at a slight depth and at relatively low temperatures has as sequel two thermal metamorphoses (anatexis 1 and 2), between which the gneissification were to be placed as a further kinematic metamorphosis, is interesting for geophysical and especially for geothermic consideration. Some information concerning different temperatures at pressures of 2–4 kilobar follows:

670–700 °C ≙ the first melting processes in bright rocks
780–850 °C ≙ a major part is melted

Fig. 8 demonstrates that in the course of geological history the crystalline basement was subjected to long processes of formation; according to chronological classification, these occured basically before the Variscan orogenesis. The Variscan belt itself has hardly left significant traces in the gneiss rocks, although large granitic plutons were pushed up in the central Moldanubic zone.

The crystalline basement is fractured. The joints are reactions of the rock masses to tectonic pressures, to unloading and cooling processes, and in deeper sections also to hydraulic force. Moreover, the rock is permeated with minuscule fissures in the individual crystals, as a result of cooling and unloading processes in the region of the grain boundary.

5. The Earth's Crust

Fig. 9 illustrates possible pressure and temperature conditions of the crustal rock in the area of the anomaly and their effect upon the mechanical properties of the crust. In this hypothetical diagramme, 4 zones are to be distinguished:

From 1.6 to 7 km of depth, the rock is slightly porous because of the cracking of crystals by unloading processes near the grain boundary the rock thereby show a minimal hydraulic conductivity ($k_f \sim 10^{-12}$ m/s). The rock is jointed; the clefts are open but often have incrustations of calcite and pyrite, upon which one finds salt water. Its mobility decreases with increasing depth despite a decreasing viscosity (reason: the higher pressure closes the gaps). The pressure of the rock is absorbed to a less extent because of the uplift geodynamics. Thermal convection currents of the salt water in the fissures have not been proven. Fundamentally, one has to expect only limited movements. Since the thermal water levels carry carbonate and sulphate in aqueous solution, a direct ascent of salty granite water up to these levels is not possible.

Between 7 and 16 km of depth, joint water loses its free mobility. The fissure frequency probably decreases; furthermore, the porosity of the rock probably disappears completely at 16 km depth. In parallel with this, the electrical resistance and the density increase, and the velocity of seismic waves attains a maximal value for granite and gneiss rocks. The critical temperature of water is reached at a depth of 12–13 km, beyond which the water as a fluid agent contains a great deal of dissolved silicic acid. One must assume that under static crustal conditions, the pressure of the fluid material in the existing fissures approaches that of the rock. The joints are separated by quartz fillings.

Below a depth of 15 km, the lithic chemistry seems to change. The moderately acid granite-gneiss rock is increasingly replaced by a more basic type of rock with dioritic and restitic character. The change is presumably not very abrupt, since acidic crystalline

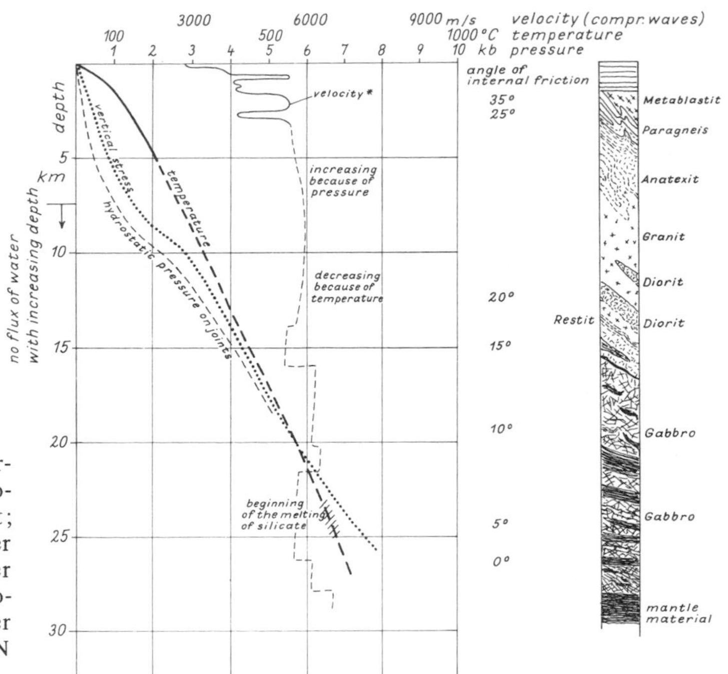

Fig. 9. Pressure, temperature and seismic velocity in the earth's crust; velocity of the upper part of the profile after WOHLENBERG (this volume) and in the lower part after BARTELSEN et al. (this volume).

rocks are still solid down to 23 km of depth. Replacing the fluid water and its silicate solutions which fill the fissures above a depth of 16 km, fused rock (pegmatic phase), abundant in water and fluxard, exudes into bands and joints (?) in spaces with reduced pressure. The rock begins to react with ever less brittleness, yielding rather more elastically to slow and constant pressure. The pegmatic phase in dioritic or restitic rock tends to move upwards, as far as it is at all capable of movement in the present geological state of slight lifting since the Tertiary.

Below about 23 km, acidic rocks, relatively "abundant in water", are no longer stable in their entire mineral composition in the solid state. A part is molten; hence they must exist as crystal mush in intergranular fusion. On the basis of the geological development, it must be assumed that acidic "water-abundant" rocks have already terminated their melting process for a long time. In the region of the lowest crustal layer, the dioritic or restitic plutonic rocks are replaced by gabbros, between which mantle material (?) is bedded; intruded in flat paths of motion into the Palaeozoic once the granite had already begun to melt out of the lower crust. Today, this crustal region is solidified; the rocks are crystalline and have a temperature which lies 200–300 °C below their melting point.

The knowledge of the lamellar structure of the lower crust is due to the results of the combined seismic refraction and reflection technique (BARTELSEN et al. this volume). From the geological point of view, it must be interpreted as a moving structure. The lower crust must have been in such a state that the internal friction was suspended for slow flow processes along the paths of motion within it. Since 12–15 km of rock was abraded during the Palaeozoic, the lower crust lay at 30–40 km of depth during that era; at this depth, a temperature of about 1100 °C can be assumed; the basic rock material was also molten.

References

Branco, W. (1894): Schwabens 125 Vulkan-Embryonen und deren tufferfüllte Ausbruchsröhren. – Jh. Ver. Vaterl. Naturkunde. Württ., **50/51**.
Büsch & Otto (1980): Endogenetic inclusions in Granites of the Black Forest, Germany. Neues Jb. f. Min. Mh.
Carlé, W. (1955): Bau und Entwicklung der Südwestdeutschen Großscholle – Beitr. geol. Jb., **16**, Hannover.
Cloos, H. (1941): Bau und Tätigkeit von Tuffschloten (Untersuchungen an dem Schwäbischen Vulkan). – Geol. Rdsch., **32**.
Geyer & Gwinner (1964): Einführung in die Geologie von Baden-Württemberg. – Stuttgart.
Gwinner, M. (1959): Die Geologie des Blattes Urach (7522) 1 : 25 000 (Schwäbische Alb). – Arb. Geol-Paläont. Inst. TH Stuttgart, N.F. **24**. Tektonik, Sedimentation und Vulkanismus im Gebiet der "Uracher Mulde". – Jber. Mitt. Oberrh. geol. Ver., N.F. **43**.
Mäussnest, O. (1969): Die Ergebnisse der magnetischen Bearbeitung des Schwäbischen Vulkans. – Jb. Mitt. Oberrh. Geol. Ver., **51**.
Rehbinder, G. (1978): Measurements of the Average Pore Velocity of Water Flowing Through a Rock Specimen. – Rock Mechanics, **11**.
Schwarz, H. (1905): Über die Auswürflinge von kristallinen Schiefern und Tiefengesteinen in den Vulkanembryonen der Schwäbischen Alb. – Jh. Ver. Vaterl. Naturkde. Württ., **61**.
Schreiner et al. (1977): Erläuterungen zur Geologischen Karte von Freiburg i. Brsg. und Umgebung 1 : 50 000, Stuttgart.
Stenger, R. (1982): Petrology and Geochemistry of the Basement Rocks of the Research Drilling Project Urach 3. – This volume.

The Urach Geothermal Project, p. 157–160;
Schweizerbart'sche Verlagsbuchhandlung, Stuttgart, 1982

The Volcanic Phenomena in the Urach Region

O. MÄUSSNEST

with 1 figure

Abstract: A great number of volcanic pipes are evident in the Urach region (centre 48°30'N, 9°27'E). These are filled with melilite-bearing olivine nephelinite tuffs, into which melilite-bearing olivine nephelinites occasionally intruded. They are of Tertiary age. This volcanic activity is unusual because its development of single vents terminated in the formation of a maar. Volcanic cones and real lava flows have never existed in the entire volcanic area. This particular type of volcanism is termed an "embryonic volcanism" throughout the literature.

Evidence of volcanic activity in the area was first given by RÖSLER in 1790. The majority of occurrences involve volcanic tuffs, into which magma has occasionally intruded, e.g. in the Jusi. The chemistry is ultrabasic with SiO_2-contents of about 33 %, and as a consequence nepheline has developed instead of feldspars; melilite is considered to replace pyroxenes in undersaturated magmas. These volcanites, derived from initial melts of the anatexis of the Upper Mantle, are with great probability from the region of the upper astenosphere at about 80 km of depth. An intermediate hearth has not yet been found by means of geophysical research work. If such a hearth exists, it must be assumed to be located at an extreme depth.

The steep escarpment of the Swabian Alb, formed by plateau recession, passes through the centre of the volcanic zone. The Swabian Alb highland and its escarpment is formed by limestones of Upper Jurassic age (Malm). It must be pointed out, however, that the denomination 'Swabian Alb' is the name of the landscape. The foreland of the plateau is built up by Lias and Dogger which consist mostly of clay-like material. On the plateau the volcanos have usually formed depressions with some maars preserved better than others. The volcanic material in the pipes includes mixtures of fragments of all strata intersected during the ascent to the surface (crystalline basement and sediments of Permian, Triassic and Jurassic age). Large amounts of fragments of Permian age are found in the volcanic tuffs near Metzingen (BRÄUHÄUSER 1918). Suspicions of a trough filled with Permian material have now been confirmed by the deep borehole Urach 3.

Although the borehole confirmed evidence of Carboniferous sediments in the subsurface, none have been detected to date in the tuffs. On the other hand fragments younger than Portlandium (Upper Malm) i.e. the strata found outcropping in the Swabian Alb highland adjacent to the maars, have not yet been found.

In naming the 'Swabian Volcano', CLOOS (1941) concluded that the large number of small and large pipes were not the result of an unsuccessful attempt to form independent volcanic cones. However, as part of the large and special Swabian Volcano, they ramified in the crust and thus were able only to form maars – possibly because of the sapping of energy in the ramification process.

In a recent explanation given by LORENZ (1979), the Swabian maars were formed by phreatic explosion of magma rising from the depths and coming into contact with ground water. According to an earlier explanation by BRANCO (1894, 1895) the pipes and maars were formed by sudden and strong explosion. This particular 'maar type' of volcanism was classified as an embryonic volcanism. However, CLOOS (1941), in his studies of the phenomenon, came to the conclusion that such a process does not apply to all of the pipes and that at least part of them were formed by fluidization. A good example is a quarry inside the pipe of Aichelberg near the motorway from Stuttgart to Ulm.

Pipes situated in the lowland contain fragments of Malm not found elsewhere. This is evidence that at the time of volcanic activity the plateau extended much farther NW than it does today. In the lowlands the pipes form hills, since the contents of the pipes are much more resistant to erosion than the mostly clayey strata of the lowlands (Lias and Dogger). Because of the recession of the precipice, fillings of the pipes can be studied down to 800 m below the old surface where one can find them as outcrops; the old surface is still preserved in the highland. One may even assume that the pipes situated in the present lowlands once ended at the surface of the now eroded plateau, forming depressions or maars.

The fillings of the pipes are not permeable and therefore the pipes cannot serve as waterways. Therefore lakes were formed in the maars; the last of these were drained by man, as shown by old local maps of the past century (e.g. Lachen 4 km E Urach or Molach 2,3 km ESE Neuffen). Fossil evidence in the sediments of the lakes give Tortonian and possibly Sarmatian ages for the pipes (SEEMANN 1926).

WEISKIRCHNER (1967) indicated a high degree of homogeneity in the chemical composition of the magma. On the basis of mineralogical considerations he came to the conclusion that the duration of the volcanic activity was of the order of 6 million years, thus comprising the period from Aquitanian to Helvetian or Tortonian. LIPPOLT et al. (1973) performed K–Ar age determinations of Urach volcanics and found ages between 30 Ma and 11 Ma ago (11 Ma for the dike near Grabenstetten and Hohenbohl pipe 1,5 km NE Owen; about 16 Ma for the pipes Sternberg 8 km WSW Münsingen, Bölle 1 km SW Owen, and Gelbenhalde [also called Hofwald] 8 km NW Urach; 20 Ma for the Dietenbühl pipe 5 km N Münsingen; 27 Ma for the Götzenbrühl pipe 1,5 km NE Owen near the Hohenbohl pipe; and about 30 Ma for the Buckleter pipe 3 km NW Urach). The ages are distributed at random throughout this volcanic province. Ages of about 11 Ma are in agreement with palaeontological evidence, while the authors have been unable to demonstrate whether the older ages of about 30 Ma ago are real or the result of excess argon.

The volcanic occurrences are concentrated in the region of Urach, and distributed over an area of 1600 km^2 (see Fig. 1). Today more than 350 occurrences (pipes and some dikes) compiled by MÄUSSNEST (1974) are known. Nearly half of them were found by magnetic field research. Most of the pipes show normal magnetization. The largest pipe is the Randecker Maar pipe with an area of approximately 1 km^2, situated 13 km NE of Urach. No causal relation has hitherto been found between the position of the pipes and the faults observed on the surface. On the other hand the great majority of the pipes is situated in the area of a syncline, the so-called 'Uracher trough'. A structural map has been published by GWINNER (1961). Three tectonic systems or their prolongations intersect here. The author's opinion is that the Uracher trough existed before volcanic activity began and possibly the Permian trough is a sign for the subsidence of the area at the end of the Palaeozoic period. Recently BINDER et al. (1978) drew attention to the late growth of 'lime precipitating sponges' in the Malm of the Urach region. Since these sponges prefer shallow water, the area is in contrast to other parts of the highland,

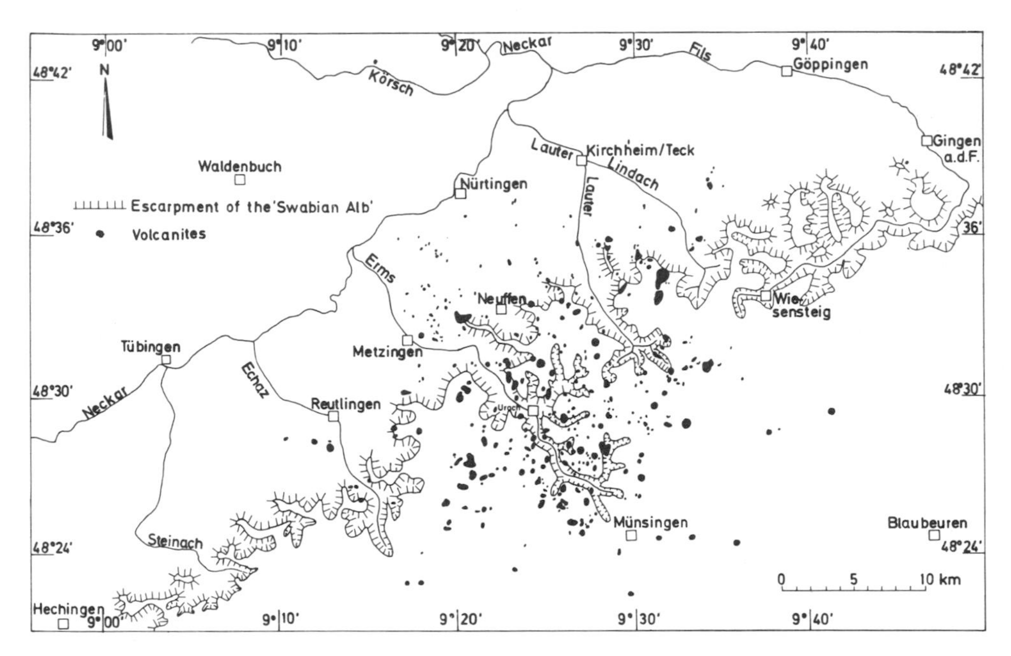

Fig. 1. General map of the Swabian Volcano.

which indicate relative subsidence at this time. Therefore, the distribution of the pipes – as far as one knows – has to be ascribed to tectonic patterns in the basement.

After deposition of Portlandian sediments the sea definitely withdrew from most parts of the Swabian Alb. One may assume that the beginning of the karstification took place at the time when Malm sediments emerged above sea level. This was about the beginning of the Cretaceous period (GWINNER 1974), since the Malm sediments consist mainly of calcareous material. There is absolute proof of the existence of karst phenomena since Eocene time i.e. Lower Tertiary, because from that date on there exist karst cleavages with fossils in the Swabian Alb, thus providing the necessary evidence (DEHM 1961 a, 1961 b). Karst cleavages, a phenomenon restricted to surface areas, show that not much overall lowering of the tableland surface can have taken place since it was found. Drainage occured underground. Karst cleavages did not contain fragments of sediments younger than those outcropping today in the highland. Therefore, very old features of early Tertiary time are still preserved in the landscape of the highland today. Of course, during the glacial epochs of the Quaternary a rejuvenescence of the surface drainage took place because karst activity was stopped by soil frost. A consequence was a partial considerable deepening of the dry valleys which then carried water again.

The preservation of a great number of maars e.g. the one at Zainingen (10,5 km E of Urach), can be attributed to the very early start of the karstification of Malm strata on the highland. There are, however, no remnants of tuff rings which may have surrounded the maars. A certain preservation of volcanic cones and lava flows has to be expected on the grounds of early karstification, if cones and lava flows ever existed. Lava flows are unlikely to have occured since lava intrusions are lacking in the pipes; i.e. most of the pipes are filled only with volcanic tuffs.

References

Binder, H., Glöckler, K. & Timmermann, G. (1978): Morphologie, Geologie, Speläologie der mittleren Schwäbischen Alb. – Kl. Schrift. Karst- u. Höhlenkde., **17**: 54–74, München.

Bräuhäuser, M. (1918): Die Herkunft der kristallinen Grundgebirgsgerölle in den Basalttuffen der Schwäbischen Alb. – Jh. Ver. vaterl. Naturkde. Württemberg, **74**: 212–274, Stuttgart.

Branco, W. (1894, 1895): Schwabens 125 Vulkanembryonen und deren tufferfüllte Ausbruchsröhren; das größte Maargebiet der Erde. – Jh. Ver. vaterl. Naturkde. Württemberg, **50**: 505–997 and **51**: 1–337, Stuttgart.

Cloos, H. (1941): Bau und Tätigkeit von Tuffschloten; Untersuchungen an dem Schwäbischen Vulkan. – Geol. Rdsch., **32**: 709–800, Stuttgart.

Dehm, R. (1961): Über neue tertiäre Spaltenfüllungen des süddeutschen Jura- und Muschelkalkgebietes. – Mitt. Bayer. Staatsslg. Pal. hist. Geol., **1**: 27–56, München.

– (1961): Spaltenfüllungen als Lagerstätten fossiler Wirbeltiere. – Mitt. Bayer. Staatsslg. Pal. hist. Geol., **1**: 57–72, München.

Gwinner, M.P. (1961): Tektonik, Sedimentation und Vulkanismus im Gebiet der Uracher Mulde. – Jber. u. Mitt. oberrh. geol. Ver., N.F., **43**: 25–40, Stuttgart.

– (1974): Erläuterungen zu Blatt 7622 Buttenhausen der Geologischen Karte von Baden-Württemberg 1 : 25 000. – Stuttgart (Landesvermessungsamt).

Lippolt, H.J., Todt, W. & Baranyi, I. (1973): K–Ar ages of basaltic rocks from the Urach volcanic district, SW Germany. – Fortschr. Miner., **50**, Beih. 3: 101–102, Stuttgart.

Lorenz, V. (1979): Phreatomagmatic Origin of the Olivine Melilitite Diatremes of the Swabian Alb, Germany. – Proc. 2 nd. int. Kimberlite Conf., **1**: 354–363, Washington (American Geophys. Union).

Mäussnest, O. (1974): Die Eruptionspunkte des Schwäbischen Vulkans. – Z. Deutsch. Geol. Ges., **125**: 23–54 and 277–352, Hannover.

Rösler, G.F. (1790): Beyträge zur Naturgeschichte des Herzogthums Wirtemberg nach der Ordnung und den Gegenden der dasselbe durchströmenden Flüsse. – Heft 2, Tübingen (Cotta).

Seemann, R. (1926): Geologische Untersuchungen in einigen Maaren der Albhochfläche. – Jh. Ver. vaterl. Naturkde. Württemberg, **82**: 81–110, Stuttgart.

Weiskirchner, W. (1967): Der Vulkanismus der Schwäbischen Alb. – Abh. geol. Landesamt Baden-Württemberg, **6**: 142–143, Freiburg i. Br. and Mém. Serv. Carte Géol. Als. Lorr., **26**, Strasbourg.

The Urach Geothermal Project, p. 161–163;
Schweizerbart'sche Verlagsbuchhandlung, Stuttgart, 1982

Remarks on the Depth of the Urach Magma Chamber

W. WIMMENAUER

Abstract: Petrographic observations and petrological considerations indicate the existence of an intermediate magma chamber in the upper crust as the source of the numerous diatremes. This model is in good accordance with the discovery of a low-velocity body, interpreted as a zone of relatively elevated temperatures, in the underground of the Urach volcanic area.

The problem concerning the depth of the magma chamber of the Urach volcanic area has been repeatedly discussed in the past (e.g. CARLÉ 1958). The present state of knowledge of the magmatic rocks and of the various ejectamenta originating from the depth allows further statements on the ascent of the magma and its possible sojourn before its definitive eruption (BERG & WEISKIRCHNER 1979, PAPENFUSS 1963, 1974, SICK 1970, WALENTA 1960, WEISKIRCHNER 1967, 1980). Certainly, the magma ultimately comes from the upper mantle; its unusually basic and undersaturated character indicates that it was formed under considerable pressure (> 25 kbar) and with a low degree of partial melting in the source region. In spite of the long distance of ascent to the surface, neither the massive magmatites nor the pipe breccias contain noteworthy amounts of mantle fragments. In contrast to many diatremes of other regions, the magma did not rise immediately from the mantle with high velocity. On the contrary, a comparatively quiet ascent has to be assumed for the massive igneous rocks (mostly dykes); thus, heavy mantle rock fragments, which eventually had been included, would not have been carried up to the level exposed at present. The pipe breccias and conglomerates, too, are free from rock fragments which could be derived from the mantle or from the lower crust. Charnockites, granulites, gabbros and similar rocks from these zones, which are quite frequent in other volcanic areas of Europe, are missing. Some rare and small peridotite nodules in the olivine melilitite of the Bölle near Owen and isolated grains of enstatite, chromian diopside and chromian spinel do not contradict the general conclusion that the eruptive processes failed to bring up many and sizable fragments from the mantle or from the lower crust. A considerable production of rocks from the underground started in the upper crust only – partly in the crystalline basement, partly in the sedimentary cover. This is evidenced by numerous ejected fragments of granite, granodiorite, gneisses and other, rarer rocks, as already described by SCHWARZ 1905; they occassionally show phenomena of partial fusion. Some of the pipe breccias, particularly those in a zone between Eningen and Nürtingen, are very rich in such constituents; other occurrences are remarkably devoid of larger ejectamenta, but single minerals from the basement are found in nearly all of the diatremes (personal communication from W. WEISKIRCHNER). Rounded pebbles of basement rocks may also originate from the Permian conglomerates (BRÄUHÄUSER 1918).

It follows from all these observations, that the explosive activity with upward transport started at a rather shallow, but not everywhere the same, level. Many details of the volcanic activity and its products have been described by CLOOS 1941, PAPENFUSS 1963, 1974, BERG & WEISKIRCHNER 1979 and others. It is suggested that the level in question may be identical with that of a magma chamber, in which the melt persisted for some time after its rise from the mantle. The concentration of gases during this sojourn finally led to the break-through to the surface and to the formation of the diatremes and maars presently exposed in different levels of erosion. The residence of the magma in the crust brought about heating of the surrounding rocks to some extent. Explosive degassing at many places would very soon effectuate the crystallization of the magma while still at a relatively high temperature. The small volume of massive volcanic material erupted and the low degree of differentiation observed with the volcanics and ejectamenta also favour the conclusion that the magma was not active for a long time. The major portion of the magmatic material belongs to the very basic olivine melilite and olivine nephelinite types. There is a certain contradiction between the assumption of a short epoch of volcanic activity on the one hand and the radiometric ages determined by LIPPOLT, TODT & BARANYI (1973), which vary between 11 and 30 Ma. The same authors, however, indicate the possibility that the higher ages may be the effect of excess argon. A Tortonian age is most likely according to palaeontological criteria.

The question of whether one single magma chamber underlies all the diatremes, or whether many smaller local magma chambers occur at different levels of the crust, cannot easily be answered by petrographic or petrological arguments. PAPENFUSS (1974) prefers the model of one single magma chamber at least for the main field of the Urach volcanic area. MÄUSSNEST (1956) emphasizes instead that no large integrate magnetic anomaly can be traced. He concludes from this, that the magma chamber should be situated at a depth of at least 4 km; even then, however, the absence of a strong magnetic effect presents a problem. Another model involving many small individual magma chambers a few kilometres in depth has been considered by CARLÉ (1958), but PAPENFUSS (1974) points to the specific difficulties associated with the idea of many separate subvolcanic intrusions, all of which were discharged to the surface in practically the same way. A model reconciling these contradictory conceptions has been proposed by MÄUSSNEST (1956). Many of the magnetic anomalies connected with single diatremes indicate a dykelike extension of the latter in downward direction and even the junction of several such bodies to larger units. This author thus assumes a rather complicated system of magma chambers, which are linked by mediating conduits in the manner of communicating tubes. It is a common feature of all these models, that they do not ascribe a large volume to the magma reservoir, although it must extend below a surface area of 30 by 40 km. Indeed, a compact body of heavy ultrabasic rock of several kilometres thickness should be perceived by gravimetry and magnetometry. The assumption that the nonappearance of the magma chamber be due to a temperature presently above the CURIE-point of titanomagnetite seems, for the time being, to be rather speculative. However, BARTELSEN & MEISSNER (1980) and KREY, SCHMOLL, MEISSNER & BARTELSEN (1980) found an important low-velocity body in the underground of the Urach volcanic area. They believe that the low seismic velocities are the consequence of higher temperatures in that region. The anomaly has a diameter of 50 km in the WSW-ENE direction; depending on the mode of calculation, its surface can be located at 5 to 10 km depth. This result corresponds quite well with the conclusions from the petrographic observations explained above. If the arguments against a large volume of the magma chamber are valid, then the low-velocity body should be, for the most part, hot country rock (e.g. granite, gneiss). The same assumption was also the basis of the geophysical calculations.

References

Bartelsen, H. & Meissner, R. (1980): Ergebnisse der reflexionsseismischen Messungen im Bereich der Wärmeanomalie Urach. – Statusreport der Projektleitung Energieforschung des Bundesministeriums für Forschung und Technologie 1980, p. 39–60.

Berg, W. & Weiskirchner, W. (1979): Petrographische Untersuchungen an vulkanischen Gesteinen des Jusi (Schwäbische Alb). – Jber. Mitt. oberrhein. geol. Ver., **61**: 337–346.

Bräuhäuser, M. (1918): Die Herkunft der kristallinen Grundgebirgsgerölle in den Basalttuffen der Schwäbischen Alb. – Jh. Ver. vaterländ. Naturkunde Württemberg, **74**: 212–274.

Carlé, W. (1958): Kohlensäure, Erdwärme und Herdlage im Uracher Vulkangebiet und seiner weiteren Umgebung. – Z. dtsch. geol. Ges., **110**: 71–101.

Cloos, H. (1941): Bau und Tätigkeit von Tuffschloten. Untersuchungen am Schwäbischen Vulkan. – Geol. Rdsch., **31**: 709–800.

Krey, T., Schmoll, J., Meissner, R. & Bartelsen, H. (1980): Variations of crustal seismic velocities derived from reflection seismic data in the Urach volcanic area. – Meeting Europ. geophys. Soc. Budapest 1980. Vortragsmanuskript.

Lippolt, H., Todt, W. & Baranyi, I. (1973): K-Ar-ages of basaltic rocks from the Urach volcanic district. – Fortschr. Miner., **50**, Beih. 3: 101–102.

Mäussnest, O. (1956): Erdmagnetische Untersuchungen im Kirchheim-Uracher Vulkangebiet. – Jber. Mitt. oberrhein. geol. Ver., **38**: 23–54.

Papenfuss, K.H. (1963): Das Schlotkonglomerat des Bürzlen bei Eningen u.d. Achalm (Schwäbische Alb). – Jh. geol. Landesamt Baden-Württemberg, **6**: 461–505.

– (1974): Mineralogisch-petrographische Untersuchungen an vulkanischen Tuffen im Uracher Vulkangebiet (Schwäbische Alb). – Jh. geol. Landesamt Baden-Württemberg, **16**: 13–34.

Schwarz, H. (1905): Über die Auswürflinge von kristallinen Schiefern und Tiefengesteinen in den Vulkanembryonen der Schwäbischen Alb. – Jh. Ver. vaterländ. Naturkunde Württemberg, **61**: 227–288.

Sick, U. (1970): Über Melilith-Nephelinite der Schwäbischen Alb. – Diss. Tübingen, 93 S.

Walenta, K. (1960): Die vulkanische Tuffbreccie von Scharnhausen bei Stuttgart. – Jber. Mitt. oberrhein. geol. Ver., **42**: 23–53.

Weiskirchner, W. (1967): Der Vulkanismus der Schwäbischen Alb. – Abh. geol. Landesamt Baden-Württemberg, **6**: 142–143.

– (1980): Der obermiozäne Vulkanismus in der mittleren Schwäbischen Alb. – Jber. Mitt. oberrhein. geol. Ver., **62**: 33–41.

The Urach Geothermal Project, p. 165–178;
Schweizerbart'sche Verlagsbuchhandlung, Stuttgart, 1982

Hydrogeological Aspects of the Geothermal Area of Urach

E. VILLINGER

with 6 figures and 1 table

Abstract: The six groundwater storeys which are developed in the area of the geothermal anomaly at the northern edge of the Middle Swabian Alb are described, together with their hydrogeological parameters and characteristics. Particular emphasis is placed on the Upper Muschelkalk, which is especially important for investigating the nature of the geothermal anomaly.
The temperature distribution at a depth of 500 m below the surface, at the top of the Upper Muschelkalk, and in the karst groundwater of the Malm, is presented.
The hydraulic potentials of the various groundwater storeys indicate that the ascent of hot water from a great depth is highly improbable, and therefore cannot serve as an explanation for the heat anomaly.
Considerations of the heat balance in the Urach region suggest the existence of a second heat source in the Upper Lias/Dogger. The extent of the thermal anomaly is strongly influenced by the flow of the hot groundwater in the northeastward direction in the Upper Muschelkalk.

1. Einleitung

Die Ergebnisse der regionalen hydrogeologischen Untersuchungen im Gebiet der Wärmeanomalie von Urach sind ausführlich bei E. VILLINGER (1982 b) dargestellt. Die vorliegende Arbeit faßt ihre wesentlichen Ergebnisse zusammen. Aufgabe der Untersuchungen war es, zu klären, ob aus den regionalen hydrogeologischenn Verhältnissen Hinweise auf die Ursachen der Wärmeanomalie abzuleiten sind.

2. Hydrogeologischer Überblick

Die Wärmeanomalie im Gebiet von Urach folgt mit ihrer Längsachse etwa der Grenze zwischen den Grundwasserlandschaften Albvorland und Schwäbische Alb (Fig. 1 sowie ZOTH in diesem Band). Im Albvorland sind die vorherrschend tonigen, untergeordnet sandigen und kalkigen Gesteine des Lias und Doggers hydrogeologisch bestimmend. Im Norden wird das Albvorland vom Neckartal begrenzt, das bis in die vorherrschend sandigen Schichten des Mittelkeupers eingeschnitten ist. Jenseits des Neckars, im Gebiet der tektonisch eingesenkten Filder, sind wieder Gesteine des unteren Lias, im Schönbuch des Keupers, verbreitet.

Der hydrogeologische und landschaftliche Charakter der Schwäbischen Alb wird durch die 200–250 m mächtige Serie verkarsteter Kalksteine und Mergel des mittleren und oberen Malms geprägt. Die zahlreichen obermiozänen Vulkanschlote des Urach-Kirchheimer Vulkangebietes besitzen jeweils nur lokale hydrogeologische Bedeutung.

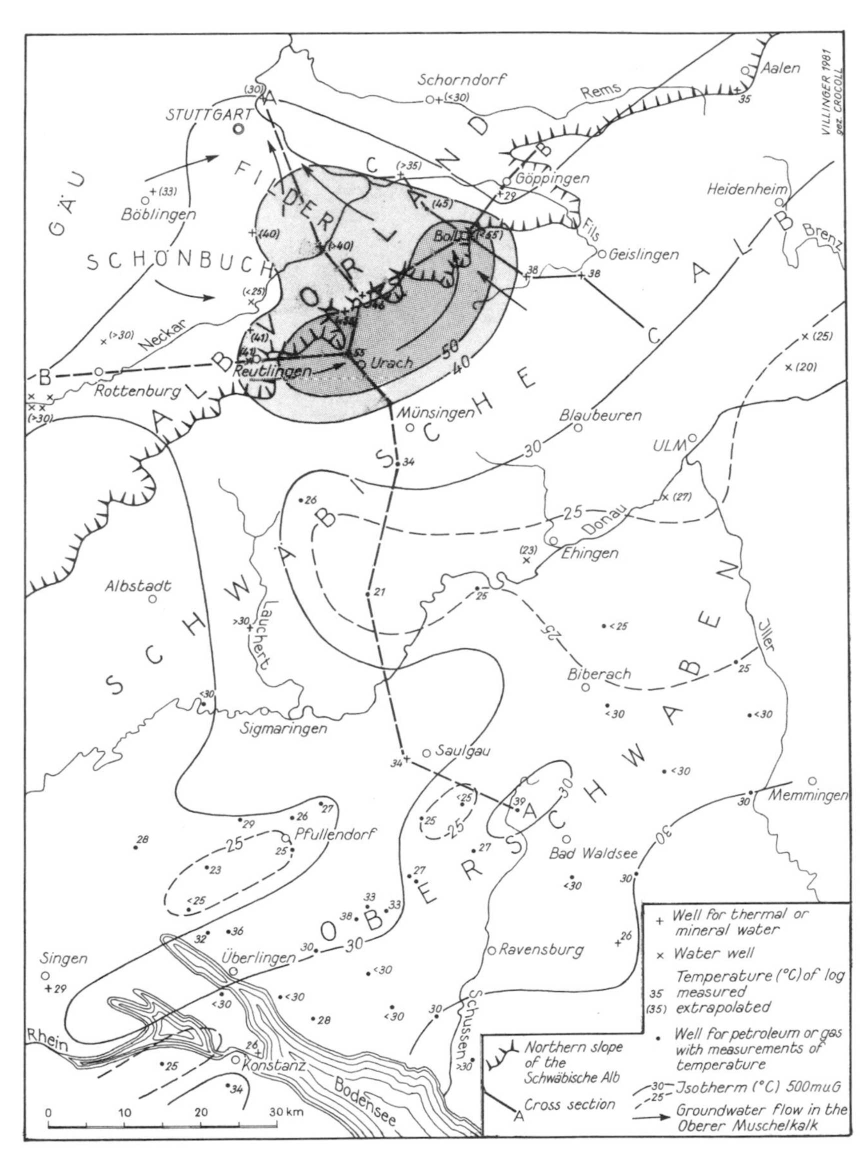

Fig. 1. Temperaturen in 500 m Tiefe, nach Temperaturlogs in Thermal-/Mineralwasserbohrungen und Einzelmessungen der Maximaltemperatur in Erdölbohrungen.

In den Tälern des Neckars und seiner größeren Seitenbäche bilden die kiesigen Talfüllungen örtliche Grundwasserleiter.

In der an der Oberfläche ausstreichenden Gesteinsfolge vom Malm bis hinab zum Keuper sind vier nennenswerte Grundwasserstockwerke vorhanden:

- der große, weiträumig zusammenhängende Karstaquifer im Malm (wβ–ζ = ox2-ti, Oxfordkalke bis Tithon), dessen Sohlschicht die Oxfordmergel (wα = ox 1) bilden; darunter die Kluft-/Porenaquifere der

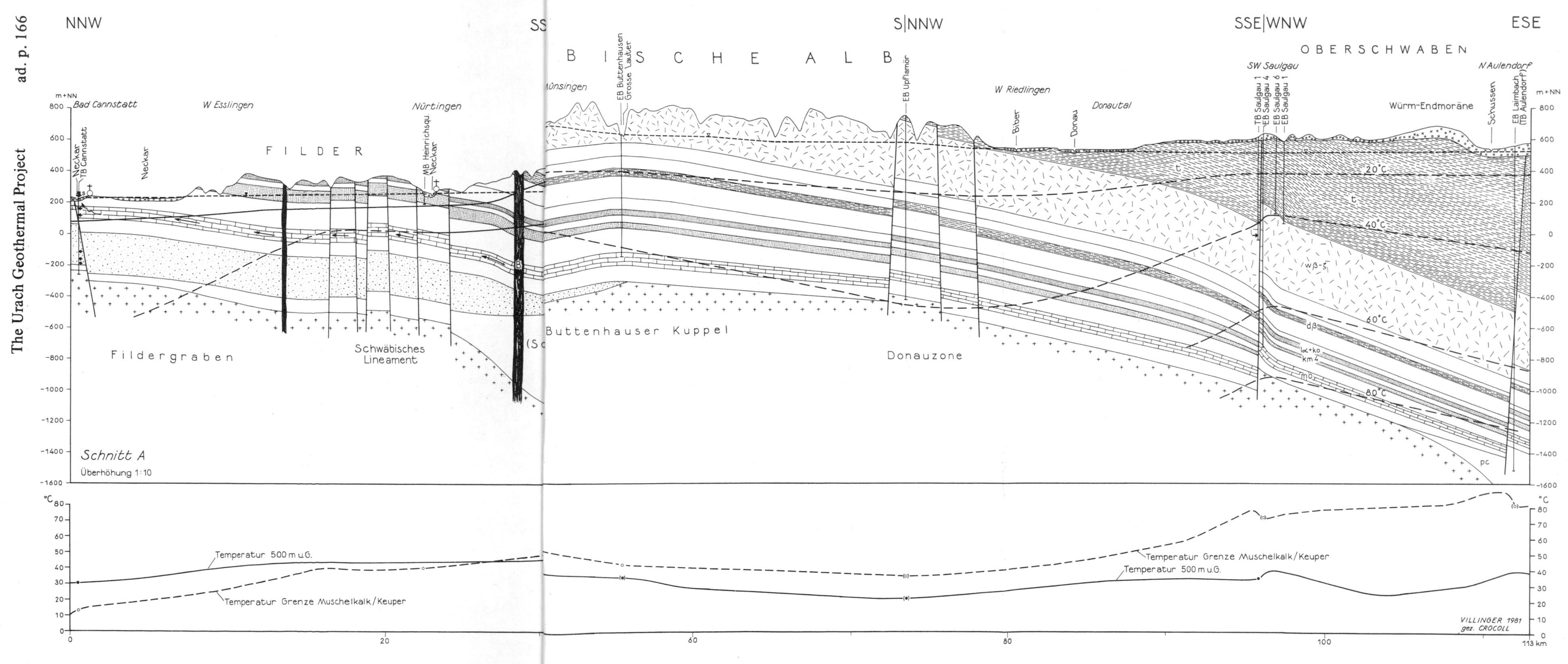

Fig. 2. Hydrogeologische und geothermische Schnitte durch die Wärme Rückseite.

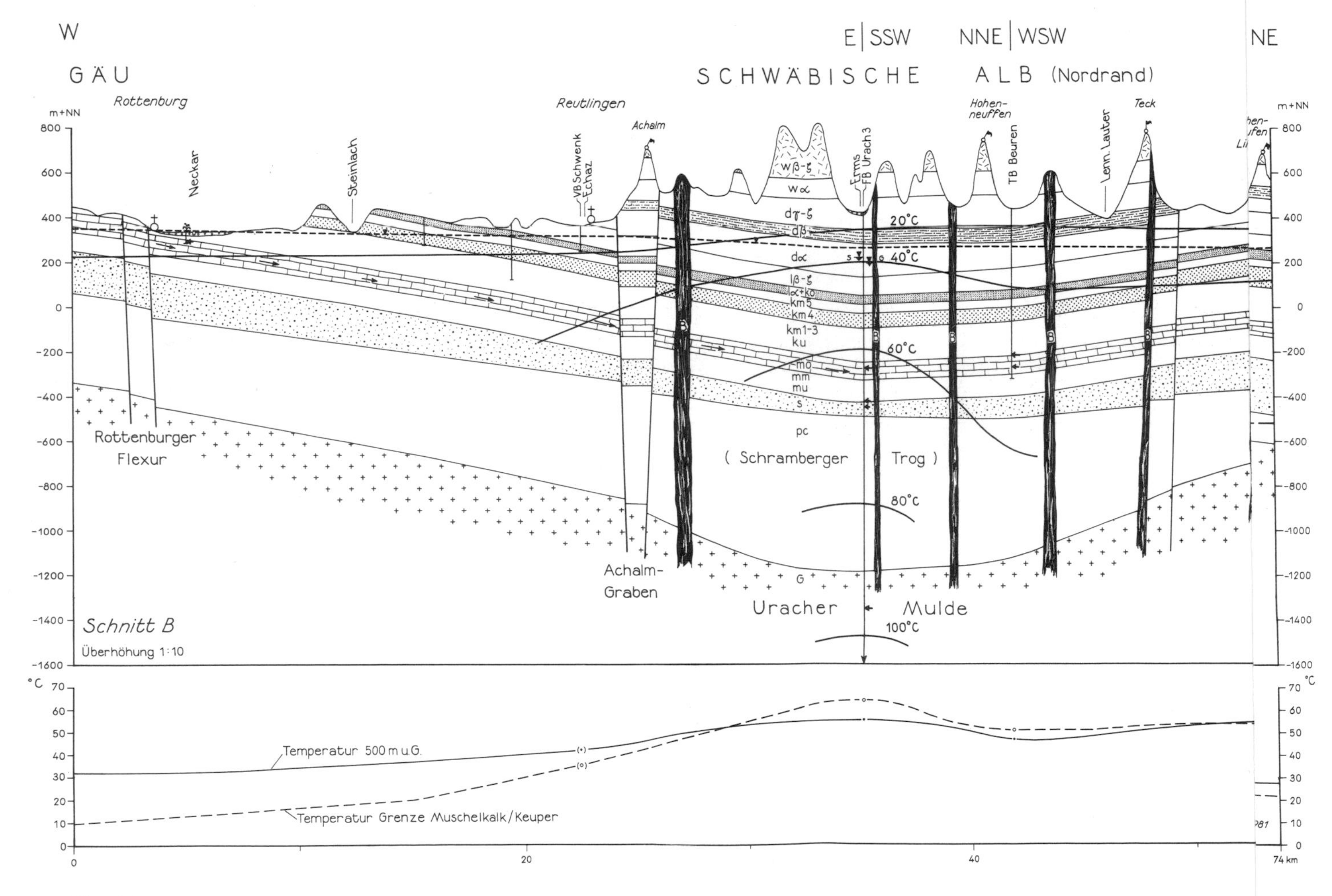
W
E | SSW
NNE | WSW
NE
GÄU
SCHWÄBISCHE ALB (Nordrand)
Rottenburg
Reutlingen
Achalm
Hohen-neuffen
Teck
Neckar
Steinlach
Erms
FB Urach 3
TB Beuren
Lenn. Lauter
Rottenburger Flexur
Achalm-Graben
(Schramberger Trog)
Uracher Mulde
20°C
40°C
60°C
80°C
100°C
Schnitt B
Überhöhung 1:10
Temperatur 500 m u.G.
Temperatur Grenze Muschelkalk/Keuper
m+NN
°C
74 km

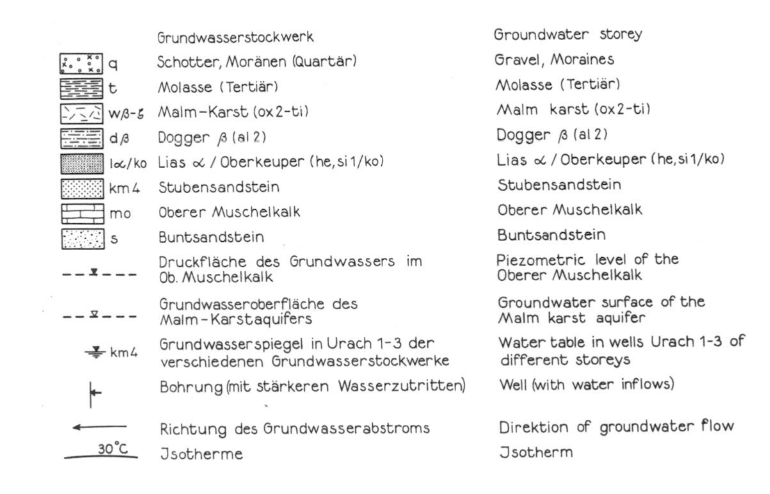
Grundwasserstockwerk
Groundwater storey
q Schotter, Moränen (Quartär)
Gravel, Moraines
t Molasse (Tertiär)
Molasse (Tertiär)
wβ-ζ Malm-Karst (ox2-ti)
Malm karst (ox2-ti)
dβ Dogger β (al 2)
Dogger β (al 2)
lα/ko Lias α / Oberkeuper (he, si 1/ko)
Lias α / Oberkeuper (he, si 1/ko)
km4 Stubensandstein
Stubensandstein
mo Oberer Muschelkalk
Oberer Muschelkalk
s Buntsandstein
Buntsandstein
Druckfläche des Grundwassers im Ob. Muschelkalk
Piezometric level of the Oberer Muschelkalk
Grundwasseroberfläche des Malm-Karstaquifers
Groundwater surface of the Malm karst aquifer
Grundwasserspiegel in Urach 1-3 der verschiedenen Grundwasserstockwerke
Water table in wells Urach 1-3 of different storeys
Bohrung (mit stärkeren Wasserzutritten)
Well (with water inflows)
Richtung des Grundwasserabstroms
Direktion of groundwater flow
30°C Jsotherme
Jsotherm

Fig. 2 B.

- Sandsteine im Dogger (dβ = al 2), vor allem Unterer und Oberer Donzdorfer Sandstein und Personatensandstein im Filsgebiet;
- Kalksteine und Sandsteine des Lias α (= he 1 + 2 und si1) und die Sandsteine des Oberkeupers (ko), die als gemeinsames Stockwerk zu betrachten sind; und
- der Stubensandstein (km 4) im Mittelkeuper.

Die übrigen Gesteine dieser Schichtenfolge spielen hydrogeologisch so gut wie keine Rolle.

Im Gebiet der Wärmeanomalie nicht an der Oberfläche anstehend, aber im Untergrund vorhanden und hydrogeologisch wichtig sind die Grundwasserstockwerke des Oberen Muschelkalks (mo) und des mit dem tiefsten Teil des Unteren Muschelkalks (mu) hydraulisch oft zusammenhängenden Buntsandsteins (s). Der Obere Muschelkalk wird erst ab Bad Cannstatt vom Neckar angeschnitten, der Buntsandstein im Unterlauf des Flusses am Rand des Odenwalds.

Die sandig-tonigen, unter Urach 700 m mächtigen Sedimente des Permocarbons (pc) in der Fortsetzung des Schramberger Trogs sind hydrogeologisch unbedeutend. Im darunter liegenden Kristallin des Grundgebirges zirkulieren, zumindest örtlich, auf Klüften im obersten Abschnitt stark salzhaltige Thermalwässer, wie in der Bohrung Urach 3 nachgewiesen wurde (Wasserzutritte in 1775 m Tiefe mit einem Feststoffinhalt über 60 g/kg, DIETRICH in diesem Band).

Nachstehend werden diese sechs Grundwasserstockwerke mit ihren hydrogeologischen Daten in stratigraphischer Reihenfolge von unten nach oben etwas näher beschrieben. Die Schnitte A, B (Fig. 2) zeigen den geologischen Bau im Gebiet der Wärmeanomalie und die Abfolge der Grundwasserstockwerke.

3. Grundwasserstockwerke

3.1 Buntsandstein/Unterer Muschelkalk

Buntsandstein und Unterer Muschelkalk sind im Bereich der Wärmeanomalie nur durch wenige Tiefbohrungen bekannt. Danach ist der Buntsandstein 62 m (Urach 3) bis 268 m (Bad Cannstatt) mächtig, die Basissande des Unteren Muschelkalks 15–22 m.

Pumpversuche haben in diesem Grundwasserstockwerk Ergiebigkeiten von 0,3–5,1 l/s bei Absenkungen des Wasserspiegels von 18–178 m erzielt. Die Höhen der Ruhewasserspiegel zeigen, daß der Neckar ab etwa Eßlingen – Stuttgart die Vorflut für das im Buntsandstein/Unteren Muschelkalk strömende Thermalwasser bildet: die Bohrung Cannstatt im Neckartal läuft mit einer Druckhöhe von 242 m ü. NN artesisch aus.

Das Thermalwasser im Buntsandstein ist hoch mineralisiert: der Feststoffinhalt beträgt 11–26 g/kg, die Temperatur im engeren Anomaliegebiet über 40 °C bis über 60 °C.

3.2 Oberer Muschelkalk

Die Mächtigkeit des Oberen Muschelkalks beträgt im Gebiet der Anomalie etwa 80 m, nach Osten und Süden nimmt sie ab. Zu dem Grundwasserstockwerk hinzuzurechnen sind die 6–11 m mächtigen Oberen Dolomite des Mittleren Muschelkalks sowie örtlich die Unteren Dolomite des Unterkeuper. Innerhalb dieses mächtigen Stockwerks strömt das Grundwasser hauptsächlich im obersten Teil des Oberen Muschelkalks, dem kavernös-klüftigen Trigonodus-Dolomit.

Pumpversuche in zahlreichen Mineral- und Thermalwasserbohrungen haben Ergiebigkeiten zwischen 1 l/s und über 12 l/s bei Absenkungen von wenigen Metern bis 115 m nachgewiesen. Als durchschnittliche Transmissivität des Grundwasserstockwerks kann man aufgrund der Pumpversuchsergebnisse etwa T = 0.0007 m^2/s ansetzen. Das hydraulische Potential ist im Oberen Muschelkalk durchweg höher als im Buntsandstein, ausgenommen im Bereich der Vorflut im Neckartal bei Stuttgart.

Die Linien gleichen Potentials im Oberen Muschelkalk zeigt Fig. 3. Da in tiefen Bohrungen die Länge der Wassersäule infolge der bei erhöhten Temperaturen verringerten Dichte des Wassers bei gleichem Aquiferdruck größer ist, sind die in Fig. 3 verwendeten Spiegelhöhen der Thermalwasserbohrungen auf die Dichte von Wasser bei 15 °C korrigiert. Die Korrektur beträgt in der Bohrung Urach 1 über 5 m, in Beuren 1 über 3 m, da dort die Temperaturen am höchsten sind.

Fig. 3 zeigt, daß die Vorflut für das Muschelkalk-Grundwasser der Neckar im Gebiet von Stuttgart ist: Dort treten in Bad Cannstatt mit die stärksten Mineralquellen Europas aus (Schüttung zusammen über 300 l/s). Die von dort weit nach Süden zurückreichende Mulde im Potentiallinienbild scheint mit den tektonischen Störungszonen von Fildergraben, Filstalmulde und Donauabbruch zusammenzuhängen. Die den Aquifer des Oberen Muschelkalks durchströmende Wassermenge Q läßt sich nach der DARCY-Gleichung Q = B · T · i wenigstens in etwa abschätzen. Der Aquiferquerschnitt entlang der gedachten 270 m-Potentiallinie vom Neckar bei Nürtingen bis zur Fils oberhalb von Göppingen (B = rd. 42 km, Fig. 3) erfaßt den gesamten ungestörten Grundwasserabstrom im Oberen Muschelkalk aus dem weiteren Bereich der Wärmeanomalie. Bei einem T-Wert von 0.0007

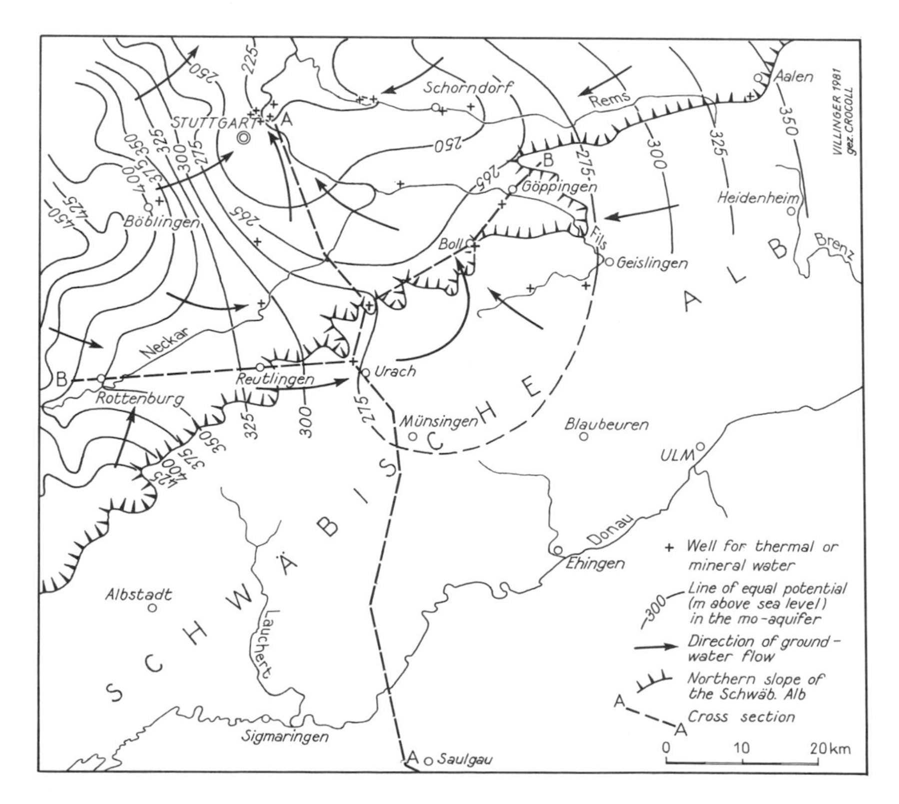

Fig. 3. Linien gleichen hydraulischen Potentials im Oberen Muschelkalk (vereinfacht nach Fig. 3 in VILLINGER 1982 b).

m^2/s und dem aus dem Abstand zwischen der 265 m- und der 275 m-Isolinie in Fig. 3 errechneten Potentialgefälle i ergibt sich ein Abstrom von insgesamt Q = 40 l/s. Davon entfallen auf den westlichen Abschnitt zwischen dem Neckar und einer Stromlinie Reutlingen – Beuren rd. 25 l/s, auf den mittleren Abschnitt von dieser Stromlinie bis zu einer zweiten, östlichen Stromlinie mit Verlauf Bad Ditzenbach – Boll etwa 10 l/s und auf den östlichen Abschnitt von hier bis zur Fils rd. 5 l/s. Der mittlere Abschnitt erfaßt den Grundwasserabstrom aus dem Zentrum der Wärmeanomalie.

Der Verlauf der Grundwassergleichen zeigt, daß der größte Teil des im Raum Urach – Beuren im Oberen Muschelkalk strömenden Thermalwassers von Westen aus dem Gebiet des oberen Neckartals und des Oberen Gäus kommt. Dort wird Grundwasser durch direkte Einsickerung von Niederschlagswasser neugebildet, wovon jedoch weitaus der größte Teil wieder in Karstquellen aus- oder ins Neckartalgrundwasser übertritt (E. VILLINGER 1982 a). Im Bereich der Schwäbischen Alb und in Oberschwaben ist die Erneuerung von Muschelkalk-Grundwasser außerordentlich gering, was wiederum das minimale Potentialgefälle dort mitbedingt. Die Neubildung geschieht hier durch extrem langsames Absickern von Grundwasser aus den überlagernden Schichten höheren Potentials (Leakance).

Obwohl sich nach Fig. 1 das Zentrum der Wärmeanomalie vom Raum Urach bis Boll erstreckt, sind die Wassertemperaturen im Oberen Muschelkalk nördlich der Donau nach den bisherigen Daten bei Urach am höchsten: am Auslauf der Bohrung Urach 1 wurden bis 60 °C, im Aquifer an der Grenze Muschelkalk/Keuper 63 °C gemessen (Temperaturlog s. Fig. 6). Dies ist eine Folge der tiefen tektonischen Absenkung der Schichtenfolge in der Uracher Mulde (Fig. 2: Schnitt A).

In Oberschwaben steigen die Temperaturen an der Obergrenze des Muschelkalk-Aquifers als Folge seiner in dieser Richtung anwachsenden Tiefenlage (Fig. 2: Schnitt A) zum Alpenvorland hin bis auf über 100 °C an. Fig. 4 zeigt die Linien gleicher Wassertemperatur, wie sie aus den im Ruhezustand gefahrenen Temperaturlogs in den Thermalwasserbohrungen abzulesen und – besonders in Oberschwaben – aus Einzelmessungen der Maximaltemperatur in Erdölbohrungen inter- bzw. extrapoliert werden können.

Hydrochemisch ist das Thermalwasser im Oberen Muschelkalk im Bereich der Wärmeanomalie durch Feststoffinhalte von 5–7 g/kg (im übrigen Gebiet ca. 1–4 g/kg) und CO_2-Gehalte von 1–1,5 g/kg gekennzeichnet (CARLÉ 1975).

3.3 Stubensandstein

Das Grundwasserstockwerk des Stubensandsteins ist im westlichen Teil der Wärmeanomalie etwa 50 m, im östlichen 50–100 mächtig. Nach Süden, unter der Schwäbischen Alb, nehmen die Mächtigkeiten ab. In Pumpversuchen wurden Ergiebigkeiten meist von 0,5–2,5 l/s bei Absenkungen von 23–80 m ermittelt.

Der Ruhewasserspiegel des Stubensandstein-Stockwerks ist durchweg höher als der des Oberen Muschelkalks, ausgenommen in der Bohrung Urach 1, was wohl auf bohrlochbedingte Einflüsse zurückzuführen ist.

Der Feststoffinhalt der Stubensandstein-Wässer liegt meist bei 2–4,5 g/kg, nur in Nürtingen und bei Reutlingen ist er mit 12 g/kg bzw. 8,6 g/kg höher. Der CO_2-Gehalt erreicht in Urach mit 0,7 g/kg das Maximum.

Die Temperaturen sind im Stubensandstein nur im zentralen, tektonisch tiefer liegenden Bereich der Wärmeanomalie hoch genug, um Thermalwässer (> 20 °C) zu erzeugen.

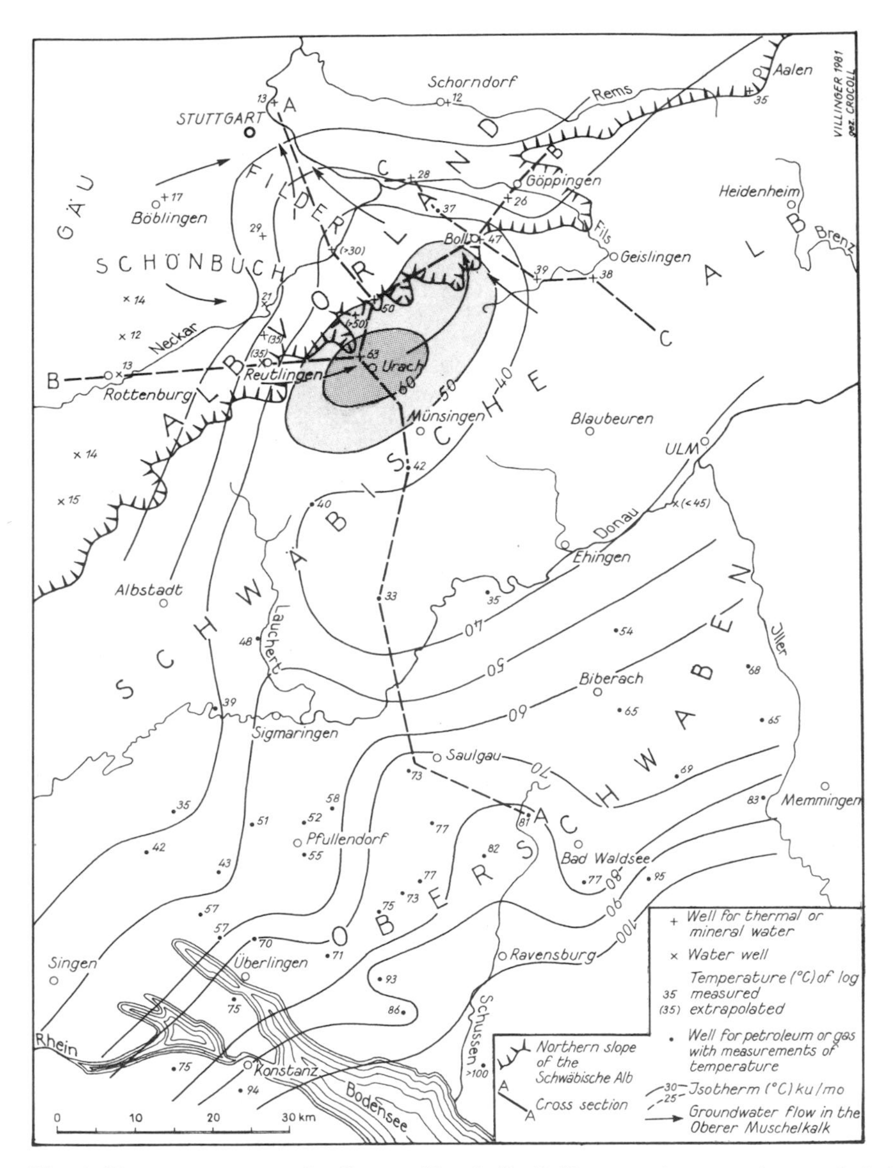

Fig. 4. Temperaturen an der Grenze Muschelkalk/Keuper. Ausgangsdaten wie bei Fig. 1.

3.4 Lias α/Oberkeuper

Arietenkalk und Angulatensandstein des Lias α bilden mit dem Oberkeuper-Sandstein (ko) – sofern er ausgebildet ist – ein gemeinsames Grundwasserstockwerk mit einer Mächtigkeit von etwa 30–40 m. Die Ergiebigkeiten betragen meist 0,5–2 l/s bei Absenkungen bis über 90 m, nur in Urach 2 erzielte man 4 l/s mit 39 m Absenkung.

Der Feststoffinhalt des Wassers beträgt meist 2–5 g/kg, der CO_2-Gehalt im Osten der Anomalie bis 1,5 g/kg. Die Thermalwassergrenze (20 °C) wird nur in den Bohrungen Urach und Beuren überschritten.

3.5 Dogger β

Die Mächtigkeit dieser Schicht beträgt etwa 50–75 m. Als Grundwasserstockwerk wirkt sie nur etwa von Urach ab nach Osten, wo mehrere Sandsteinhorizonte auftreten.

Die Ergiebigkeiten betragen dort i.a. 0,1–0,5 l/s, vereinzelt bis 1 l/s. Der Feststoffinhalt der Dogger β-Wässer liegt bei 2–5 g/kg, der CO_2-Gehalt bei 0,2–1,5 g/kg. Die Temperaturen bleiben durchweg unter 20 °C.

3.6 Malm-Karst

Das etwa 200–250 m mächtige, von verkarsteten Kalksteinen und Mergeln aufgebaute Grundwasserstockwerk des mittleren und oberen Malms in der Schwäbischen Alb enthält die weitaus größten Grundwassermengen. Sie strömen zu starken Karstquellen (Schüttungen bis mehrere hundert l/s), die am Nordrand der Alb und in tief eingeschnittenen Tälern entspringen. Das nutzbare Hohlraumvolumen des Karstaquifers beträgt 1–2 %, die regionale Durchlässigkeit etwa 0,0001–0,0005 m/s. Die durchschnittlichen Abstandsgeschwindigkeiten des Karstgrundwassers liegen im oberen Aquiferbereich bei etwa 100 m/h (E. VILLINGER 1977). Der Feststoffinhalt des Malm-Karstwassers beträgt 0,4–0,6 g/kg.

Die Temperaturverhältnisse im Malm-Karst lassen sich durch Messung der Quelltemperaturen in Zeiten mit niedrigen Schüttungen erkennen: Fig. 5 zeigt diese Temperaturen nach Messungen vom 25.9. bis 6.10.1980. Auffallend ist eine sich von SE Reutlingen über Urach bis SE Beuren erstreckende positive Anomalie. In der engeren Umgebung von Urach sind die Quelltemperaturen bis 1 °C höher als zu erwarten wäre. Dies dürfte eine Folge der Uracher Wärmeanomalie sein (s. Kap. 4.3).

4. Die Wärmeanomalie aus hydrogeologischer Sicht

4.1 Wärmeaufstieg

Für den Mechanismus des Wärmetransports von unten nach oben kommen grundsätzlich zwei Möglichkeiten in Frage: Aufstieg durch Wärmeleitung (Konduktion) und/oder Aufstieg von heißen Tiefenwässern bzw. Gasen (Konvektion). Nach den Modellberechnungen von HOFFERS (1980) und D. WERNER (in diesem Band) kann die Anomalie nicht allein durch Wärmeleitung im Gestein erklärt werden, wenn sich der erkaltende Magmenherd in 30–40 km Tiefe oder noch tiefer befindet. Die seismischen Ergebnisse von MEISSNER et al. (in diesem Band) lassen einen Wärmeherd in 5–25 km Tiefe erkennen.

CARLÉ (1974: 211) deutet an, die Anomalie könnte durch aufsteigendes heißes Tiefenwasser verursacht werden. HAENEL & ZOTH (in diesem Band) nehmen aufgrund der Meßdaten in der Bohrung Urach 3 aufsteigende Wässer aus 1600–5000 m Tiefe an. Sie errechneten eine Aufstiegsgeschwindigkeit des Wasser von 10^{-8} cm/s = ca. 3 mm/Jahr.

Damit Wasser aus größerer Tiefe aufsteigen kann, müssen vor allem zwei Voraussetzungen erfüllt sein:

- ausreichende Wasserwegsamkeit des Gesteins und
- ein hydraulisches Potentialgefälle von unten nach oben.

Die erste Voraussetzung ist in der tektonisch und vulkanisch erheblich beanspruchten Region der Anomalie sicherlich gegeben, weil das sedimentäre Deckgebirge wie auch das Grundgebirge ausreichend geklüftet sind. Dies hat auch die Bohrung Urach 3 bestätigt.

Table 1. Water levels of the different groundwater storeys in the area of the heat anomaly.

Groundwater storey	Wells with hdydrostatic water levels (m above sea level)							
	Urach 1	Urach 3	Beuren 1	Boll 1	Ditzenbach 2 + 3	Überkingen 1	Bonlanden 1	Cannstatt
Dogger β	–	–	–	–	ca. 485[1]	um 435[2]	–	–
Lias α/Oberkeuper	288	–	275	–[3]	(237)[4]	–	–	–
Stubensandstein	(256)[4]	–	–	296	287	304	–	–
Ob. Muschelkalk	277,5	265	264	267,5	274	276	270	224[5]
Buntsandstein/Unt. Muschelkalk	–	ca. 226 (?)	–	–	–	268	308	242[6]
Basement	–	ca. 204[7]	–	–	–	–	–	–

1) Marienquelle
2) Helfensteinquelle and Adelheidquelle
3) Water level of Lias ϵ about 390 m above sea level (Alter Schwefelbr.)
4) Real water level probably higher
5) Artesian well
6) Buntsandstein without Unt. Muschelkalk
7) Water from depth at 1775 m

Ob auch die zweite Bedingung erfüllt ist, läßt sich an der Höhenlage der Grundwasseroberfläche in den verschiedenen Stockwerken erkennen. In Tab. 1 sind die in den Tiefbohrungen angetroffenen Ruhewasserspiegel der verschiedenen Grundwasserstockwerke zusammengestellt. Die Spiegelhöhen wurden nicht temperaturkorrigiert, da sich dadurch für diese Fragestellung nur unwesentliche Änderungen ergeben hätten. Es zeigt sich, daß im Kerngebiet der Anomalie die Wasserspiegel von oben nach unten kontinuierlich immer tiefer stehen. Lediglich die im Stubensandstein in Urach 1 und im Lias β/Oberkeuper in Ditzenbach gemessenen Spiegelhöhen machen davon eine Ausnahme: sie sind wahrscheinlich durch ungenügenden Anschluß des Bohrlochs an den Aquifer oder andere bohrlochbedingte Effekte zu niedrig.

Es ist somit festzuhalten: die hydraulischen Potentiale der Stockwerke nehmen im Untergrund der Wärmeanomalie bis mindestens 1800 m Tiefe von oben nach unten ab und nicht umgekehrt. Ein Aufstieg von erhitztem Tiefenwasser, z.B. aus dem kristallinen Grundgebirge oder gar dem Permokarbon-Trog, ist somit höchst unwahrscheinlich.

Das Modell des Tiefenwasseraufstiegs hätte nämlich einen Aufstieg der heißen Wässer – mindestens im Kerngebiet der Anomalie um Urach – Neuffen – bis in die Tongesteine des oberen Lias/Dogger als Konsequenz. Die heißen Wässer müßten alle tieferen Grundwasserleiter durchqueren, um schließlich ausgerechnet in einem Nichtleiter ihre Aufwärtswanderung zu beenden und seitlich, schichtparallel aus der Anomalie abzuströmen. Dies wäre aus der Beobachtung zu schließen, daß die Temperaturkurven in diesem Abschnitt der Schichtfolge einen deutlichen Knick zeigen, d.h. zur Erdoberfläche hin einen überdurchschnittlich hohen geothermischen Gradienten besitzen. Nur die Bohrung Neuffen macht hiervon eine Ausnahme: vermutlich war sie nicht tief genug.

Sofern die reine Wärmeleitung tatsächlich nicht ausreichen sollte, kämen als Transportmedium für den Wärmeaufstieg vielleicht Gase in Frage, was auch HAENEL & ZOTH (in diesem Band) als Möglichkeit andeuten. Im Uracher Raum böte sich hierfür das reichlich aufsteigende Kohlendioxid an. Seine Wärmekapazität reicht aus,um die Anomalie zu erklären (D. WERNER in diesem Band). Auf den augenscheinlichen Zusammenhang zwischen der Wärmeanomalie und dem Kohlendioxidaufstieg hat auch CARLÉ (1974: 120) hingewiesen.

4.2 Geothermischer Gradient

In Fig. 6 sind alle verfügbaren Temperaturlogs – soweit sie einigermaßen zuverlässig sind – sowie die Grundwasserstockwerke aufgezeichnet. Zwar wurden einige Logs nur relativ kurze Zeit nach Pumpversuchen gefahren. Doch kann davon ausgegangen werden, daß allenfalls im oberen Teil der Logs geringe Verschiebungen der Kurven durch das zuvor gepumpte wärmere Wasser aus den tieferen Aquiferen hervorgerufen wurden. Fig. 5 zeigt, daß sich die Bohrungen Göppingen und Aalen bereits außerhalb der Wärmeanomalie befinden, auch Ditzenbach und Überkingen liegen schon an ihrem Rand. Der Knick im Log von Überkingen 1 im Grenzbereich Buntsandstein/Muschelkalk geht auf vertikale Wasserbewegungen zurück. Bei Urach 1 und 2, besonders aber in Boll 1, ist der Gradient innerhalb des Oberen Muschelkalks stark verringert. Auch dies ist auf Konvektion innerhalb des hier stark von Grundwasser durchströmten Stockwerks zurückzuführen.

Der Knick in Urach 3 in 1600 m Tiefe beruht auf der unterschiedlichen Wärmeleitfähigkeit von sedimentärem Deckgebirge (λ = 2–2,5 W/m · k) und Grundgebirge (λ = 2,5–3 W/m · k nach HAENEL & ZOTH 1980, Anl. 11). Die Ursache für den wesentlich stärkeren Knick in etwa 300 m Tiefe, der wie oben erwähnt auch in den anderen Bohrungen dieses Gebiets auftritt, ist bisher noch nicht geklärt (s. unten).

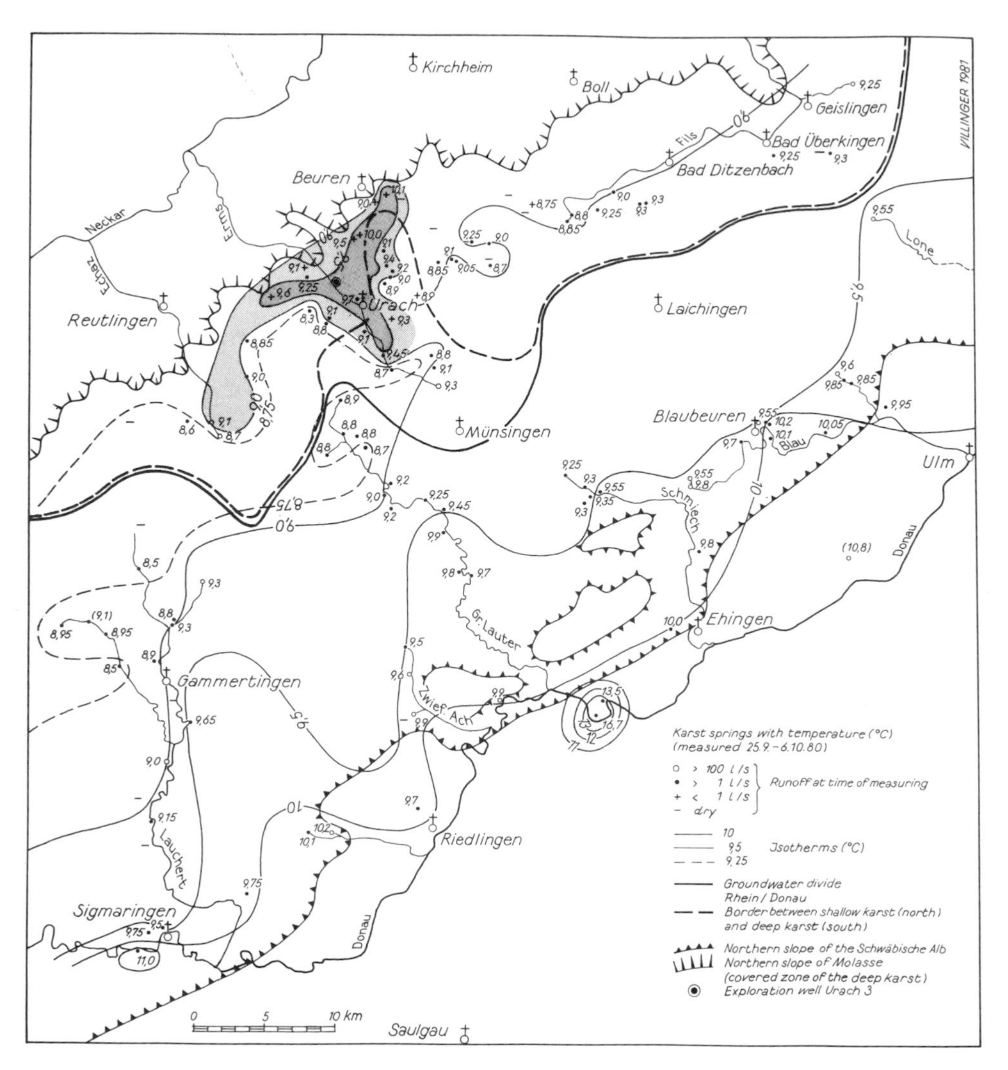

Fig. 5. Temperaturen der Karstquellen im Malm der mittleren Schwäbischen Alb nach Messungen vom 25.9. bis 6.10.1980 bei niedrigen Abflüssen.

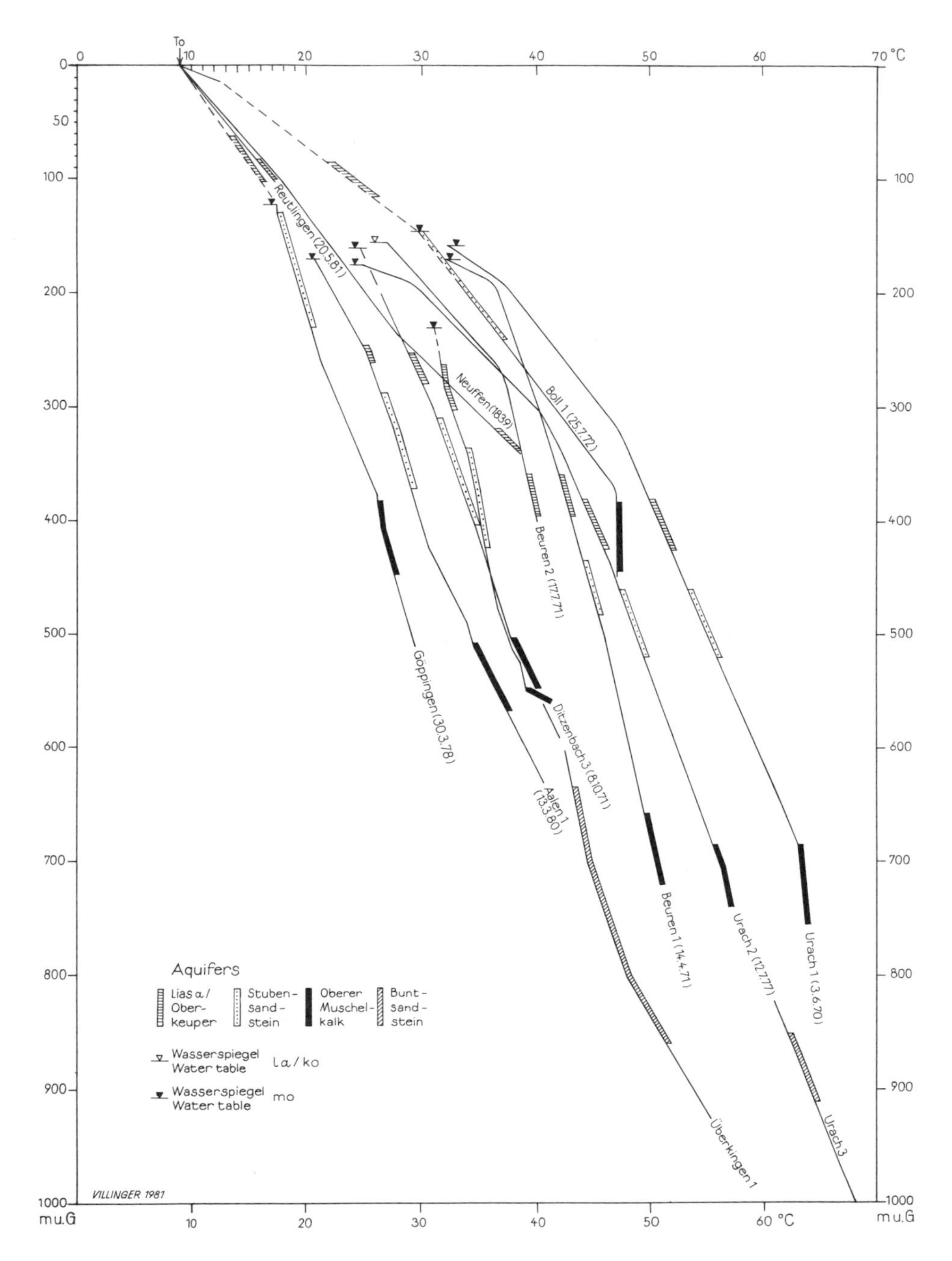

Fig. 6. Temperaturverlauf und Grundwasserstockwerke in den tiefen Bohrungen im Bereich der Wärmeanomalie und ihrer Umgebung.

Die kuppelartige Aufwölbung der Isothermen im Raum Urach und ihre, die Erhöhung des geothermischen Gradienten verursachende Scharung zur Oberfläche hin zeigen die Schnitte (Fig. 2).

4.3 Wärmestromdichte

Um einen so viel höheren Gradienten (in Urach von 0–200 m gemessen 11 °C/100 m, nach HAENEL & ZOTH in diesem Band, mit topographischer Korrektur 9,1 °C/100 m) zu erzeugen, müßte die Wärmeleitfähigkeit der Gesteine in diesem Abschnitt (oberer Lias und unterer Dogger), auf den extrem niedrigen Wert von etwa $\lambda = 0.9$ W/m · k absinken. Geht man in diesem relativ oberflächennahen Bereich von einem realistischeren Wert von 1,5–1,7 W/m · k aus – wie er durch die Untersuchungen von BEHRENS et al. (in diesem Band) in allerdings nur 50 m tiefen Bohrungen ermittelt wurde –, dann würde sich dort mit dem topographisch korrigierten Gradienten grad T = 9,1 °C/100 m ein Wärmestrom von q = 135–155 mW/m^2 ergeben, d.h. ein Plus von 50–70 mW/m^2 zusätzlich zu dem aus der Tiefe aufsteigenden, ebenfalls schon um 10–15 mW/m^2 erhöhten Wärmestrom.

Dies könnte auf eine zweite Wärmequelle hinweisen, die in etwa 200–350 m Tiefe, also im oberen Lias bis unteren Dogger zu lokalisieren wäre. Als Energielieferant kämen am ehesten die in diesen Schichten, vor allem im Lias ϵ angereicherten Kohlenwasserstoffe (Bitumengehalt im lϵ bis 20 %), vielleicht auch Schwefelkies in Betracht: Zu denken wäre an einen exothermen mikrobiellen Abbau im anaeroben Milieu, wie er z.B. auch zur Entstehung der Schwefelwässer im Lias ϵ führt, oder an Inkohlungsvorgänge.

Daß dieses „sekundäre" Wärmephänomen nur im Uracher Raum auftritt, obwohl die gleichen Gesteine entlang der ganzen Alb ausgebildet sind, könnte auf eine Art Kettenreaktion zurückgehen: der um 15–20 % erhöhte Tiefenwärmestrom wirkt als eine Art Vorheizung, durch die – vielleicht unter Beteiligung aus der Tiefe aufsteigender Gase – exotherme Umwandlungsprozesse induziert oder intensiviert werden, so daß der zur Erdoberfläche aufsteigende Wärmestrom erheblich verstärkt wird.

Am Beispiel der Eckisquelle, die in Urach am nördlichen Rand des Ermstales, an der Karstbasis (Basis Oxfordkalk) entspringt und in einem tief in den Berg getriebenen Horizontalstollen gefaßt ist, läßt sich eine quantitative Wärmebilanz versuchen. Diese Quelle und ihr Einzugsgebiet liegen im Kern der im Malm-Karstgrundwasser festgestellten Wärme-Insel (Fig. 5): Die Eckisquelle hatte bei den Temperaturmessungen im Herbst 1980 eine um etwa 1 °C erhöhte Quelltemperatur von 9,7 °C. Die Schütterung der Quelle hat zur Meßzeit etwa 20 l/s betragen, im Mittel dürfte sie, zusammen mit Nebenaustritten, bei etwa 40–50 l/s liegen.

Daraus läßt sich ein Wärmemehraustrag in der Eckisquelle von

$$Q_W = (40 \text{ bis } 50) \cdot 4{,}2 = 170 \text{ bis } 210 \text{ kW}$$

errechnen. Bezogen auf das unterirdisch sich nach Nordosten erstreckende, etwa 3,5 km^2 große Einzugsgebiet ergibt das eine gegenüber normalen Gebieten um

$$q = 50 \text{ bis } 60 \text{ mW/m}^2$$

erhöhte Wärmestromdichte, was mit dem aus dem Gradienten abgeschätzen Wert der mutmaßlichen Zusatzwärme aus dem Lias/Dogger gut übereinstimmt.

4.4 Wärmetransport

Für die Erklärung der auffallend hohen Temperaturen in der Thermalwasserbohrung Boll scheidet das Modell der zweiten Wärmequelle im Lias/Dogger aus, weil diese Schichten hier bereits an der Oberfläche ausstreichen. Der Verlauf der Isothermen im Schnitt B (Fig. 2) deutet an, daß im Boller Gebiet zudem der von unten aufsteigenden Wärmestrom gegenüber Urach reduziert ist: Von Urach nach Beuren sinkt die 60 °C-Isotherme steil ab, während die beiden oberen Isothermen in etwa gleicher Höhe bis Boll bleiben, um dann aber in Richtung Göppingen steil abzutauchen. Aus diesem Bild kann – in Verbindung mit den Strömungsrichtungen des Grundwassers im Oberen Muschelkalk (Fig. 3) – geschlossen werden: Die hohe Temperatur in dem in Boll erbohrten Tiefenbereich wird im wesentlichen durch seitlich aus dem Zentrum der Anomalie bei Urach herangeführte Wärme hervorgerufen. Dies geschieht innerhalb des Oberen Muschelkalks durch das Grundwasser auf seinem von den Potentiallinien (Fig. 3) vorgezeichneten Strömungsweg: von Urach zunächst nach Osten der tektonischen Filstalmulde folgend, dann im Bogen westlich Wiesensteig nach Norden bis Nordwesten an Boll vorbei in die Richtung des Fildergrabens einschwenkend, um neckarabwärts der Vorflut bei Bad Cannstatt zuzuströmen. Göppingen liegt im Osten außerhalb dieser thermischen Konvektionsströmung und partizipiert deshalb nicht mehr an der Uracher Wärme, Bad Ditzenbach liegt gerade noch in ihrem Randbereich.

Die Größenordnung der durch die Warmwasserströmung aus dem Anomaliezentrum weggeführten Wärmemenge läßt sich überschlägig berechnen: am Westrand der Anomalie, etwa bei Reutlingen (325 m Linie in Fig. 3), strömen in einem etwa 6 km breiten Streifen 10 l/s Grundwasser mit einer Temperatur von 30 °C in den Raum Urach und – wie in Kap. 3.2 beschrieben – im Nordosten durch die (vertikal im Aquifer gedachte) 270 m Potentialfläche mit 50 °C wieder ab. Daraus ergibt sich eine mit dem Grundwasserstrom abfließende Wärmeleistung von etwa

$$Q_W = 10 \cdot 20 \cdot 4{,}2 = 840 \text{ kW}.$$

Mit ihren Ausläufern wirkt sich die Warmwasserströmung stromabwärts im Muschelkalk und in den darüberliegenden Schichten bis Bad Cannstatt aus (Schnitt A in Fig. 2 u. Fig. 4). Auch die Temperaturverteilung in 500 m Tiefe wird von ihr beeinflußt (Fig. 1).

Am westlichen Ende der Wärmeanomalie, etwa in der Reutlinger Gegend und nördlich davon, ist auf Fig. 4 umgekehrt eher der Einfluß des von Westen aus dem oberflächlichen Neubildungsgebiet heranströmenden kühleren Muschelkalk-Grundwassers zu erkennen: die 40 °C-Isotherme sinkt dadurch nach Westen zu vermutlich etwas rascher in die Tiefe ab, soweit dies aus den Temperaturmessungen in der Reutlinger Versuchsbohrung geschlossen werden kann (Fig. 2: Schnitt B).

Im Südwesten scheint die Uracher Anomalie über das Lauchertgrabensystem eine – allerdings bisher wenig belegte – Verbindung nach Süden zur Saulgau – Überlinger-Anomalie zu besitzen. Fig. 4 zeigt dies deutlicher als Fig. 1, weil der Muschelkalk in diesem Graben bis über 100 m tiefer als in der Umgebung liegt. Wasserströmungen innerhalb des Aquifers sind somit nicht verantwortlich für diese Anomalienbrücke.

Der kühlende Einfluß des stark durchströmten Malm-Grundwasserstockwerks geht aus Schnitt A hervor, wo die 20 °C-Isotherme – den wenigen interpolierten Temperaturdaten zufolge – unter der Schwäbischen Alb in größere Tiefen herabgedrückt zu sein scheint. Allerdings sinkt auch die 40 °C-Isotherme weit ab, um erst wieder im Bereich der Saulgauer Anomalie an der antithetischen Verwerfung hoch aufzusteigen.

References

Carlé, W. (1974): Die Wärme-Anomalie der mittleren Schwäbischen Alb (Baden-Württemberg). – Approaches to Thaphrogenesis, p. 207–212, Stuttgart (Schweizerbart).
– (1975): Die Thermalwasserbohrung von Stuttgart-Bad Cannstatt. – Jh. Ges. Naturk. Württ., **130**: 87–155, Stuttgart.
Haenel, R. & Zoth, G. (1980): Thermische Untersuchungen. – In: Endbericht über Geophysikalische Untersuchungen in der Forschungsbohrung Urach. – Report, p. 5–21. – NLfB Hannover, Archive No. 86 453.
Hoffers, B. (1978): Modellrechnungen zur Uracher Wärmeanomalie. – Oberrhein. geol. Abh., **27**: 33–39, Karlsruhe.
Villinger, E. (1977): Über Potentialverteilung und Strömungssysteme im Karstwasser der Schwäbischen Alb (Oberer Jura, SW-Deutschland). – Geol. Jb., **C 18**: 3–93, Hannover.
– (1982 a): Grundwasserbilanzen im Karstaquifer des Oberen Muschelkalks im Oberen Gäu (Baden-Württemberg). – Geol. Jb., **C**, Hannover (in press).
– (1982 b): Hydrogeologische Aspekte zur geothermischen Anomalie im Gebiet Urach – Boll am Nordrand der Schwäbischen Alb (SW-Deutschland). – Geol. Jb., **C**, Hannover (in press).

The Urach Geothermal Project, p. 179–186;
Schweizerbart'sche Verlagsbuchhandlung, Stuttgart, 1982

Geological and Hydrochemical Investigations in the Area of Urach-Kirchheim

BALKE, K.-D., EINSELE, G., ERNST, W., FRIEDRICHSEN, H., GREGAREK, R., KOLLER, B., KOZIOROWSKI, G., LOESCHKE, J., MAIER, U., METZKER, S. and STAFFEND, H.-J.

with 7 figures and 1 table

Abstract: The tectonic situation in the area of the geothermal anomaly of Urach-Kirchheim is characterized by a generally SW-NE-strike and SE-dip of the layers. The occurrence of numerous bendings and faults indicates a high tectonic stress. The most significant structures are synclines and anticlines. Furthermore, the layers are penetrated by a great number of volcanic pipes. The tectonic pattern contains many vertical and steeply dipping elements.
The hydrochemical type and the mineralization of the groundwater vary in correspondence with the petrological condition and the depth of the different aquifers. In some areas in the region of Reichenbach-Göppingen and at the Achalm near Reutlingen hydrochemical and hydrothermal anomalies which may be caused by rising deeper groundwater were found.
Isotopic research and geochemical temperature determinations show that there are mixtures of deep circulating hot water and cooler water situated near the surface.
The tectonic, hydrochemical and isotope geological investigations prove that in the area of the geothermal anomaly of Urach-Kirchheim vertical groundwater convection takes place. This can be considered as the most important reason for the development of the anomaly.

The development of the geothermal anomaly, which is observed in the area of Urach-Kirchheim, can be explained in two ways:
– by the flow of heat from a magma centre below the anomaly,
– by vertical convection of deep hot groundwater to the vicinity of the surface.
In the latter case it is to be expected that paths of ascent have been created by tectonic activity, and that hydrochemical and thermal anomalies occur within the higher aquifers.

The investigations should clarify to what extent the convection theory can be proved by tectonic and hydrochemical facts. The area of investigation is situated in the southwestern German cuesta landscape at the northwestern edge of the Schwäbische Alb, (Fig. 1). Except for Quatenary formations – especially sandy-pebbly vally sediments – the outcrops show the layers from Keuper to Upper Malm (Fig. 2). The sediments of the Muschelkalk, Buntsandstein and Rotliegendes underly these layers and are encountered only by drilling. The basement is found at a depth of 1606 m and consists of gneiss and syenite.

This rock sequence is cut by more than 350 volcanic channels, which appear to be very frequent in the area of Urach-Kirchheim. Mainly gaseous explosions, which left behind channel breccias with tuff content, took place. Basaltic lava emerged only in a few

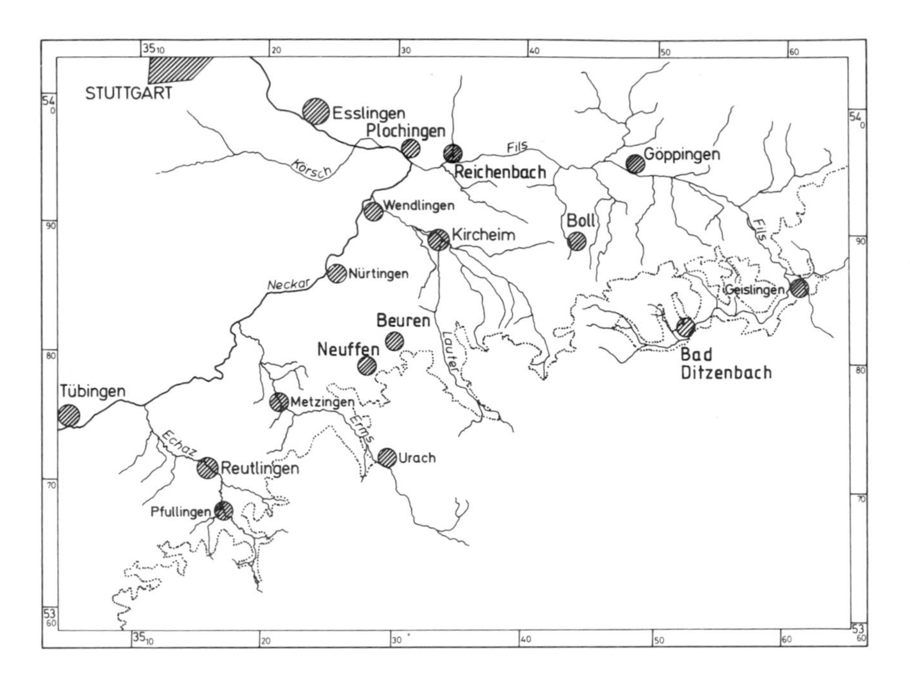

Fig. 1. Area of investigation (dotted line: edge of the Schwäbische Alb).

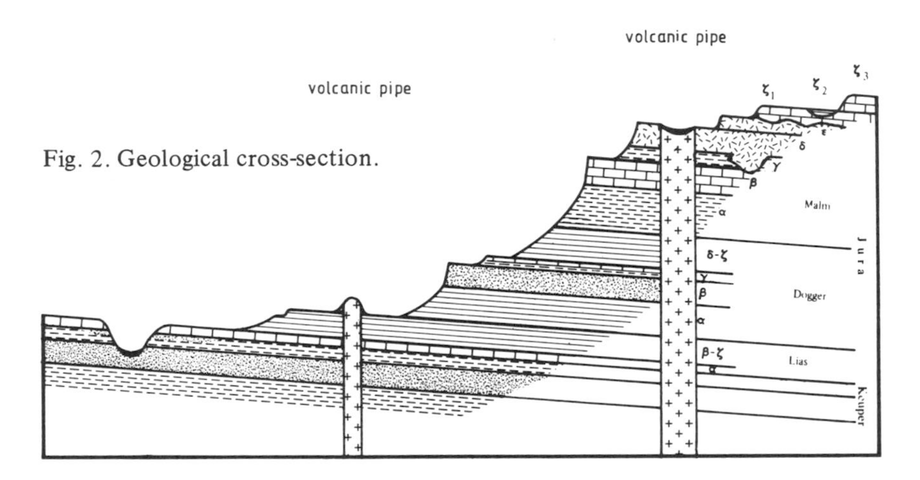

Fig. 2. Geological cross-section.

cases. The "Upper Miocene Alb volcanism" is closely related to the regional tectonics. The area of volcanism is situated at the extension of the Filder system of faults, the Lauchert trench, and the Filstal fault. Furthermore, it is touched by the Bebenhäuser fault zone.

The Urach-Kirchheim heat anomaly has its centres at Neuffen and Bad Boll. It extends over the area of Tertiary volcanism.

For clarifying the question of whether paths of ascent for vertically circulating water could occur in an area with such tectonic structures the following working methods were used: production of an off-lap map, interpretation of aerial photographs and geological orbital photography (ERTS), measurement of joints, measurement of soil gas for the discovery of hidden faults.

The off-lap map (Fig. 3) shows that there is a generally SE-NE strike and SE-dip in the area. The occurrence of a great number of bendings and faults indicates a high tectonic stress. The most significant structures are synclines and anticlines, for example in the area of Wendlingen, Reutlingen, Pfullingen and Urach. Other significant structures are scalariform bendings where the layers show a steeper dip than in the surrounding areas. Areas like these are found at Neuhausen, Dettingen, Grabenstetten, Ochsenwang (see Fig. 3). Furthermore, there are several fault zones which cut the bent "plate" into different blocks. The strike of these fault zones is Hercynian (about NW-SE, 120–135°), Rhenish (NNE-SSW, 15°), and Swabian (ENE-WSW, 75°).

The measurements of the joints prove that the directions of the joints are very different in distinct areas. In the whole region there is no direction which shows a significant development. The Hercynian, the "erzgebirgische" (about NE-SW, 50–70°) and the Rhenish directions are represented rather often, whereas the Swabian and the "eggische" (NNW-SSE, 160°) directions are subordinate. A conspicuous feature is that the directions of the joints are similar to those of the faults which are situated in the surrounding area near the measurement points.

The above-mentioned directions can be seen in aerial photographs, too. Especially the Swabian lineament and the Filstal zone show significant structures. The strike of both tectonic elements is parallel to the longitudinal direction of the Urach heat anomaly.

A clear relation between the anomaly of Urach, whose major axis is parallel to the Swabian and the "erzgebirgische" direction, and the tectonic direction of the region cannot be proved because of the lack of definitely predominating tectonic directions, (Fig. 4). The underground shows a great number of faults. Hence, the conception that tectonic activity created paths of ascent for vertical groundwater streams in joint and fault zones is a plausible starting point. Furthermore, the great number of volcanic channels can serve as paths of ascent for the water. The more or less impervious tuffites of the funnels probably do not provide paths for the stream of water, but the surroundings of the funnels, which are mechanically and thermally stressed, can do so.

For clarifying the question of whether or not the rising of deeper groundwater could be noticed from hydrochemical and/or thermal exchange in the upper aquifers, 170 wells and perennial springs were investigated within the period from August, 1977 to March, 1979. Furthermore, a study of the literature provided a great number of data about water analysis. Measurements in summer and winter should show whether there is a difference in the state of the water at different seasons. The investigations were made with groundwater from aquifers in the following strata: Muschelkalk, Keuper, Lias, Dogger, Malm and Alluvium.

The mineral water from the area of investigation, which is obtained from the Muschelkalk aquifer by drilling, is of the Na-Ca-Cl-SO_4- and the Na-Ca-Cl-SO_4-HCO_3-types. The chemism is very similar for both types; conspicuous differences were not found.

The Stubensandstein water which occurs in springs corresponds to the Ca-Mg-HCO_3-/ Ca-Mg-HCO_3-SO_4-type but shows a very high Mg-content. A different chemism was found at the Stöcklicht spring in Reichenbach/Fils. The type of water is Na-Ca-Mg-HCO_3 with a very high Na-content. The spring is located above a fault where rising Na-HCO_3 water is mixed with the Stubensandstein-groundwater. This interpretation is supported by the observed higher content of fluoride, which is characteristic in lower Stubensandstein layers.

In the Stubensandstein aquifer of this area mineral water is also obtained. It is of the

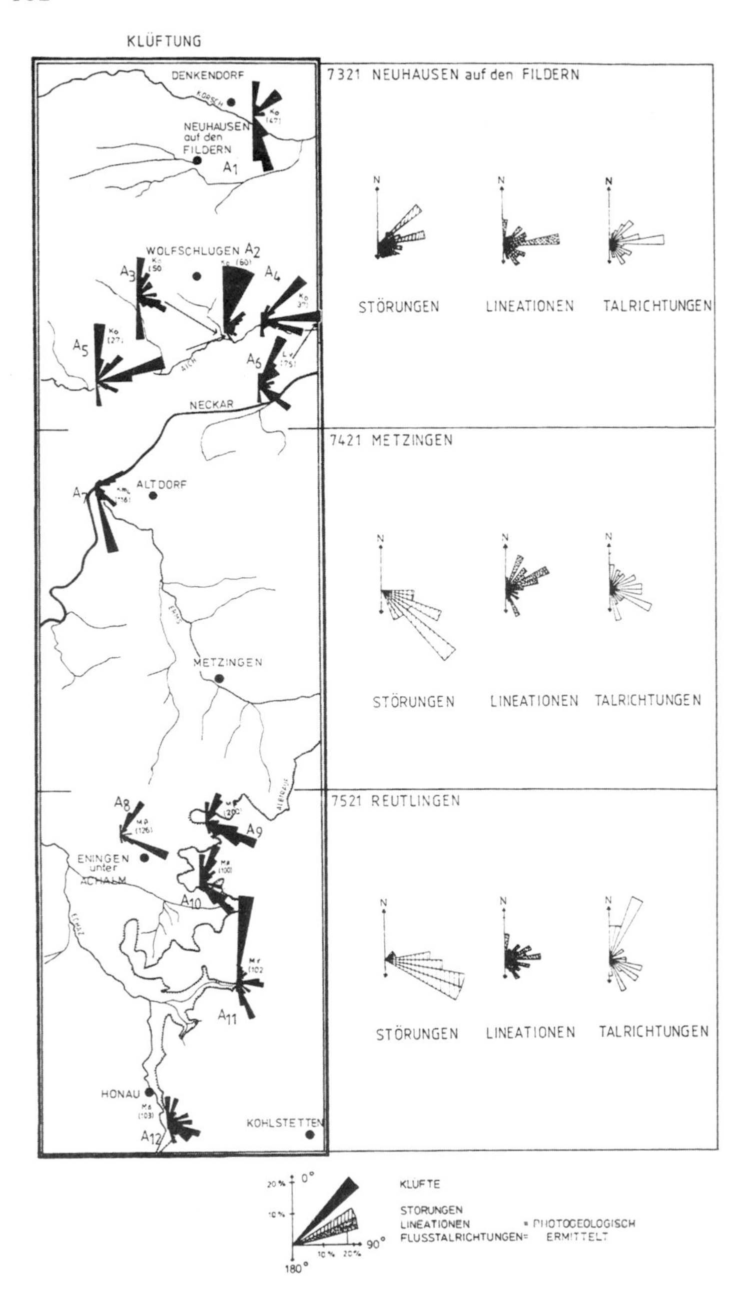

Fig. 4. Directions of joints, faults, lineations and valleys (Klüftung – jointing, Klüfte – joints, Störungen – faults, Lineationen – lineations, Talrichtungen – directions of valleys).

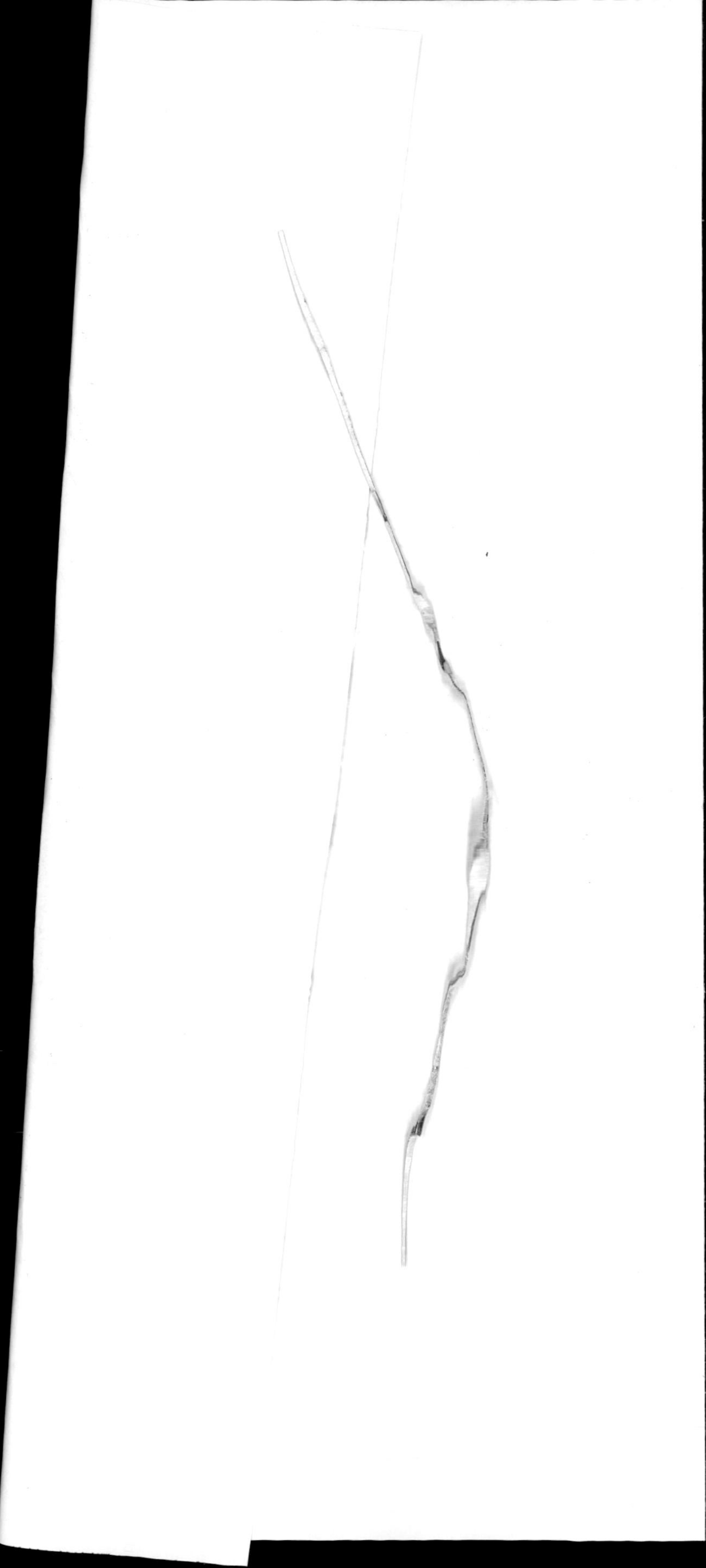

Na-HCO_3-, Na-HCO_3-SO_4-, and Na-SO_4-HCO_3-types. An exception is the „Bitterquelle" in the area of Göppingen, where water which has a higher temperature and a higher Ca-, Mg-, and SO_4-content is obtained. This irregularity can be explained by rising roundwater from aquifers of the Gipskeuper or by water which rises from yet deeper layers and crosses those of the Gipskeuper. The faults situated in the area of the Filstal edge can be viewed as a plausible path for the infiltration of such vertical convections streams (Fig. 5).

The layers of the Lias contain some aquifers. The most important aquifer is the Angulatensandstein. The chemism of the water of the Angulatensandstein is very different. In particular, higher temperatures and extreme differences of the salt content in the water of the Badquelle 1, Staufenbrunnen, Mörickebrunnen, Neuer Brunnen, Barbarossabrunnen and the Uhlandbrunnen (all situated in the area of Göppingen) suggest the occurrence of admixtures.

In the region of Reichenbach-Göppingen, areas where rising groundwater affects the

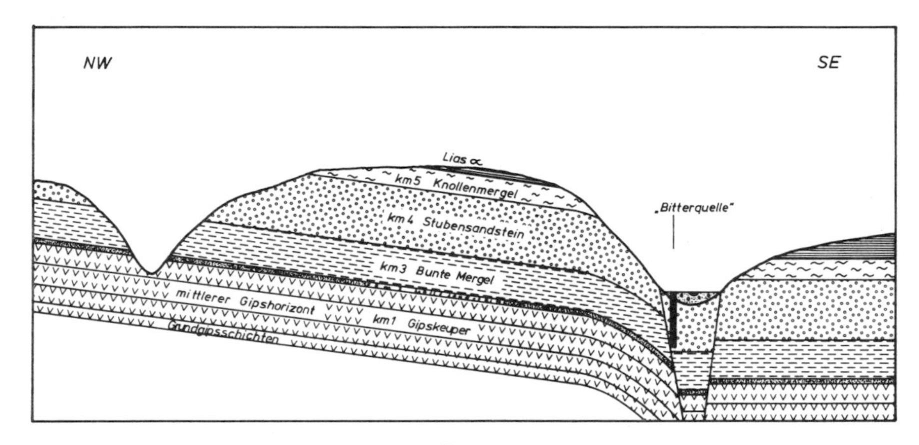

Fig. 5. Geological cross-section of the Fils-valley.

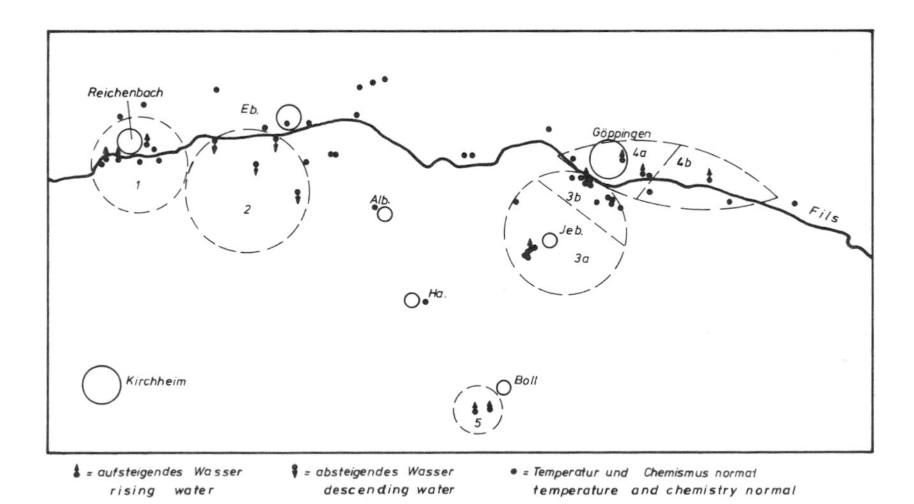

Fig. 6. Areas with rising and descending waters.

groundwater of the Stubensandstein and the Angulatensandstein can be determined. Descending water probably occurs only within a limited area (Fig. 6). Rising groundwater can be expected because of the data of the Bitterquelle, Stöcklicht- and Weberbrunnen in the area 1. The Ca-HCO_3 Stubensandstein water which occurs in the area 2 – this groundwater normally has a high Na-content at a relatively low temperature – indicates an admixture of descending groundwater. The region is tectonically very much disturbed; hence there probably are paths for the ascent of the water. The area 3 is nearly unaffected; only the Schloßquelle could receive inflows of the rising groundwater. Higher contents of Na^+, Cl^-, F^- and elevated temperatures of the water indicate an admixture of deep groundwater in the areas 4 and 5.

The layers of the Dogger do not have an important function as aquifers because of their mainly clayey composition. Nevertheless, some wells and springs with little discharge, where water flows from intercalated sandstone and limestone sequences, are of a certain local importance. In the area of the Achalm (a small mountain near Reutlingen), several shallow wells are situated within aquifers of the Dogger. Their water temperature is raised to 5–9 °C and they show a higher degree of mineralisation. These facts can be regarded as a clear indication of an inflow of deeper groundwater, (Tab. 1). This water

Table 1. Temperature and mineralisation of the groundwater of the Achalm area.

	Depth of well m	Temperature of the sample in summer °C	Temperature of the sample in winter °C	TDS in summer mg/l	TDS in winter mg/l
51	13.5	18.2	12.6	804.3	810.5
52	10.5	17.2	14.5	881.9	876.9
54	17.5	–	12.1	–	1109.1
58	8.6	13.9	12.1	932.0	916.3
46	13.0	–	10.8	–	1248.5
47	7.0	11.7	4.5	898.1	794.7
48	7.0	10.2	7.7	899.9	762.2
49	spring	–	9.5	–	952.2
50	7.0	11.6	7.6	781.9	821.2
55	5.0	–	9.2	–	758.7

is of the Ca-HCO_3-category with in part very high Na and Cl-contents up to 40 respectively 31 mval %. The paths of ascent by the water probably are the numerous faults of the Achalm-trench; most of the wells are situated above the faults or in the proximata area, (Fig. 7).

Most of the calcareous deposits of the Malm contain Karst-groundwater. Rising deep groundwater could not be proved in the investigated water.

The groundwater of the valley plains of the rivers Neckar, Echaz, Erms, Lauter and Fils does not give a clear indication of vertical groundwater convection, probably because it is affected by anthropogenetic influences in most of the areas.

To obtain information about the origin of the deeper groundwater and any mixtures of rising deep groundwater with higher groundwater, isotopic researche and geochemical temperature determinations were carried out. Within a period of two years the isotopic ratios of hydrogen and oxygen in thermal water and surface water of the area of the Urach heat anomaly were investigated. The surface water shows variations of its isotopic

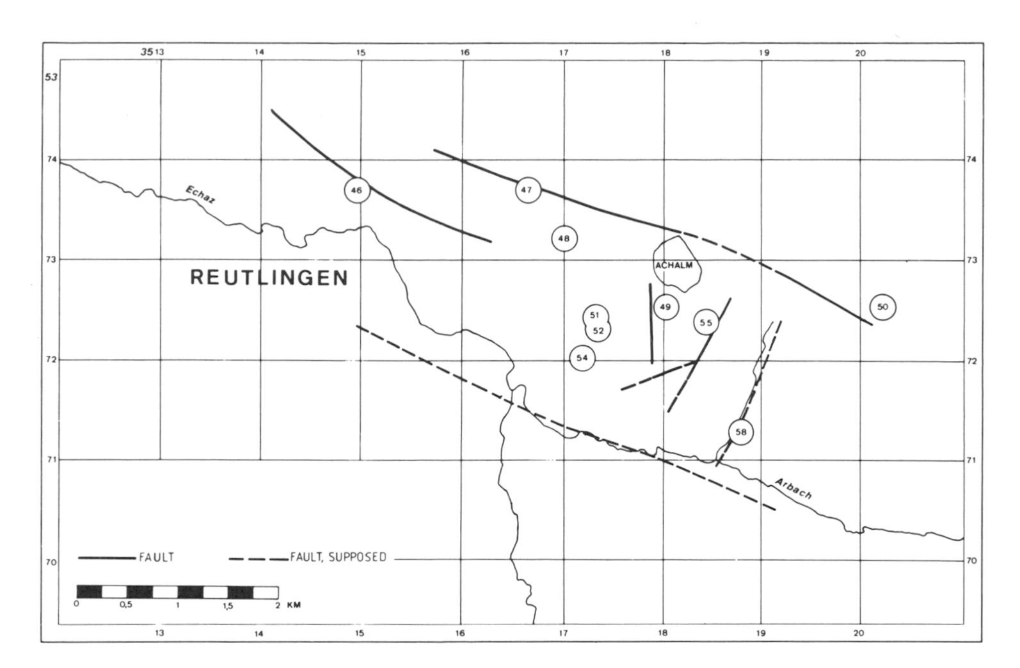

Fig. 7. Positions of the wells and the tectonic pattern in the area of Achalm.

composition. In summer heavier ^{18}O- and D-data were measured than in winter. The mean isotopic data of the surface water yield $\delta^{18}O = -10{,}0\ ^0/_{00}$ and $\delta D = -71{,}0\ ^0/_{00}$. The water complyies exactly with the Craig-curve.

Samples of thermal water were at first taken in a 14-day and later in a 4-week cycle. This water can be classified into 5 categories on the basis of the differences in chemism and isotopic composition. In the Muschelkalk aquifer, water with three different isotope ratios is found; the thermal water of Beuren 2, Bad Ditzenbach-Canisiusquelle and Bad Boll has ^{18}O- and D-contents which are lower than those of the local rainwater.

The thermal water can be viewed as a mixture of deep circulating hot water and cooler water which already is situated near the surface. The portions of cold water were estimated to be 40–77 % according to different models of mixing. Measurements of tritium yielded data which indicate higher age of the thermal water. Nevertheless it is supposed that admixtures, probably episodical, took place, because thermal water from one source showed a higher concentration of tritium.

The use of different geochemical thermometers showed highly divergent results at distinct temperatures. The cation thermometers show higher temperatures than the SiO_2-thermometer. The highest SiO_2-temperatures were measured in the water of the deepest aquifers (Urach and Beuren 1). The SiO_2-temperatures are interpreted as the probable data for water which is circulating and rising at the transition zone of sedimentary rocks and basement. While rising it mixes with cooler water penetrating from above, and this process results in modified thermal water. The temperatures which were measured by the SO_4-thermometer correspond quite well with the temperatures of the aquifer for the specific thermal water. With a high probability, the SO_4-ions of this water originate from the gypsum dissolved from the Gipskeuper layers.

The $\delta^{18}O$- and $\delta^{13}C$ data of the CO_2 samples which were taken in Urach have to be viewed in close correlation with the carbonates of that area. The $\delta^{18}O$ and $\delta^{13}C$ of the gas samples showed only slight variations within the period of measurement.

Gas samples which originated from the research borehole Urach 3 also contained CO_2 gas with $\delta^{18}O$ data and $\delta^{13}C$ data which correspond to those of the CO_2 from the thermal aquifer. A gas sample, which was taken from the Urach 3 borehole at a depth of 1453 m, showed an unusually low ^{13}C concentration. A probable reason for this observation is the replacement by CH_4 also present at greater depth.

The Urach Geothermal Project, p. 187–196;
Schweizerbart'sche Verlagsbuchhandlung, Stuttgart, 1982

The Thermal Water of the Urach-Kirchheim Heat Anomaly, Germany: An Isotopic and Geochemical Study

B. KOLLER and H. FRIEDRICHSEN

with 6 figures and 3 tables

Abstract: The oxygen and hydrogen isotopic composition, and the Na, K, Ca and SiO_2 contents of various samples of thermal and surface water in the area of the Urach-Kirchheim heat anomaly were investigated during the years 1975–1979. The thermal water, which is recovered from drilled wells, has temperatures up to 60 °C. Isotopic and chemical analyses clearly indicate that most of this water has been derived directly from meteoric water, and that none has experienced significant oxygen isotopic exchange with the host rocks.

In most cases, two or more aquifers, which can be distinguished isotopically and chemically, were tapped in each borehole. The depths of the aquifers range from 350 to 780 m below the surface, and the host rock consists mainly of Triassic limestone and sandstone. The different samples of thermal water vary in their isotopic composition, with $\delta^{18}O$ values ranging from – 10.0 ‰ to – 11.4 ‰ and δ D values from – 66.6 ‰ to – 78.2 ‰. In comparison, surface springs, which represent average groundwater conditions, have $\delta^{18}O$ values of – 8.7 ‰ to – 11.3 ‰, and δ D values of –65.7 ‰ to – 72.7 ‰. Within one formation, such as the Upper Muschelkalk, as many as three isotopically different types of thermal water have been found. These differences are interpreted as being the result of mixing of a deep circulating component with a shallower, colder component. Two samples of thermal water with unusually light $\delta^{18}O$ values may have been derived from a different, more elevated recharge area, or from older precipitation formed during a cooler period. The application of chemical thermometers yields considerably different estimates of the temperatures. Na-K and Na-K-Ca yield temperatures which are too high. The sulphate geothermometer, however, yields temperatures which are generally in very close agreement with the measured reservoir values. The estimated quartz temperatures are somewhat higher than the measured reservoir values, but in reasonable agreement for temperatures inferred from geophysical methods for the lower part of the hole. This fact, together with calculations based on chemical mixing models, suggests that a hot water component at about 90°–110 °C rises from the bottom of the sedimentary column and mixes with cooler, shallower meteoric water to produce the thermal water of the reservoir.

Introduction

The chemical composition of geothermal fluids and the isotopic composition of certain dissolved ions and gases are closely related to the thermal history. The world-wide compilation of hot spring water by TRUESDELL (1976) indicates that the Na, K, Ca and SiO_2 contents, as well as the 0-isotopic composition of dissolved sulphate, can yield information on reservoir temperatures. Such data are fundamental tools in the exploration and exploitation of geothermal energy. Until now nearly all investigations have been concentrated on high-enthalpy geothermal resources with reservoir temperatures higher than 100 °C. Only limited work has been done on low-enthalpy resources with reservoir temperatures between 50° and 100 °C.

The purpose of this study was to obtain information on water-rock exchange processes in low-temperature geothermal reservoirs in the Urach-Kirchheim area, and to test the applicability of geothermometers which have been used successfully in high-temperature geothermal fields.

$\delta^{18}O$ and δD studies on geothermal fluids from different regions of the earth show that they are of local meteoric origin. Juvenile components are generally not detectable, and at most make only a minor contribution to the thermal water. Since the work of CRAIG (1963), we know that the isotopic composition of meteoric water (and therefore of hot-spring water) varies with the mean annual temperature of different climatic regions. Cold areas have isotopically light precipitation, whereas tropical areas have 0- and H-isotopic values near SMOW. Furthermore, within a given region, mean 0- and H-isotopic values of precipitation (and groundwater) are lower in topographically elevated areas than in adjacent valleys. On this basis, it is possible to detect recharge areas for the different reservoirs.

Experimental Procedure

Na, K, Ca, Mg and SiO_2 contents were determined by standard AAS techniques on filtered water samples. The standard deviation for replicates is ± 1 % for Na, K, Ca and Mg, and ± 2 % for SiO_2. δD and $\delta^{18}O$ values were measured on H_2 and CO_2 gases with the use of a double collector Nier-type mass spectrometer. H_2 was prepared by reduction of water with hot (800 °C) uranium; CO_2 was equilibrated with water at 18 °C.

Dissolved sulphate was separated with an ion exchange column, and precipitated as $BaSO_4$. After drying at 120 °C, sulphate oxygen was allowed to reacted with carbon at 1050 °C, yielding CO and CO_2. The CO was then converted to CO_2 by high voltage discharge between two platinum plates. Total CO_2 was analysed for its ^{18}O-isotopic content. Replicate analyses yielded a precision of ± 1.0 ‰ for δD and ± 0.2 ‰ for $\delta^{18}O$. Isotopic ratios are reported by using common δ-notation relative to SMOW (CRAIG 1961).

Geothermometry

Various geothermometers have been applied to estimate reservoir temperatures of geothermal fluids. The thermometers are based on Na, K and Ca contents, on silica solubility, and on the 0-isotope fractionation between water and dissolved sulphate (for details see TRUESDELL 1976; MIZUTANI & RAFTER 1969, and ROBINSON 1978). The following equations have thereby been used:

Na-K thermometer (Na, K contents in ppm):

$$T\,(^{\circ}C) = \frac{777}{\lg_{10}\,(Na/K) + 0.70} - 273.15$$

Na-K-Ca thermometer (Na, K, Ca concentrations in mole/l):

$$T\,(^{\circ}C) = \frac{1647}{\lg_{10}\,(Na/K) + \beta \cdot \lg_{10}\,(\sqrt{Ca}/Na) + 2.24} - 273.15$$

where $\beta = 4/3$ for $\sqrt{Ca}/Na > 1$ and $T < 100\ ^\circ C$
$\beta = 1/3$ for $\sqrt{Ca}/Na < 1$ or $T_{4/3} > 100\ ^\circ C$

Silica thermometers (Si content in ppm):

$$T\,(^\circ C) = \frac{1315}{5.205 - \lg_{10} SiO_2} - 273.15$$: Dissolution of quartz, and conductive cooling of water

$$T\,(^\circ C) = \frac{1015.1}{4.655 - \lg_{10} SiO_2} - 273.15$$: Dissolution of chalcedony

Sulphate isotope-thermometer:

$$T\,(^\circ C) = \sqrt{\frac{2.88 \cdot 10^6}{10^3\ 1\ \ln\alpha + 4.1}} - 273.15$$: Sulphate-water isotope fractionation

$$\text{where } \alpha = \frac{(^{18}0/^{16}0)_{SO_4^{\ 2-}}}{(^{18}0/^{16}0)_{H_2O}}$$

Generally, geothermometers can be applied to thermal water only under the following conditions: (1) temperature-dependent partitioning occurs in the reservoir, and the elements involved in the reactions are present at a sufficient concentration; (2) equilibration is attained between the water and host rock in the reservoir, and no re-equilibration takes place during ascent of the water to the surface. It should also be noted that if hot water mixes with cooler, shallower water, the quartz thermometer is affected and yields lower apparent temperatures. Thermometers based on K-Na-Ca ratios are not affected by such dilution and can be used, provided the cooler water is free of these ions. However, valid results are obtained from these thermometers only if reservoir temperatures are higher than 120 °C (TRUESDELL 1976), and if feldspars are the only minerals which have exchanged cations with the water. Since the first condition is not met, and the second condition is probably not met in the reservoirs under study, erroneous apparent temperatures can be expected.

Results and Discussion

Stable Isotopes

The oxygen and hydrogen isotopic composition of surface and thermal water at the Urach-Kirchheim heat anomaly were measured during the years 1975 to 1979. One hundred and forty-three meteoric water samples from rain, snow, rivers and surface springs were analysed for their $\delta^{18}0$ composition, and 44 samples for their δD composition. Seasonal variations of surface springs, which are representative of average groundwater, range from $\delta^{18}0 = -8.7$ ‰ to -11.3 ‰, and $\delta D = -65.7$ ‰ to -72.7‰. Variations in the isotopic composition of precipitation are considerably greater. However, all these values closely fit the meteoric water line of CRAIG (1963) (Fig. 2). The average $\delta^{18}O$ and δD values for all samples of surface water are -10.1‰ and -71‰ respectively.

Various thermal wells, which yielded water at temperatures ranging from 35 °C to 60 °C, have been drilled in the Urach-Kirchheim geothermal area. They are: Therme Urach (TU), Beuren 1 (Be1), Beuren 2 (Be2), Bad Boll (BB), Bad Ditzenbach-Canisiusquelle (BD-C), Bad Ditzenbach-Theresienquelle (BD-T), Bad Überkingen-Renatatherme (BU-RT), Bad Überkingen-Ottotherme (BU-OT) (Fig. 1). In most cases, each borehole

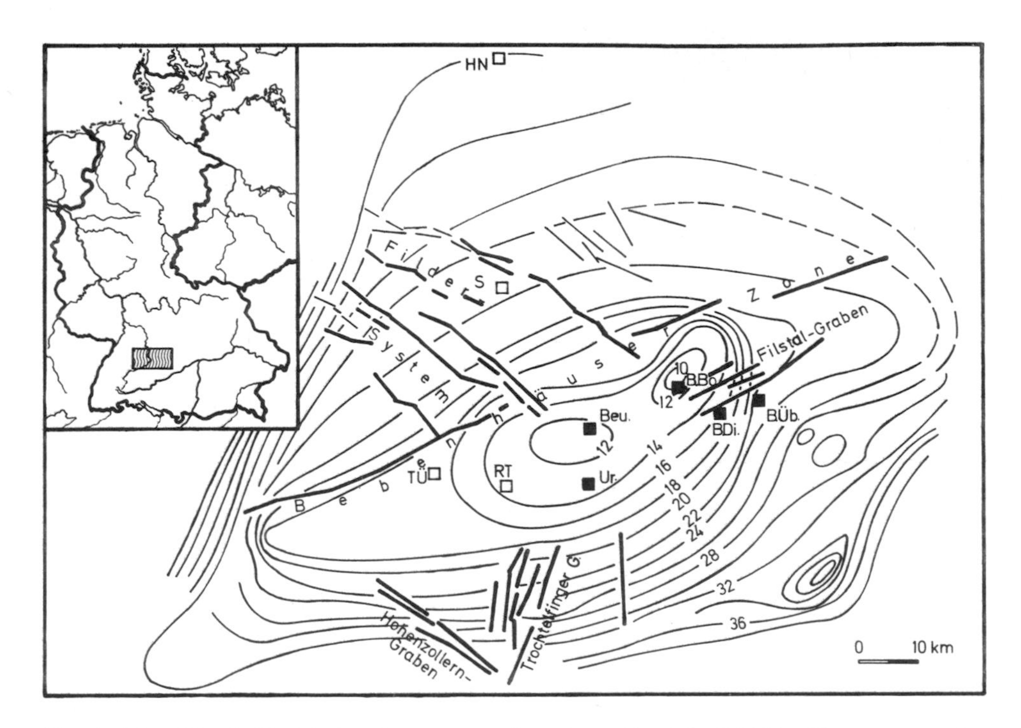

Fig. 1. The Urach-Kirchheim heat anomaly, after CARLÉ (1974). The contour lines indicate the vertical distance in metres for a temperature increase of 1 °C. The solid squares indicate locations of drilled thermal wells. Ur = Urach, Beu = Beuren, BDi = Bad Ditzenbach, BBo = Bad Boll, Büb = Bad Überkingen. Inset map shows location of study area, in the region of Baden-Württemberg.

tapped two and sometimes more aquifers. The thermal water was sampled over intervals of 14 days to one month. The mean values of all isotopic data and Na, K, Ca and SiO_2 contents are compiled in Tab. 1. Only minor variations in the $\delta^{18}O$ and δD values (± 0.4 and 3‰, respectively) have been detected during the sampling period; these variations are only marginally greater than the analytical uncertainty.

In Fig. 2, the average oxygen and hydrogen isotopic compositions of water from the different aquifers are plotted. It can be seen that the composition of most of the thermal water samples does not differ significantly from the average value of local meteoric surface water ($\delta^{18}O$ = – 10.1‰, δD = – 71.0‰). This indicates that extensive oxygen isotopic exchange between the host rock and the thermal water, which would lead to heavier $\delta^{18}O$ values, did not take place.

Nevertheless, five different thermal water categories can be distinguished on the basis of their oxygen isotopic and chemical compositions. Of these, three are located in the same formation, the Upper Muschelkalk, which consists mainly of Triassic marine limestones. These three categories, which represent the hottest and deepest water, are: Group 1, with $\delta^{18}O$ values of – 10.0‰ to – 10.1‰ and δD values of – 67.7‰ to – 68.4‰; Group 2, with $\delta^{18}O$ values of – 10.9‰ and δD values from – 77.7‰ to – 78.2‰; and Group 4, with a $\delta^{18}O$ value of – 10.5‰ and a δD value of – 73.4‰. Two aquifers located in the Stubensandstein formation (Group 3) have $\delta^{18}O$ values of – 10.0‰ to – 10.1‰ and δD values of – 66.6‰ to – 66.9‰. The thermal aquifer in the

Table 1. Chemical and isotopic data for the different wells in the Urach-Kirchheim area. In each average, 25 analyses are involved, except for TU where 43 samples were measured. Isotopic values in $^0/_{00}$.

Group	Well	Na (ppm)	K (ppm)	Ca (ppm)	SiO_2 (ppm)	$\delta^{18}O$ H_2O	δD H_2O	$\delta^{18}O$ Sulphate
1	TU	802	70	689	31	– 10.0	– 67.1	+ 11.7
	Be1	1020	83	741	28	– 10.1	– 68.4	+ 12.9
2	BB	1383	143	596	24	– 10.9	– 78.2	+ 13.0
	BD-C	1390	117	673	26	– 10.9	– 77.7	+ 12.0
3	BD-T	965	19	28	26	– 10.0	– 66.6	+ 13.0
	BU-RT	1117	15	21	19	– 10.1	– 66.9	+ 13.4
4	BU-OT	888	58	455	22	– 10.5	– 73.4	+ 13.7
5	Be2	540	8	6	18	– 11.4	– 75.8	+ 13.0

Rät-formation (Group 5) has an average $\delta^{18}O$ value of $- 11.4^0/_{00}$ and a δD value of $- 75.5^0/_{00}$, and thus contains the lightest water in the area of investigation, with $\delta^{18}O$ values even more negative than those of local meteoric surface water. The isotopically light nature of this water, as well as that from Group 2, requires that it be derived either from a more elevated recharge area, or from older precipitation formed during a cooler period.

The ^{18}O and ^{13}C isotopic composition of CO_2 gas discharged at the TU site has also been investigated. It has an average $\delta^{18}O$ value of $+ 25.5\ ^0/_{00}$ and a $\delta^{13}C$ value of $- 5.2\ ^0/_{00}$ relative to PDB. Both $\delta^{18}O$ and δD values are relatively constant with time. From the oxygen isotopic fractionation between H_2O and CO_2, the equilibration temperature between thermal water and CO_2 has been determined to be 58 °C. Since the temperature of the water has been measured as 59 °C (CARLÉ 1975), it is apparent that carbon

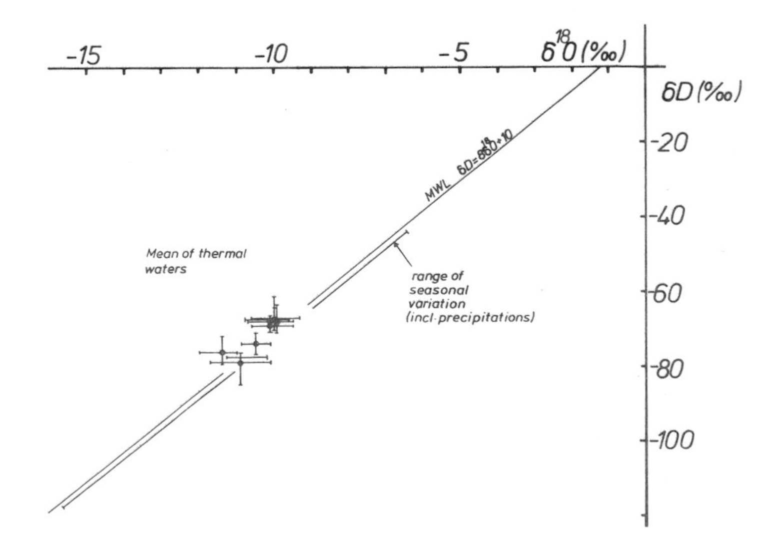

Fig. 2. $\delta^{18}O$ and δD values of thermal water. Crosses indicate the isotopic range of each aquifer, and dots the mean value. The bar parallel to the meteoric water line (MWL) indicates the range of seasonal variations of surface water and precipitation in the Urach-Kirchheim area.

dioxide and water equilibrated at the reservoir temperature, without re-equilibration during ascent to the surface.

Gas samples from the nearby research borehole Urach 3, taken at depths of 834 m and 1453 m below the surface, have also been analysed. They show $\delta^{18}O$ values of + 26.9 $^0/_{00}$ and +28.1$^0/_{00}$, and $\delta^{13}C$ values of –6.7$^0/_{00}$ and –15.7$^0/_{00}$ (PDB), respectively. The unusally light $\delta^{13}C$ values probably contain a component of CO_2 derived from oxidized organic matter.

Temperatures Estimated by Chemical Thermometers

The results of temperature estimates using the geothermometers mentioned above are listed in Tab. 2. It can be seen that considerably different temperatures result from the application of the various thermometers. Na-K-Ca temperatures range from 116 °C to 189 °C, and Na-K temperatures from 30 °C to 187 °C. Both thermometers yield temperatures which are too high in almost all cases for the reasons discussed in the section on geothermometry. Temperatures estimated from the quartz thermometers are considerably lower and closer to measured reservoir temperatures. The chalcedony thermometer yields temperatures that are lower than the measured reservoir temperatures, probably because insufficient chalcedony was present in the source to saturate the thermal water.

Fig. 3 shows that there is a close positive correlation between the reservoir temperatures and the estimated quartz temperature. The fact that the quartz temperatures are some-

Table 2. Depth of aquifers from which the samples were collected, directly measured reservoir temperatures, and temperatures estimated by using the different chemical thermometers.

Well	Depth (m)	T_{quartz} (°C)	T_{chalc} (°C)	$T_{Na\text{-}K}$ (°C)	$T_{Na\text{-}K\text{-}Ca}$ (°C)	T_{SO_4} (°C)	$T_{Reservoir}$ (°C)
TU	760	82	42	168	169	61	59
Be1	750	76	41	161	169	49	48
BB	440	70	31	187	189	45	49
BD-C	560	74	39	164	177	53	46
BD-T	427	74	43	52	129	42	38
BU-RT	350	62	27	30	116	49	35
BU-OT	520	67	29	139	160	44	41
Be2	381	60	29	35	118	44	37

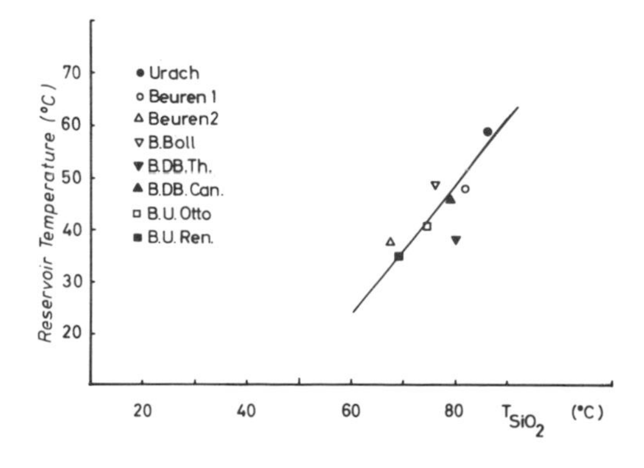

Fig. 3. Estimated quartz temperature versus reservoir temperature.

what higher than the reservoir temperatures indicates that a component of the reservoir water is derived from greater depths in the sedimentary column. The bottom of the sedimentary column is located at a depth of 1604 m, and has a temperature of 92.5 °C.

A plot of quartz temperatures versus depth (Fig. 4) suggests that the water becomes heated with increasing depth, and attains progressively higher SiO_2 concentrations. Subsequently, this hot water rises and mixes with cooler water which contains little or no SiO_2. Temperatures estimated from the oxygen isotopic fractionation between water and dissolved sulphate are, within analytical uncertainty, in general accordance with the respective reservoir temperatures. Estimated temperatures and actual reservoir temperatures agree best for the hot water of the TU and Be1 wells. The temperatures of the cooler water do not agree so well.

The fact that different temperature estimates result from different chemical thermometers is probably due to the circumstance that the reservoirs contain a mixture of hot deeper water and cooler shallower water (cf. TRUESDELL & FOURNIER 1976). Under favourable conditions, the warm-spring-mixing-model (WSMM) of FOURNIER & TRUESDELL (1974) provides a means of estimating the fraction of hot and cold water of mixed thermal water, as well as the maximal temperature reached by the hot water fraction. The dissolved-silica-versus-enthalpy-model (DSVE) of TRUESDELL & FOURNIER (1977) also can be used to estimate these parameters. Tab. 3 lists the fractions of cold water and the maximal temperatures reached by the hot component, calculated on the basis of these models.

From these models, cold water fractions of about 40 % to 70 % are estimated. The maximal temperature reached by the deep circulating water is about 110 °C (not considering BD-T), thus being about 25 °C higher than the maximal temperatures calculated

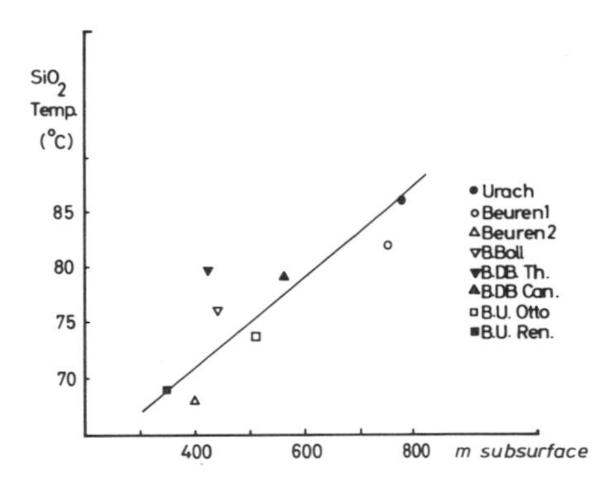

Fig. 4. Estimated quartz temperature versus depth.

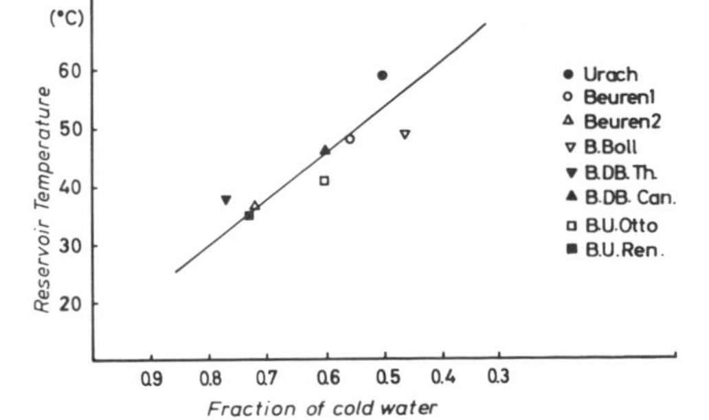

Fig. 5. Fraction of cold water component versus reservoir temperature.

Table 3. Cold water fractions and maximal temperatures of the hot component, as calculated for the warm-spring-mixing-model (WSMM) and the dissolved silica-versus-enthalpy-model (DSVE).

Well	Cold Water Fraction		Max. Temperature of Hot Water Component (°C)	
	WSMM	DSVE	WSMM	DSVE
TU	0.50	0.40	105	100
Be1	0.64	0.60	115	110
BB	0.54	0.42	105	80
BD-C	0.65	0.60	112	110
BD-T	0.76	0.77	132	138
BU-RT	0.73	0.74	108	105
BU-OT	0.65	0.60	105	90
Be2	0.68	0.68	105	100

for the quartz thermometers. A plot of cold water component versus reservoir temperature (Fig. 5) supports the idea that a hot water component at 90 °C to 110 °C rises from the bottom of the sedimentary column and subsequently mixes with cooler, shallower meteoric water.

The low tritium concentrations measured for this water (analysed by Dr. W. RAUERT, GSF, München) indicate that this is generally recycled water, older than several hundred years. However, the elevated tritium concentrations for one aquifer (BU-OT) suggests episodic influx of young meteoric water.

Fig. 6 illustrates the various geothermal gradients which can be calculated from the temperature data of the thermal water. The high geothermal gradient of up to 10 °C/

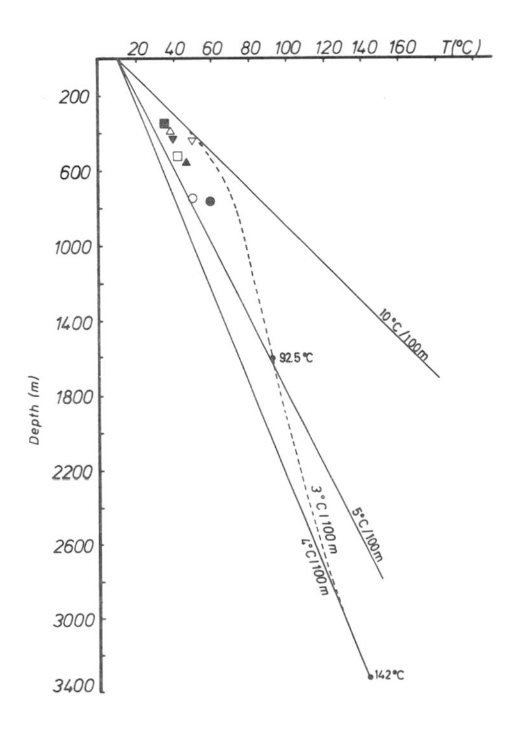

Fig. 6. Various geothermal gradients.

100 m (CARLÉ 1974) appears to be appropriate only for shallow depths, and for the BB thermal water. Geothermal gradients for the other cases of thermal water are between 5° to about 7 °C/100 m. However, temperature measurements at depths of 1604 m (92.5 °C) and 3334 m (142 °C) indicate that below the aquifers only a normal gradient of 3 °C/100 m exists. The Urach-Kirchheim heat anomaly is consequently interpreted primarily as the result of convective heat transport. The circulation of meteoric water heated by a normal temperature gradient has been made possible by the extensive fracture system which prevails in the area (Fig. 1).

Conclusions

(1) The isotopic composition of most Urach-Kirchheim thermal water is essentially the same as that of local surface spring water. There is no evidence for an oxygen isotopic shift towards heavier values, which would be indicative of water-rock isotopic exchange.

(2) Different reservoirs are characterised by different isotopic and chemical compositions. Three isotopically different categories of thermal water have been found within the same geological formation, the Upper Muschelkalk.

(3) A few samples of thermal water have isotopic compositions lighter than those of average local surface water. This water may have been derived either from a more elevated recharge area, or from older precipitation formed during a cooler period.

(4) Apparent temperatures derived from the different thermometers vary considerably. Cation thermometers cannot be applied to this water. The quartz thermometers agree best with temperatures inferred from geophysical data for the lower part of the sedimentary column. Temperatures calculated with the sulphate thermometer are generally in the range of the reservoir temperatures.

(5) Mixing models indicate cold water fractions of 40 %–70 % for the various samples of thermal water, and a maximal temperature of about 90° to 110 °C for the deep hot component.

Acknowledgements

We are indebted to M. SCHUMANN, C. SCHWARTZ and I. RUPP for their assistance with the isotope analyses; to A. SCHNECK for her assistance with the chemical analyses; and to Dr. T. J. BARRETT (Tübingen) for his critical review of the manuscript. We also acknowledge the financial support from the Deutsche Forschungsgemeinschaft.

References

Carlé, W., (1974): Die Wärmeanomalie der mittleren Schwäbischen Alb, Baden-Württemberg. – In: Approaches to Taphrogenesis (J.H. Illies & K. Fuchs, Eds.), p. 207–212, E. Schweizerbart, Stuttgart.

– (1975): Die Mineral- und Thermalwässer von Mitteleuropa: Geologie, Chemismus, Genese. – Bü. d. Zeitsch. Naturw. Rdsch. – Wissensch. Verlagsges., Stuttgart.

Craig, H., (1961): Standard for reporting concentrations of deuterium and oxygen-18 in natural waters. – Science **133**: 1833–1834.

– (1963): The isotopic geochemistry of water and carbon in geothermal areas. – In: Nuclear Geology on Geothermal Areas (E. Tongiorgi, Ed.), (Spoleto), p. 17–54, Consiglio Nazionale delle Ricerche Laboratoria di Geologia Nucleare, Pisa.

Dietrich, H.-G., Haenel, R., Neth, G., Schädel, K. & Zoth, H., (1980): Deep investigation of the geothermal anomaly of Urach. – In: Advances in Europa Geothermal Research (A.S. Strub & P. Ungemach, Eds.), p. 253–266, Proc. of the 2nd Int. Seminar on the Results of EC Geothermal Energy Research, Strasbourg.

Epstein, S. & Mayeda, T., (1953): Variations of ^{18}O content of waters from natural sources. - Geochim. et Cosmochim. Acta **4**: 213–224.

Fournier, R.O. & Truesdell, A.H., (1974): Geochemical indicators of subsurface temperature: part 2, estimation of temperature and fraction of hot water mixed with cold water. – Jour. Res. U.S.G.S., p. 263–270.

Friedmann, I., (1953): Deuterium contents of natural waters and other substances. – Geochim. et Cosmochim. Acta **4**: 89–103.

Koller, B., (1980): Geochemische und isotopengeochemische Untersuchungen an Thermalwässern aus der Urach-Kirchheimer Wärmeanomalie. – Dis. Univ. Tübingen.

Mizutani, Y., Rafter, T.A., (1969 a): Oxygen isotopic composition of sulphates; 3. Oxygen isotopic fractionation in the bisulphate ion-water system. – N.Z. Jour. Sci. **12**: 54–59.

Truesdell, A.H., (1976): Geochemical techniques in exploration. – Proc. 2nd U.N. Symposium on the Development and Use of Geothermal Resources (San Francisco, 1975), 1. U.S. Govt. Printing Office, Washington, D.C.

Truesdell, A.H. & Fournier, R.O., (1976): Calculation of deep temperatures in geothermal systems from the chemistry of boiling spring waters of mixed origin. – Proc. 2nd U.N. Symposium on the Development and Use of Geothermal Resources (San Francisco, 1975), 1. U.S. Govt. Print. Office.

(1977): Procedure for estimating the temperature of a hot water component in a mixed water by using a plot of dissolved silica versus enthalpy. – Jour. Res. U.S.G.S. 5, **1**: 49–52.

The Urach Geothermal Project, p. 197–203;
Schweizerbart'sche Verlagsbuchhandlung, Stuttgart, 1982

Geothermal Investigations in the Area of the Urach Anomaly

BEHRENS, J., ROTERS, B. and VILLINGER, H.

with 5 figures and 1 table

Abstract: During geothermal investigations in the area of the Urach anomaly five heat flow values have been determined with measurements made in shallow boreholes with a depth of about 35 m. The heat flow in two of them – situated north of Urach – is significantly higher than the measured mean value for Germany, which is about 70 mW/m^2. In comparison with these two values the other three have a normal heat flow of about 75 mW/m^2. The error of all values amounts to about 30 %. The reliability of four values is low. Only the value at Urach is very reliable because it is in good accordance with the heat flow value determined in the deep borehole Urach 3, which gives a heat flow values of 86 mW/m^2.
In boreholes where no cores were available the thermal conductivity has been measured with a newly developed probe for in situ thermal conductivity measurements. The in situ values could be checked in some cases where cores were available and the agreement of the in situ values with the laboratory values is remarkably good.

Introduction

The aim of the geothermal investigations in the area of the Urach anomaly was to find out whether reliable heat flow determinations could be made with measurements in shallow boreholes with a depth down to 35 m.

For this purpose the measurement of the undisturbed temperature gradient in the borehole and the thermal conductivity of the sorrounding rock are indispensable. Normally, the thermal conductivity is determined on core samples in the laboratory. Problems often arise when the shallow borehole is in unconsolidated sediments, as was frequently the case with shallow boreholes. If no cores are available the thermal conductivity has to be determined in situ. Several attempts to solve this problem have been made in the past, but no satisfactory working probe has been developed up to now. Therefore, the development of a probe for in situ determination of the thermal conductivity was necessary (BEHRENS et al. 1980).

After the determination of the thermal conductivity and the undisturbed temperature gradient it is possible to calculate the heat flow as the product of the thermal conductivity and the temperature gradient.

The positions of the shallow boreholes mentioned are listed in Tab. 1 and are shown in Fig. 1. They are arranged along a profile perpendicular to the strike of the anomaly. All of them are cased with a PVC-casing with an inner diameter of 2". The total length of the core samples amounts to 36 m. Reliable temperature gradients could be determined only in five boreholes. In the other boreholes the temperature field was severely disturbed, and so they could not be used for heat flow determinations.

Table 1. Results of heat flow measurements.

Borehole	Position west/north	Depth- (m)	Measured gradient (°C/m)	Topogr. correct. (°C/m)	Thermal con-ductivity (W/m K)	heat flow (mW/m²)
Plattenhardt	9°12'/49°39'	20–30	0.0725	+ 0.0137	1.65	142.2
Frickenhausen	9°21'/48°35'	16–35	0.108	– 0.0125	1.31	125.1
Urach	9°22'/48°30'	16–36	0.0707	– 0.021	1.54	76.5
Hechingen	8°57'/48°20'	20–30	0.0457	+ 0.0027	1.58	76.5
Hausen	9° 4'/48°17'	20–35	0.080	– 0.0238	1.31	73.6

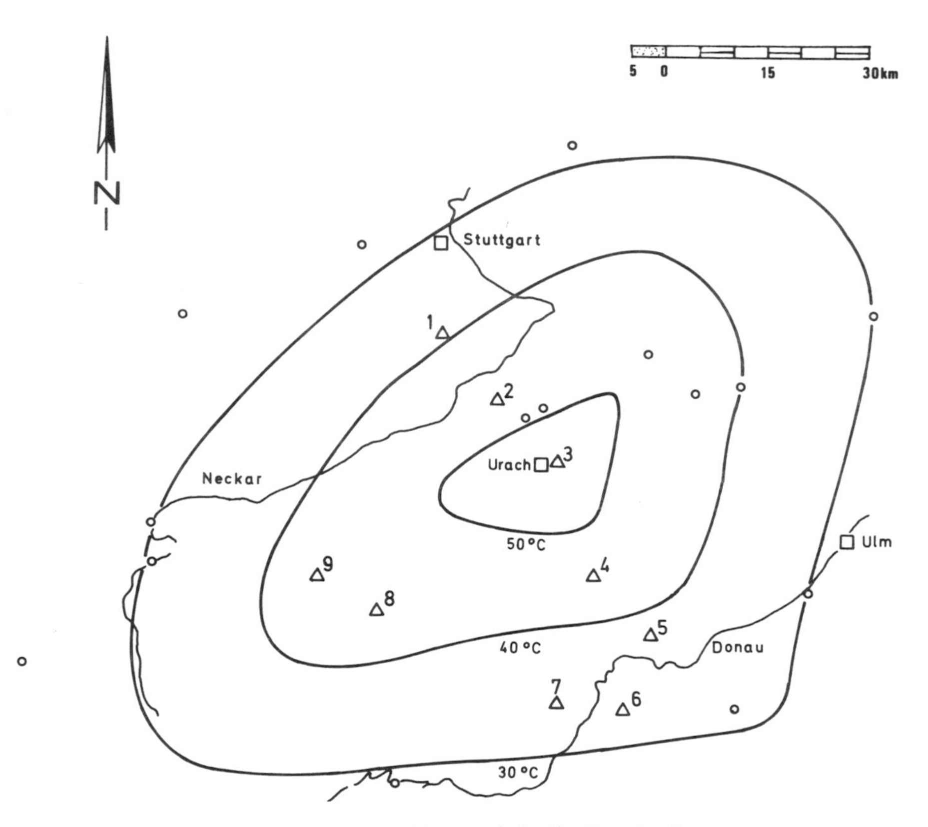

Fig. 1. Geothermal anomaly after Haenel (1974). The isolines represent temperatures at 500 m depth. △: positions of shallow boreholes.

Determination of Temperature Gradient and Thermal Conductivity

After the boreholes had attained thermal equilibrium several temperature measurements were performed during one year. The relative accuracy of the measurements is ± 0.01 °C. Figs. 2 and 3 show the resulting temperature-depth curves of two boreholes together with the stratigraphy.

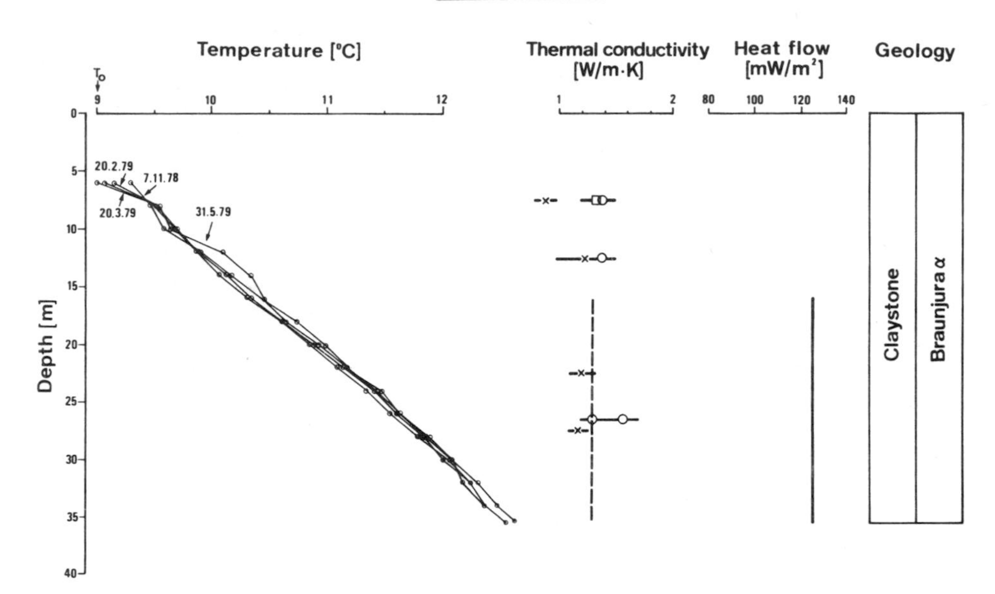

Fig. 2. Borehole Frickenhausen. Temperature and thermal conductivity as a function of the depth and the calculated mean heat flow.

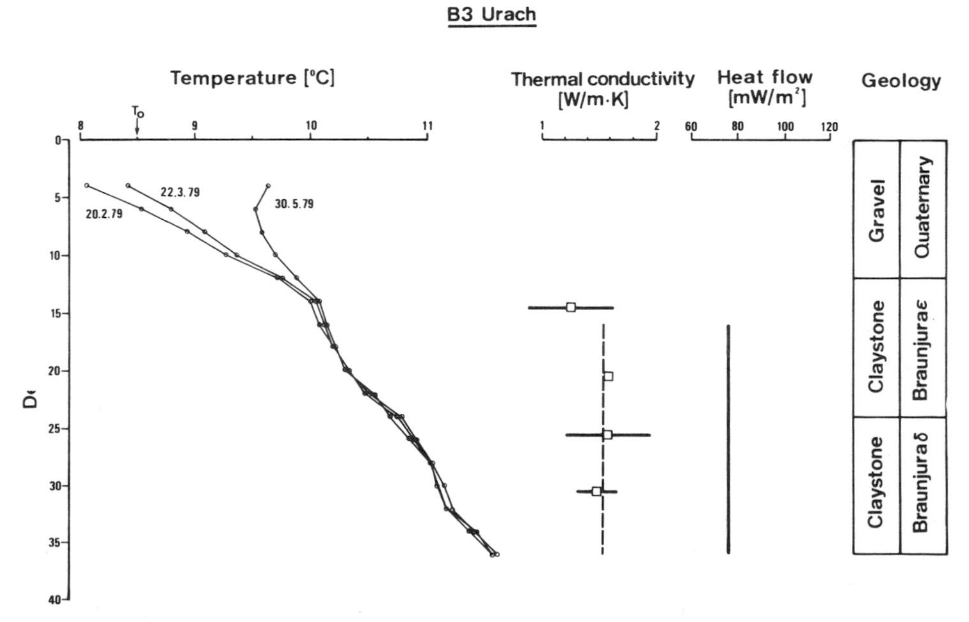

Fig. 3. Borehole Urach. Temperature and thermal condutivity as a function of the depth and the calculated mean heat flow.

The estimation of temperature disturbances due to seasonal variations of the temperature at the earth's surface shows that this influence is less than 0.01 °C below a depth of 15 m. Therefore the temperature gradient has been calculated only from measurements performed below 15 m. The extrapolation of the temperature depth curve to the earth's surface is supposed to give the mean annual temperature T_0 of the earth's surface. This value can also be obtained from the mean air temperature (Klimaatlas von Baden-Württemberg, 1953) by correcting for the vegetation in the surroundings of the hole (KAPPELMEYER & HAENEL 1974). The agreement between the calculated T_0 and the extrapolated T_0 is quite good. This fact indicates that the temperature field in the investigated depth interval is not severely disturbed.

The topography in the immediate surroundings of the boreholes disturbs the normal temperature field down to a certain depth. This disturbance calculated after Bullard (KAPPELMEYER & HAENEL 1974) and the resulting corrections of the gradients are tabulated in Tab. 1. In two cases (Hausen and Urach) where the drill holes are situated in quite narrow valleys, their corrections amount to about 33 % of the measured gradient. All other corrections are less in magnitude.

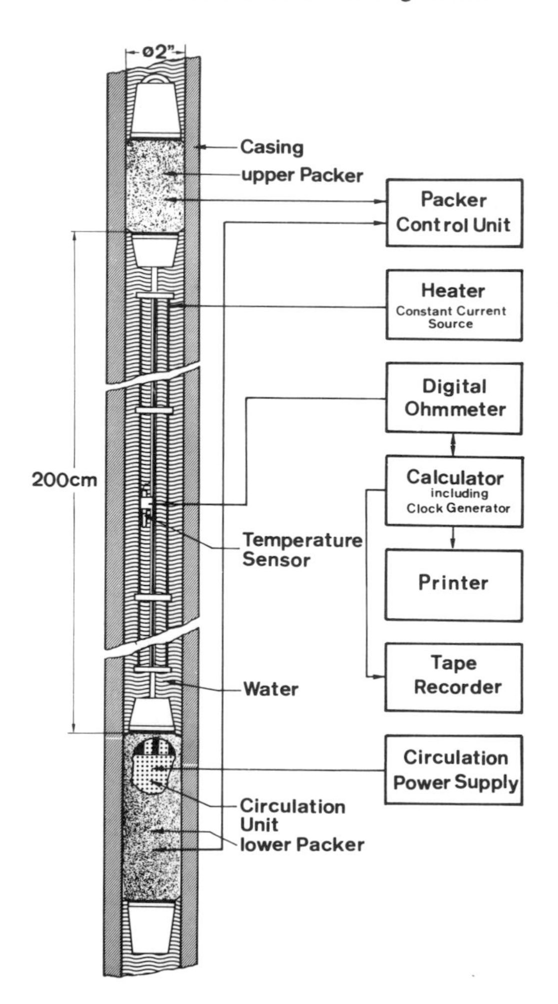

Fig. 4. Schematic cross section of the thermal conductivity probe (left) and block diagram of the control- and measuring unit (right).

The calculated mean temperature gradient, the topographic corrections and the corrected gradients are listed in Tab. 1. Thermal conductivity measurements were made on core samples in the laboratory of the Niedersächsisches Landesamt für Bodenforschung (Hannover). Two different instationary methods were used, according to the mechanical conditions of the cores. One instrument was the well known needle probe (von HERZEN & MAXWELL 1959) the other was the "Quick Thermal Conductivity Meter" (QTM), described in detail by SUMIKAWA & ARKAWA (1976). Several measurements were carried out on each core, which normally had a length of one metre. The mean values representing the average over one metre of core and the standard deviations are shown for the boreholes Frickenhausen and Urach in Figs. 2 and 3. The large standard deviations are due to the inhomogeneities of the cores.

After the basic investigations of BECK et al. (1956), the measuring principle for the in situ determination of thermal conductivity consists of a cylindrical source heated in a water-filled borehole. The source is realized in a separate section of the water-filled borehole. If the length to diameter ratio exceeds a value of about 20 and the surrounding rock has a uniform and isotropic thermal conductivity, the heat source may be regarded as an infinite cylindircal source. According to this theory the temperature rise measured by means of a temperature sensor at the centre of the source is a linear function of the logarithm of time after some time of heating. The slope of the straight line thus obtained is inversely proportional to the thermal conductivity of the surrounding rock.

The realization of the measuring principle is shown schematically in Fig. 4. To separate a section of the water-filled borehole two packers, either working with air pressure or based on the bottle brush principle are used. The thermally insulated water column is heated with a constant current input to four wire heating elements. To fulfill the condition of a perfectly conducting probe (BECK et al. 1956) the water column is stirred with the aid of a small pump installed in the lower packer. Because of the stirring it is sufficient

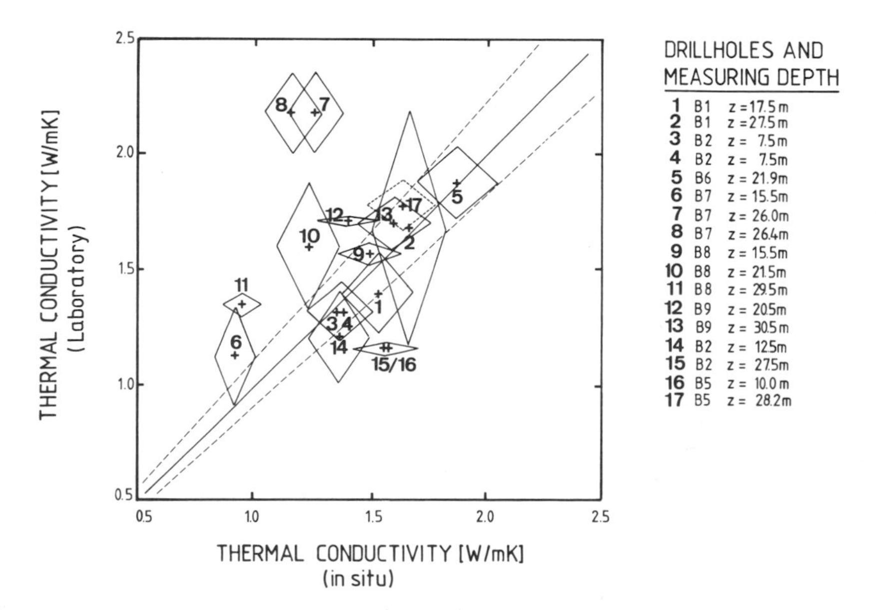

Fig. 5. Comparision of thermal conductivities (in situ) and thermal conductivities (laboratory) for the same depth section.

to measure the temperature at the centre of the water column. The temperature sensor used is a calibrated thermistor.

The recording and control unit (see Fig. 4) is located at the earth's surface. It consists of a unit for handling and control of the packers and a device for measuring and storing the data with the aid of a computer controlled data logging system. To test the reliability of the in situ values, thermal conductivities were measured in the laboratory on core samples from the same depth interval. Fig. 5 shows the comparison of in situ and laboratory values. The in situ values agree remarkably well with the laboratory values and are mostly within a range of ± 10 %. This comparison allows only a very rough estimate of the accuracy of the proposed new method because of the different spatial dimensions in which the thermal conductivities were measured in situ and in the laboratory.

Heat Flow Calculation

The terrestrial heat flow can be calculated as the product of the undisturbed vertical temperature gradient and the thermal conductivity of the surrounding rock. The temperature gradients used for the calculation of the heat flow are listed in Tab. 1. The thermal conductivities used are the mean values of all thermal conductivity values measured in the investigated depth sections. The mean heat flow values are also listed in Tab. 1 and are shown for the boreholes Urach and Frickenhausen in Figs. 2 and 3. Since the error of the thermal conductivities is about ± 10 % and the error of the gradient is about ± 20 %, the relative accuracy of the calculated heat flow is about ± 30 %.

Conclusion

The results of the geothermal investigations in the Urach area can be summarized as follows:

1. Five mean heat flow values have been obtained from measurements in shallow boreholes with a depth down to 35 m. The reliability of four of these values is uncertain because it is very hard to exclude disturbance of the temperature field which cannot be detected over such a small depth interval. An error of about ± 30 % therefore has to be assumed for these values. The reliability of the heat flow value of Urach is quite good because this shallow borehole is situated very close to the deep borehole Urach 3. The value obtained in the shallow borehole is 76.5 mW/m^2 and agrees quite well with the value of 86 mW/m^2, obtained with measurements in the deep borehole.

2. The proposed new method for the in situ determination of thermal conductivity in boreholes gives encouraging results, as the comparison of in situ and laboratory values shows.

3. The mean heat flow values along a profile perpendicular to the strike of the anomaly do not reflect the geothermal anomaly. This may not be caused by the errors of the heat flow values but may be due to the limited number of boreholes along this provile. Geothermal anomalies can be detected with heat flow determinations in shallow boreholes only if a great number of such shallow boreholes is available in order to minimize the influence of disturbed values on the resulting heat flow profile.

Acknowledgement

The authors are indebted to Dr. HAENEL and Mr. ZOTH, both of the Niedersächsisches Landesamt für Bodenforschung (Hannover, FRG) and Messrs. GEISER, SCHENKLUHN and SCHMARSOW, as well as to Mr. TÖPPER and Mrs. CRAMER (all TU Berlin) for their valuable help and advice. The work was kindly supported by the Commission of the European Communities (Brussels, Belgium) and the Ministry of Research and Technology of the Federal Republic of Germany (Bonn, FRG).

References

Beck, A., Jager, J.C. & Newstead, G. (1956): The measurement of the thermal conductivities of rocks by observations in boreholes. – Austr. J. Physics, **9**: 286–296.

Behrens, J., Roters, B. & Villinger, H. (1980): In situ determination of thermal conductivity in cased drill holes. – In: Seminar on Geothermal Energy, Straßburg.

Kappelmeyer, O. & Haenel, R. (1974): Geothermics. – Bornträger, Stuttgart, Berlin.

Sumikama, S. & Arakawa, Y. (1976): Quick thermal conductivity meter. – Instrumentation and Automation, **4**: 60–66.

von Herzen, R. & Maxwell, A.E. (1959): The measurement of thermal condictivity of deep-sea sediments by a needle probe method. – J. Geophys. Res., **64**: 1557–1563.

The Urach Geothermal Project, p. 205–221;
Schweizerbart'sche Verlagsbuchhandlung, Stuttgart, 1982

The Temperature Field of the Urach Area

G. ZOTH

with 12 figures and 2 tables

Abstract: Ten new temperature maps have been constructed for depths between 50 m and 5000 m. Existing and recently measured temperature data have been incorporated from the Swabian Alb and its surroundings. Four shallow boreholes were drilled down to 50 m and one down to 80 m. From the temperature measurements in these boreholes and the measured thermal conductivity from rock cores, the heat flow density was determined to be 67–135 mW m^{-2}. And additional three deep boreholes yielded values between 64 mW m^{-2} and 126 mW m^{-2}. Further, it is shown that near the surface, the high temperature areas correspond to ascending water and low temperature areas with descending water.

1. Introduction

Additional temperature and heat flow density measurements were necessary to more precisely define the limits of the Urach temperature anomaly. The most recent representation of the Urach temperature dates from 1976 (HAENEL 1976). Since then, 37 new boreholes were drilled where the thermal conditions were not well known. The cost for the five shallow boreholes were partly covered by the EG Project 1 (Geothermics) 077–76 and partly by a common fund of the geological surveys of the Federal Republic of Germany.

The temperature and the thermal conductivity measurements were carried out by the Geological Survey of Lower Saxony, Hannover.

2. Shallow Boreholes and their Use for Heat-Flow-Density Determinations

The sites of the shallow boreholes were selected in cooperation with Dr. K. SCHÄDEL from the Geological Survey of Baden-Württemberg.

The following aspects were taken into consideration:

– The boreholes should be drilled near the expected transition from the high temperature to the normal temperature distribution.

– The sites of the boreholes should be selected so that karst rock containing circulating water is not drilled through, i.e. stratigraphically below any karstified rock.

– The non-consolidated cover should be not too thick, so that samples of consolidated rock for thermal conductivity measurements can be obtained as close to the surface as possible.

– For reasons of cost, the boreholes should not be deeper than 50 m.

The positions of the boreholes are shown in Fig. 1 and the coordinates are listed in Tab. 1.

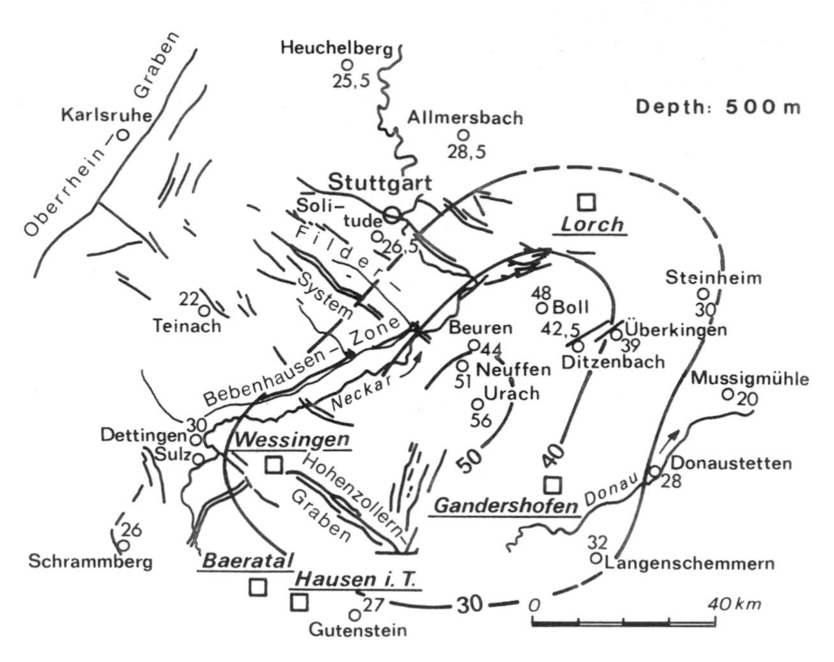

Fig. 1. Position of the shallow boreholes in the Urach area with temperature isolines calculated by HAENEL (1976).

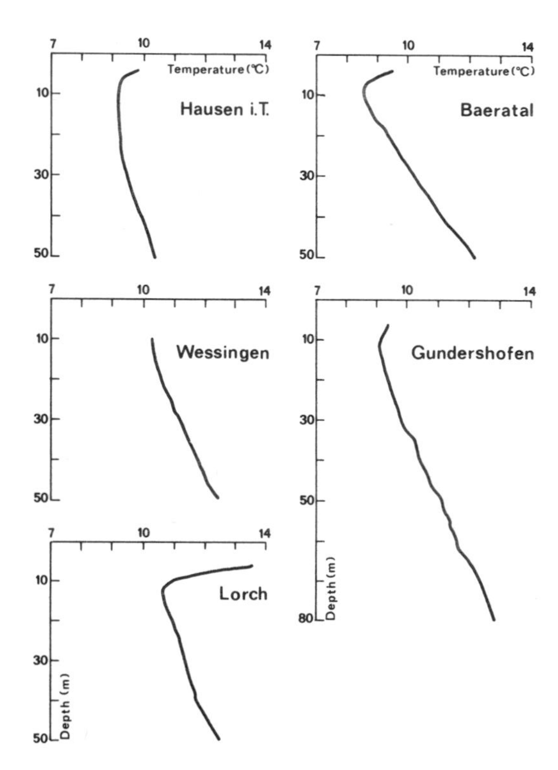

Fig. 2. Temperature measurements versus depth.

A 3-m core was taken every 10 m drilling depth. The description of the rock samples and the geological profile of the five shallow boreholes have already been published by ZOTH (1981).

The boreholes Hausen i. Tal and Baeratal were drilled in autumn 1978 and the borehole Wessingen, Gundershofen and Lorch in spring 1979. All of the boreholes have a diameter of 86 mm and 2" plastic pipe which is open at the bottom. The temperature measurements were repeated in each boreholes several months apart to assure an undisturbed temperature gradient. The most recent temperature measurement in each borehole is shown in Fig. 2.

To calculate the heat flow density, only the section deeper than 30 m can be considered, because the annual temperature variation on the earth's surface penetrates down to a depth of 25–30 m (KAPPELMEYER & HAENEL 1974). The extrapolation of the uninfluenced measured temperature to the earth's surface of each borehole is in good agreement with the mean annual air temperature T_0 at the depth $z = 0$ m. This value is in agreement with T_0 from the Climate Atlas of Baden-Württemberg (Klima Atlas, 1964).

Assuming that water movements in the part of the boreholes used for the determination has not falsified the temperature gradients, it is necessary to eliminate the influence of topography.

The differences in elevation up to 12 km from the borehole were taken from a topographic map of the area. Correction factors were then calculated according to BULLARD (1940) and HAENEL (1971). The measured and corrected temperature gradients are listed in Tab. 2.

The thermal conductivity of the rock samples was determined on about 200 samples using the electrical line source method described by ARAKAWA & SHINOHARA (1980), and checked by the absolute method described by CREUTZBURG (1964). The mean values are also given in Tab. 2.

The heat flow density q can be calculated by the corrected temperature gradient grad T and the thermal conductivity λ, both from the same depth range:

$$\vec{q} = -\lambda \cdot \text{grad } T$$

The value for q are given in Tab. 2 and the average of the terrestrial heat flow density of the boreholes are as follows:

Hausen i.T.	70 mW m^{-2}
Baeratal	67 mW m^{-2}
Wessingen	135 mW m^{-2}
Gundershofen	78 mW m^{-2}
Lorch	80 mW m^{-2}

Three further heat flow density values were calculated from the deeper boreholes Hegau 3, Dinkelsbühl and the thermal well Bad Cannstatt. For a description of the data for these wells, see ZOTH (1981). The coordinates are listed in Tab. 1, whereas the thermal conductivity, the temperature gradient and its correction are listed in Tab. 2.

The calculated heat flow density values are

Hegau 3	64 mW m^{-2}
Dinkelsbühl	82 mW m^{-2}
Bad Cannstatt	126 mW m^{-2}

Table. 1. The co-ordinates of the measured boreholes.

Name		Top. Map. 1 : 25 000	Latitude N	Longitude E	Altitude (m)	Depth (m)	End of Drilling
Hausen i. T.	Bo 1	7920	48° 4.9'	9° 2.1'	600	51.8	Nov. 18. 1978
Baeratal,	Bo 2	7919	48° 5.8'	8° 54.5'	680	52	Nov. 28. 1978
Wessingen,	Bo 3	7619	48° 19.9'	8° 56.1'	541	52	Apr. 24. 1979
Gundershofen,	Bo 4	7623	48° 22.5'	9° 37.2'	590	82.2	May 08. 1979
Lorch,	Bo 5	7224	48° 47.9'	9° 40.5'	275	52.2	May 19. 1979
Cannstatt		7121	48° 49.0'	9° 13.5'	219	470	Juli 26. 1974
Hegau 3		8218	47° 47.4'	8° 46.4'	559	220	Dec. 06. 1979
Dinkelsbühl		6928	49° 2.2'	10° 22.7'	441	560	Nov. 27. 1979

Table. 2. The results of measurements and calculations.

	Depth interval (m)	Measured gradient (°C/100 m)	Topographic correction (°C/100 m)	Corrected gradient (°C/100 m)	Mean thermal conductivity ($W m^{-1} K^{-1}$)	Number of samples	Heat flow density ($mW m^{-2}$)	Mean Heat flow density ($mW m^{-2}$)
Hausen im Tal	12– 13				2.31	2		
	23– 24				2.50	2		
	33– 34	4.73	– 0.93	3.8	1.899	6	72.16	
	43– 51	4.63	– 0.93	3.67	1.818	42	66.72	
	44.8	4.50	– 0.94	3.56	2.017	1	71.81	70.23
Baeratal	20– 24	9.16	– 3.65	5.51	1.297	16	71.46	
	51– 53	9.16	– 3.61	6.00	1.093	17	65.58	
	22.25	9.16	– 3.65	5.51	1.151	1	63.42	66.82

Wessingen	33– 34	7.93	+ 0.52	8.45	2.11	2	(178.29)	
	40– 44	7.93	+ 0.41	8.34	1.75	29	145.95	
	50– 52	7.00	+ 0.36	7.36	1.55	2	114.08	
	41.9	7.93	+ 0.41	8.34	1.752	1	146.12	135.38
Gundershofen	22– 26	(5.95)	– 2.55	(3.40)	1.995	8	(67.83)	
	35– 36	5.95	– 2.33	3.62	2.00	2	72.40	
	39– 43	5.95	– 2.22	3.73	1.75	7	65.27	
	70– 71	5.95	– 1.85	4.10	2.05	16	84.05	
	74– 76	5.95	– 1.80	4.15	2.18	5	90.47	
	80– 84	5.95	– 1.72	4.23	2.10	13	88.83	78.14
Lorch	20– 22				1.404	15		
	30– 32	6.0	– 0.80	5.2	1.465	14	76.18	
	42– 43	7.83	– 0.80	7.03	1.336	6	93.92	
	30.8	6.0	– 0.80	5.2	1.362	1	70.8	80.30
Bad Cannstatt	458–460	3.92	0.119	3.801	3.305	2	125.6	125.6
Hegau 3	19– 20	5.8	0.31	5.49	1.105	1	60.66	
	28– 30	5.6	0.28	5.32	1.276	1	67.9	
	46– 47	6.2	0.216	5.84	1.073	1	64.2	
	59– 60	5.5	0.188	4.81	1.219	1	64.75	
	77– 78	5.4	0.142	5.26	1.182	1	62.17	
	88– 89	5.7	0.125	5.58	1.110	1	61.90	63.60
Dinkelsbühl	60– 80	2.2	0.025	2.175	3.400	1	73.90	
	140–145	4.6	0.055	4.545	2.026	1	92.08	
	185–190	3.2	0.030	3.170	3.041	1	96.38	
	190–200	2.5	0.025	2.475	3.286	1	81.33	
	220–230	2.9	0.025	2.875	2.529	1	72.71	
	260–270	5.5	0.035	5.465	1.397	1	72.63	81.51
	349				2.912	1		
	480				2.955	1		
	513				3.099	1		
	547				3.164	1		

3. Determination of Underground Temperature Calculated by means of Heat Flow Density

The temperature at depth z can be calculated if
- the heat flow density is known,
- the values for heat production in the rock are known,

and if the following assumptions can be made (HAENEL 1977):
- The temperature field does not change with time.
- The thermal conductivity of certain geological layers (Quartary, Tertiary, etc.) is constant.
- Anisotropy effects in the horizontal direction can be neglected.

If these conditions are fulfilled, the temperature at depth z can be calculated using the following equation:

$$T(z) = T_o + \frac{q \cdot z}{\lambda} - \frac{H \cdot z^2}{2 \cdot \lambda},$$

Where
$T(z)$ (°C)	= temperature at depth z
T_o (°C)	= average annual temperature at z = O
$\vec{q}$ (mW m^{-2})	= heat flow density with components in the x, y, and z directions
q (mW m^{-2})	= heat flow density only in z direction
z (m)	= depth
λ (W m^{-1} K^{-1})	= thermal conductivity
H (W m^{-3})	= heat production (see HAENEL & ZOTH, this volume).

The results depend on the quality of the heat flow density data, as well as on the above-mentioned assumptions. Even very slow water movements connected with heat transport can influence the results considerably. In shallow boreholes small water movements cannot be recognized in every case. Where water movement was recognized the data was not used for temperature calculation.

4. Temperature Maps

The temperature maps in Figs. 3–12 are based on the temperature values calculated as described in section 3. and on temperature measurements from 37 new boreholes. Further, the maps were constructed on the basis of data from the Atlas of Subsurface Temperatures (HAENEL 1980), as well as data published by WOHLENBERG (1979).

The temperature isoline intervals were chosen as follows:

down to 250 m depth	: 2.5 °C
down to 1000 m depth	: 5 °C
deeper than 1000 m	: 10 °C

The northern boundary of the Urach geothermal anomaly has not been clearly defined so far; further temperature data are necessary. In the southern part the anomaly is bounded by a region of low temperature isotherms parallel to the Donau valley. Further south, the temperatures are again higher; this region is called the Lake Constance anomaly. The low temperature field parallel to the Donau valley had been observed only down to 1000 m by borehole temperature measurements. Therefore, it is not clear whether the

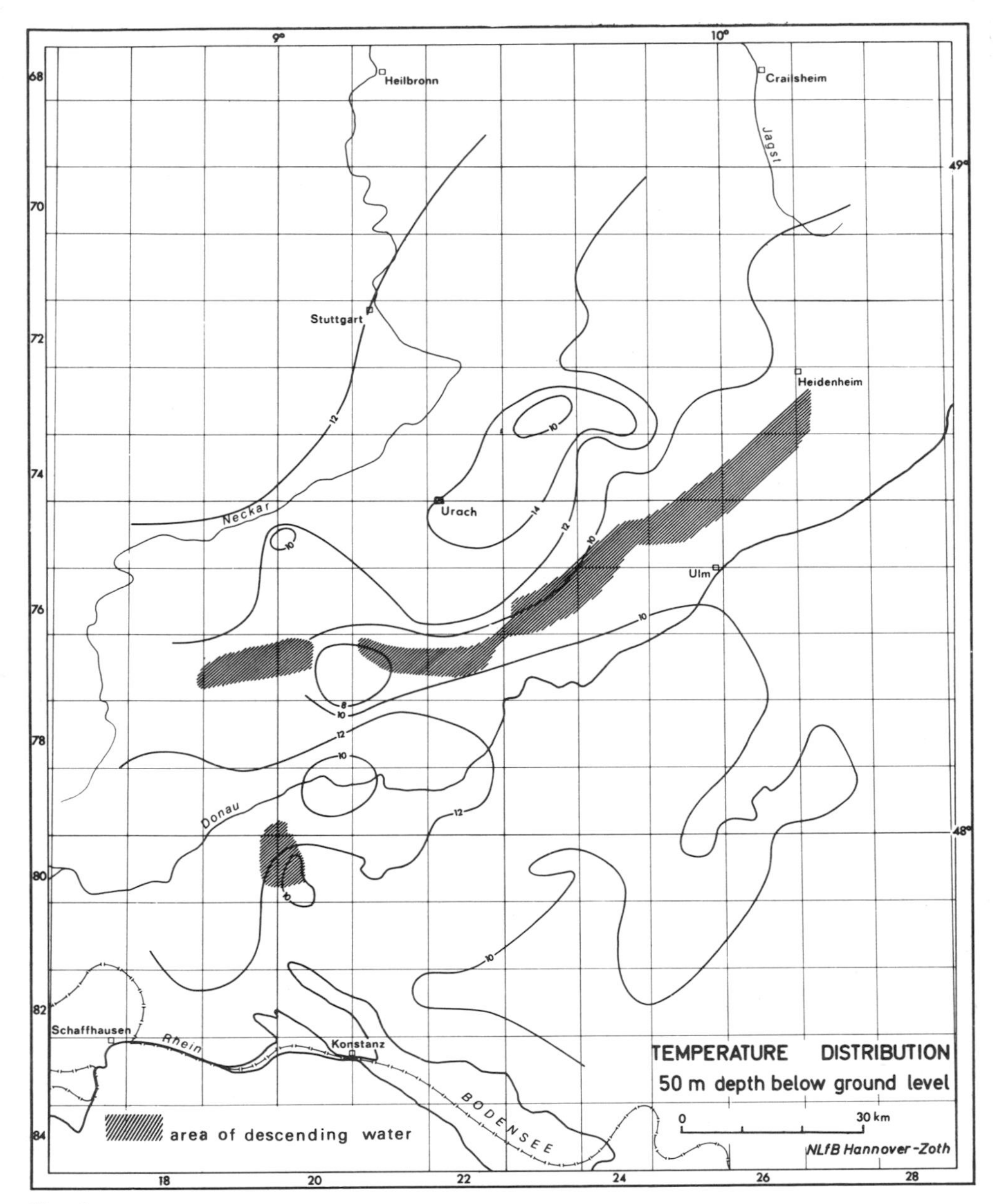

Fig. 3.

Fig. 3–12. Temperature distribution at different depths below ground level. The hatched areas in Fig. 3 are regions of descending water. The points in Fig. 7 indicates the used boreholes for temperature measurements and heat flow density calculations.

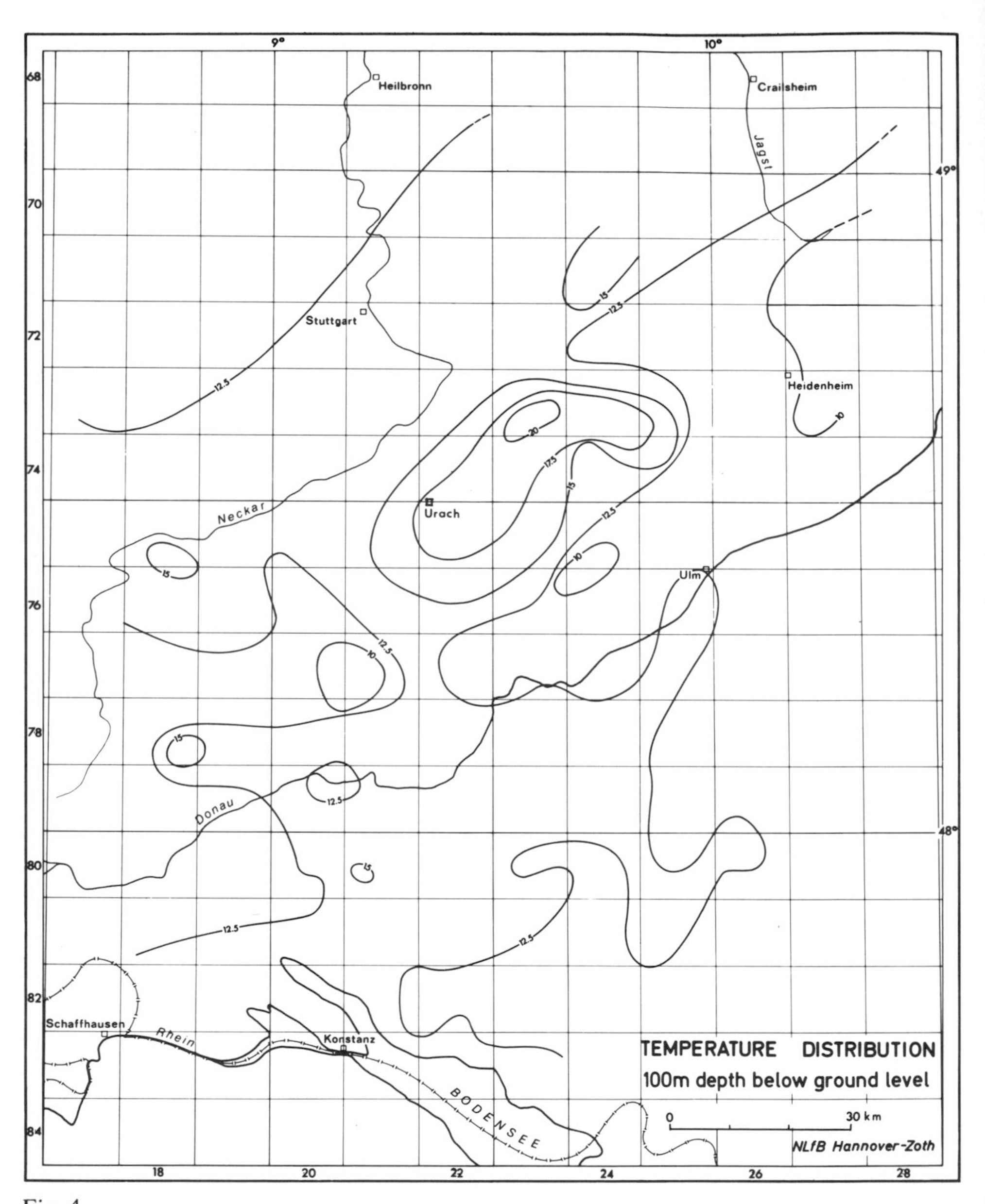
9°
10°
49°
48°
68
70
72
74
76
78
80
82
84
18
20
22
24
26
28
Heilbronn
Crailsheim
Jagst
Stuttgart
Heidenheim
Neckar
Urach
Ulm
Donau
Schaffhausen
Rhein
Konstanz
BODENSEE
12.5
15
17.5
20
10
TEMPERATURE DISTRIBUTION
100m depth below ground level
0
30 km
NLfB Hannover-Zoth

Fig. 4.

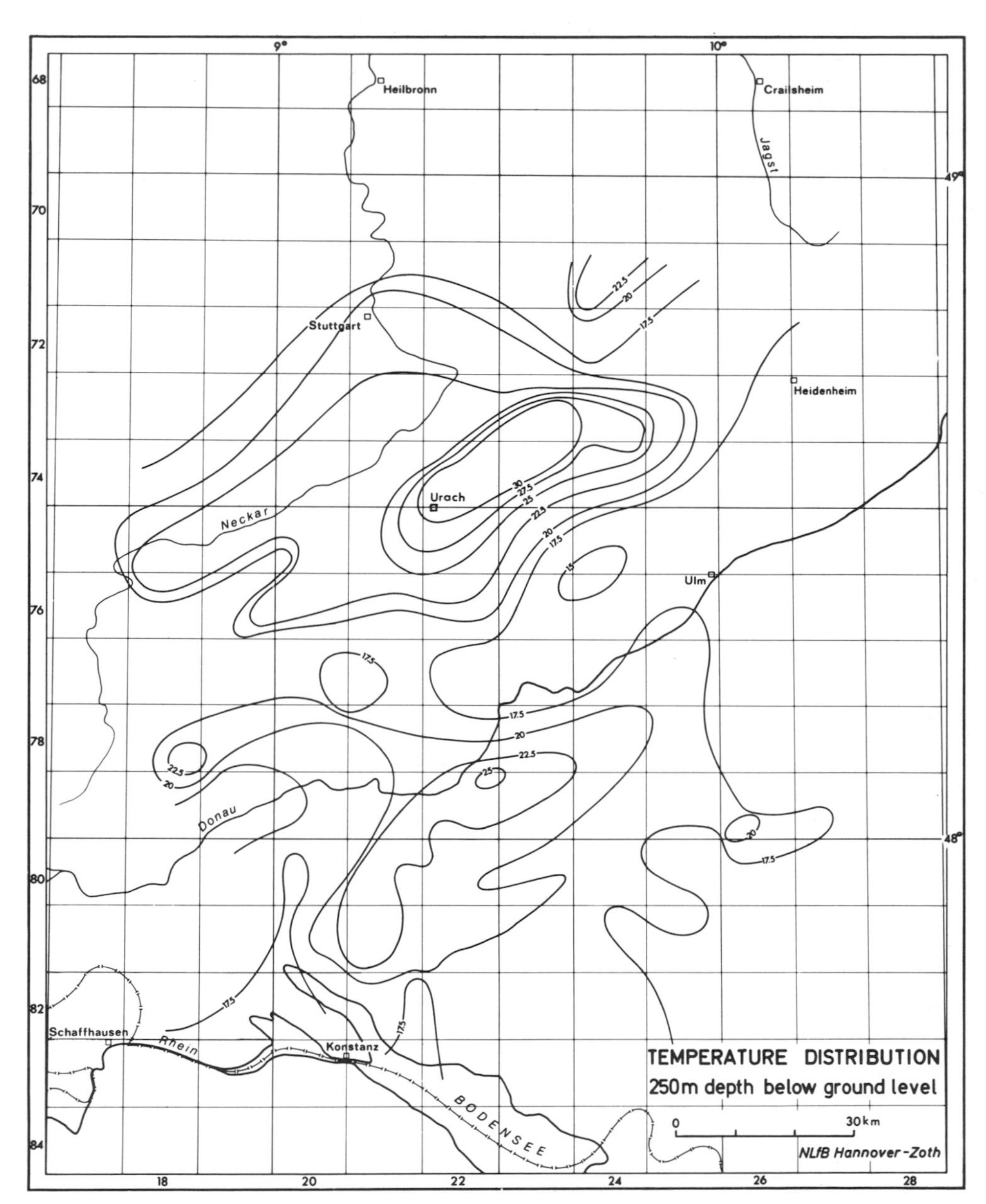

9°
10°
Heilbronn
Crailsheim
Jagst
Stuttgart
Heidenheim
Urach
Neckar
Ulm
Donau
Schaffhausen
Rhein
Konstanz
BODENSEE
49°
48°
TEMPERATURE DISTRIBUTION
250m depth below ground level
0
30km
NLfB Hannover-Zoth

Fig. 5.

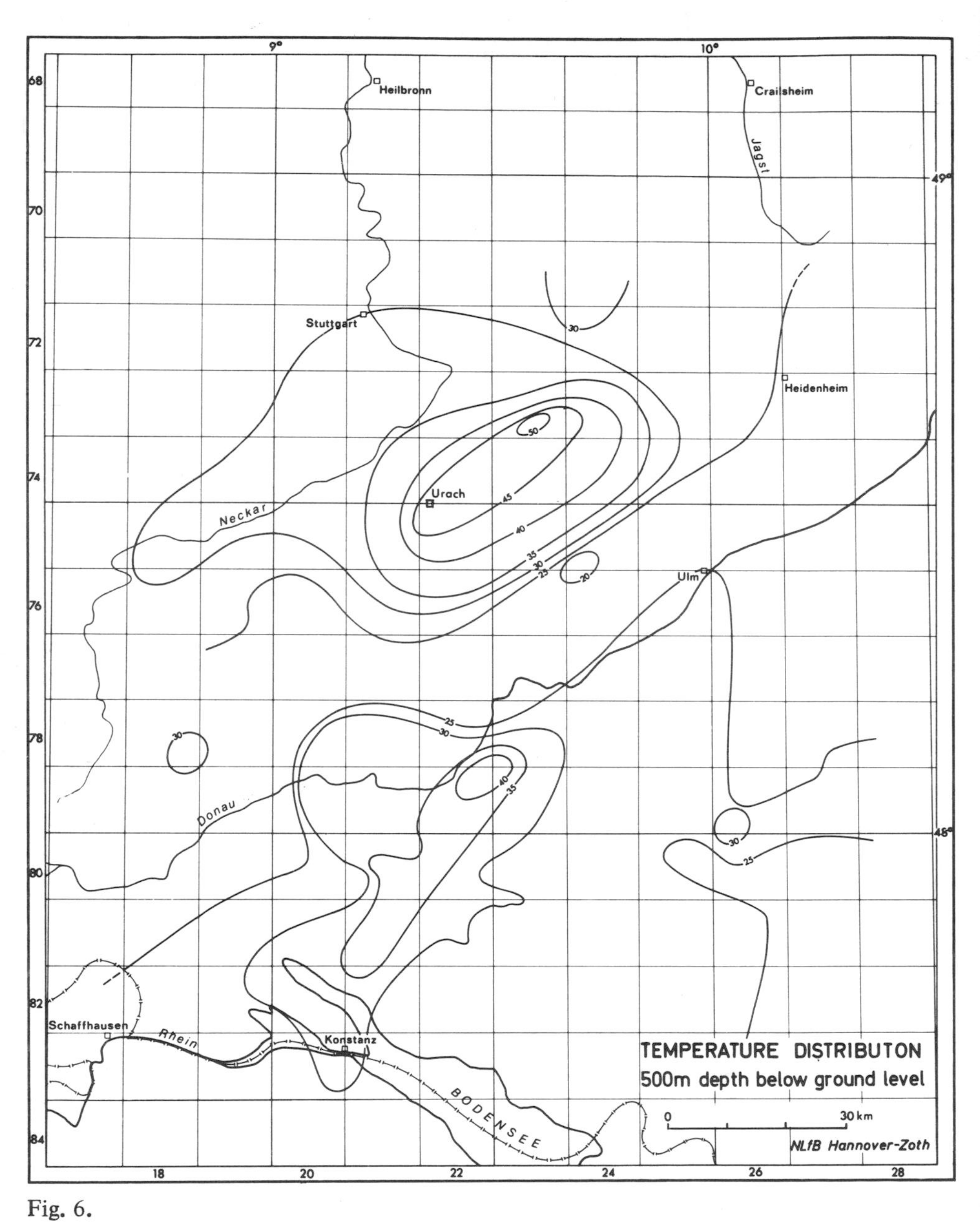
9°
10°
Heilbronn
Crailsheim
Jagst
49°
Stuttgart
Heidenheim
Urach
Neckar
Ulm
Donau
48°
Schaffhausen
Rhein
Konstanz
BODENSEE
TEMPERATURE DISTRIBUTON
500m depth below ground level
0
30 km
NLfB Hannover-Zoth
68
70
72
74
76
78
80
82
84
18
20
22
24
26
28

Fig. 6.

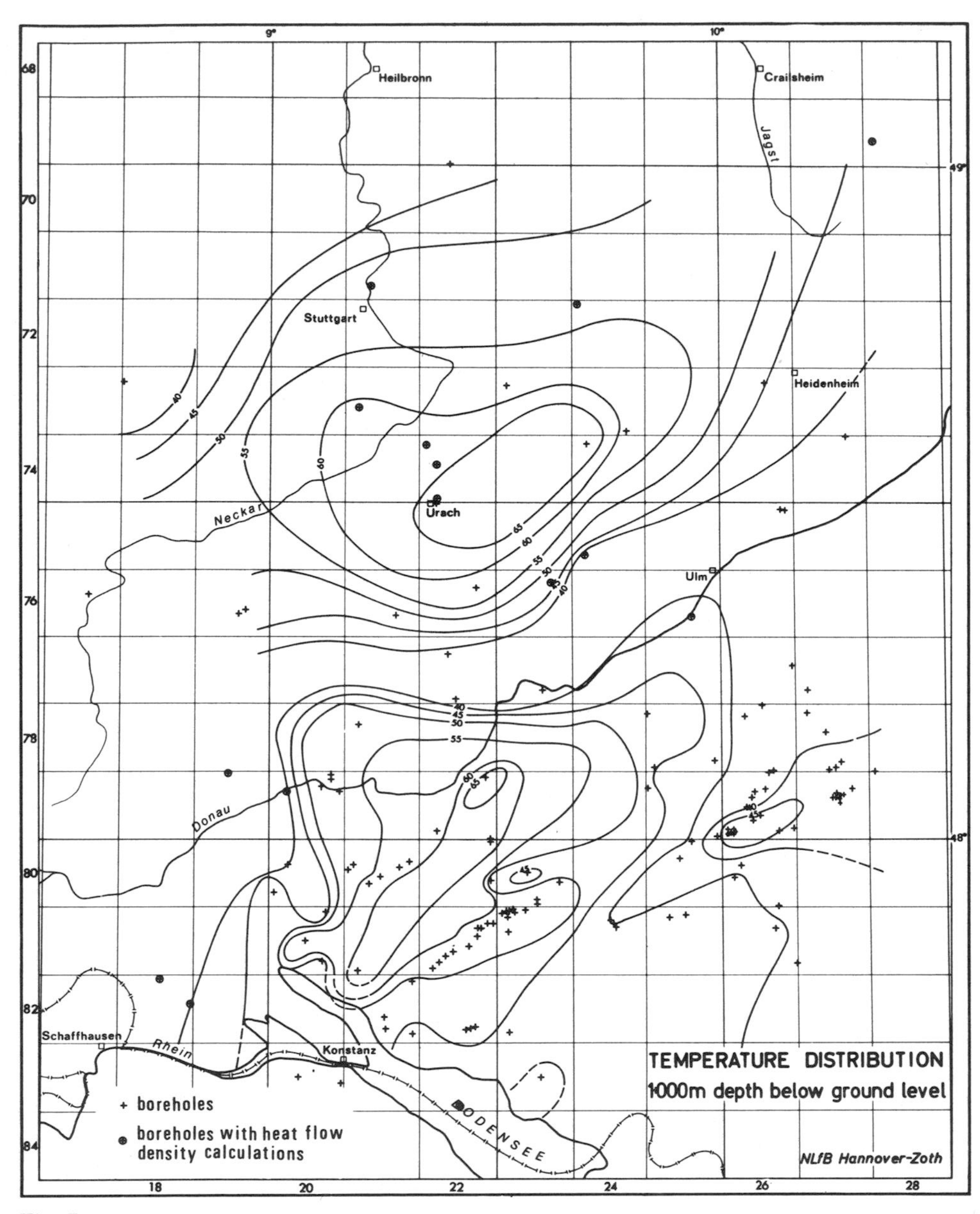

Heilbronn
Crailsheim
Jagst
Stuttgart
Heidenheim
Neckar
Urach
Ulm
Donau
Schaffhausen
Rhein
Konstanz
BODENSEE
TEMPERATURE DISTRIBUTION
1000m depth below ground level
+ boreholes
boreholes with heat flow density calculations
NLfB Hannover-Zoth

Fig. 7.

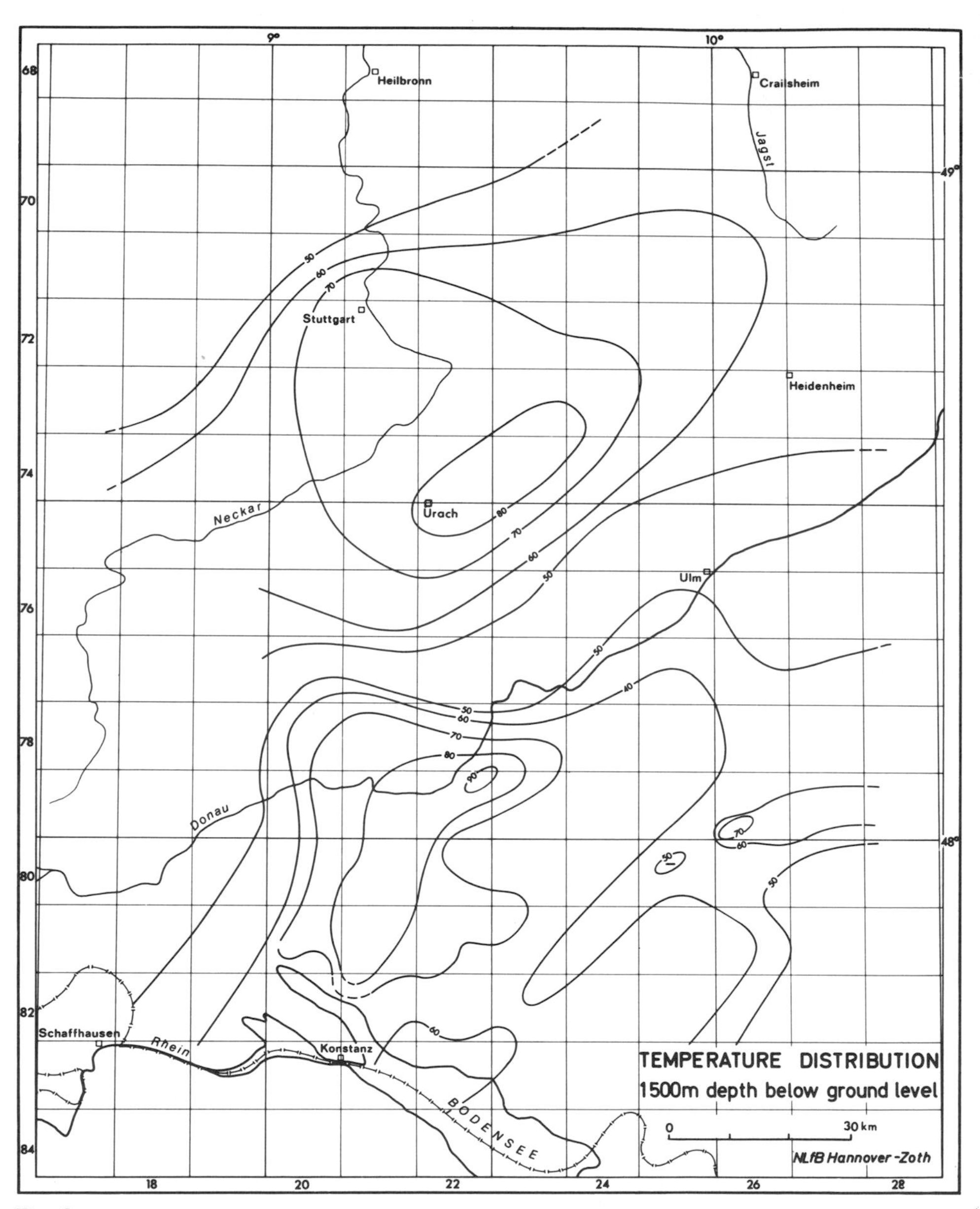
9°
10°
Heilbronn
Crailsheim
Jagst
49°
Stuttgart
Heidenheim
Neckar
Urach
Ulm
Donau
48°
Schaffhausen
Rhein
Konstanz
BODENSEE
TEMPERATURE DISTRIBUTION
1500m depth below ground level
0
30 km
NLfB Hannover-Zoth
18
20
22
24
26
28
68
70
72
74
76
78
80
82
84

Fig. 8.

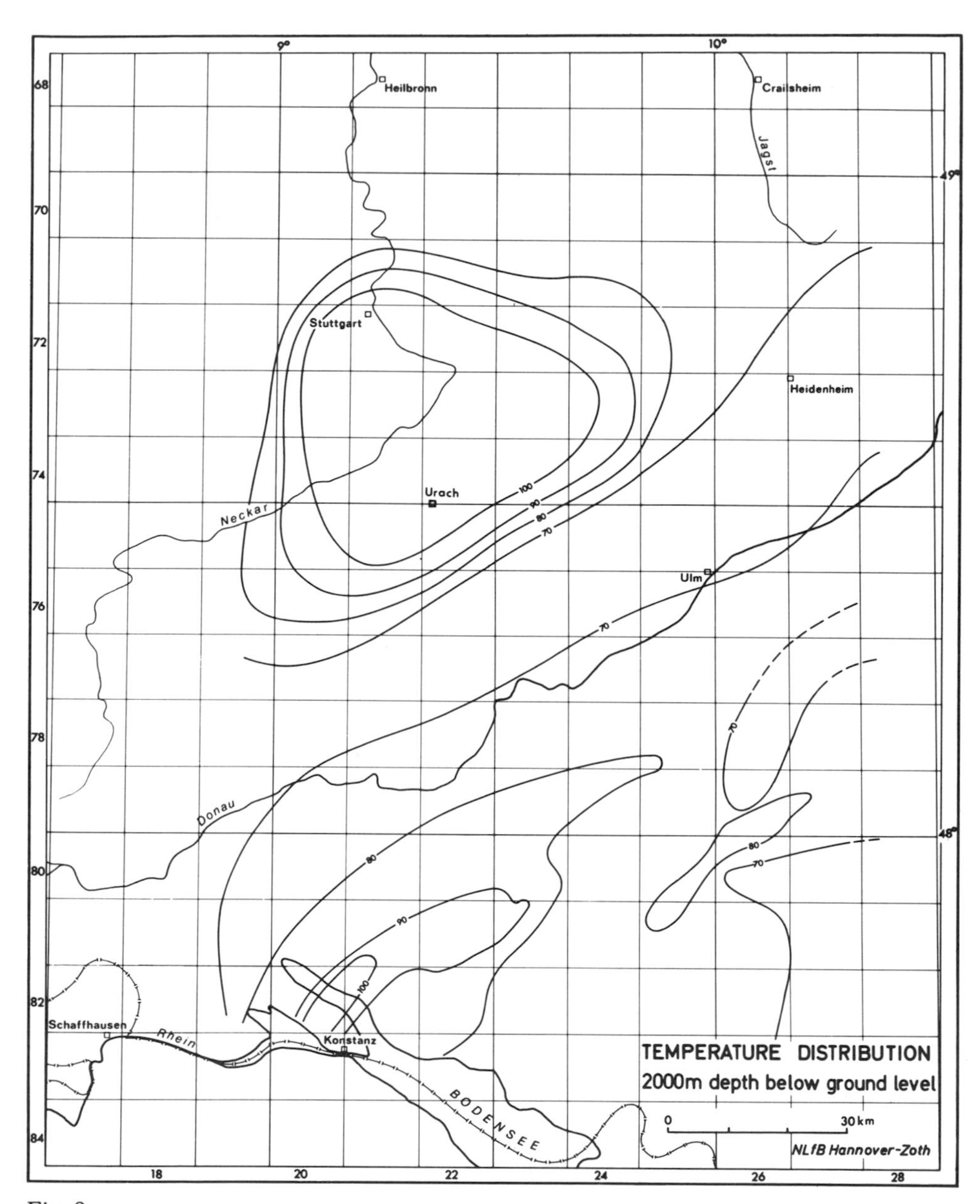
9°
10°
68
70
72
74
76
78
80
82
84
49°
48°
18
20
22
24
26
28
Heilbronn
Crailsheim
Jagst
Stuttgart
Heidenheim
Urach
Neckar
Ulm
Donau
Schaffhausen
Rhein
Konstanz
BODENSEE
100
90
80
70
TEMPERATURE DISTRIBUTION
2000m depth below ground level
0
30 km
NLfB Hannover-Zoth

Fig. 9.

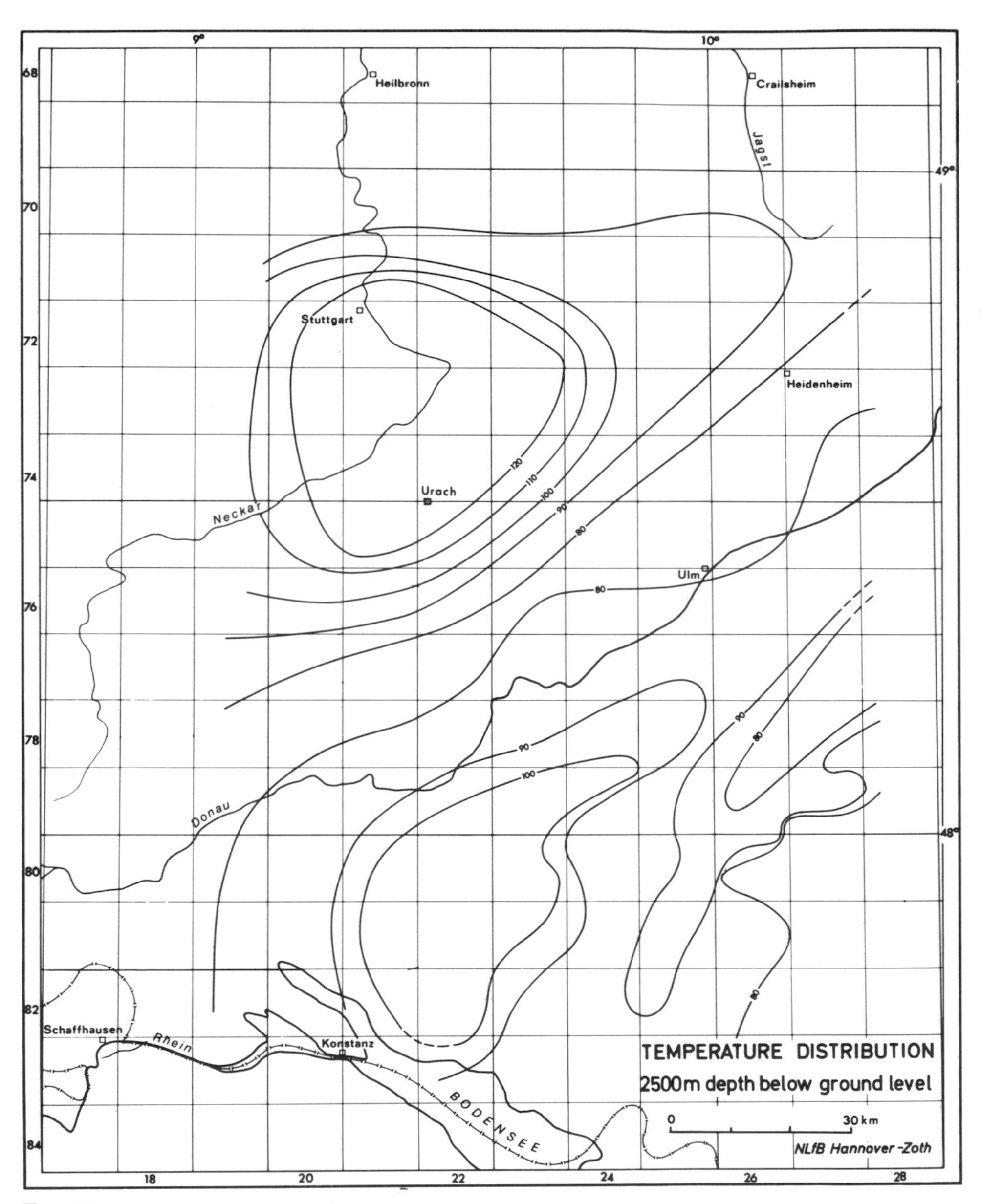
9°
10°
49°
48°
68
70
72
74
76
78
80
82
84
18
20
22
24
26
28
Heilbronn
Crailsheim
Jagst
Stuttgart
Heidenheim
Urach
Neckar
Ulm
Donau
Schaffhausen
Rhein
Konstanz
BODENSEE
120
110
100
90
80
TEMPERATURE DISTRIBUTION
2500m depth below ground level
0
30 km
NLfB Hannover-Zoth

Fig. 10.

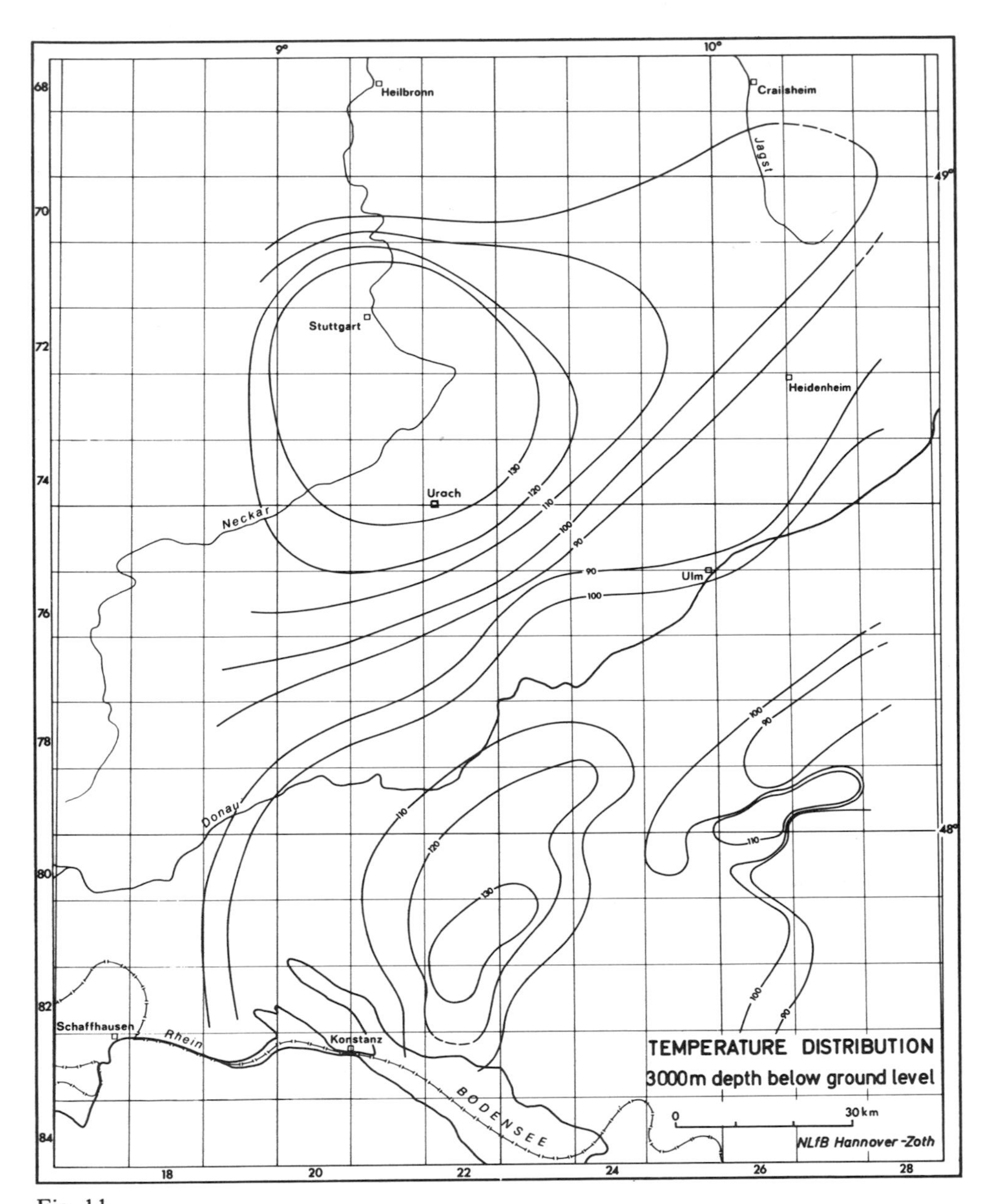
9°
10°
Heilbronn
Crailsheim
Jagst
Stuttgart
Heidenheim
Urach
Neckar
Ulm
Donau
Schaffhausen
Rhein
Konstanz
BODENSEE
49°
48°
68
70
72
74
76
78
80
82
84
18
20
22
24
26
28
TEMPERATURE DISTRIBUTION
3000m depth below ground level
0
30km
NLfB Hannover-Zoth

Fig. 11.

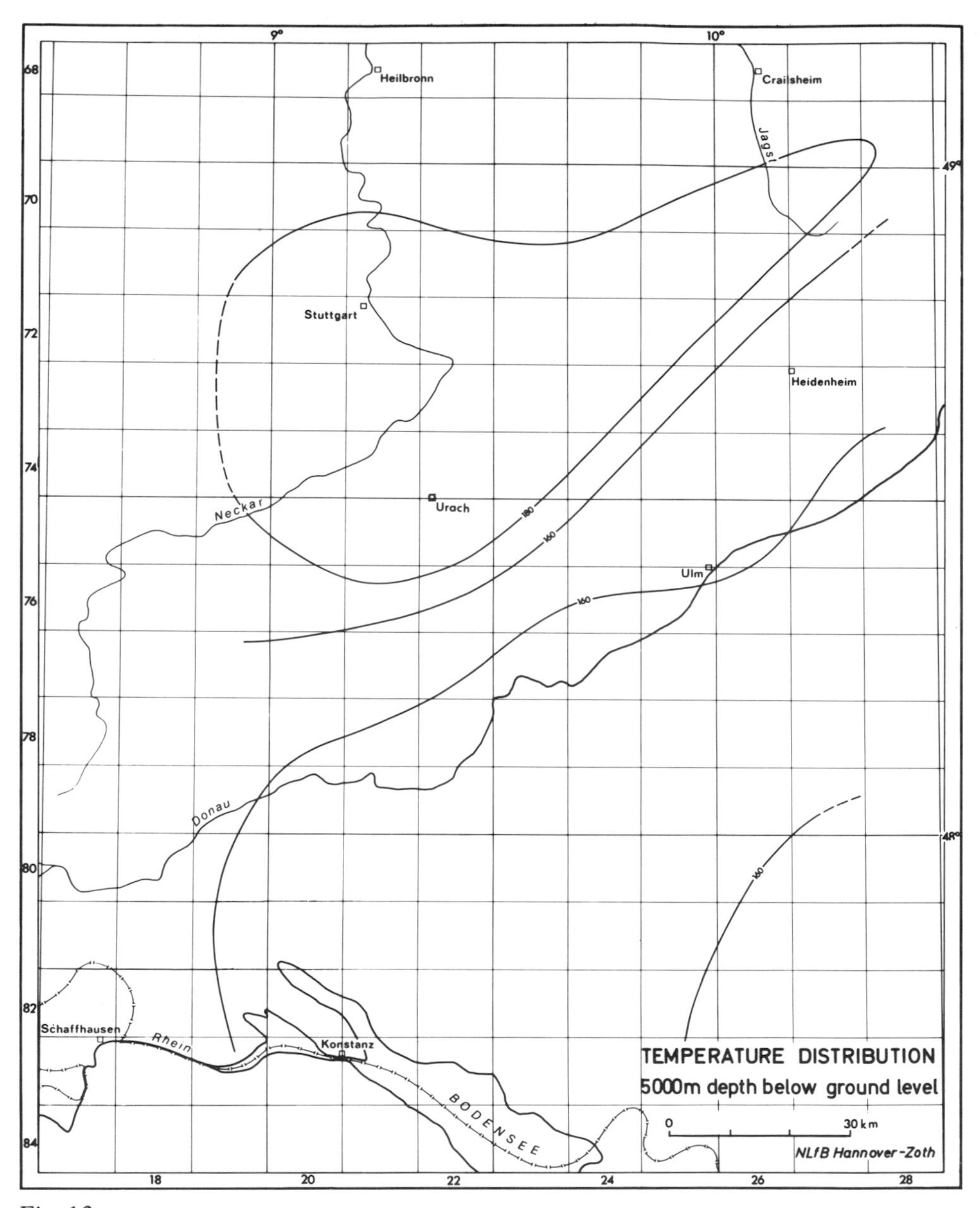

Fig. 12.

low temperature field continues below this depth or whether the Urach and Lake Constance anomalies are only separated near the surface.

VILLINGER (1977) has published a map of the Swabian Alb with isolines of the piozemetric level of the water in the Malm sediments. This map also shows areas with descending and ascending water.

The regions with descending water are shown on the map for temperatures at 50 m depth (see Fig. 3, hatched areas). It can be seen that these areas correspond to the low temperature areas. The same correspondance is observed for areas with ascending water and high temperature areas. From the isoline map of piozemetric level, another region of high temperature in the Donau valley between Sigmaringen (48° 10' N, 9° 10' W) and Ulm (48° 24' N, 10°W) is to be expected. But unfortunately, there are not enough temperature measurements available for comparison.

Altogether, isotherm maps have been constructed for ten different depths. It can be seen that, in principal, the trend of temperature distribution near the surface is also found in deeper regions. Further, it seems that the Urach anomaly covers a larger area at depth than at the surface and that the centre of the anomaly shifts slightly to the NNW with increasing depth.

References

Arakawa, Y. & Shinohara, A. (1980): Quiek Thermal Conductivity Meter. – EGS-Meering, Budapest.

Carle, K. (1975): Die Thermalwasser-Bohrungen am Fildergraben-Rand (Stuttgart, Bad Cannstatt und Bonlanden). – Heilbad und Kurort, **27**: 7–17, 15.1.1975.

Creutzburg, H. (1964): Untersuchungen über den Wärmestrom der Erde in Westdeutschland. – Kali u. Steinsalz, p. 73–108.

Bullard, E.C. (1940): The disturbance of the Temperature Gradient in the Earth's Crust by Inequalities of Height. – Month. Astron. Roy. Soc. Geoph., 1940, **4**: 360–362, London.

Haenel, R. (1971): Determinations of the Terrestrial Heat Flow in Germany. – Physica **37**: 119–134.

– (1976): Möglichkeiten der Nutzung geothermischer Energie in der Bundesrepublik Deutschland. – Report, NLfB Hannover, Archive No. 70 469.

– (1977): Die Bestimmung der Temperaturen im Untergrund der Bundesrepublik Deutschland aus Wärmestromdichtedaten. – Report, NLfB Hannover, Archive No. 78 680.

– (1980): Atlas of Subsurface Temperatures in the European Community. – Printed by Th. Schäfer, Hannover, Feb. 1980.

Haenel, R. & Zoth, G. (1976): Geothermische Untersuchungen im Rahmen des Schwerpunktsprogramms Geologische Korrelationsforschung im nordwestdeutschen Becken. – Report, NLfB Hannover, Archive No. 75 690.

– (1980): Endbericht über Geophysikalische Untersuchungen in der Forschungsbohrung Urach, 2. Thermische Untersuchungen. – Report, NLfB Hannover, Archive No. 85 805.

Kappelmeyer, O. & Haenel, R. (1974): Geothermics with Special Reference to Application. – Borntraeger, Berlin-Stuttgart.

Klima-Atlas von Baden-Württemberg, Deutscher Wetterdienst. – Selbstverlag des Deutschen Wetterdienstes, Offenbach, 1964.

Villinger, E. (1977): Über Potentialverteilung und Strömungssysteme im Karstwasser der Schwäbischen Alb (Oberer Jura, SW-Deutschland). – Geol. Jb., C, **18**: 3–93, Hannover.

Wohlenberg, J. (1979): The Subsurface Temperature Field of the Federal Republic of Germany. – Geolog. Jb., E, **15**, Hannover.

Zoth, G. (1981): Das Temperaturfeld im Bereich der Uracher Wärmeanomalie. – Report, NLfB Hannover, Archive No. 88 011.

The Urach Geothermal Project, p. 223–227;
Schweizerbart'sche Verlagsbuchhandlung, Stuttgart, 1982

The Relationships Between the Volcanism and the Geothermal Anomaly in the Urach Region*)

D. WERNER

with 1 figure

Abstract: In order to explain the thermal anomaly in the Urach region three models are discussed. A cooling magma body cannot be considered as a realistic model. As a second interpretation it is assumed that uprising deep ground water produces the thermal anomaly. This model seems to be much more realistic. Its reference to the Swabian Volcano, however, is doubtful. The third model, which is a speculative one, leads to a better relationship between the Tertiary volcanism and the present thermal anomaly. It is assumed that after the period of active volcanic eruptions (about 13 Ma b.p.) a continuously decreasing gas exhalation has been taking place up to the present. Calculations show that the extraordinarily small heat capacity of gases can be compensated under the condition that a time span of a few million years is available for such a gas exhalation process.

Introduction

It is an interesting phenomenon that both the geothermal anomaly near Urach and the Tertiary Swabian volcano are located within exactly the same region. If the map of the thermal anomaly (CARLÉ 1958, 1974, HAENEL 1976, ZOTH 1981) is examined in connection with the distribution of the more than 300 volcanic chimneys (MÄUSSNEST 1970) the geographical conformity of these two findings must be considered as significant. It cannot be assumed that this observation is an accidental one. On the other hand, it is not easy to find convincing explanations for that case.

The apparently most plausible suggestion would be that the thermal anomaly represents residual heat from a cooling magma body. Such a suggestion, however, is inadequate. A process involving the ascent of deep ground water seems to be much more realistic. The idea involved here is that deep thermal water can follow paths which have been opened earlier within the crustal material as a result of volcanic activity. Another suggestion is to assume a postvolcanic gas exhalation. These three possibilities of interpretation will be briefly discussed in the following sections.

*) Contribution no. 340 of the Institut für Geophysik, ETH Zürich

Cooling Model for a Magma Body

In this model the existence of a very large magma chamber (about 15 km x 15 km x 15 km), which may be located partly in the lower crust and partly in the uppermost mantle, is postulated. The cooling process of this body may have been taking place since about 13 million years. The question arises whether such a cooling body can influence the near-surface temperature gradient (or heat flow) at the present time. It can be shown that this simple model or a similar one, based on poor thermal conductivity effects, cannot explain the thermal anomaly (WERNER 1976), because the calculated gradients near the earth's surface are nearly the same as the corresponding undisturbed gradient. That means that the effect of the cooling magma body as mentioned above could not possibly be observed in a borehole.

The same result has been published by HOFFERS (1978). We can state that such a model is far from realistic. Furthermore, the condition of the model (existence of a large magma body) is more than doubtful.

Uprising Deep Groundwater

A better model may be based on the assumption that uprising deep ground water produces the thermal anomaly in the Urach region. In order to relate this possible process to the Swabian volcano, the idea that the broken crustal material due to the volcanic activity could permit water circulation is invoked (WERNER 1976, see also HAENEL & ZOTH in this volume). It may be that the volcanic chimneys can be considered as paths for uprising thermal water.

This case can be mathematically simulated in a relatively simple way, if we assume that the vertical flow of water is laterally constant within the surroundings of Urach. Thus the problem becomes one-dimensional with respect to the z-axis (depth). This model is a simple one but it is good enough for an estimation of the heat transfer effect. The case considered here can be described by the following differential equation:

$$\frac{\partial T}{\partial t} + \gamma u \frac{\partial T}{\partial z} = \kappa \frac{\partial^2 T}{\partial z^2} + \frac{A}{c\rho} \tag{1}$$

with:

- T = temperature
- t = time
- u = effective flow of uprising water (DARCY velocity)
- z = depth
- κ = effective thermal diffusivity
- A = density of radiogenic heat sources
- $c\rho$ = heat capacity of rocks
- γ = ratio of heat capacity between water and rocks.

The equation (1) can be evaluated numerically by means of a finite difference method. By starting with a defined undisturbed temperature distribution (which at the same time is the reference temperature curve related to the surroundings of the Urach anomaly), the time-dependent temperature changes can be calculated step by step.

By means of this method quantitative relationships will be obtained between such parameters as:

- the magnitude u of the uprising flow,
- the depth of origin of the thermal water,
- the history of the thermal anomaly, especially its age.

In order to simulate the Urach case with the help of this model, the following input data have been used; this example leads to a good conformity with the observations:

- the uprising flow (DARCY velocity) must be about 0.5 cm/a;
- the uprising water originates from the crystalline basement; its original depth is at least 4000 m;
- the age of the anomaly is of the order of magnitude of 100 000 years.

Accordingly a temperature of 105 °C should be observed at 2000 m depth; this corresponds in fact to the measurements performed in the Urach 3 borehole (see HAENEL & ZOTH this volume).

It can be stated that the hydrothermal model outlined here is suitable to explain the Urach anomaly in a relatively simple way. A quite similar explanation has been found for the geothermal anomaly of Landau/Pfalz, located within the Rhinegraben area (WERNER & PARINI 1980). It should be noted, however, that the Urach anomaly cannot be compared with the geothermal situation in the Rhinegraben area. Uprising deep ground water within the Rhinegraben seems to be plausible because of the different mean altitude of the graben and of the surrounding mountain ranges. In the Urach region, however, such a hydraulic argument has not yet been found.

There is another fact which may be taken into consideration: the age of the thermal anomaly (order of magnitude: 100 000 years) as related to the age of the Swabian volcanism (order of magnitude: 13 million years). This casts doubt on the idea that the shattered postvolcanic crust could favour deep water circulation. If it is assumed that the age of the thermal anomaly be much more than 100 000 years, a further geological argument must be taken into account. Hydrothermal processes are limited by chemical processes, i.e. small cavities within the crustal material will be filled by minerals. After that it must assumed that the crustal cavities are "closed".

In a view of this consideration, one cannot believe that the deep water hypothesis could be the best interpretation for the Urach geothermal problem. We must state that the relationship between the deep water model and the Swabian volcanism is rather doubtful.

Gas Exhalation as a Possible Postvolcanic Process

The Swabian volcanism at the end of the Miocene period (Tortonian) was characterized mainly by large gas eruptions. Therefore, it seems possible that a considerable gas exhalation, decreasing with time, may also have occured during a long period after the active volcanism had ceased. It is known that postvolcanic exhalation processes occur. For the Urach case, however, it would be necessary to assume that such a process should remain in progress during a time span of 13 million years, i.e. up to the present. Such a speculative suggestion would have to be accepted in order to explain the thermal anomaly from this point of view. On the other hand, a gas exhalation model would agree with the deep gas hypothesis after GOLD & SOTER (1980). These authors have discussed numerous findings indicating the probability that the upper mantle contains considerable quantities of hydrocarbon gases.

The heat transfer effect by means of gases is extremely small because of the small magnitude of their heat capacity (which appears as the ratio γ in the equation (1)). On the other hand the long time span mentioned above was available. In relation to the time span, but also in relation to the depth range, the model of uprising deep ground water and the model of uprising deep gases are extremely different. If the gas model is accepted it must be assumed that the whole crust in the Urach region has been permeable to gases since the times of the active Swabian volcanism.

The effect of uprising gases originating from the upper mantle can be calculated on the basis of the heat transport equation (1). It should be noted, however, that such a model is a very simple one which must be considered as a first approximation in order to explain the Urach anomaly from this speculative point of view. It is obvious that this gas model contains unanswered questions. Nevertheless, an attempt has been made to estimate the present temperature distribution, as derived from an assumed history of gas exhalation. The assumed history consists of 4 periods with decreasing gas exhalation as follows:

uprising gas flow u (m^3/m^2 a)	length of period (10^6 years)
100	1
50	5
5	5
0.5	2

Furthermore, it is assumed that the gases originate from a depth of 40 km. For the calculation, a value of $\gamma = 2.5 \times 10^{-4}$ has been used. The result can be seen in Fig. 1. The curve T_3 indicates the calculated temperature distribution at present. For comparision the curve T_1, which is the corresponding undisturbed temperature distribution is shown.

The question arises what kind of gas could be responsible for such a process. The gas which passes through the whole crustal thickness is probably methane (or other hydrocarbon gases). During its passage to the earth's surface the methane probably reacts to form carbon dioxide. As previously mentioned many questions have to be answered before the gas model can be viewed as a realistic one.

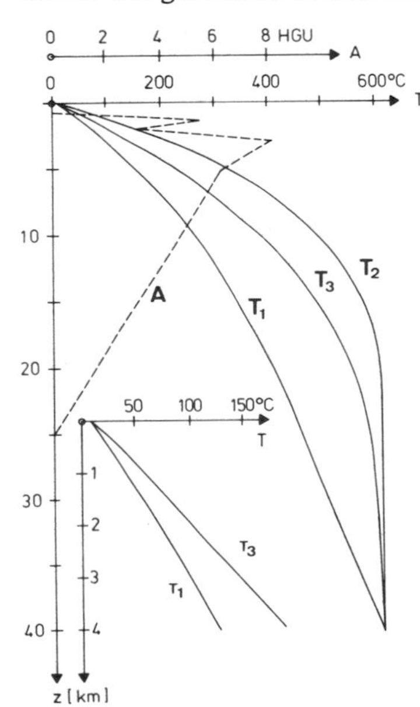

Fig. 1. One-dimensional model of deep gas exhalation.
T_1 : original temperature distribution with depth z or the reference temperature outside the anomaly;
T_2 : temperature distribution after 10 million years;
T_3 : present temperature, i.e. after 13 million years;
A : assumed distribution of the radiogenic heat sources within the crust (1 HGU = 4.19×10^{-7} W m^{-3}).
Assumptions: uprising gas flow with decreasing amount during 13 million years (see text); original depth of gas: 40 km.

On the other hand, there are arguments favouring such a model. The geographically significant conformity between volcanism and thermal anomaly could be explained in an unconstrained way. The Urach anomaly might perhaps provide an example verifying the deep gas hypothesis after GOLD & SOTER (1980). In contrast to uprising water the uprising gases must be considered to be juvenile, i.e. it is not necessary to establish a hydraulic circulation process.

Seismic findings could be used in order to prove the validity of the model. BARTELSEN et al. (this volume) have determined a strong low velocity body within the crust underneath the region in question. One can assume that this seismic result could be explained as a thermally induced effect. The difference between the curves T_1 and T_3 in Fig. 1 would produce a reduction of the seismic velocity (v_p-waves) of the order of 1 %. Observations after BARTELSEN et al. show a reduction between 1 % and 10 %.

Another possibility of verifying the gas model could be to carry out detailed investigations of the gases in the Urach borehole.

References

Carlé, W. (1958): Kohlensäure, Erdwärme und Herdlage im Uracher Vulkangebiet und seiner weiteren Umgebung. – Z. dt. geol. Ges. **110**: 71– 110.

– (1974): Die Wärmeanomalie der mittleren Schwäbischen Alb (Baden-Württemberg). – In: H.J. Illies & K. Fuchs, Approaches to Taphrogenesis, p. 207–212, Schweizerbart, Stuttgart.

Gold, T. & Soter, S. (1980): The Deep Earth-Gas Hypothesis. – Scientific American **242**, 6: 130–137.

Haenel, R. (1976): Möglichkeiten der Nutzung geothermischer Energie in der Bundesrepublik Deutschland. – Report, NLfB Hannover, Archive No. 70 4 69.

Hoffers, B. (1978): Modellrechnungen zur Uracher Wärmeanomalie. – Oberrhein. geol. Abh. **27**: 33–39.

Mäussnest, O. (1970): Regionalmagnetische Vermessung Mittelwürttembergs. – Geol. Jb. **88**: 567–586.

Werner, D. (1976): Zur Wärmeanomalie bei Urach. – 36. Tagung der Dt. Geophys. Ges. Bochum, Abstract.

– (1977): Erdwärme und Lithosphäre im Raum Südwestdeutschland/Schweiz. – Die Naturwissenschaften, **64**: 626–631.

Werner, D. & Parini, M. (1980): The Geothermal Anomaly of Landau/Pfalz: An Attempt of Interpretation. – J. Geophys. **48**: 28–33.

Zoth, G. (1981): Das Temperaturfeld im Bereich der Wärmeanomalie Urach. – Report, NLfB Hannover, Archive No. 88 011.

The Urach Geothermal Project, p. 229;
Schweizerbart'sche Verlagsbuchhandlung, Stuttgart, 1982

A Possible Explanation of the Geothermal Anomaly of Urach

K. SCHÄDEL

Abstract: A few results are not in agreement with an assumed ascent of water from great depth. Heat production can be assumed to occur also at shallow depth by oxidation of hydrocarbons and pyrite.

In the following a brief comment will be given on the question of whether the origin of the geothermal anomaly is to be sought at greater depth (more than about 1500 m) or at shallow depth (at about 500 m). Of course, no detailed investigation was possible to prove the idea presented below.

From the hydrogeological point of view an ascent of water from great depth seems to be unrealistic with regard to:

- the observed free water level of the different aquifers in the Urach 3 borehole (DIETRICH, this volume),
- the increasing salinity in the aquifers with depth (DIETRICH, this volume), and
- the possible discrepancy between the determined heat flow density of 86 mW m^{-2} in the Urach 3 borehole (HAENEL & ZOTH, this volume) and the estimated heat flow density of about 140 mW m^{-2} for the depth range of 0 m to 600 m (Jurassic).

Another possible explanation of the origin of the geothermal anomaly could be within the rocks of the Lias and Dogger. The rocks contain hydrocarbons (especially kerogen in the Lias epsilon) and pyrite. The oxidation involves exothermic reaction which can produce enough heat during a geological time span to maintain the geothermal anomaly of Urach; see also KAPPELMEYER & HAENEL (1974). How far the natural condition for an oxidation of pyrite and hydrocarbons is given still has to be investigated.

References

Kappelmeyer, D. & Haenel, R. (1974): Geothermics with special reference to application. – Geoexploration Monographs, Series 1, No. 4, Gebr. Borntraeger, Berlin-Stuttgart.

The Urach Geothermal Project, p. 231–245;
Schweizerbart'sche Verlagsbuchhandlung, Stuttgart, 1982

A Seismic-Refraction Investigation of the Basement Structure in the Urach Geothermal Anomaly, Southern Germany

M. JENTSCH, D. BAMFORD, D. EMTER and C. PRODEHL

with 11 figures and 1 table

Abstract: In 1978 and 1979 a seismic-refraction survey was carried out in the Urach geothermal anomaly, Southern Germany. All available P_g data have been subjected to an analysis with standard refraction-seismic methods, i.e. plus-minus analysis, delay-time analysis using the MOZAIC method of BAMFORD and two-dimensional raytracing. The objective of these analyses was to obtain detailed information on the topography of the crystalline basement, the velocity variations within the crystalline basement, and their possible relation to the temperature anomaly.
No evidence was found for either a velocity gradient or for velocity anisotropy within the basement. Evidence for lateral velocity variations related to the anomaly was also not found. An acceptable hypothesis is that the seismic velocity of the basement is a uniform velocity of 5.66 ± 0.02 km/s.
The resulting map of basement depths shows reasonable agreement with available borehole data and displays the variation of sedimentary thicknesses in and around the thermal anomaly. Maximal basement depths and temperature maxima show clear offsets. The results suggest that any possible heat source has to be located below 4–5 km depth.

1. Introduction

In the years 1978 and 1979 a seismic-refraction survey was carried out in the geothermal area of Urach, Southern Germany, as part of a multidisciplinary effort to get a better understanding of this area with the use of geological, geochemical and geophysical methods (HAENEL 1978).

The thermal area of Urach is a low-temperature, water-dominated system (SCHÄDEL 1977) reaching temperatures of about 130–140 °C at 3 km depth (WOHLENBERG 1978). This seismic survey was part of the overall exploration efforts which culminated in the drilling of the deep borehole Urach 3. The borehole reached a maximal depth of 3334 m and intersected the crystalline basement at a depth of 1604 m (SCHÄDEL 1978, WOHLENBERG 1978) below surface.

In this study only the first-arrival data concerning the depth and structure of the crystalline basement are interpreted. Studies of seismic parameters in areas of higher geothermal potential, as compared to that under investigation here, revealed clear effects on seismic velocities (MAJER & McEVILLY 1979, ACKERMANN 1979). However, in a low temperature anomaly such as Urach the changes in compressional wave velocities are expected to be much smaller. Laboratory measurements place them at about 0.05–0.1 km/s (VOLAROVICH & GURVICH 1957, HUGHES & MAURETTE 1956, 1957, TIMUR

1977). Hence it will be quite difficult, but not generally impossible, to detect such changes of velocities with temperature with standard refraction methods.

Thus the primary objective of the seismic-refraction survey was to define the compressional wave velocity structure and the topography of the crystalline basement and to relate possible lateral velocity variations to the thermal anomaly as a basis for a more complete understanding of the area. This will be shown to be a straightforward process. The second question to be discussed is the extent to which state-of-the-art refraction methods can provide information about changes of seismic velocities in a low temperature anomaly.

2. Geological Setting

The Urach geothermal area lies on the northern boundary of the Swabian Alb. It reveals an oval form with its axis running WSW-ENE. This is illustrated in Fig. 1 by isolines of the geothermal step after CARLÉ (1974).

The area of the thermal anomaly is divided into two parts by the cliff of the White Jurassic (Malm) plateau of the Swabian Alb which runs about parallel to the axis of the anomaly: the northern part 300–450 m, the southern part 700–850 m above sea level (SCHÄDEL 1977). The basement in the whole area consists mainly of the granitic and gneissic rocks of the Variscan orogene which under the area of investigation are overlain by Upper Permian sequences (Rotliegendes). The latter fill a Palaeozoic trough in the basement, the so-called Schramberger trough, whose existence has already been suggested by earlier geological and geophysical investigations (BREYER 1956). On top of these layers which thin out towards the north and south, the sequences of the Keuper, Lias and Dogger (mainly limestones, marls and sandstones) follow in their normal age sequence; they thin out in the northern foreland of the Swabian Alb and are overlain in the south by the well bedded limestones of the Alb Plateau. The centre of the anomaly comprises about 300 Upper Miocene volcanic chimneys of the Urach volcanic area, known also as the Swabian Volcano (GEYER & GWINNER 1968).

The main fracture zones in the vicinity of the anomaly (Fig. 1) are the Filder-Graben system (NW strike), the Bebenhausen fracture zone (NNE strike) in the north and the Lauchert-Hohenzollern-Graben systems in the south (CARLÉ 1974), the latter being the strongest earthquake centre in Southwest Germany.

3. Survey and Data

Seismograms were recorded in analog form with 35 (1978) and 30 (1979) mobile stations of the German MARS type (BERCKHEMER 1970, 1976). In 1978 mainly two vertical components of ground motion, about 500 m apart along the profile, and one radial horizontal component were installed at each recording. In 1979 three components, vertical, radial and transverse horizontal, were observed by reoccupying the same sites as in the previous year and filling gaps between them so as to obtain reversed and condensed subsurface coverage at the centre of the anomaly. Thus shots from 7 specially designed shotpoints and 3 commercial quarries were recorded on 10 reversed or partly reversed and 10 unreversed profiles and one fan (Fig. 1). Inside the anomaly observation distances reached a maximum of 40 km while some profiles were recorded well beyond the anomaly with maximal distances of 155 km. Fig. 2 shows all recording and shotpoint positions.

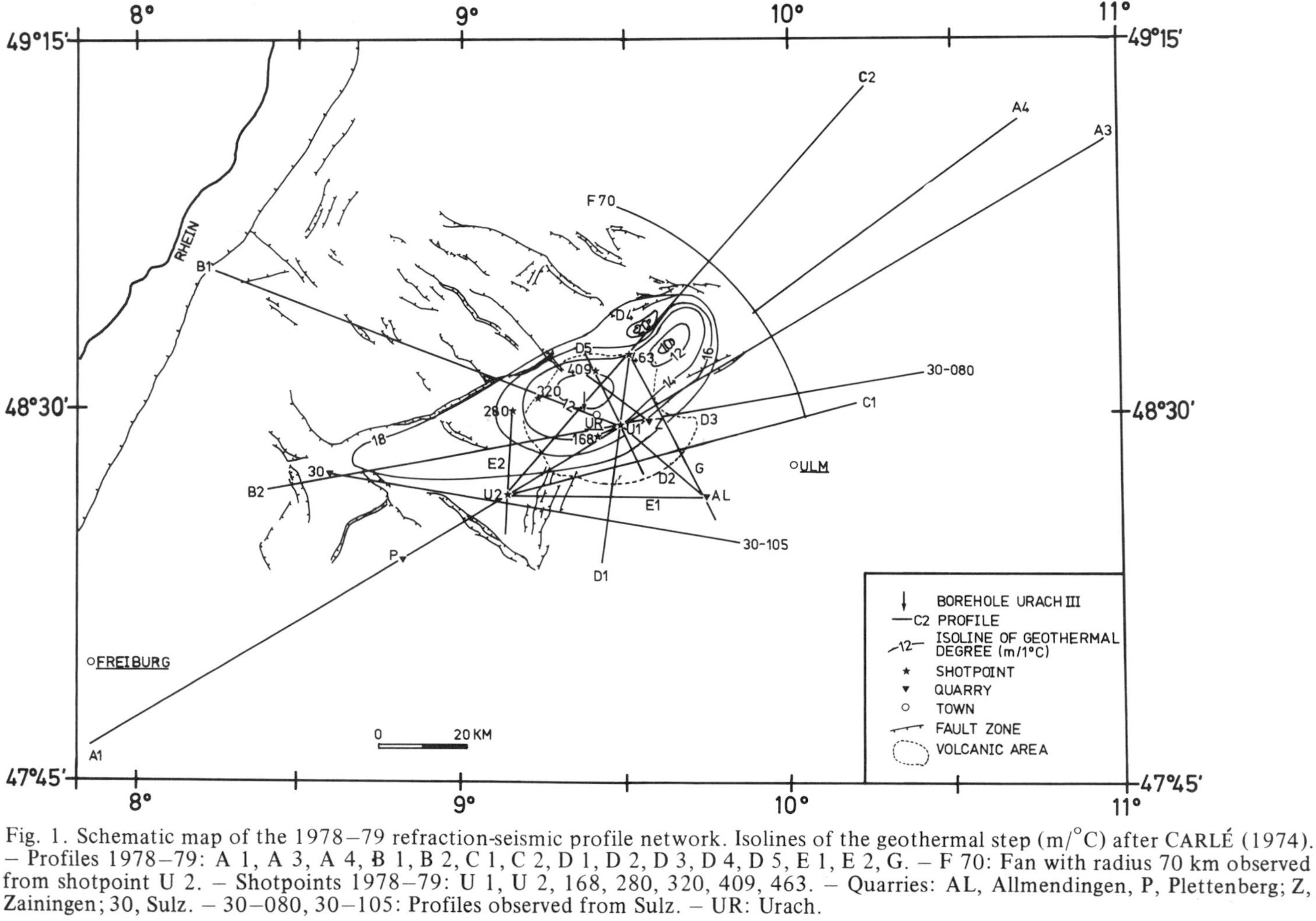

Fig. 1. Schematic map of the 1978–79 refraction-seismic profile network. Isolines of the geothermal step (m/°C) after CARLÉ (1974). – Profiles 1978–79: A 1, A 3, A 4, B 1, B 2, C 1, C 2, D 1, D 2, D 3, D 4, D 5, E 1, E 2, G. – F 70: Fan with radius 70 km observed from shotpoint U 2. – Shotpoints 1978–79: U 1, U 2, 168, 280, 320, 409, 463. – Quarries: AL, Allmendingen, P, Plettenberg; Z, Zainingen; 30, Sulz. – 30–080, 30–105: Profiles observed from Sulz. – UR: Urach.

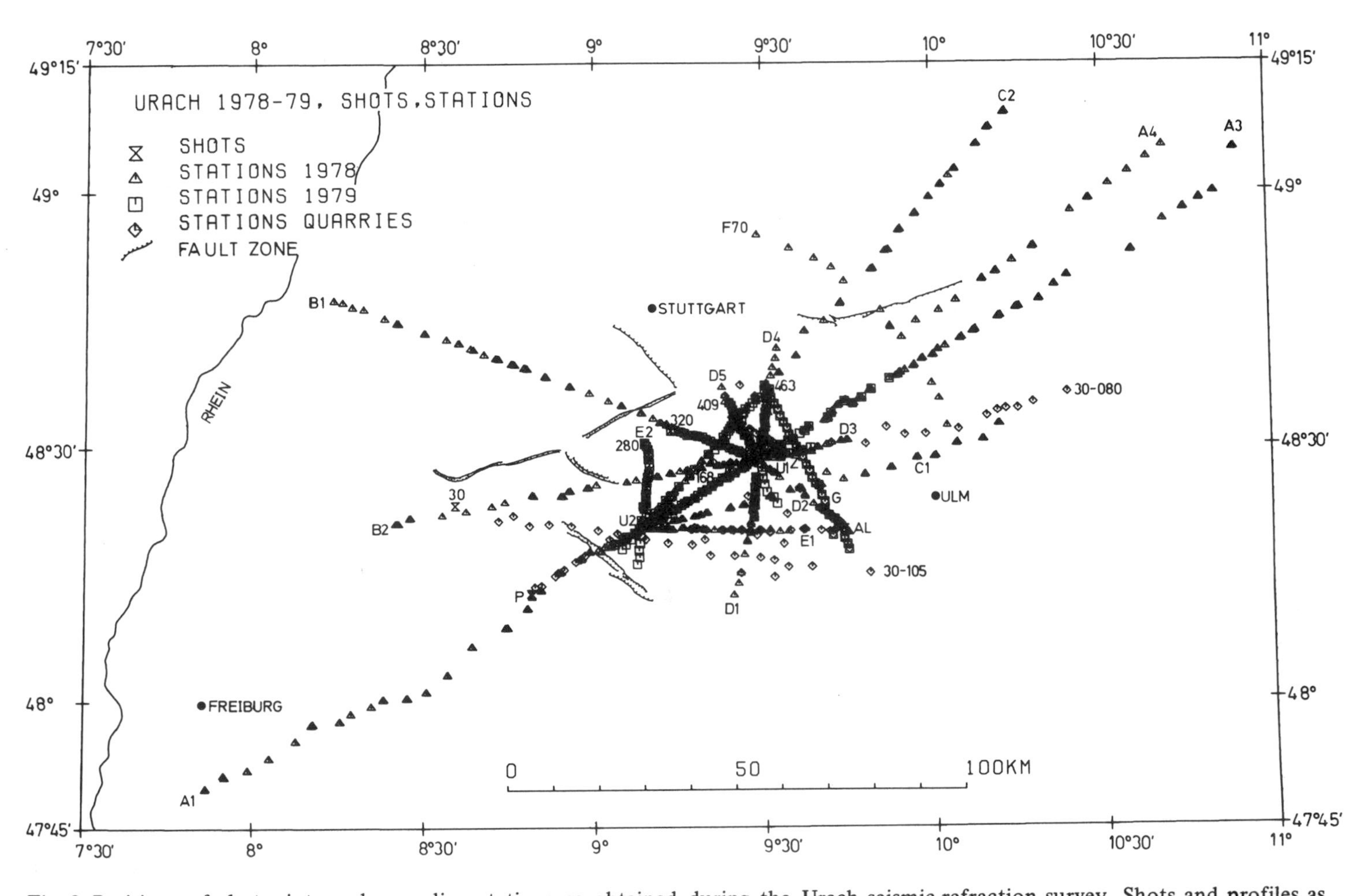

Fig. 2. Positions of shotpoints and recording stations as obtained during the Urach seismic-refraction survey Shots and profiles as in Fig. 1.

Charges between 75 and 800 kg of explosives were fired in clusters of boreholes with a mean depth of 50 m and a maximal load of 100 kg per hole.

Shot times were determined with an accuracy better than 0.01 s by recording the cap break and the output of a small array of three seismometers of one refraction unit with one seismometer placed immediately at the shothole.

In addition the data recorded between 1971 and 1977 from quarry blasts at shotpoint 30 (Fig. 1) were incorporated in this investigation (EMTER & PRODEHL 1977). All data were plotted in the form of computer-generated record sections after digitizing the analog magnetic tapes. Seismograms in general were of good quality and first arrivals could be read with confidence from the unfiltered sections. Figs. 3 and 4 show examples of bandpass filtered record sections (4–16.5 Hz). Seismograms are plotted in normalized form, i.e. true amplitudes are not portrayed. Locations, travel time and a complete set of vertical record sections are compiled in an open-file report (JENTSCH 1980).

4. Interpretation

As mentioned above, the present study concentrates only on the interpretation of waves refracted in the crystalline basement (P_g). The essential point in the analysis is therefore to identify all P_g-first arrivals. This was done by reading all first arrivals off the unfiltered record sections and constructing reduced time distance diagrammes (Fig. 5). All phases with apparent velocities between 5.2 and 6.4 km/s were then tentatively accepted as P_g.

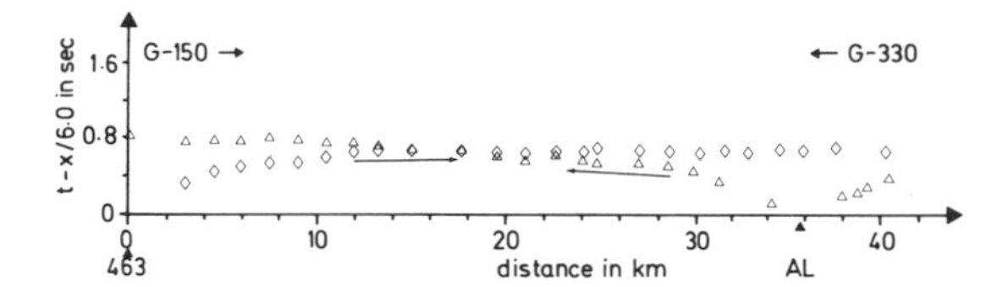

Fig. 5. Reduced time-distance plot of first arrivals as observed on profile G. Shotpoint positions are marked with ▲. Arrows indicate the phase P_g refracted in the basement. t: overall traveltime; x: distance.

4.1 Plus-Minus Method

To obtain a first idea of the refractor velocity HAGEDOORN's plus-minus method (1959) was then applied to all reversed profiles. The results of this analysis are shown in Tab. 1. With the exception of profile E2 (denoted by * in Tab. 1) all values are in a reasonable range for upper crustal crystalline rocks (SMITHSON & SHIVE 1975). In fact, no systematic relation of higher or lower velocities with regard to the thermal anomaly is present, i.e. profiles clearly lying outside the anomaly (A1/P, D3) give similar velocities to those

Profiles	Velocity (km/s)
A1-240-U1 / A3-060-U2	5.62 ± 0.03
E1-270-Al / E1-090-U2	5.81 ± 0.02
G-330-AL / G-150-463	5.75 ± 0.02
C2-040-U2 / C2-220-463	5.67 ± 0.02
D4-010-U1 / D4-185-463	5.62 ± 0.13
B1-290-U1 / B1-110-320	5.85 ± 0.17
D3-135-U1 / D3-315-AL	5.62 ± 0.08
E2-185-280 / E2-005-U2	5.20 ± 0.07 *
A1-240-U2 / P-060-P	5.75 ± 0.04

Table 1. Velocity values as derived from plus-minus analysis Profile code: Profile Identification-Azimuth (E of N)-Shotpoint. For positions see Figs. 1 and 2.

* : Explanation see text.

within (C2, G, Fig. 1). The anomalously low velocity on profile E2 can be explained with strong variations in refractor topography, such that a basic assumption of the plus-minus method is violated.

4.2 Delay-Time Analysis

In the following all available P_g data were subjected to delay-time analysis using the MOZAIC method of BAMFORD (1976 a). The basic principles of delay-time analysis are not reviewed; the interested reader is referred to BAMFORD (1976 a, b, 1977). However, a few words about the MOZAIC method are appropriate because it served as the main interpretation method here. It was designed to deal especially with the type of composite, heterogeneous and, for the classic time-term approach, possibly poorly conditioned observation scheme as available for this study. Its central idea is that for any refractor there exist small areas which can be considered to have constant delay time and that the distribution of such areas can be reasonably guessed by using ancillary information such as geological maps etc. Thus in order to design suitable MOZAICs for further analysis the available P_g data of Fig. 2 were used in conjunction with geological (Geologische Übersichtskarte von Baden-Württemberg, 1 : 200 000, 1962) and gravity (SCHLEUSENER 1945) maps of the area.

At an early stage in the analysis it already had to be concluded that not all of the data could be reasonably combined into a stable MOZAIC: some of the data were wrongly identified as P_g from the travel-time plots but in reality originated from regions deeper within the crust. The final set of some 545 time-distance data then gave stable solutions.

One of the great advantages of the MOZAIC method is that it allows, in addition to the determination of refractor topography, a consideration of variations in refractor velocity (gradients, anisotropy etc.). The various solutions are considered in the following.

4.2.1 Uniform Velocity/Velocity Gradient

In pursuit of this question the refractor velocity was modeled either as uniform or as having a small gradient. As is normal in such studies (BAMFORD et al. 1979) the effects of including the possibility of a gradient as opposed to assuming a uniform velocity can be assessed statistically by comparison of the variances of the different solutions. In fact, on a statistical basis, the uniform velocity solutions were always to be preferred to those with a velocity gradient: variances and hence standard errors did not improve for the gradient solutions.

Therefore a uniform velocity (5.66 ± 0.02 km/s) had to be accepted for the crystalline basement in the Urach area: the resulting delay times, which will be considered in detail in a later section, are shown in Fig. 9.

4.2.2 Velocity Anisotropy

The possibility of velocity anisotropy in the upper crust can be relevant to thermally anomalous areas in that fissure distributions of preferred orientation serving as aquifers for hydrothermal convective systems will probably result in a bulk anisotropic effect on seismic velocities (BAMFORD & NUNN 1979).

Techniques for studying refractor velocity anisotropy have been described by BAMFORD (1977). The principle requirement of such studies is that the azimuthal distribution must be fairly even. As may be seen from Fig. 6 the azimuthal distribution of the available data set is not ideal. However, on the base of experience – especially model studies – which allowed detecting quite weak anisotropy at great depths with much weaker data sets (2–3 % anisotropy at 40 km depth, BAMFORD et al. 1979), it was assumed that the data set at hand would permit tests for the presence of anisotropy.

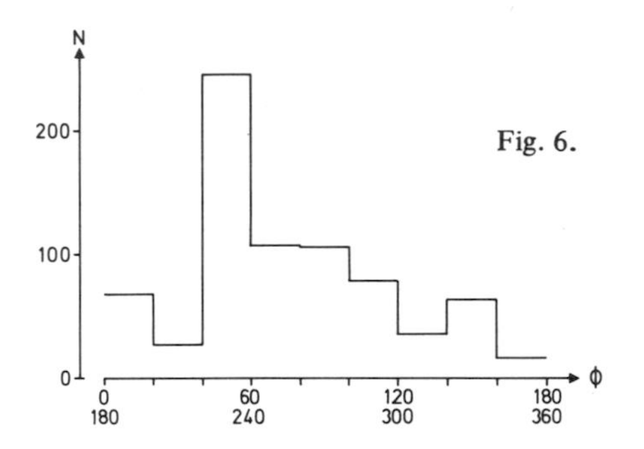

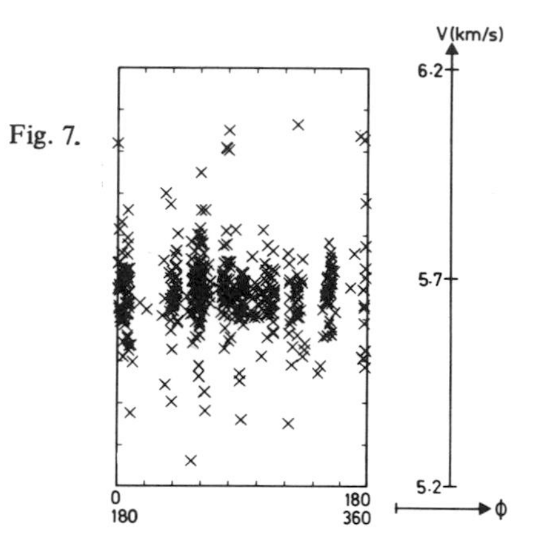

Fig. 6. Number (N) of available P_g observations versus azimuth (in deg E of N) ϕ.

Fig. 7. Scattergram with velocity (V) versus azimuth (ϕ).

In fact, both statistically and on the basis of the velocity scattergram shown in Fig. 7 (BAMFORD 1977), where a velocity value was calculated for each observation point and then plotted as a function of azimuth, solutions allowing for the possibility of velocity anisotropy were less satisfactory than those that did not. Thus the hypothesis that the basement velocity in the Urach area is isotropic within the limits of measurement error (~1 %) has to be accepted.

4.2.3 Lateral Velocity Variations

The available data sampled subsurface points inside the thermal anomaly as well as outside. Hence, tests could be made to determine whether changes of the compressional wave velocity are due to temperature increases within the anomaly.

This was pursued by plotting the single velocity values as a function of position at the midpoint between shot and receiver (Fig. 8) which are plotted as a function of azimuth in the scattergram (Fig. 7). Simple contouring was then attempted in order to identify possible lateral velocity changes. However, no systematic relation of higher or lower velocities to the thermal anomaly was found.

5. Topography of the Seismic Basement

To convert the delay times τ (Fig. 9) obtained in the MOZAIC analysis to depths H to the basement the simple relation

$$H = \frac{V \cdot \overline{V}}{(V^2 - \overline{V}^2)^{1/2}} \, \tau$$

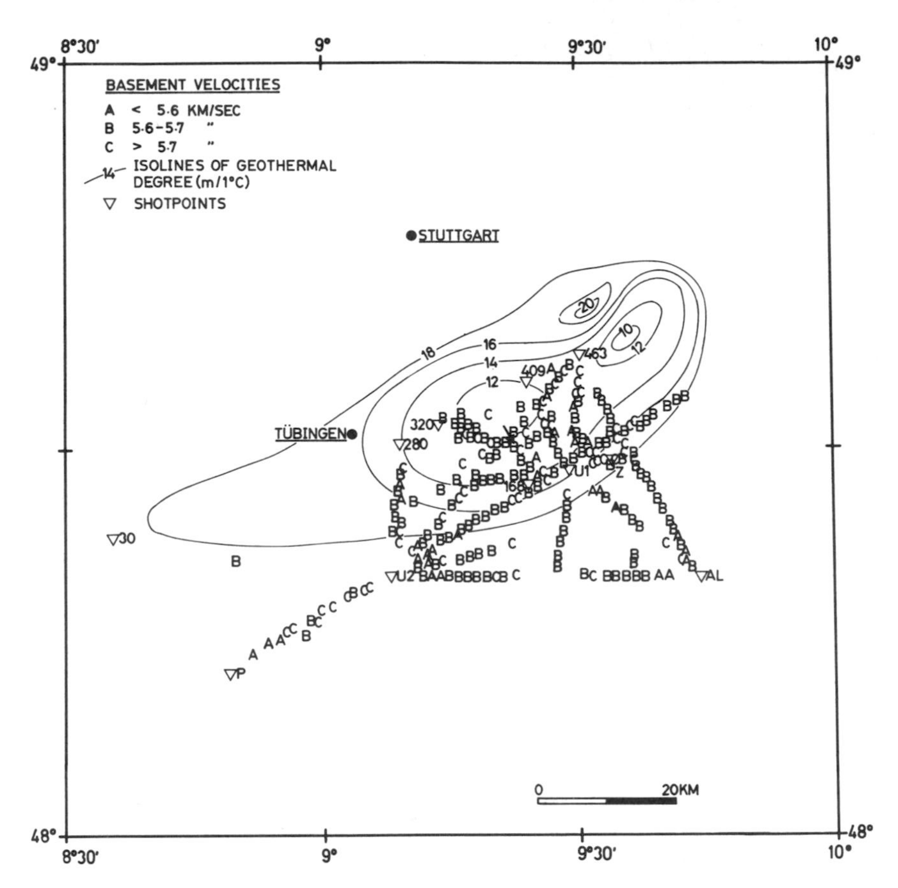

Fig. 8. Graded velocities plotted as a function of position. Isolines of inverse temperature gradient (geothermal degree) after CARLE (1974).

can be used, where V is the refractor (basement) velocity and $\overline{V}$ a mean overburden velocity. With this approximation the general difficulty of not knowing the exact velocity-depth function in the overburden at each point where a delay time is available can be avoided: SMITH et al. (1966) tested this approximation and showed that with a correctly derived mean velocity $\overline{V}$, errors in H will be 1 % or less.

With the use of the above formula a depth value for each observation point beneath the surface can be calculated and presented in the form of a map of depth isolines. Such a map has been published by JENTSCH et al. (1980, Fig. 4). However, this presentation can be misunderstood if strong variations of the topography are present, because the lines of equal depth are not related to a constant reference depth (e.g. sea level). Such strong variations are present in the area of investigation, which is traversed by the WSW-ENE trending rim of the Swabian Alb where a sudden change in elevation of 400 m occurs.

Therefore the delay times shown in Fig. 9 have been corrected to a constant station elevation of + 500 m above sea level, the elevation of the top of the Urach 3 borehole. For the correction a velocity of 3.9 km/s was used, since this represents an average velocity for the Jurassic formations.

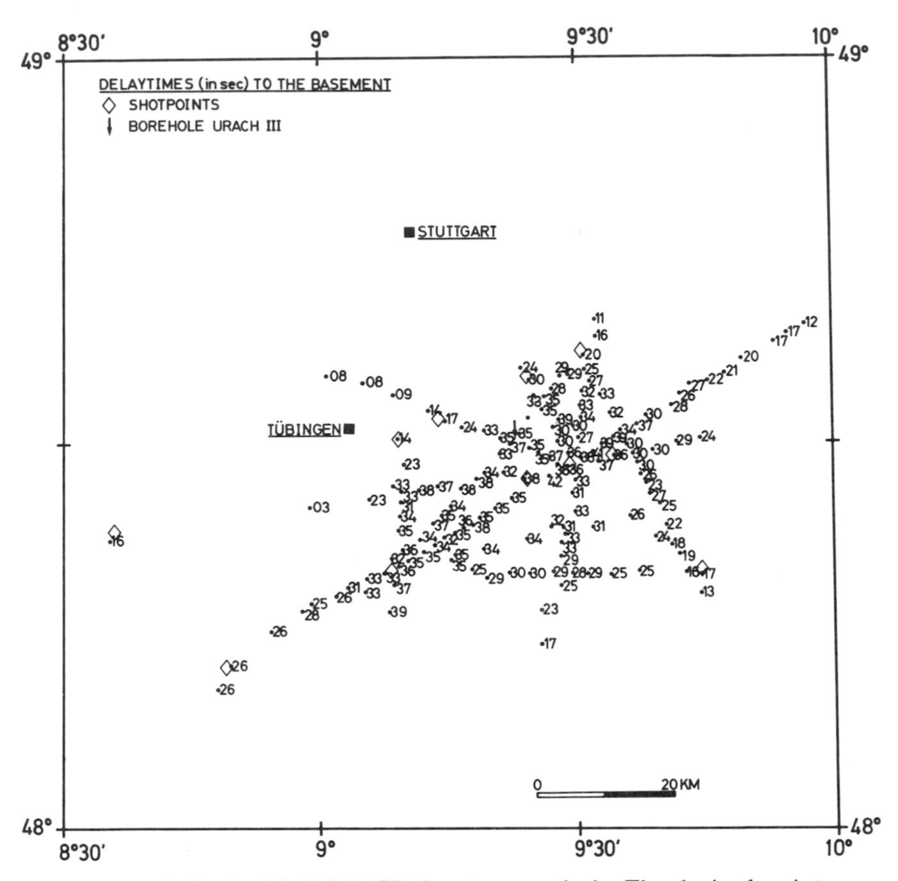

Fig. 9. Delay times as derived with MOZAIC time-term analysis. The decimal points are plotted at the position where the delay-time was obtained. For shotpoint codes see Fig. 8.

To convert the delay times thus corrected to depth below the reference level a mean velocity of 3.6 km/s for the remaining overburden was used on the basis of velocity values for the sediments given by BREYER (1956) and down-hole velocity measurements in the well Urach 3 (WOHLENBERG 1978). To check this value the delay times close to the borehole were converted and gave basement depths which agree in principle with that encountered in the well, as can be seen from the sonic log in Fig. 10. It should be noted that the mean overburden velocity derived from the borehole compensated sonic log and the well shooting turns out to be about 3.9 km/s. However, no values in the first 340 m are given; these are probably very low and would therefore lower the mean velocity. Thus, on the basis of the other measurements mentioned above and the good agreement of depths at the borehole, the lower value of 3.6 km/s was judged to be the best estimate for the overburden velocity and could be applied to the whole area. The resulting contour map is shown in Fig. 11. Allowing for the 0.02–0.05 s uncertainty of the delay times, which corresponds to a depth uncertainty of 100–200 metres, there seems to be reasonable agreement with borehole data (GEYER & GWINNER 1968, SCHÄDEL pers. comm., WOHLENBERG 1978) and earlier measurements (BREYER 1956). However, it should be noted that the delay times computed for points on the edge of the survey are prone to systematic errors and therefore the corresponding depths have not been used.

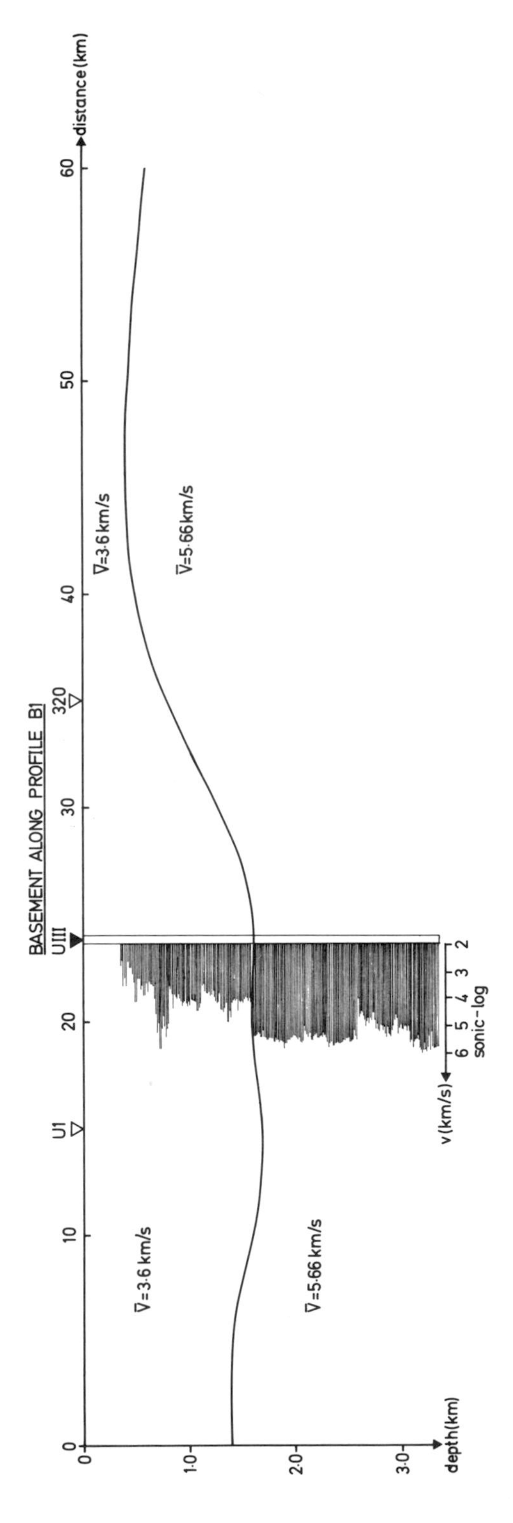

Fig. 10. Basement depth below surface (heavy line) as obtained with two-dimensional ray tracing from observations on profile B 1. V̄: mean compressional wave velocities. The sonic log as measured in the borehole URACH 3 is shown for comparison (WOHLENBERG 1978). △: shotpoints; ▲: position of borehole.

The cross section in Fig. 10 shows a further check on the results obtained so far. Here, the data obtained on profile B 1, which runs through the centre of the anomaly and passes the borehole URACH 3 have been inverted with two-dimensional ray tracing (CERVENY et al. 1974) without applying elevation corrections. A comparison with the sonic-log data plotted over the basement surface as derived from the refraction observations illustrates the good agreement between them.

A depth limit of the lower boundary of the 5.66 km/s layer could only be roughly estimated from unreversed profiles where phases originating from depths below the basement were observed. The basic formulae for the 2-layer case with plane dipping interfaces were thereby employed. This limit can be placed at 4–5 km depth.

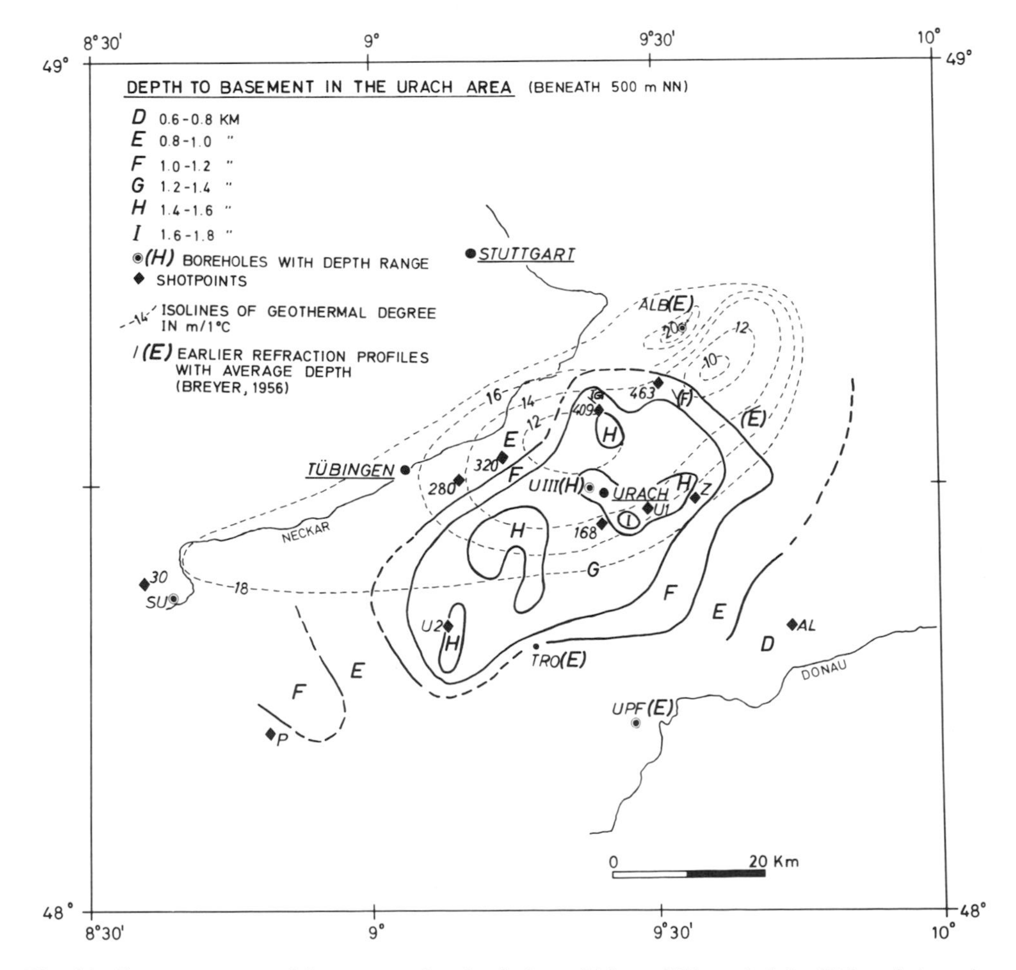

Fig. 11. Contour map of basement depths below 500 m NN graded in 200 m intervals (heavy lines). Isolines of geothermal degree after CARLE (1974). – Explanation of shotpoints see Fig. 1; Boreholes: ALB Albershausen, SU Sulz, TRO Trochtlfingen, U III Urach 3, UPF Upflamör.

6. Discussion

The basic result of this study is the basement contour map shown in Fig. 11. The overall feature of this map is that it provides an outline of the extension and the borders of the Rotliegend trough for the area of investigation, as was already suggested by a few earlier data (BREYER 1956). The greatest depths of 1.6–1.8 km below + 500 m NN are obtained SE of Urach. In principle such an anomaly can arise from incorrect delay times which can result at stations with observations over only a limited azimuth window if, for example, the actual refractor velocity is locally different from that determined by the model (BAMFORD 1973). However, at the particular stations involved here the azimuthal coverage is quite good. In the area of the Hohenzollerngraben the depth to the basement seems to decrease slightly. However, this area is covered by only one profile and more detailed investigations have to be made here.

It should be noted that maximal basement depths do not coincide with the maxima of the thermal anomaly (Fig. 11). The results of the analysis of possible velocity varations within the basement can be summarized as follows:

– If one compares the delay time solutions allowing for a uniform and for a gradient velocity distribution there is no reason to assume that the basement compressional wave velocity is not uniform (5.66 ± 0.02 km/s) in the area under study.

– No significant lateral velocity changes can be found within the resolving power of the method applied here when comparing the thermal anomaly with adjacent areas (Fig. 8).

– Both on a statistical basis and on the basis of the velocity scattergram shown in Fig. 7 there is no indication of velocity anisotropy within the limits of measurement error (~1 %) for the basement in the Urach area.

This leads to the following conclusions:

If any velocity changes in the basement in the Urach area due to thermal effects are present they must be smaller than 2 % on the basis of laboratory measurements (VOLAROVICH & GURVICH 1956; HUGHES & MAURETTE 1956, 1957), which is at the limit of the resolving power of our method. This agrees with the temperature measurements in the Urach 3 borehole where less than 50 °C temperature increase, as compared to normal, occurs within the basement. This corresponds to less than 1 % change in P-velocity (MEISSNER et al., 1980).

If the reason for the geothermal anomaly is a high-temperature source it will certainly be located below 4–5 km depth; otherwise it should have stronger effects on the seismic velocities.

Any crack or fissure systems that can serve as aquifers for hydrothermal convection within the basement are likely to have random distributions, eventually connected with fractures associated with the Upper Miocene volcanic activity. Regular crack patterns should have had a bulk anisotropic effect on the seismic velocities (BAMFORD & NUNN 1979).

As to the question to what extent standard refraction-seismic methods can be used as an exploration tool for areas of geothermal potential we think the following answers can be given: With the type of experiment and the methods described in this paper a straightforward and comparatively cheap way is given to derive refractor velocity structure and topography as related to compressional waves over a large area and thus provide the basic information to aid in defining where other, more detailed methods should be applied.

In low temperature areas such as in Urach the changes of seismic parameters due to temperature are at the limit of the resolving power of refraction methods. Although they would eventually just about be able to resolve such small effects they would probably fail to detect areas of low thermal potential not known prior to the investigation. As other studies in high temperature fields have shown (e.g., MAJER & McEVILLY 1979), however, refraction seismic methods can well be used to outline the extent of such areas.

Hence well controlled seismic-refraction experiments should be used as a first step in assessing the structure of a geothermally anomalous area as an aid to defining areas of special interest where more detailed methods, e.g., reflection seismic studies, should be applied and other geophysical methods such as electric or magnetic are likely to add significant information.

Acknowledgements

The investigation was funded by a joint project of the Commission of the European Communities (EG) under contract no. 071-76 EG G/A-23-5 and the German Federal Ministry for Research and Technology (BMFT) under contract no. EG 4027 A as part of a multidisciplinary project headed by Dr. R. HAENEL.

We are indebted to Prof. K. FUCHS for his aid during the organisation of the experiments. W. KAMINSKI supplied continuous improvements of the data processing system. Without his efforts we would not have been able to complete this study. Data processing was done on the RAYTHEON RDS 500 of the Geophysical Institute and most of the calculations were carried out at the Computing Centre, both of the University of Karlsruhe. We thank all participants in the field experiments for their personal engagement and effort which contributed so much to the success of the experiments. The following institutions participated: Universities of Bochum, Frankfurt, Göttingen, Karlsruhe, Kiel, Stuttgart, ETH Zürich, the Niedersächsisches Landesamt für Bodenforschung and the Geologisches Landesamt Nordrhein-Westfalen. For the organisation of the shots the help of the Gewerkschaften Brigitta and Elwerath Betriebsführungsgesellschaft m.b.H., Hannover, Mr. C. BEHNKE of Niedersächsisches Landesamt für Bodenforschung, Hannover and of Mr. KLEINMANN and Mr. DUDENHÖFFER of the Landesbergamt Baden-Württemberg at Freiburg is greatly acknowledged. The personal engagement of Dr. DIETRICH, Urach, and Dr. SCHÄDEL, Freiburg, is highly appreciated. The communities of Burladingen-Melchingen (1978) and Urach (1979) kindly provided the headquarters for the field experiments and much other support. The company Lupold, Vöhringen, enabled the use of many blasts in the quarry Sulz (shotpoint no. 30).

References

Ackermann, H.D. (1979): Seismic refraction study of the Raft River geothermal area, Idaho. – Geophysics, **44**: 216–225.

Bamford, D. (1973): Refraction data in Western Germany – a time-term interpretation. – Z. Geophys., **39**: 907–927.

– (1976 a): MOZAIC time-term analysis. – Geophys. J. R. astr. Soc., **44**: 433–466.

– (1976 b): An updated time-term interpretation of P_n data from quarry blasts and explosions in Western Germany. – **In** Explosion seismology in Central Europe, data and results, ed. by P. Giese, C. Prodehl, A. Stein, Springer Verlag, Berlin-Heidelberg-New York, p. 215–220.

– (1977): P_n velocity anisotropy in a continental upper mantle. – Geophys. J. R. astr. Soc., **49**: 29–48.

Bamford, D., & Nunn, K. (1979): In situ seismic measurements of crack anisotropy in the carboniferous limestone of northwest England. – Geophys. Prospect., **27**: 322–338.

Bamford, D., Jentsch, M., & Prodehl, C. (1979): P_n anisotropy studies in northern Britain and the eastern and western United States. – Geophys. J.R. astr. Soc., **57**: 397–430.

Berckhemer, H. (1970): MARS 66 – a magnetic tape recording equipment for deep seismic sounding. – Z. Geophys., **36**: 501–518.

– (1976): Standard equipment for deep seismic sounding. – **In** Explosion seismology in Central Europe, data and results, ed. by P. Giese, C. Prodehl, A. Stein, Springer Verlag, Berlin-Heidelberg-New York, p. 115–118.

Breyer, F. (1956): Ergebnisse seismischer Messungen auf der Süddeutschen Großscholle besonders im Hinblick auf die Oberfläche des Varisticums. – Z. deutsch. geol. Ges., **108**: 21–36.

Carlé, W. (1974): Die Wärme-Anomalie der mittleren Schwäbischen Alb (Baden-Württemberg). – **In** Approaches to taphrogenesis, ed. by J.H. Illies, K. Fuchs, Schweizerbart'sche Verlagsbuchhandlung, Stuttgart, p. 207–212.

Červený, V., Langer, J., & Pšenčik, I. (1974): Computation of geometric spreading of seismic body waves in laterally inhomogeneous media with curved interfaces. – Geophys. J.R. astr. Soc., **38**: 9–20.

Emter, D., & Prodehl, C. (1977): Use of explosion seismics for the study of the Urach area. – Proc. Seminar on geothermal energy, Brussels, December 6–8, **1**: 351–366.

Geologische Übersichtskarte von Baden-Württemberg (1 : 200 000) (1962). – Geologisches Landesamt Baden-Württemberg.

Geyer, O.F., & Gwinner, M.P. (1968): Einführung in die Geologie von Baden-Württemberg. – Schweizerbart, Stuttgart.

Haenel, R. (1978): Die Erkundung des Temperaturfeldes bis in größere Tiefen im Bereich von Urach sowie Erprobung geophysikalischer und geochemischer Methoden. – Report, NLfB Hannover, Archive No. 81 370.

Hagedoorn, J.G. (1959): The plus-minus method of interpretation of seismic refraction sections. – Geophys. Prospect., **7**: 158–182.

Hughes, D.S., & Maurette, C. (1956): Variation of elastic wave velocities in granites with pressure and temperature. – Geophysics, **21**: 277–284.

– – (1957): Variation of elastic wave velocities in basic igneous rocks with pressure and temperature. – Geophysics, **22**: 23–31.

Jentsch, M. (1980): A compilation of data from the 1978–79 Urach, Baden-Württemberg, seismic-refraction experiment. – Open file report 80–1, Geophysical Institute, University of Karlsruhe.

Jentsch, M., Bamford, D., Emter, D., & Prodehl, C. (1980): Structural study of the Urach area by deep refraction seismics. – Commission of the European Communities, Second International Seminar: Advances in European Geothermal Research, 4–6 March, 1980, Strasbourg, p. 223–230.

Majer, E.L., & McEvilly, T.V. (1979): Seismological investigations at the Geysers geothermal field. – Geophysics, **44**: 246–269.

Meissner, R., Bartelsen, H., Krey, Th., Lüschen, E. & Schmoll, I. (1980): Combined reflection and refraction measurements for investigating the geothermal anomaly of Urach. – Commission of the European Communities, Second International Seminar: Advances in European Geothermal Research, 4–6 March, 1980, Strasbourg, p. 231–234.

Schädel, K. (1977): Die Geologie der Wärmeanomalie Neuffen-Urach am Nordrand der Schwäbischen Alb. – Proc. Seminar on geothermal Energy, Brussels, December 6–8, 1977, **1**: 53–60.

– (1978): Bohrung Urach 3, Erdwärmebohrung, vorläufiges Schichtprofil (1 : 1 000). – Geologisches Landesamt Baden-Württemberg, Freiburg.

Schleusener, A. (1945): Bouguer Isoanomalenkarte von Südwestdeutschland (1 : 200 000). – Report, Seismos GmbH, Hannover.

Smith, T.J., Steinhart, J.S., & Aldrich, L.T. (1966): Lake Superior crustal structure. – J. Geophys. Res., **71**: 1141–1172.

Smithson, S.B., & Shive, P.N. (1975): Field measurements of compressional wave velocities in common crystalline rocks. – Earth Planet. Sci. Lett., **27**: 170–176.

Timur, A. (1977): Temperature dependence of compressional and shear wave velocities in rocks. – Geophysics, **42**: 950–956.

Volarovich, M.P., & Gurvich, A.S. (1957): Investigation of dynamic moduli of elasticity of rocks in relation to temperature. – Bull. Acad. Sci. USSR, Geophys. Ser., engl. transl., **1**,4: 1–9.

Wohlenberg, J. (1978): Geophysikalische Untersuchungen in der Forschungsbohrung Urach. – Report, NLfB Hannover, Archive No. 81 388.

The Urach Geothermal Project, p. 247–262;
Schweizerbart'sche Verlagsbuchhandlung, Stuttgart, 1982

The Combined Seismic Reflection-Refraction Investigation of the Urach Geothermal Anomaly*

H. BARTELSEN, E. LUESCHEN, TH. KREY, R. MEISSNER, H. SCHMOLL and CH. WALTER

with 17 figures

Abstract: The combined reflection-refraction measurements with extended spread lengths and 8-fold coverage across the Urach geothermal anomaly have provided a high resolution of crustal structure with significant velocity anomalies.
From the (refraction) evaluation of the first arrivals of the large reflection spread a minimum in (horizontal) velocity up to about 2 % was found in the upper 3 km depth in the area of the geothermal anomaly. The average velocity decrease of about 1 % can perfectly be explained by a geothermal effect of + 50 °C as measured in the borehole Urach 3.
The evaluation of the reflection data indicates a large body with a 4 % deviation in stacking velocities and a maximum of 10 % deviation of the (vertical) interval velocity in the middle and deep crust below the Urach anomaly. The existence of such a large low-velocity body is also compatible with data from the wide angle stations and is partly indicated by a reflector of reversed polarity along its upper boundary. The interpretation of this body requires a combination of a thermal effect and a large-scale crustal alteration.

1. Introduction

The objective of the combined reflection-refraction measurements in the area of the Urach geothermal anomaly was to find out how seismic parameters, especially horizontal and vertical velocities, are related to the anomaly. In order to get some idea how an anomalous temperature might affect seismic velocities first the data of special experimental work were investigated. The expected small velocity variations of a few percent required a special field arrangement which was never tested before. The necessary accuracy of about 1 % for the seismic stacking velocity required a huge spread length of more than 20 km in order to get a high resolution in move-out times, i.e. about 1 sec for 10 sec 2-way travel time. Also a high coverage for the reflection and the refraction work was necessary for increasing the signal to noise ratio by stacking and for a higher accuracy of the refraction work which should provide a high lateral and depth resolution. In addition to the combined large reflection-refraction spread portable refraction stations with in-line offsets of up to 120 km should provide an extension to larger distances and permit a comparison between steep angle and wide angle reflections.

Hence, the basic question about the detectibility of a weak geothermal anomaly by seismic methods contains many elements which have not been used for seismic crustal

* Publ. No. 217 from Institut für Geophysik Kiel.

studies before, and much weight was devoted to the best possible method for velocity determination. First results on the investigations were published by MEISSNER & BARTELSEN (1979, 1980), BARTELSEN & MEISSNER (1980) and MEISSNER et al. (1981).

2. Consequences of Laboratory Experiments

In order to determine the effect of temperature on the seismic velocity under high temperature-high pressure conditions various experiments were carried out using a 3 axial, 200 ton press of the Mineralogisch-Petrologisches Institut in Kiel (KERN, 1978; MEISSNER & FAKHIMI 1977). 14 Gneiss samples were investigated

(i) at constant pressure and variable temperatures,
(ii) at constant temperature and variable pressure, and
(iii) at variable pressure and temperatures along an appropriate geotherm.

In order to estimate anisotropy and natural scattering of velocity values, Fig. 1 presents examples from 4 Gneiss samples, using the method (iii). Up to 20 % deviations in P-velocity are found in these Gneisses, a large part resulting from anisotropy as also seen from Fig. 1 by comparing the different directions of the same samples.

Temperature effects on velocity are much smaller. Fig. 2 shows the sensitivity of the V_p-velocity versus temperature for granitic material. It strongly depends on the pressure and on the average temperature level. For one kb or about 3.5 km depth and for 150 °C the value for $\Delta V_p/\Delta T$ is about $-$ 100 ms^{-1}/100 °C. For 2 kb and 200 °C average temperature $\Delta V_p/\Delta T$ is about $-$ 60 ms^{-1}/100 °C a value similar to that recently published by CHRISTENSEN (1979) for 2 bar and sialic rocks. Fig. 3 gives another representation of

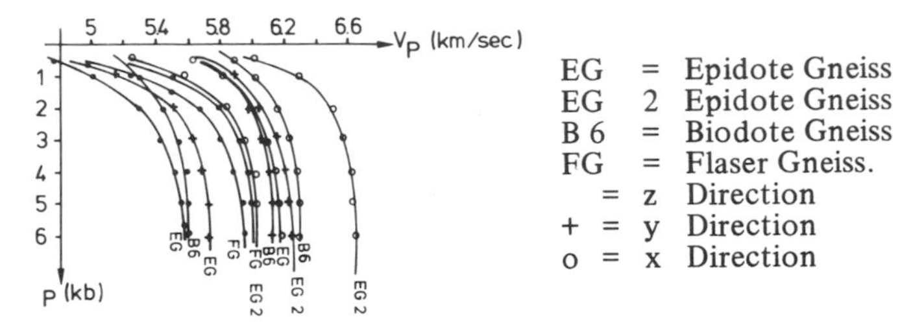

Fig. 1. Velocity-pressure relationship for 4 different Gneiss samples. (Velocity determined in 3 directions).

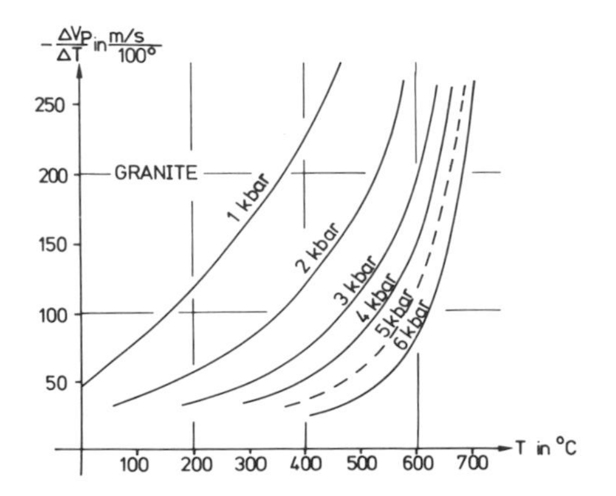

Fig. 2. Sensitiveness of V_p-velocity towards temperature at selected pressure levels.

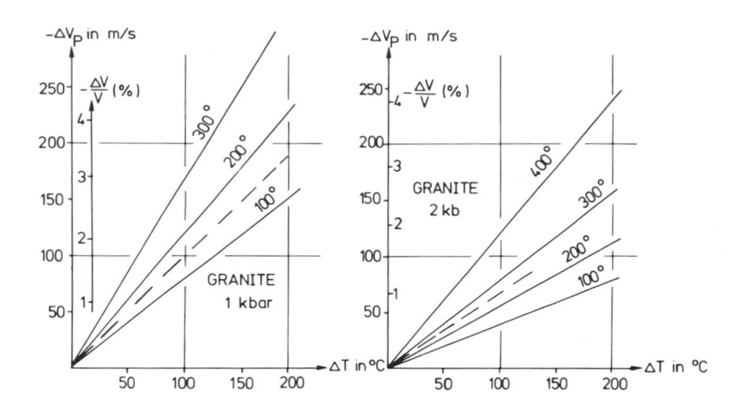

Fig. 3. Change in V_p-velocity as a function of a temperature difference at constant pressure and at selected temperature levels.

the temperature influence on velocities, showing the accuracy which is needed in the determination of velocity values for varying temperatures at constant pressure (depth) in order to measure temperature effects. As seen in the left diagram, a resolution of 50 m/s is needed for a change of 50 °C, a situation representing the Urach geothermal anomaly in the upper 3 kilometers. The whole temperature effect gives only a 1 % change in velocity (for Δ T = 50 °C) which is certainly small compared to effects of 16–20 % caused by inhomogeneity with anisotropy. There is some hope that lateral variations in temperature by their very nature cause smooth and continuous changes in the isotaches (= lines of constant velocities) whereas the material of rocks may change more discontinuously and abrupt, preferably along fault zones, and stay constant over large areas. Moreover, anisotropy, as measured in crustal studies is generally very small (MEISSNER 1967). The pure thermal effect on the other hand, may be increased by hydrothermal alterations, generally directed towards a decrease of velocities for increasing temperature. These effects, however, are very hard to control. For the investigations in the Urach area, changes of about 50 m/s representing 50 °C temperature variations were expected from the pure thermal effect in the center of the Urach anomaly for the first 3 to 4 kilometers. If the anomaly in temperature would not increase with depth, velocity variations of less than 50 m/s or about 1 % would be expected for the lower crust.

3. The Field Experiment

In order to measure small variations and details of the velocity structure with the highest accuracy possible, both horizontal and vertical velocities had to be gathered by the field set-up. Measuring the two horizontal directions in velocity on two perpendicular profiles would provide some additional boundary conditions on the anisotropy and possibly on lateral changes in anisotropy. The field set-up on line Ul consisted of a geophone line of about 23 km in length consisting of three independent reflection units each with 48 traces. An 8-fold coverage for reflection and a 16-fold coverage in refraction was achieved by shooting at 1.3 km intervals. For the picking of the first refracted arrivals and some wide-angle arrivals the reflection line was extended to more than 100 km by 27 portable refraction stations. A slightly different arrangement was made for line U 2 crossing line

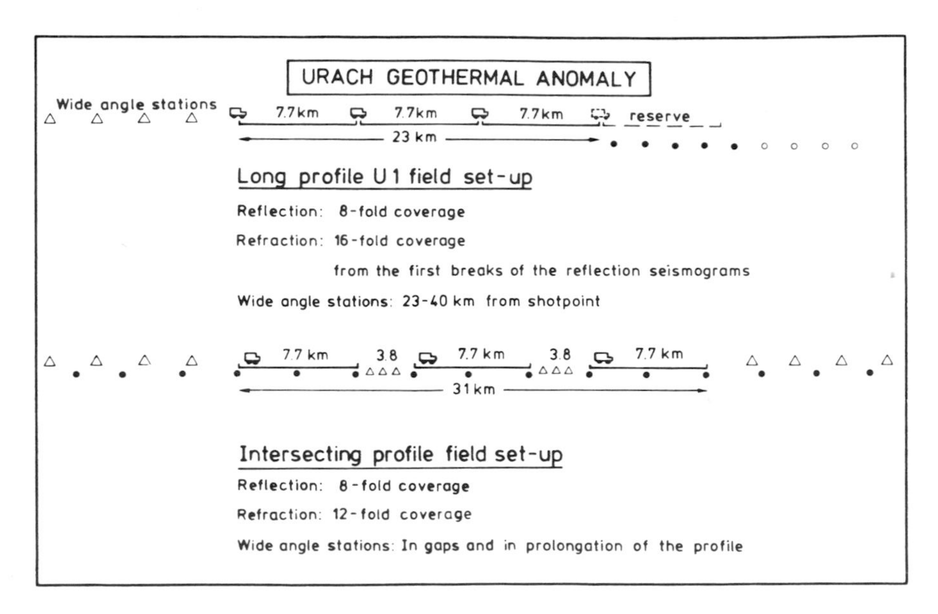

Fig. 4. Field set-up of the two reflection seismic profiles in the area of the Urach geothermal anomaly.

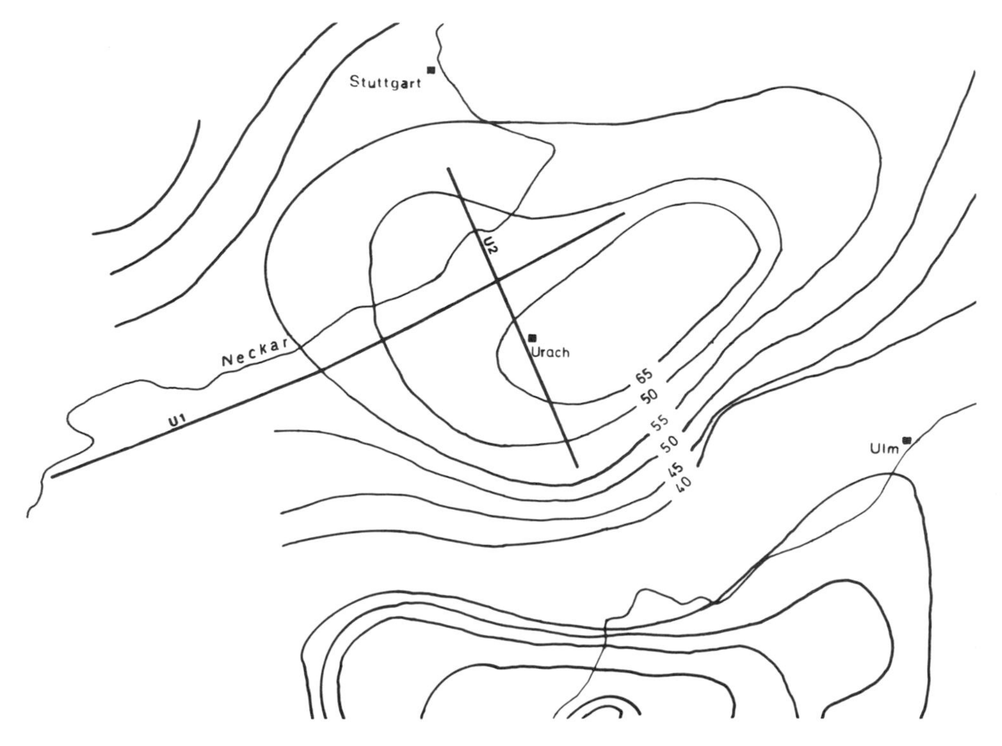

Fig. 5. Situation map of the profiles together with the temperature distribution at 1000 m depth below ground level (ZOTH this volume).

U 1 nearly perpendicularly. Fig. 4 shows both kinds of field set-up. The location of the two lines with regard to the known temperature maps (ZOTH, this volume) is shown in Fig. 5.

4. Interpretation of Refraction Work

The refraction work consists of the evaluation of the first arrivals of the reflection set-up and the evaluation of the additional refraction stations. On the two profiles up to 144 traveltime points from each of the 63 shotpoints form a system of refraction branches with 16-fold coverage. These traveltimes were first corrected to a datum plane at + 300 m above sea level. In another step, additional weathering corrections were determined, later used also for the processing of the reflection data. For the evaluation of (horizontal) velocities within the crystalline basement, additional corrections for the basement topography were applied. The depth to basement was computed from the time delays of the traveltime branch of the basement refractor. The full system of corrected traveltimes together with some additional reciprocal traveltimes at the shotpoints (vertical lines) is shown in Fig. 6.

The apparent velocities along these traveltime branches were transformed into true velocities by inverse averaging. An additional smoothing process with regard to time and position was applied. The smoothed data were then transformed by the Wiechert-Herglotz integral method (GIESE et al. 1976) into individual velocity-depth functions along the profile U 1. The resulting two-dimensional velocity distribution is shown in Fig. 7 together with the geological model above basement. The velocities range from about 5.2 km/s in the upper part of the basement to about 5.8 km/s in 2–3 km depth.

From the first arrivals of the intersecting profile U 2 the basement topography was computed in a similar way. The northern flank of the Rotliegend trough with a maximum thickness somewhat south of the deep borehole Urach 3 is found near the intersection U 1/U 2. The scatter of velocity values and the short profile length of U 2 allowed only the determination of a representative one-dimensional velocity-depth-function (Fig. 8). In general, the basement velocities on profile U 2 are about 0.3 km/s higher than the velocities on U 1 in the uppermost part of the basement in the region of the profile intersection. This observation is in agreement with the direction of maximum tectonic stress which originates from the Alpine orogeny and is generally NW-SE (GREINER & ILLIES 1977), in the direction of U 2.

5. Interpretation of Reflection Work

Because of the large spread length and the strong variations of the sediments along the profiles several correction procedures were applied to the data. Besides the normal and automatic residual static corrections, bandpass filtering (8–35 Hz), deconvolution and dynamic corrections, also a special stripping correction after KREY (1978) was used. Single coverage and stacked sections were available for time control and evaluation. Fig. 9 shows the long profile U 1 in an 8-fold stacked version with a picking of the correlatable events. The profile intersection with U 2 is marked by a vertical line. Fig. 9 a shows profile U 1 with somewhat different filter technique and without interpretation. The upper crust seems to be seismically transparent, and possible inhomogeneities have to

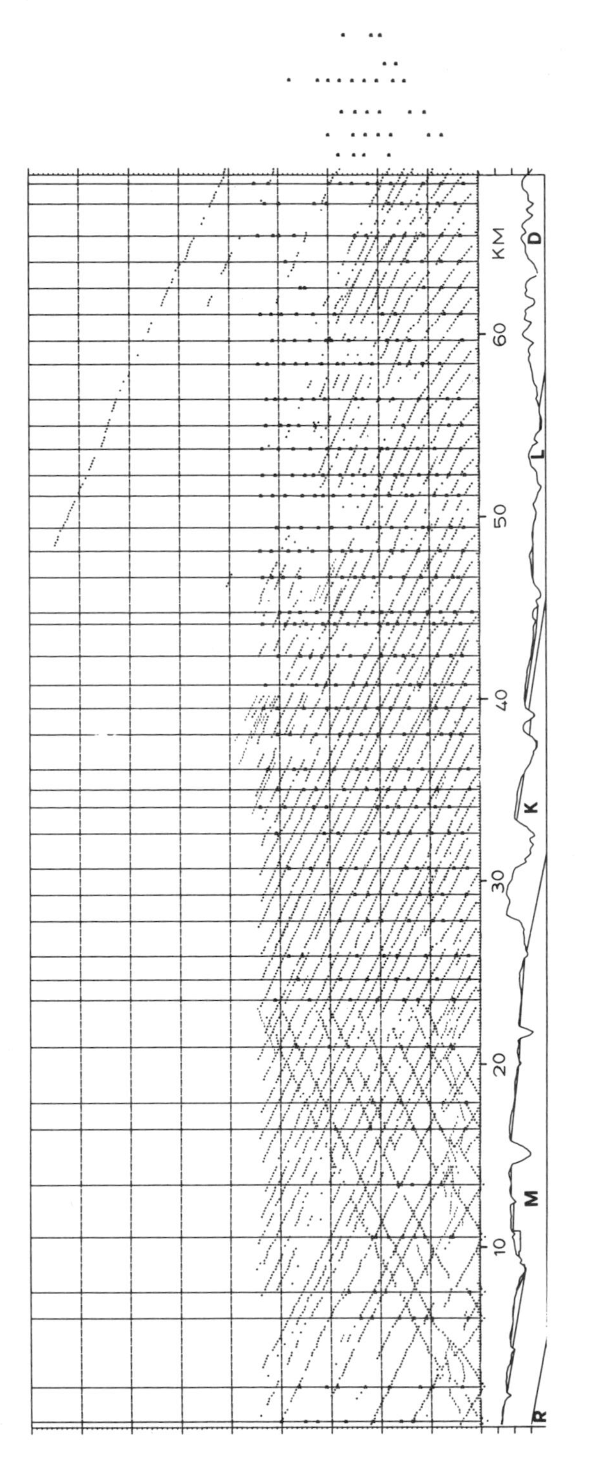

Fig. 6. The system of first arrivals and reciprocal times (at the shotpoints = vertical lines) on the long profile U 1. Traveltimes are corrected to datum plane at + 300 m above sea level. The datum plane corresponds to the baseline of the geological profile. R = Rotliegendes and Buntsandstein, M = Muschelkalk, K = Keuper, L = Lias, D = Dogger.

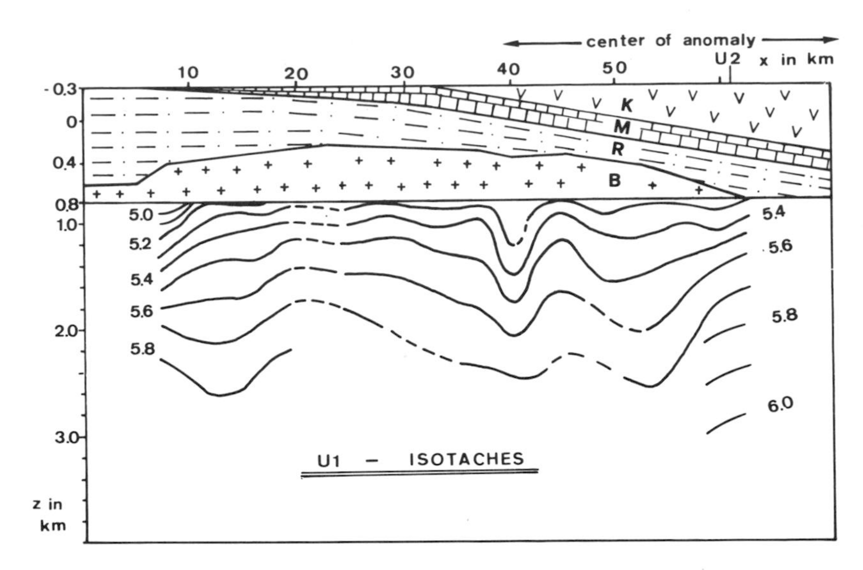

Fig. 7. Isotaches (= lines of constant velocities) in the basement below a datum plane of 800 m below sea level and geological profile. Symbols as in Fig. 6. B = Crystalline basement.

be of small size. The lower crust on the other hand, shows a generally high reflectivity, especially in the lowermost part, the transition zone to the mantle – the "Moho".

The mantle itself seems again rather homogeneous. These observations are in agreement with those of other crustal reflection data in Germany (MEISSNER et al. 1980). The Moho dips down from WSW (left hand side) to ENE. The distance from the western end of the profile to the eastern flank of the Rhinegraben is only 100 km so that the Moho-dip might be part of a large scale uplift of the shoulders of the Rhinegraben (GIESE 1976). The remarkable updoming in the middle crust decreases after migration. The Moho in the righthand part of the profile shows some signs of faulting. The stacked section of the shorter intersecting profile U 2 after similar data processing is shown in Fig. 10; Fig. 10 a shows profile U 2 with a somewhat different filter technique and without interpretation; the intersection with profile U 1 is marked by a vertical line. The band reflections in the lowermost crust shows a down-dip towards the Alps where crustal thickening ("mountain root") was to be expected. This spatial distribution of reflections on U 2 seems to be more irregular than on U 1. This might be an effect of lower seismogram quality and degree of coverage.

In order to reduce and control the influence of interpretational effects, the correlatable events were marked not only in the stacked section but also in the corresponding 8 single coverage sections. Events were then digitized and plotted in a common section in the form of an optical stack with no attenuation of events by interference. The comparison with the independently marked events of Fig. 9 shows a very good agreement and some additional features. All digitized events from the single coverage sections were then migrated and plotted into a common x-z-section (Fig. 11). The extension of the updoming structure at kms 20 to 38 in Fig. 11 is now reduced to about 9 km by migration.

As mentioned before, the special field set-up with the unusually long geophone spreads permitted a high accuracy in the determination of stacking (and interval) velocities. For the determination of optimum stacking velocities an initial uniform velocity model was changed systematically from 91 % to 110 % in steps of 1 %. Stacking was thus repeated

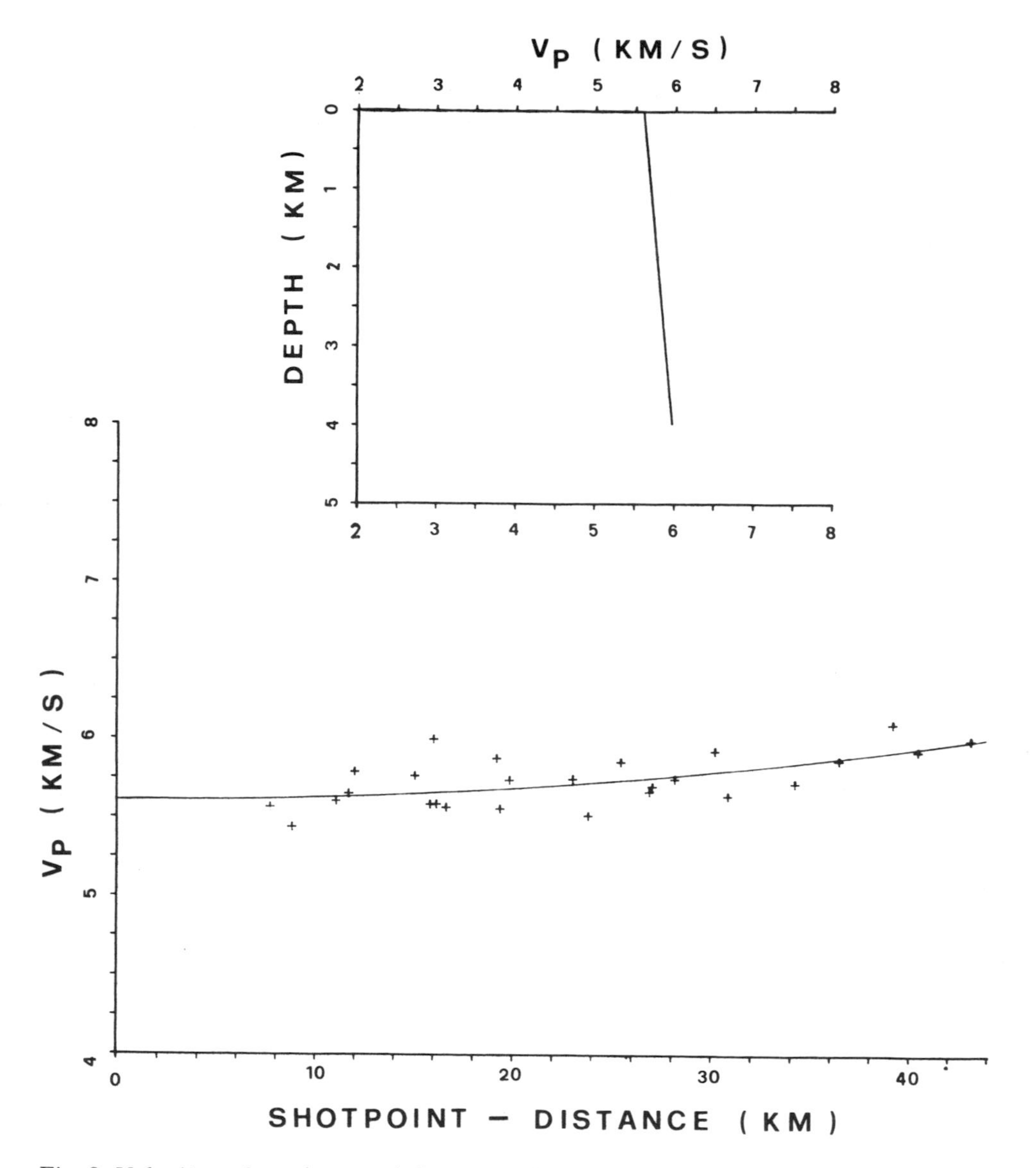

Fig. 8. Velocity values (averaged from traveltime branches and corresponding reversed traveltimes) on profile U 2. The regression polynomial of 2nd order was transformed into a velocity-depth function by the Wiechert-Herglotz integral method.

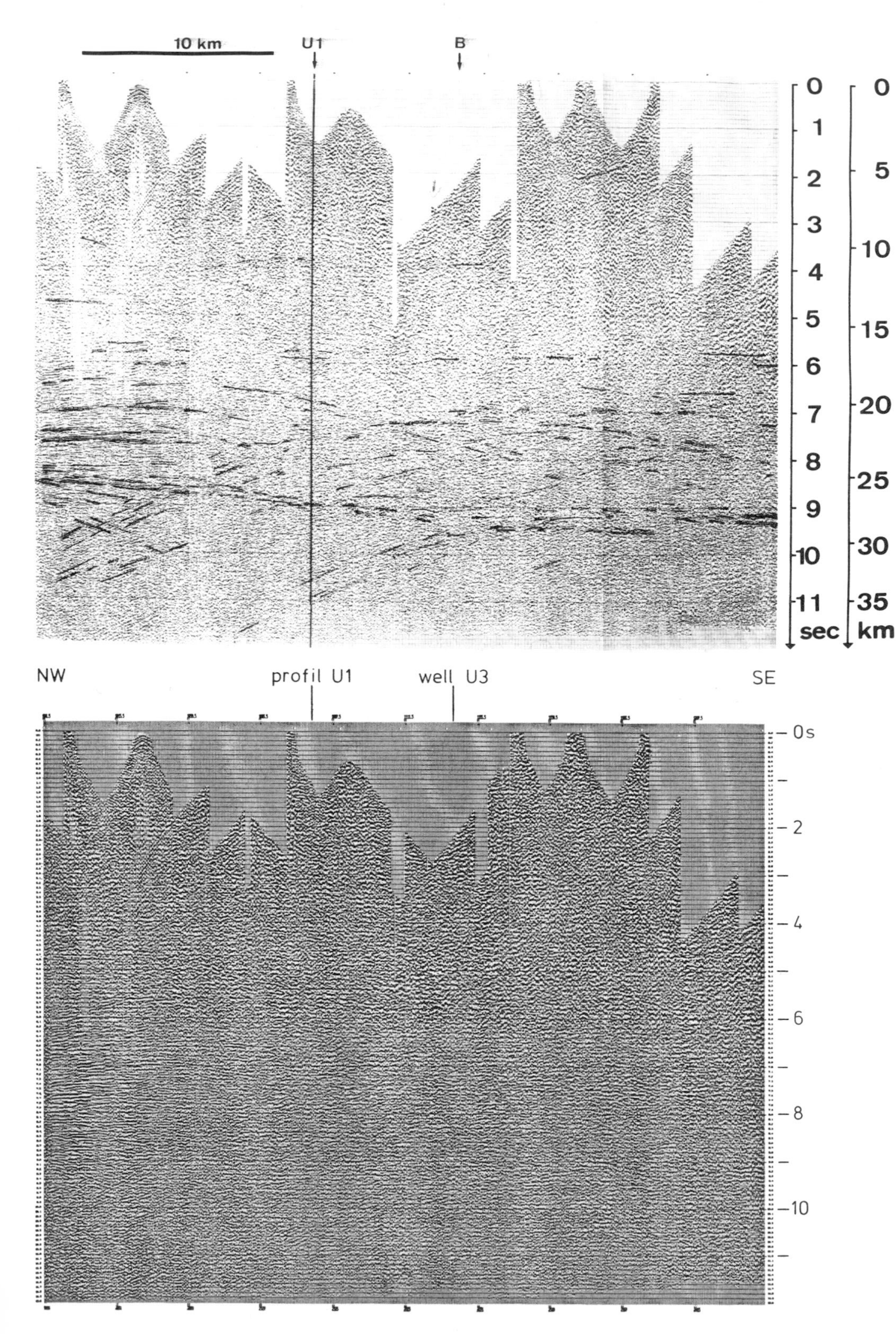

Fig. 10. Stacked section of profile U 2. The intersection with profile U 1 is marked by a vertical line. B = deep borehole Urach 3.

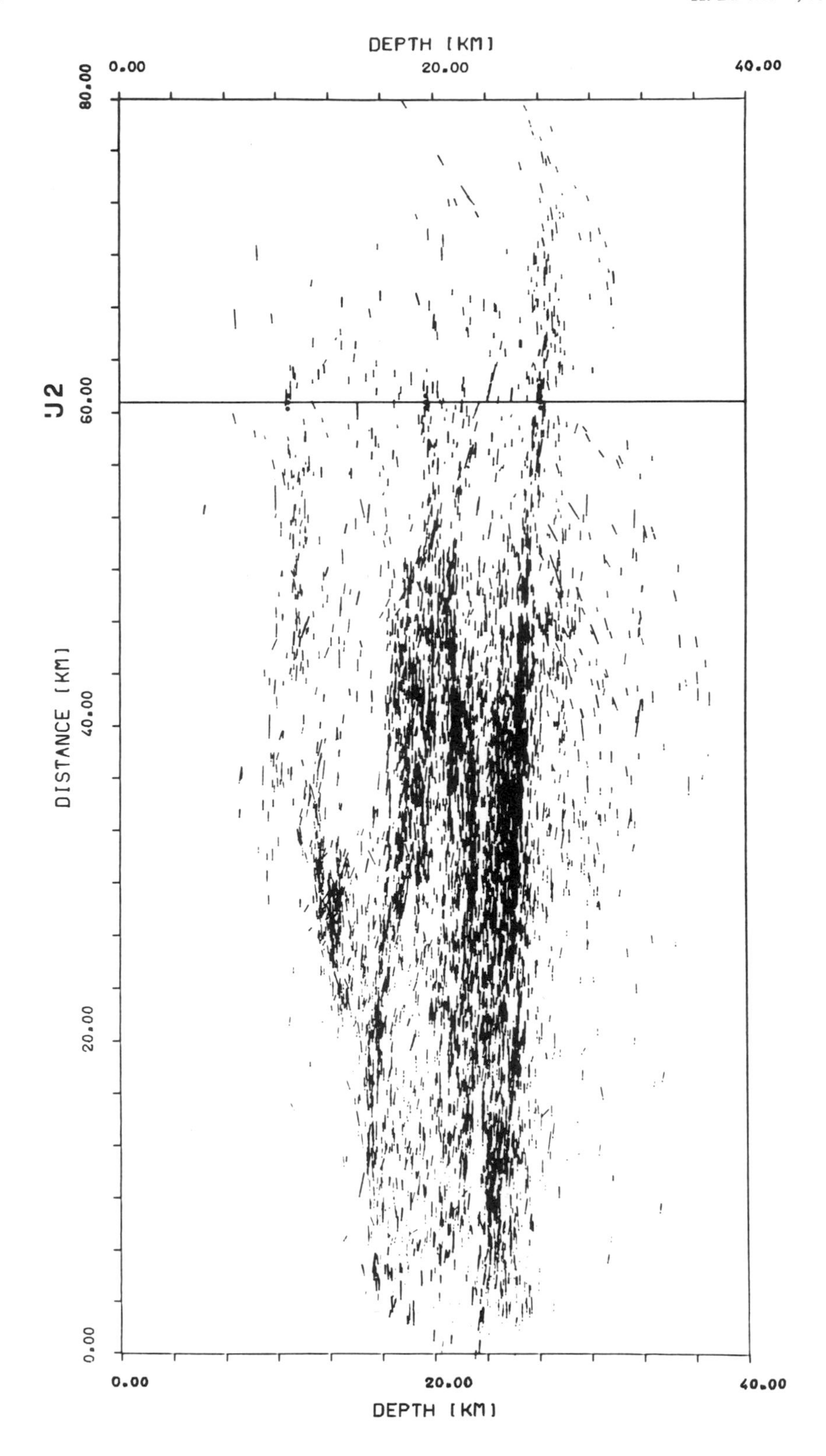
DEPTH [KM]
0.00
20.00
40.00
80.00
J2
60.00
DISTANCE [KM]
40.00
20.00
0.00
0.00
20.00
40.00
DEPTH [KM]

20 times for all data points. In most cases the percentages yielding optimum stacking can easily be decided upon. Fig. 12 shows the result of this procedure for reflections of two different time intervals along U 1. The points contain root-mean-square error bars and are connected by a smoothed line. For the deeper reflections a distinct low velocity body in the region of the Urach geothermal anomaly is clearly defined. For the shallower reflections a corresponding trough is less clear.

It is well known from KREY (1976, 1978, 1980) that the lateral variations of the optimum stacking velocity are not identical with the same variations of the average velocity or of interval velocities. Using Krey's method (1978) different velocity models and their fit with the observed variations of optimum stacking velocities were computed. Fig. 13 presents the final velocity model after 4 steps of iteration. It shows a large body of reduced velocity (hatched area) in the region of the Urach anomaly. The Moho reflectors and some other prominent reflectors were re-constructed and migrated with the velocities of. the final model. The influence of the new, more precise velocities on the position and even on the dip of reflectors is most remarkable. The fit in stacking velocities is now greatly improved and lies within the precision range of the method (Fig. 14), especially if a certain regional velocity trend is admitted. This regional trend would somewhat

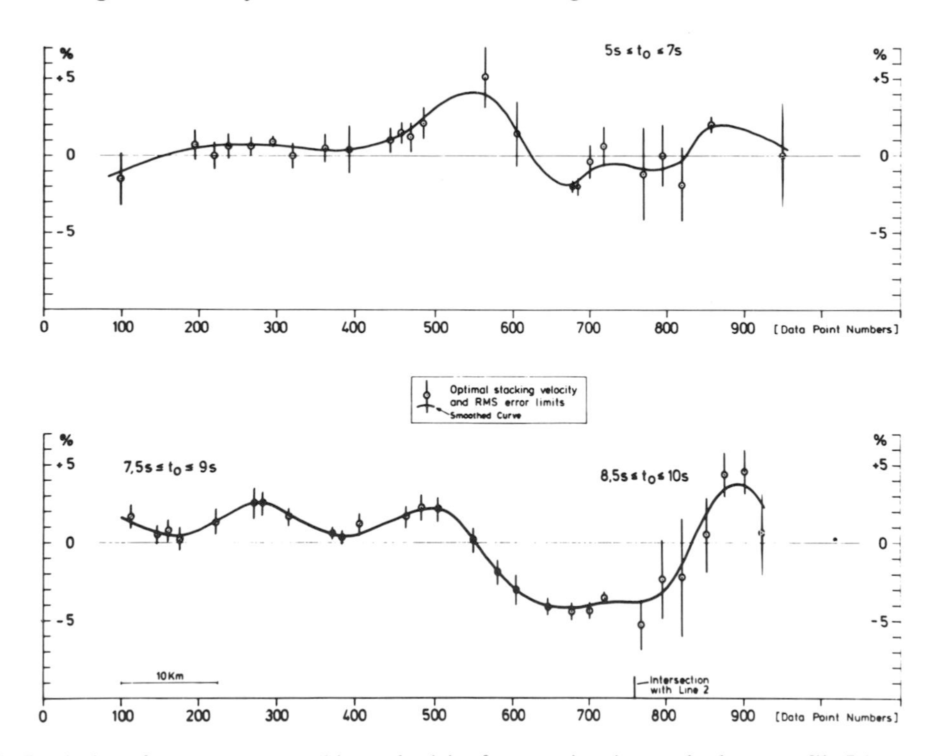

Fig. 12. Deviations from average stacking velocities for two time intervals along profile I 1.

Fig. 11. Profile U 1: Digitized events of single coverage sections after migration in x-z-representation.

increase the general updip towards the Upper Rhinegraben rift. The reliability of optimum stacking velocities on the short profile U 2 is poorer than on U 1, and a joint interpretation is not yet finished.

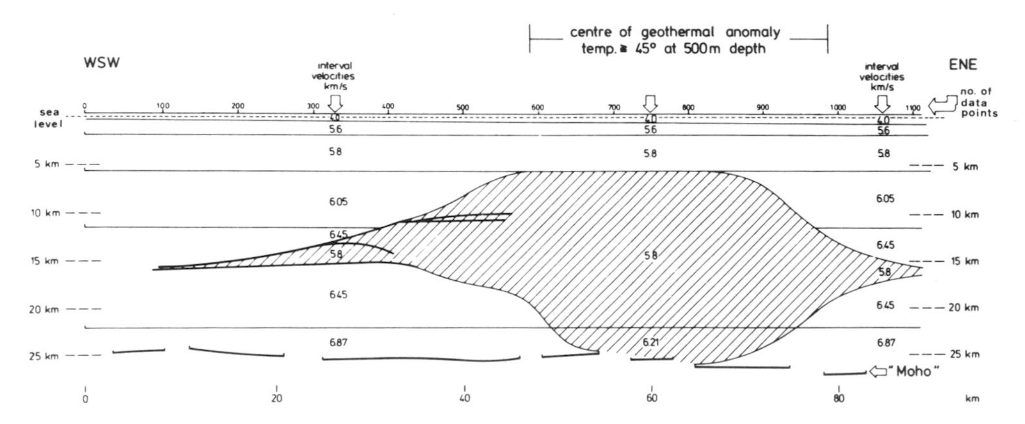

Fig. 13. Final velocity model along U 1 (4th iteration) showing a body of reduced velocity (hatched area). Numbers are V_p in km/s.

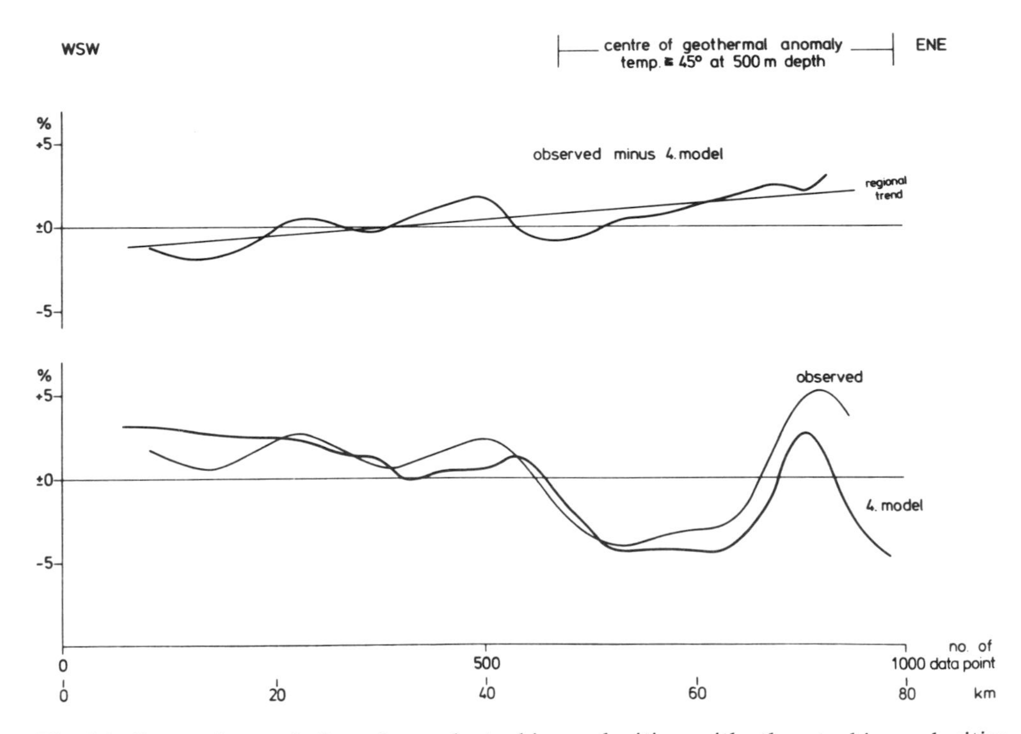

Fig. 14. Comparison of the observed stacking velocities with the stacking velocities resulting from the velocity model of Fig. 13. The differences are shown in the upper part.

6. Comparison of the Steep Angle Reflection Results with those of Wide Angle Stations

The final velocity model for profile U 1 was used for the computation of theoretical traveltimes for the wide angle range. The recordings of the portable stations provide some control of the velocity model by comparison of observed and calculated traveltimes even if dominant frequencies and the resolution of wide angle recordings are lower. The velocity model of Fig. 15 corresponds to the final model of Fig. 13. It contains

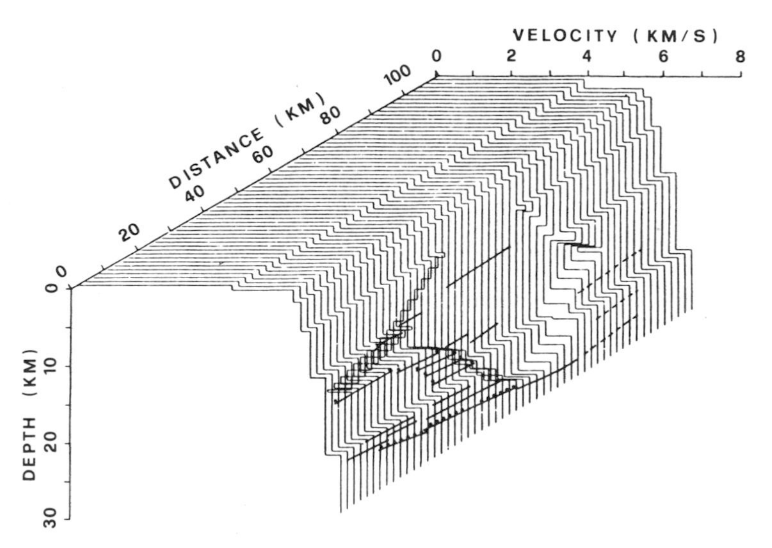

Fig. 15. Final velocity model for the computation of theoretical traveltimes and reflectors corresponding to Fig. 13. Dots mark agreement with observed events (see text).

only some additional reflecting elements on the NE extension of U 1. Only one example of these wide angle recordings is shown in Fig. 16. It contains the data of different shots recorded at station B, at the western end of the profile. The calculated traveltimes are marked by broken lines within the seismogram. Much to our surprise, these traveltime curves gave a good fit to the observed arrivals along the profile without any modification of the low-velocity body of Fig. 13 and without applying any trial and error methods. Only the additional reflectors between 70 and 90 km were introduced according to single coverage observations. The result of the comparison of all wide angle stations with the velocity model is also shown in Fig. 15. Reflectors in the model that show agreement between calculated traveltimes and strong correlatable events in the wide angle recordings are dotted. One dot only at the end of a reflector signifies agreement with observed events of weaker amplitudes. The broken lines in Fig. 15 mark reflectors at the ENE end of the reflection profiles that are indicated only in some single coverage sections and postulated from wide angle data in the ENE.

7. Reflection Polarity

From the data of the previous chapter the good coincidence between prominent steep-angle and wide angle reflectors was obvious. It is stressed that these reflectors, because of

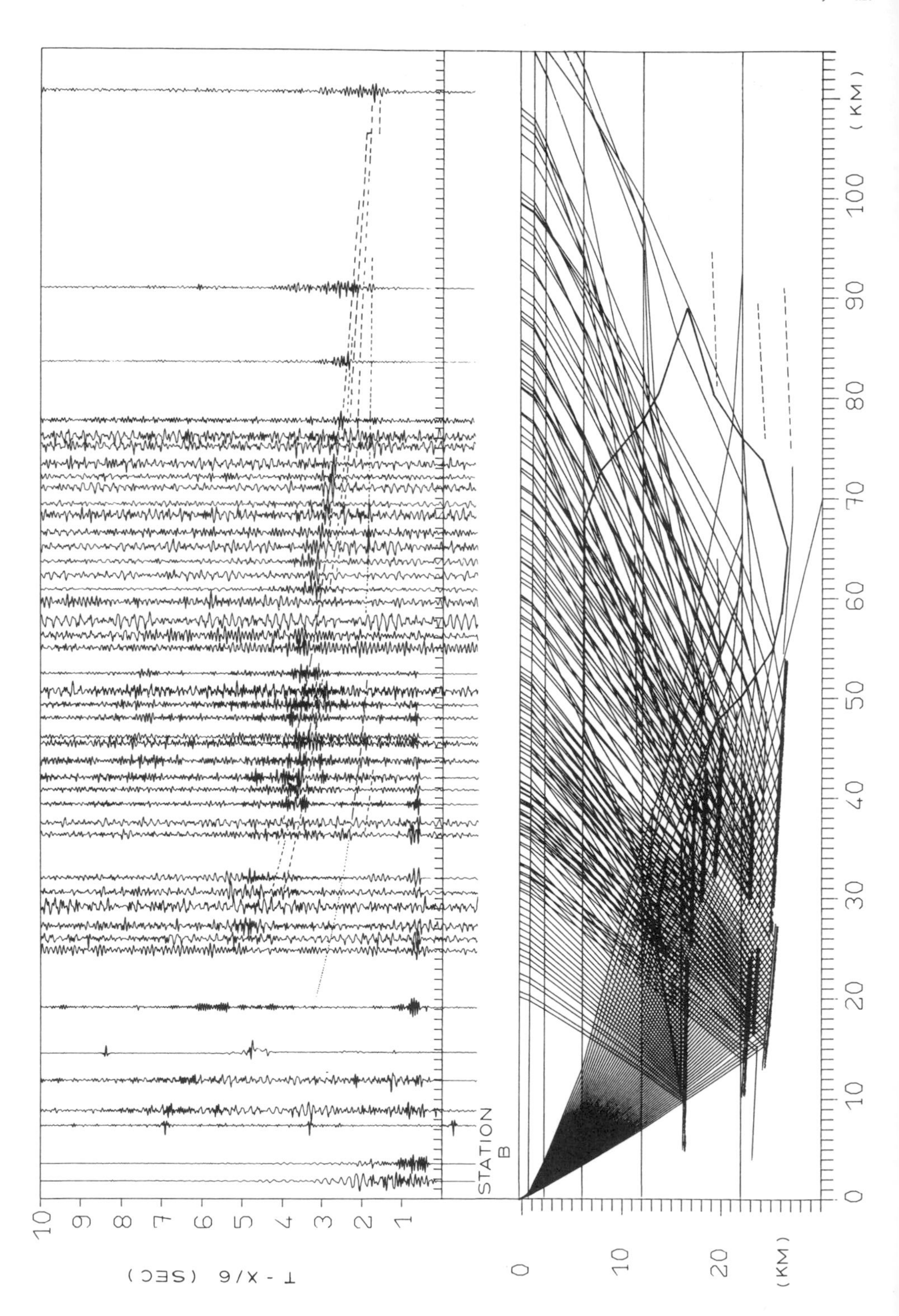

STATION B
T - X/6 (SEC)
(KM)
(KM)

their strongly enlarged amplitudes in the wide angle (= critical angle) area must represent relatively sharp boundaries from lower to higher impedance, resp. velocity (= normal polarity). It is well known that boundaries with an inverse polarity (i.e. those from higher to lower velocities) do not show up in the wide angle area but may well be observed by steep angle observations. Such boundaries are also required by the lamellation model of the crust mantle boundary (MEISSNER 1967, 1973). Moreover, the abundance of reflectors in the lowermost crust strongly suggest such a layering. In order to detect these reflectors with an inverse reflection polarity a cross correlation of a uniform wavelet with high quality single coverage sections of U 1 was performed. An example of this procedure is shown in Fig. 17. The weak but reliable boundary in the middle of the reflection band represents a boundary of inverse polarity. This is the first time that such an inversion of reflection coefficients is observed directly, providing a strong support for the lamellation concept. Also the top of the low velocity body of Fig. 15 should consist of a reflector with inverse polarity. At least on the left-hand side of this body there are some indications for such a velocity inversion while the main part of the body is not marked by reflections, probably because of a rather continuous transition zone.

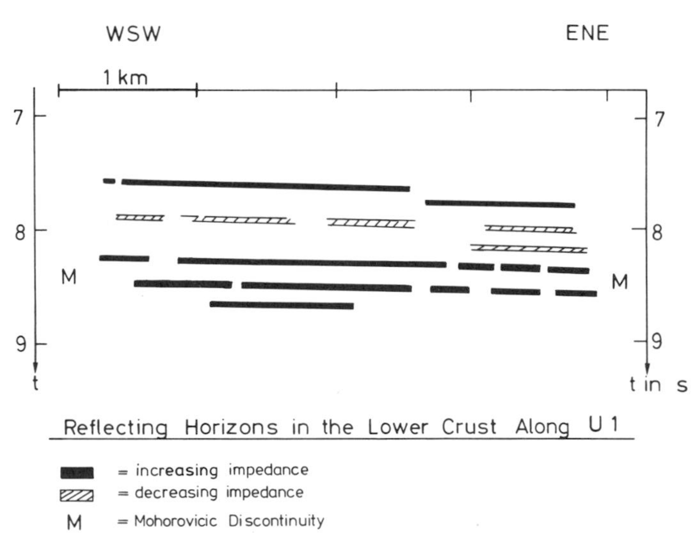

Fig. 17. Example of a reflector of inverse polarity (hatched bar) between layers of normal polarity (dark bars).

8. Conclusions

Below the area of the Urach geothermal anomaly a large body of low velocities – its center shifted only slightly to the West of the center of the heat flow anomaly – was detected by the evaluation of the reflection data. The mapping of such a body was made possible only by using extended spread lengths, providing an accuracy of 1 % in velocities down to depths of 30 km. The careful investigation of the refraction data shows only a small velocity anomaly in the upper 3 km which agrees with the small temperature anomaly of the Urach 3 borehole compared to its vicinity. The velocity anomaly in the middle and lower crust is surprisingly large and reaches up to 10 %. The determination

Fig. 16. Upper part: Example of a reduced record section in the wide angle range. The velocity model of Fig. 15 was used for the computation of rays and traveltimes (broken lines). Lower part: Ray pattern in the crust.

of such a strong low velocity body imposes serious problems and doubts on the accuracy of depth and structure as obtained by ordinary refraction work and studies of teleseismic residuals. Structure, depths and dips of reflections are modified by using the improved, laterally varying velocity data. In addition to the accurate velocities along the profile U 1, information about structure and reflection polarity are useful for the understanding of the geothermal anomaly.

The reasons for the large velocity anomaly area are not yet fully understood. It may be a combination of a pure thermal effect and a chemical alteration of crustal material, both effects possibly induced by the rise of basaltic material from the mantle or – generally – by a strong interaction between crust and mantle.

References

Bartelsen, H. & Meissner, R. (1980): Ergebnisse der reflexionsseismischen Messungen im Bereich der Wärmeanomalie Urach. **In**: BMFT/PLE: Programm Energieforschung und Technologien, Statusreport 1980, p. 39–59.

Christensen, N.J. (1979): Compressional wave velocities in rocks at high temperatures and pressures, critical thermal gradients and crustal low-velocity zones. – J. Geoph. Res. **84**: 6849–6858.

Giese, P. (1976): The basic features of crustal structure in relation to the main geologic units. In: Explosion Seismology in Central Europe, P. Giese et al. (eds.), Springer Verlag Berlin, Heidelberg, New York.

Giese, P., Prodehl, C. & Stein, A. (1976): Explosion Seismology in Central Europe, data and results. – Springer Verlag Berlin, Heidelberg, New York.

Greiner, G. & Illies, J.H. (1977): Central Europe: Active or residual tectonic stresses. – Pageoph. **115**: 11–26.

Kern, H. (1978): The effect of high temperature and high confining pressure on compressional wave velocities in quartz-bearing and quartz-free igneous and metamorphic rocks. – Tectonophysics, **44**: 185–203.

Krey, Th. (1976): Computation of interval velocities from common reflection point moveout times for layers with arbitrary dips and curvatures in three dimensions when assuming small shot-geophone distances. – Geoph. Prosp. **14**: 91–111.

– (1978): Seismic stripping helps unravel deep reflections. – Geophysics, **43**, 5: 899–911.

– (1980): Mapping non-reflecting velocity interfaces by normal moveout velocities of underlying horizons. – Geoph. Prosp. **28**: 359–371.

Meissner, R. (1967): Zum Aufbau der Erdkruste. – Gerl. Beitr. z. Geophys. **76**: 211–254 and 295–314.

– (1973): The Moho as a transition zone. – Geophys. Surv. **1**: 195–216.

Meissner, R. & Fachimi, M. (1977): Seismic anisotropy as measured under high pressure, high temperature conditions. – Geoph. J.R. astr. Soc. **49**: 133–143.

Meissner, R. & Bartelsen, H. (1979): Reflexionsseismische Untersuchungen struktureller und lithologischer Parameter in geothermisch anomalem Gebiet am Beispiel Urach. In: BMFT/PLE: Programm Energieforschung und Technologien, Jahresbericht 1979, p. 1255–1266, ET 4207 A.

– (1980): 2. Zwischenbericht über die reflexions- und refraktionsseismischen Messungen im Gebiet der Wärmeanomalie Urach, Projekt ET 4207 A.

Meissner, R., Bartelsen, H. & Murawski, H. (1980): Seismic reflection and refraction studies for investigating fault zones along the Geotraverse Rhenoherzynikum. – Tectonophysics **64**: 59–84.

Meissner, R., Bartelsen, H., Krey, Th. & Schmoll, J. (1981): Detecting velocity anomalies in the region of the Urach geothermal anomaly by means of a new seismic field arrangement. – Proc. EGS-ESC meeting Budapest 1980, in press.

The Urach Geothermal Project, p. 263–270;
Schweizerbart'sche Verlagsbuchhandlung, Stuttgart, 1982

Seismological Study of the Urach Geothermal Anomaly

G. SCHNEIDER

with 3 figures and 2 tables

Abstract: Seismograms recorded by the permanent seismological stations in Baden-Württemberg and by three mobile PCM-stations have been used to study the physical structure of the Urach geothermal anomaly (Fig. 1). From travel time and amplitude investigations some crustal pecularities in the Urach area could be found. Body wave signals from near-by and distant hypocentres are guided when travelling in a north-south direction through the crust of the studied area. An increase of absorptivity within the anomaly indicates a temperature increase of about 100 K.

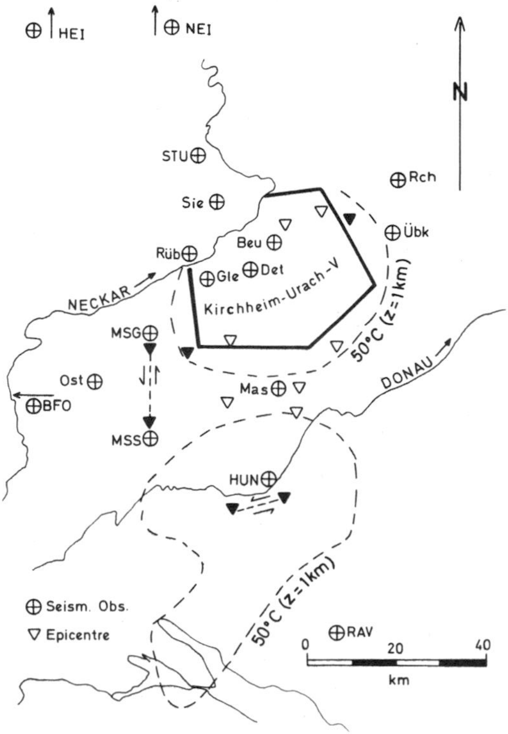

Fig. 1. Seismological observation sites used in the study of the Urach geothermal anomaly.
▽ : epicentres of local earthquakes
⊕ : seismological stations
--- : 50 °C-isotherm (HAENEL 1980)
—— : border of the Urach-Kirchheim volcanic area (MÄUSSNEST 1978)
▼---▼ : seismic focal zone

1. Travel Time Methods

1.1 Near-by Earthquake Travel Time Curves

Signals from near-by earthquake foci can be used to detect anomalous structures in the earth's crust and mantle. In comparison with refraction and reflection seismic methods an earthquake seismological study gives only a very coarse picture of the substructure. However, the different focal depths of natural seismic events allow the study of seismic wave propagation along paths which are different from those used by waves radiated from near-surface sources. For epicentral distances below 100 km the following epicentral areas could be used:

- Western and central Swabian Jura
- Oberschwaben
- Northern border of the Alps.

Along profiles having the same direction as the main strike of the Swabian Jura small time delays in the arrival of body wave pulses are observed in the travel time curves when the arrival times are compared with average travel time curves. These delays diminish with increasing focal depth from 0.3 to about 0.1 s when the focal depth increases from 2 to 10 km. The delays can therefore be explained by the influence of the P-velocity gradient in the uppermost part of the crystalline basement. Borehole measurements show the shape of such a velocity depth-variation (WOHLENBERG 1980).

When the body waves observed along profiles with a north-south orientation, forerunners can be seen on the seismograms. They arrive up to 0.4 s earlier than the direct P-waves propagating through the upper crust and are characterized by a P-velocity of ± 6.0 km/s. These onsets are interpreted as waves propagated along north-south striking "dykes" with a higher P-velocity of about 6.5 km/s. For details see SCHNEIDER (1980).

1.2 Deep Reflections

For epicentral distances below 100 km it is typical that reflections of strong amplitude appear after the direct ray phases. The Conrad- and Mohorovičić-discontinuities are the major reflectors in the South-western part of the German triangle and are normally found at depths of 20 and 30 km, respectively. In the western part of the studied area an upwarping of the Conrad-discontinuity from the "normal" depth of 20 km to a minimal depth of 14 km could be observed. Refraction seismic studies made in the same area show similar effects (SCHICK 1968, MEISSNER et al. 1980, EMTER 1980). For details see SCHNEIDER (1980).

2. Amplitude Methods

2.1 Spectral Methods

In the case of wave propagation problems the physical system can be divided into three main parts: source, propagation medium and receiver. The influence of the receiver, that is the seismograph, can be eliminated by idential calibration of the instruments (KEPPLER & SCHNEIDER 1977).

Differences in the shape of the seismic signals and their spectra due to source radiation pattern effects can be avoided by recording at an identical radiation angle. In the case of distant earthquakes the aperture of the array (maximal extension about 30 km) is related to very small differences in radiation angle. The damping of seismic signals during transmission is a function of the frequency of the particular spectral components. On propagation paths of 30 km length, as used for the study of the Urach anomaly, it is not possible to find any measurable amplitude differences at frequencies lower than 1 Hz even if the differences in propagation quality (Q) are quite high. This explains why pulse from distant earthquakes ($\Delta > 1000$ km) are almost identical in shape on records of the time function and the spectrum. At frequencies above 7 to 10 Hz the relation between propagation quality and absorption coefficient is no longer linear (Fig. 2). This is caused by the water content and the polycrystalline structure of the rocks in the upper crust (BORN 1941, POSTNIKOV 1974, JOHNSTON et al. 1979). Only in the frequency band from 1 to 7 Hz can interpretations of differences in spectral amplitudes caused by anomalous absorptivity be made. The analysis of a P-wave pulse from the Swabian Jura epicentral area is shown in Fig. 2.

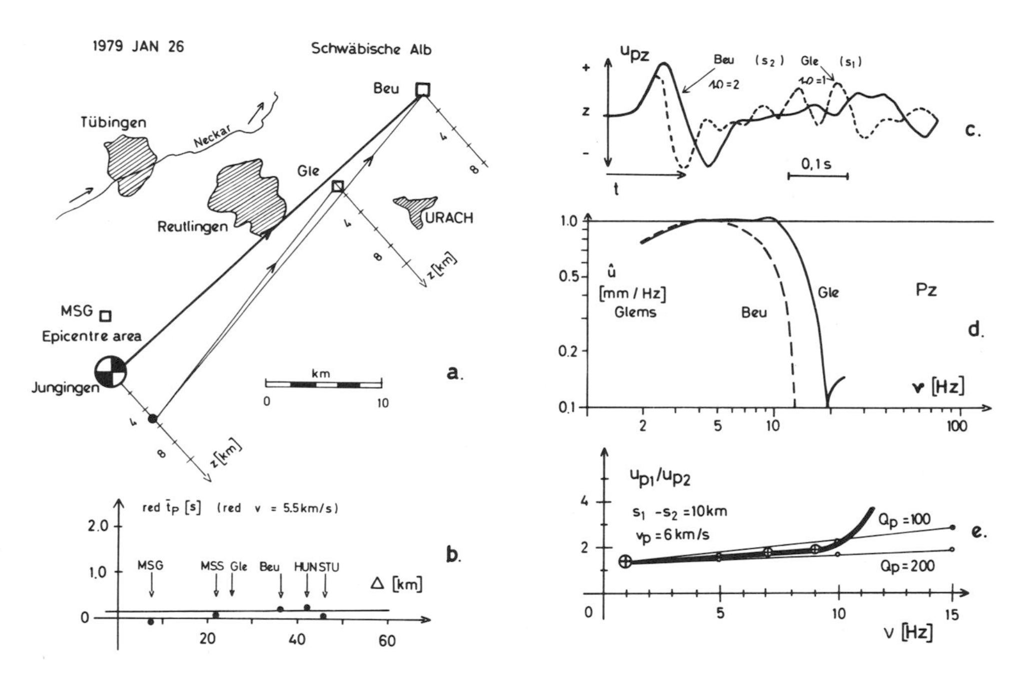

Fig. 2. Analysis of the records made during a Swabian Jura earthquake.
a) Epicentre location and observation points
b) Reduced travel time curve; red ≙ reduced
c) Time functions recorded at the stations Beuren (Beu) and Glems (Gle)
d) P-pulse amplitude spectra from Beuren and Glems
e) Interpretation of the spectra for differences in the propagation quality factor Q.

2.2 Time Function Method

The time function method for determining the absorptivity uses rise time and width of seismic pulses (GLADWIN & STACEY 1974, MINSTER 1978, REITER & MONFORT 1977). This method has been used to compare the amplitude and Q-variations along different propagation paths inside the anomalous area.

3. Interpretation of the Seismological Effects

3.1 Temperature Differences

Most of the models which attempt to describe the mechanism of mechanical wave absorption involve frequencies much higher than those observed in seismic signals. For frequencies between 1 and 10 Hz the mechanism of high temperature internal friction seems to be very probable (JACKSON & ANDERSON 1970). For the exponential relation between Q and temperature, as given in Fig. 3, a decrease in Q, amounting to about 50 % (see Tab. 1) is related to a temperature increase ascribed to the anomaly of about 100 K. Since rays from the local earthquake sources propagate directly through the upper crust, the anomaly must be located there.

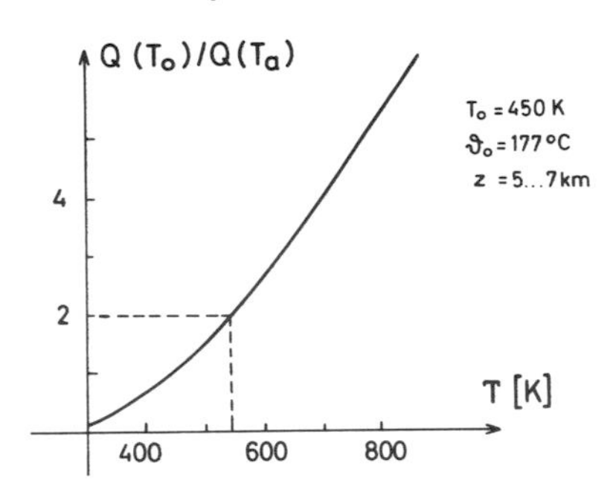

Fig. 3. Interpretation of variations in the propagation quality Q as caused by temperature differences in the earth's interior. T_o and ϑ_o are the reference temperature values. z is the depth interval for which the temperature T_a has been determined.

3.2 Structure Anomalies

In the case of P-pulse propagating in a north-south direction through the Urach area an amplitude increase at 1 Hz is found. In the case of guiding the channel width would be about 2 km (REDWOOD 1960). This value is in agreement with the average distance between volcanic pipes of the Kirchheim-Urach volcano (MÄUSSNEST 1978). It is also possible that metamorphic material or basic crystalline rocks can cause such an effect. Electromagnetic and gravimetric interpretations also show north-south striking positive anomalies in electric conductivity and density, respectively (BERKTOLD & KEMMERLE 1980, RICHARDS et al. 1980, MÜLLER 1979).

Data on the most important events used in this study as well as the basic physical equations are compiled in Tab. 1 and 2, respectively.

4. Possibilities of the Seismological Method

Seismological measurements can help to detect temperature and structural anomalies in the earth's crust when earthquake signals are recorded at epicentral distances between 0 and about 1000 km. The best information is obtained from registrations of local events distributed over a wide interval of focal depth. Such signals give important additional information when seismic waves from surface foci have been studied in the same area.

Acknowledgements

The author wishes to express his thanks for active help to the following colleagues: Dipl.-Phys. H. KEPPLER, Stuttgart (now Hannover); Mech.-Mstr. P. KREBS, Stuttgart; Frl. J. TURNOVSKY, M. Sc., Stuttgart; Dr. J. WIECK, Stuttgart.

Table 1. Seismic events and observations.

a) Fast forerunners in the case of local earthquakes

1976 Nov 29	1965 . . . 1978
Bosler	Oberschwaben
48° 36' N	48° 02 ± 5' N
9° 36' E	9° 21 ± 5' E
ho = 14 km	ho = 10 ± 5 km
δ Beu = 14 km	δ Beu = 61 km

b) Reflexions from Conrad- and Mohorovičic-discontinuities

1961 Apr 28	1960 . . . 1979
Südschwarzwald	Western Swabian Jura
47° 43' N	48° 17 ± 5' N
7° 53' E	9° 02 ± 5' E
ho = 10 km	ho = 6 ± 5 km
δ Beu = 155 km	δ Beu = 42 km

c) Guiding of distant earthquake signals in north-south direction

1979 Dec 68	1980 Feb 03	1979 Nov 16
Sicily	Tonga-Islands	Fiji-Islands
38.24⁶ N	17.641 S	16.512 S
11.843 E	171.271 W	179.923 W
ho = 16 km	ho = 33 km (N)	ho = 33 km (N)
δ Beu = 1150 km	δ Beu = 16 800 km	δ Beu = 16 300 km
1979 Nov 08	1977 Dez 30	
Kermadec-Islands	Tyrrhenian Sea	
32.376 S	40.000 N	
179.249 E	15.4191 E	
ho = 450 km	ho = 281	
δ Beu 18 000 km	δ Beu = 1000 km	

d) S-Wave spectral relations

1978 Apr 30	MSG → Beu	: Q_S = 180 ± 50
Kaiserstuhl		
46° 06.5' N		
7° 38.2' E		
ho = 12 km		
δ Beu = 142 km		
1978 Feb 23	MSG	: Q_S = 200 (fan)
Moloja-Pass	Beu	: Q_S = 100
46.45 N		
9.76 E		
ho = 3 km		
δ Beu = 225 km		
1979 Jan 27	MSG → Beu	: Q_S = 200 ± 50
Waldkirch		
48° 07.5' N		
7° 57.5' E		
ho = 16.4 km		
δ Beu = 119 km		
1979 Jul 03	MSG → STU	: Q_S = 120
Fribourg/CH		
46° 57.7' N		
7° 11.8' E		
ho = 5 km		
δ Beu = 230 km		

e) Comparison of rise times and spectra fo isolated pulses by using near-by earthquakes

Western Swabian Jura	Central Swabian Jura
as given under b)	1980 Mar 12–14
	48° $22 \pm 1'$ N
	9° $11 \pm 1'$ E
	ho = 12 ± 3 km

Observations along profiles between the border and the centre of the anomaly: Q_p = 150 ± 50; $Q_s = 75 \pm 25$. At the centre of the anomaly: $Q_p = 70 \pm 30$.

f) Q-values inside and outside the anomaly using isolated pulses from alpine P_n-phases. Important decrease of amplitudes at about 7 Hz (see also EMTER, 1971). No change in amplitude below the anomaly at about 2 Hz.

1978 Dec 12	Records of permanent stations in the distance range between 200 and 600 km.
Venetian Alps	
46.444 N	
12.742 E	
ho = 4 km	
δ Beu = 358 km	

Earthquake data published by the following institutions have been used: C.S.E.M. Strasbourg; Seismological Central Observatory, Erlangen; Geological Survey, USA; Geophysical Institute, Karlsruhe; Institute of Geophysics, Zürich; Institute of Geophysics, Stuttgart.

Table 2. Relationships used in sections 2 and 3.

Section 2.1.:

a) Determination of Q along a profile

$$Q = \frac{\pi \cdot \nu}{v \cdot \alpha} = -\frac{\pi \nu (s_2 - s_1)}{v \cdot \ln \frac{u_2}{u_1} \frac{s_2}{s_1}}$$

ν = frequency [Hz]
v = P or S-velocity [m/s]
α = absorption coefficient $[m^{-1}]$
s_1 = hypocentral distance to the first recording station along the profile [m]
s_2 = hypocentral distance to the second recording station along the profile [m]
u_1, u_2 = amplitudes and amplitude density at stations 1 and 2 [m]

b) Determination of δ Q on a fan

$$\delta Q = \frac{\pi \nu s}{v} \left(\frac{1}{\ln u_1} - \frac{1}{\ln u_2} \right)$$

s = effective path, for which a difference in propagation quality δQ has to be measured [m]

c) Relationship between Q_p and Q_s

$$\frac{Q_s}{Q_p} = \xi \left(\frac{v_s}{v_p} \right)^2$$

$$\xi = \frac{4}{3} \ldots \frac{7}{3} \text{ (limiting values)}$$

$Q_p \approx 1.71\ Q_s$ for upper earth crust.

d) Reference values for Q, WIECK (1974), MEISSNER & VETTER (1979)

Section 2.2:

e) Determination of Q from the rise time

$$Q = \frac{c\,\overline{t}}{\tau_2 - \tau_1}$$

c = empirical constant ≈ 0.5
$\overline{t}$ = travel time [s]
τ_1, τ_2 = rise times at stations 1 and 2 [s]

f) Guiding of wave phases in channels

$$T = 2\,d\,\sqrt{\frac{1}{v_s^2} - \frac{1}{v_p^2}} \quad [s]$$

T = period [s]
d = width of the channel [m]
v_s = S-velocity of the channel-material [m/s]
v_p = P-velocity of the channel-material [m/s]

Section 2.3:

g) Relation between differences in Q and in T

$$\frac{Q(T_o)}{Q(T_a)} = \exp\left[\frac{E^*(T_a - T_o)}{R\,T_a\,T_o}\right]$$

$Q(T_o)$ = propagation quality at reference temperature T_o
$Q(T_a)$ = propagation quality at temperature T_a
T_o = reference temperature [K]
T_a = temperature [K]
E^* = activation energy [J]
R = gas constant [J/kg · mol]
E^*/R = $1.76 \cdot 10^3$ K

References:

Berktold, A. & Kemmerle, K. (1980): Magnetotelluric measurements and geomagnetic depth sounding in the area of the Urach geothermal anomaly. – Sem. Strasbourg, 4–6 march 1980, p. 352–355.

Born, W.T. (1941): The attenuation constant of earth materials. – Geophysics **6**: 132–148.

Emter, D. (1971): Ergebnisse seismischer Untersuchungen der Erdkruste und des obersten Erdmantels in Südwestdeutschland. – Dissert. Uni. Stuttgart, 108 pp.

Gladwin, M.T. & Stacey, F.D. (1974): Anelastic degradation of acoustic pulses in rock. – Phys. of the Earth and Planet. Int. **8**: 332–336.

Haenel, R. (1980): Atlas of subsurface temperatures in the European Community. – Commission of the European Communities, Th. Schäfer, Hannover, 36 pp, 43 maps.

Jackson, D.D. & Anderson, D.L. (1970): Physical mechanisms of seismic wave attenuation. – Rev. of Geophys. and Space Phys. **8**: 1–63.

Johnston, D.H., Toksöz, M.N. & Timur, A. (1979): Attenuation of seismic waves in dry and saturated rocks: II. Mechanisms. – Geophysics **44**: 691–711.

Keppler, H. & Schneider, G. (1977): Seismological methods applied to Urach geothermal anomaly. – Sem. Geothermal Energy, Brussels 1977. EUR 5920, I: 367–381.

Mäussnest, O. (1978): Karte der vulkanischen Vorkommen der mittleren Schwäbischen Alb und ihres Vorlandes (Schwäbischer Vulkan) 1 : 100 000. – Landesvermessungsamt Baden-Württbg., Stuttgart.
Meissner, R., Bartelsen, H., Krey, Th., Lüschen, E. & Schmoll, J. (1980): Combined reflection and refraction measurements for investigating the anomaly of Urach. – Sem. Strasbourg, 4–6 march 1980, p. 231–234.
Meissner, R. & Vetter, U. (1979): Relationship between the seismic quality factor Q and the effective viscosity. – J. Geophys. **45**: 147–158.
Minster, J.B. (1978): Transient and impulse responses of a one-dimensional linearly attenuating medium. II. A parametric study. – Geophys. J.R. astr. Soc **52**: 503–524.
Müller, K. (1979): Gravimetrische Untersuchungen im Gebiet der thermischen Anomalie Urach (Arbeitsgruppe MAKRIS, Hamburg). – Vortrag 10. Arbeitssitzung Geothermik, Urach 1979.
Postnikov, V.S. (1974): Innere Reibung bei Metallen (russ.). – Verlag Metallurgie, Moskau, 320 pp.
Redwood, M. (1960): Mechanical waveguides. – Pergamon Press, Oxford u.s.w., 300 pp.
Reiter, L. & Monfort, M.E. (1977): Variations in initial pulse width as a function of anelastic properties and surface geology in Central California. – Bull. Seism. Soc. Am. **67**: 1319–1338.
Richards, M.G., Steveling, E. & Watermann, J. (1980): Low frequency magnetic and magnetotelluric soundings in the Rheingraben and the Black Forest. – Sem. Strasbourg, 4–6 march 1980, p. 339–341.
Schick, R. (1968): Die Tiefenlage der Mohorovičić- und Conrad-Diskontinuität des Schwäbischen Juras. – Veröffentl., Landeserdbebend. Bad.-Württbg., Stuttgart, 5 pp.
Schneider, G. (1980): Seismological investigations in Urach. – Proc. 2nd Intern. Sem. on the Results of EC Geothermal Energy Res., march 1980, Strasbourg, D. Reidel Publ. Company, Dordrecht (Holland), p. 567–575.
Wieck, J. (1974): Ereignisgesteuerte Erdbebenaufzeichnung mit hohem Dynamikumfang. – Dissert. Uni. Stuttgart, 70 pp.
Wohlenberg, J. (1980): Seismische Untersuchungen. Endbericht über geophysikalische Untersuchungen in der Forschungsbohrung Urach. – Report, NLfB, Hannover, Archive No. 8505.

The Urach Geothermal Project, p. 271–277;
Schweizerbart'sche Verlagsbuchhandlung, Stuttgart, 1982

Microseismic Investigation of the Urach Geothermal Area

M. STEINWACHS

with 7 figures

Abstract: Seismic noise was measured above the Urach geothermal anomaly on two profiles with six mobile, telemetric seismic stations from September 26 to October 3, 1977, and October 24 to November 2, 1978. Only the seismic noise recorded between 2 and 4 a. m. was evaluated. The average amplitude of the seismic noise in the central part of the anomaly, i.e. within the 45° isotherm for a depth of 500 m, was larger than that outside the anomaly (25° isotherm for 500-m depth) by a factor of 2.2.

1. Introduction

Microseismic recordings were made in the search for a geophysical method that can be used for the exploration for geothermal anomalies at the beginning of the seventies, primarily in the geothermal areas of the USA and New Zealand. Some of these studies showed that above the anomalies, the amplitude of the seismic noise was greater than outside the anomalies. The explanation for the origin of this "geothermal noise" (according to the models that have been made) seems plausible only for high enthalpy anomalies. The conversion of hot water to steam in permeable rock or the percolation of water and steam through fissures in the cap rock releases seismic energy in the form of "noise", which can be measured at the surface. Since amplitudes of 1 μm in the short-period range (1–30 Hz) are large in comparison to the average amplitudes of seismic noise outside geothermal anomalies, the use of suitable recording equipment and evaluation procedures make it possible to detect at the earth's surface very-low-energy seismic sources in the subsurface (DOUZE & SORRELS 1972). This is also demonstrated by the experience gained in the use of microseismics for monitoring gas storage reservoirs in porous rock (HARDY 1980).

Based on the results of a study of short-period seismic noise in the Federal Republic of Germany (STEINWACHS 1974), two consecutive research programs for the microseismic investigation of geothermal anomalies were carried out between 1977 and 1980, commissioned by the Commission of the European Community (contract no. 177–77 and 630/78 EGD). The anomalies in the Urach (southern Germany) and Torre Alfina (central Italy) areas were chosen for these studies. Results of the work at the Urach anomaly are discussed in this paper. The results of the investigations at Torre Alfina have already been published (STEINWACHS 1980).

2. Planning and Execution of the Recordings

2.1 The Recording Stations

Six recording stations were used. Each station is equipped with a vertical seismometer (Geotech S 13). The signals for all six stations were transmitted by radiotelemetry to a mobile receiving and recording station up to 30 km away. The recordings were made analog on magnetic tape in the frequency range from 0.2 to 17 Hz. (The analog equipment has since then been replaced with digital equipment, which has already been used at Torre Alfina). The calibration of the stations was done on a shaking table in the laboratory. The calibration was checked in the field by a simultaneous recording by all the stations set up at the same site (Fig. 1).

2.2 Method and Planning of Station Sites

Recordings were made along two profiles. The profiles were chosen on the basis of the isotherms at 500 m depth, as far as these were known at that time (HAENEL 1974). The 50 °C isotherm marked the central part of the anomaly (Fig. 2). To determine whether there were characteristic differences in the seismic noise between the anomaly and the area around it, the six stations were first set up in 1977 along profile 1, which stretches about 30 km from near Urach towards the SSE (Fig. 2).

In 1978, recordings were made along a second profile. Stations 5 and 6 were left at the same sites as the year before, the other stations were set up along profile 2, stretching about 25 km to the ENE. The choice of the profiles and the placement of the stations

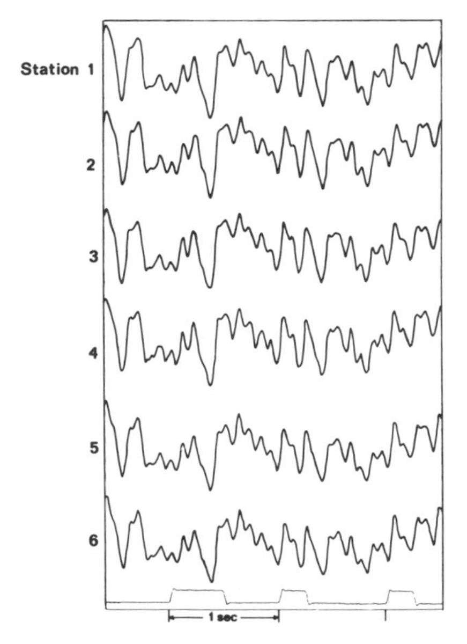

Fig. 1. Recordings made simultaneously at a single site for comparison and checking of the calibration of the seismometer stations.

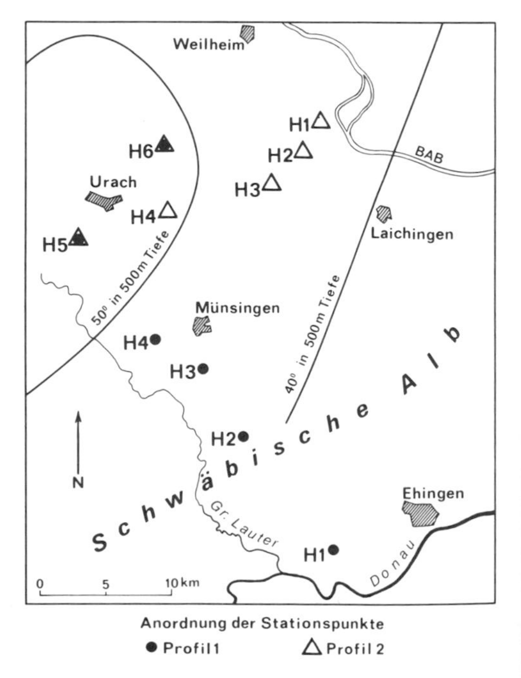

Fig. 2. Location of the station sites.

hade to take into consideration that there are numerous sources of seismic noise other than geothermal sources. If one considers the seismic noise N (t) recorded at a particular station as the superposition of noise from various sources, it can be represented in the time domain in the following manner:

$$N(t) = G(t) + D(t) + H(t) + M(t)$$

where G (t) is the term for geothermal noise,
D (t) is the term for noise from distant sources,
H (t) is the term for noise produced by local human acitivities, either directly or indirectly (i.e. machinery, especially motor vehicles), and
M (t) is the term for noise from local meteorological or other natural sources (e.g. the influence of wind on trees, buildings, and other topographic hindrances to the wind, rapidly flowing streams, etc.).

In the terminology of communications engineering, the term G (t) is called the "signal" and the sum D (t) + H (t) + M (t) is called "noise".

The station sites were selected to keep the "noise" as low as possible.

It is well known from refraction and reflection seismic studies, as well as from numerous seismic noise studies, that the amplitudes of seismic noise in urban and industrial areas are greater than in thinly settled areas with no industry. The propagation of seismic noise produced by industry has been investigated, especially by WILSON (1953) at Cambridge.

There is no especially noise-intensive industry in the Swabian Alb. Thus, distant noise D (t) was very low.

Owing to greatly reduced industrial activity, road and rail traffic, and use of agricultural machinery after midnight, the human noise term H (t) varies in a distinct diurnal cycle. The amplitudes of the seismic noise during the night can be more than a factor of 6 less than during the day (STEINWACHS 1974) with a minimum between 2 and 4 a.m. To keep the term H (t) as low as possible, recordings of seismic noise were made only between these times. In addition, the stations were not placed near houses or heavily travelled roads, with the exception of station 1 on profile 2, which was 1 km from the Stuttgart-Munich motorway. The transient noise produced by a lone vehicle (at night) could, with a little experience, be easily recognized and eliminated during the evaluation of the data.

To keep the M (t) term as low as possible, the stations were placed only in open fields. The distance to buildings, trees, and bushes was always at least 30 m. The seismometers were installed in pits to prevent direct influence of wind and rain. Recordings made when the wind strenght was greater than 2 (according to the weather report) were deleted from the evaluation.

A further factor that had to be considered in the selection of station sites was the effect of the subsoil at the site known as "seismic soil amplification". This can be important particularly in the case of thin, surface-near, unconsolidated layers above consolidated rock. The conditions in the Swabian Alb were favorable for setting up seismometers. Except for station 1 on profile 1, all of the pits for the stations could be dug down to bedrock (limestone).

2.3 Recording

Observations of the conditions given above for keeping to a minimum the unwanted terms D (t), H (t), and M (t), as well as the seismic soil amplifications, resulted in the siting of some of the stations several kilometers from the originally planned profiles.

Recordings were made on profile 1 from September 26 to October 3, 1977, on profile 2 from October 24 to November 2, 1978. On each profile, continous recordings were made simultaneously with all 6 stations for at least a week.

3. Evaluation of the Recordings and Interpretation of the Results

The recordings made during the day had irregular fluctuations in the amplitude. These were due primarily to vehicular traffic and agricultural machinery. There are no systematic differences between the recordings made at different stations. In contrast, there are distinct differences between the recordings made at the different stations on profile 1 at night between 2 and 4 o'clock. Fig. 3 shows as an example the recording between 2 : 35 and 3 : 00 a.m. on October 2, 1977. Fig. 4 shows a similar example from profile 2.

The seismic noise at stations 4, 5, and 6 on profile 1, which are close to the center of the anomaly, has a considerably greater amplitude than that of stations 1, 2, and 3 (Fig. 3). The ratio of the mean amplitude at station 6 to that at station 1 averages 2.2 with a variation of less than 20 % between different days.

On profile 2 (Fig. 4), there is little difference between the amplitudes at the different stations. The average amplitudes of the seismic noise are shown as columns on the profiles in Fig. 5. The bottom of each column is at the site of the station projected into the profile line. The height of the column represents the average amplitude of the seismic noise at night. The average was calculated from 10-minute recordings made between 2 and 4 a.m. on three different days.

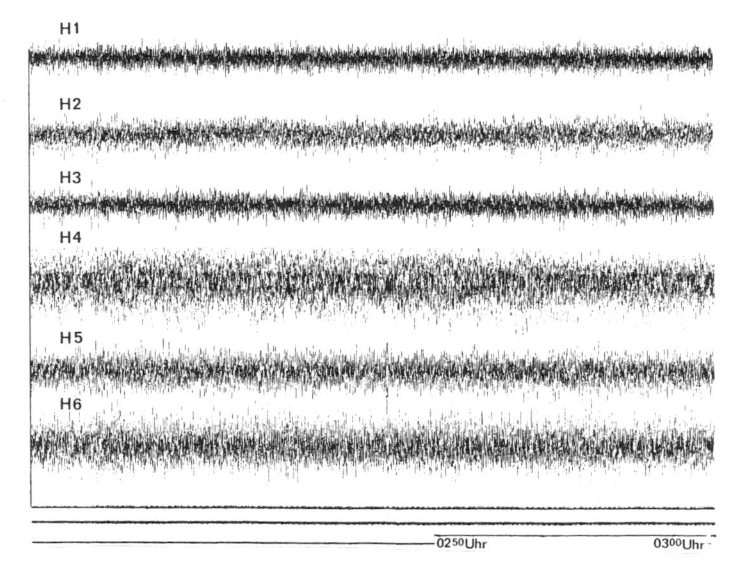

Fig. 3. Example of a play-back from magnetic tape of seismic noise at night on profile 1.

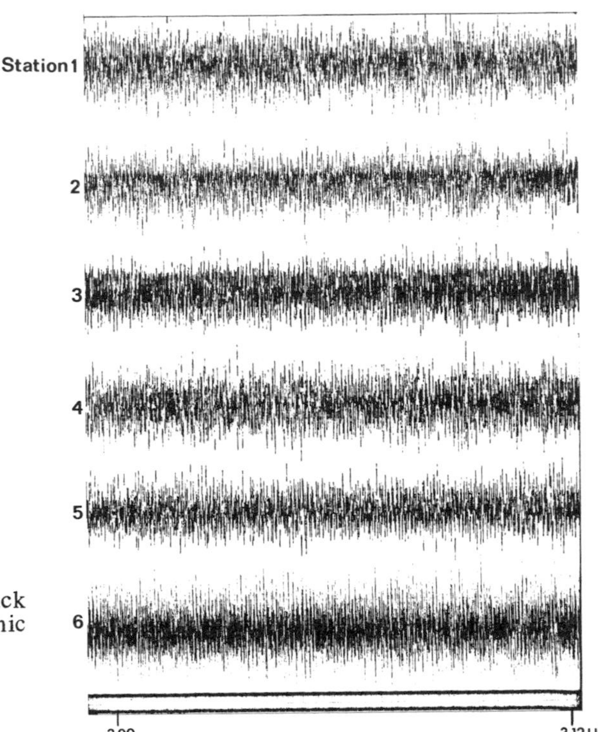

Fig. 4. Example of a play-back from magnetic tape of seismic noise at night on profile 2.

The amplitudes on profile 1, after a slight increase from station 5 to station 4, decrease rapidly with increasing distance from the center of the anomaly. There is only a slight indication of this trend on profile 2 from station 6 to station 3. This general result and the jump in amplitude between stations 3 and 4 on profile 1 was at first difficult to interpret on the basis of the isotherms (STEINWACHS 1978). Only after comparison of the seismic noise profiles with the isotherm map obtained from the most recent temperature measurements (ZOTH this volume) could the results of the recordings be confirmed (Fig. 6). The large decrease in the amplitudes south of Münsingen to the "normal level" as found at station 1 outside of the anomaly correspond to the decrease in the distance between isotherms that marks the edge of the anomaly. On the basis of this new isotherm map, a decrease in amplitude with increasing distance from Urach on profile 2 similar to that on profile 1 is not to be expected, which was also the case. The slight increase in amplitude on profile 2 from station 3 to station 2 and from station 2 to station 1 is probably due to the closeness of the motorway on which even at night there was a constant stream of traffic that could not be recognized on the recordings as transient noise.

A fourier analysis was done to determine which frequencies predominate. The Fourier spectra for stations 2 and 6 of profile 1 are shown in Fig. 7 as examples. The curves are the result of averaging 15 individual curves. The peaks at 3.5–4.5 Hz and 5.5–6.5 Hz are quite distinct on the curve for station 6 (dotted line). This peaks are up to 5 times greater for the curve for station 6 than for the curve for station 2 (solid line).

In conclusion, it can be stated that the results of the microseismic study of the Urach anomaly are in good agreement with isotherms drawn from recent data.

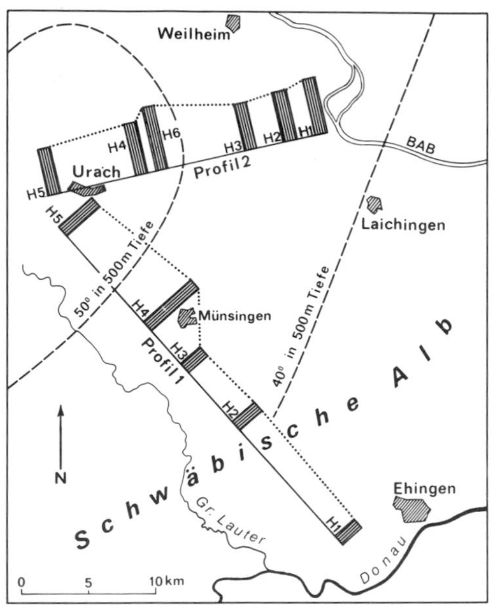

Fig. 5. Average amplitudes of the seismic noise at night and isotherms at 500 m depth from HAENEL (1974).

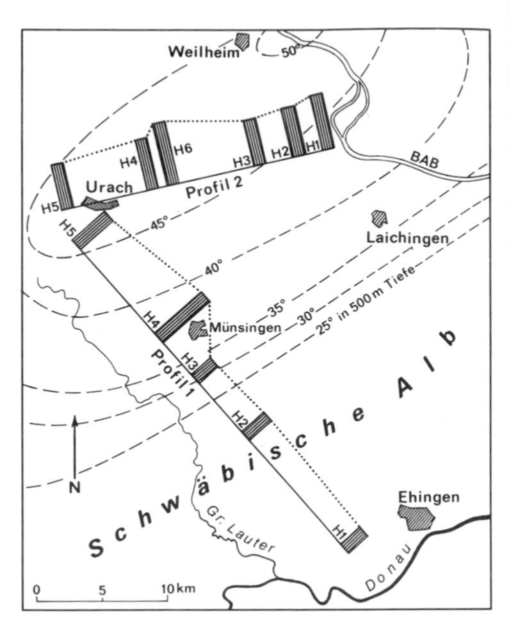

Fig. 6. Average amplitudes of the seismic noise at night and isotherms at 500 m depth based on the newest temperature data (ZOTH, this volume).

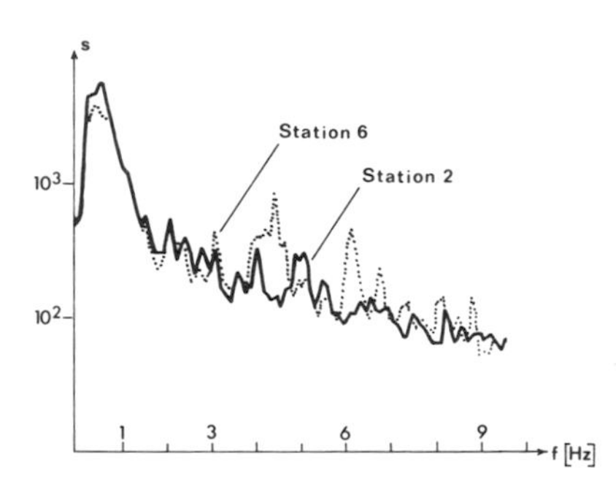

Fig. 7. Fourier spectra of the seismic noise at station 2 (solid line) and station 6 (dotted line) on profile 1. Each of the curves are the result of averaging 15 individual spectra.

As mentioned in the introduction, after the recordings were made at Urach, microseismic recordings were made at the Torre Alfina geothermal anomaly in central Italy. The temperature gradient there is 20° per 100 m and is considerably greater than the one at Urach. The seismic noise amplitudes at the center of the Torre Alfina anomaly were greater than those outside the anomaly by a factor of over 10. On the basis of the model developed for Torre Alfina, with the low-temperature gradient at Urach only noise of relatively low seismic energy can be expected.

The autor wishes to express his appreciation to Mr. J. LIPS, W. FESCHE, and K. BECKHAUS for their work on the preparation and implemention of the field work necessary for this project.

References

Douze, E.J. & G.G. Sorrels (1972): Geothermal Ground-Noise Survey. – Geophysics **37**, 5: 813–825.

Haenel, R. (1974): Bericht über geothermische Messungen in der mittleren Schwäbischen Alb (Uracher Vulkangebiet). – Report, NLfB Hannover, Archive No. 31 292.

Hardy, H.R. (1980): Stability Monitoring of an Underground Gas Storage Reservoire. – Proc. Sec.. Conf. Acoustic Emission/Microseismic Activity in Geologic Structures and Materials, Trans. Tech. Publ.

Steinwachs, M. (1974): Systematische Untersuchungen der kurzperiodischen seismischen Bodenunruhe in der Bundesrepublik Deutschland. – Geol. Jb. **E 3**: 1–59.

– (1978): Mikroseismische Untersuchungen geothermischer Anomalien. – Abschlußbericht über den Forschungsauftrag Nr. 177–77 EGD, Report, NLfB Hannover, Archive No. 80 041.

– (1980): Investigation of Microseismic Noise of the Geothermal Anomaly at Torre Alfina (Italy). – Sec. Seminar on the Res. of EC Geothermal Energy Research, Strasbourg.

Whitefort, P.C. (1975): Studies of the Propagation and Sources Location of Geothermal Seismic Noise. – Proc. Sec. UN Symposium, San Francisco, 1263–1272.

Wilson, C.D.V. (1953): An Analysis of Vibrations Emitted by Some Man-made Sources of Microseisms. – Proc. Royal Society of London, 188–202, London.

The Urach Geothermal Project, p. 279–284;
Schweizerbart'sche Verlagsbuchhandlung, Stuttgart, 1982

Geomagnetic Aspects of the Proposed Geothermally Disturbed Area at Urach

R. PUCHER and A. HAHN

with 2 figures and 2 tables

Abstract: The attempt to detect a geothermally disturbed area by means of magnetic measurements is based on considerations regarding the Curie temperature and Curie isotherm for different crustal compositions and on magnetic model calculations for different upwardly directed concavities of the Curie isotherm.
An application to the Urach area resulted in the statement that in the magnetic anomalies there is no positive indication for the presence of a large concavity of the Curie isotherm. However, an amplitude up to 5 km might be hidden within the anomalies. If such a concavity were present one should expect that the CURIE isotherm (550–560 °C) – normally positioned at 25 km depth – would rise to 20 km. This would correspond to a positive temperature difference of 100 °C.

1. Introduction

In zones of the earth's crust where the temperature is higher than normal, the distribution of magnetization can be affected mainly in three ways:

1.1 If the temperature anomaly reaches the depth of the Curie isotherm of magnetite (585 °C) and if at that depth there is a sufficient content of magnetite, the space below the bump-shaped Curie isotherm represents a hole in a more or less uniformly magnetized layer which should produce a detectable magnetic anomaly at the earth's surface.

1.2 High temperature may alter rock minerals, possibly by means of circulating water or steam, which may result in a change in the bulk magnetization in those parts of the crust. Such an alteration by hot water and steam was observed by SKINNER & BHOGAL (1977).

1.3 Moreover one should take into consideration that the creation of a geothermal anomaly within the crust is most probably caused by or connected with tectonic movements involving consolidated rock units and/or magmatic masses of considerable extent. If these movements result in the displacement of rock units with different magnetization against a smoother previous configuration one can again expect to find magnetic anomalies.

Although each of the described mechanisms easily can yield detectable anomalies, these anomalies can be attributed to high temperature zones only if there is a clear idea of the "normal" distribution of magnetic mass without the alteration.

Table 1. CURIE temperatures of crystalline rocks from Southern Sweden (geograph. coord. 57° N/14° E Gr.) Analyses with a magnetic field of 560 kA/m, a vacuum of 10^{-1} torr, maximal temperature of 680 °C, 10 °C/min.

Sample no.	Grade of metamorphism	Basic/ acid	Rock type	CURIE temperature °C during heating/cooling	Age	Locality
6	none	basic	red gneisses	561/551	Tertiary	Pallate
7	none	basic	red gneisses	557/555	Tertiary	Pallate
8	none	basic	red gneisses	539/523	Tertiary	Pallate
9	none	basic	red gneisses	563/544	Tertiary	Pallate
10	none	basic	red gneisses	555/537	Tertiary	Pallate
11	none	basic	hyperite	549/534	Precambrian	Sutareboda
12	none	basic	hyperite	(400) 557/535	Precambrian	Sutareboda
13	none	basic	hyperite	551/537	Precambrian	Sutareboda
14	none	basic	hyperite	(400) 539/539	Precambrian	Sutareboda
15	none	basic	hyperite	(400) 553/537	Precambrian	Sutareboda
35	none	acid	granite	557/543	Precambrian	Hanaskog
36	none	acid	granite	556/539	Precambrian	Hanaskog
37	none	acid	granite	557/539	Precambrian	Hanaskog
38	none	acid	granite	559/547	Precambrian	Hanaskog
1	low	basic	hyperite	531/520	Precambrian	Smal. Taberg
2	low	basic	hyperite	543/522	Precambrian	Smal. Taberg
3	low	basic	hyperite	539/520	Precambrian	Smal. Taberg
4	low	basic	hyperite	541/518	Precambrian	Smal. Taberg
5	low	basic	hyperite	529/512	Precambrian	Smal. Taberg
40	low	acid	ignimbritic rhyolite	561/547	Precambrian	Lenhovda
41	low	acid	ignimbritic rhyolite	563/547	Precambrian	Lenhovda
42	low	acid	ignimbritic rhyolite	(400) 559/541	Precambrian	Lenhovda
43	low	acid	ignimbritic rhyolite	561/547	Precambrian	Lenhovda
44	low	acid	ignimbritic rhyolite	563/539	Precambrian	Lenhovda
45	low	acid	quartzite	559/547	Precambrian	Almvik-Gamleby
46	low	acid	quartzite	559/547	Precambrian	Almvik-Gamleby
48	low	acid	quartzite	565/553	Precambrian	Almvik-Gamleby
49	low	acid	quartzite	559/539	Precambrian	Almvik-Gamleby
50	low	acid	Finnlands granite	559/545	Precambrian	NE Växjo
51	low	acid	Finnlands granite	(400) 554/539	Precambrian	NE Växjo
53	low	acid	Finnlands granite	559/543	Precambrian	NE Växjo
54	low	acid	Finnlands granite	(400) 555/541	Precambrian	NE Växjo

Sample no.	Grade of metamorphism	Basic/ acid	Rock type	CURIE temperature °C during heating/cooling	Age	Locality
55	low	acid	Smalandporphyry	400, 555/539	Precambrian	Tingsryd
56	low	acid	Smalandporphyry	559/541	Precambrian	Tingsryd
57	low	acid	Smalandporphyry	400, 559/543	Precambrian	Tingsryd
58	low	acid	Smalandporphyry	400, 555/541	Precambrian	Tingsryd
59	low	acid	Smalandporphyry	400, 553/535	Precambrian	Tingsryd
60	low	basic	uraliteporphyry	558/543	Precambrian	Madesjö
61	low	basic	uraliteporphyry	557/541	Precambrian	Madesjö
62	low	basic	uraliteporphyry	558/541	Precambrian	Madesjö
63	low	basic	uraliteporphyry	400, 554/537	Precambrian	Madesjö
16	high	acid	red gneisses	547/529	Precambrian	Sondrum
17	high	acid	red gneisses	551/537	Precambrian	Sondrum
18	high	acid	red gneisses	551/537	Precambrian	Sondrum
19	high	acid	red gneisses	541/531	Precambrian	Sondrum
20	high	acid	red gneisses	549/533	Precambrian	Sondrum
21	high	acid	granitic gneiss	559/543	Precambrian	Morrum
22	high	acid	granitic gneiss	555/537	Precambrian	Morrum
23	high	acid	granitic gneiss	549/538	Precambrian	Morrum
24	high	acid	granitic gneiss	549/535	Precambrian	Morrum
25	high	acid	granitic gneiss	549/539	Precambrian	Morrum
26	high	basic	charnockite	553/539	Precambrian	Varberg
27	high	basic	charnockite	547/533	Precambrian	Varberg
29	high	basic	charnockite	549/533	Precambrian	Varberg
31	high	acid	charnockite	559/541	Precambrian	Varberg
32	high	acid	biotite gneiss	559/545	Precambrian	Fjäras
33	high	acid	biotite gneiss	559/547	Precambrian	Fjäras
				Max. 563/Max. 555		
				Min. 529/ Min. 512		

2. Magnetic Model Calculation

As an approach to the solution of the question whether there are geomagnetic anomalies connected with the geothermal anomaly in the Urach area, the magnetic field of the following model body was calculated: a horizontal layer with a finite thickness but infinite extension is homogeneously magnetized. Its upper surface is flat and lies parallel to an infinite reference plane which represents e.g. the measuring plane of an airborne survey. The lower surface of the layer has a concavity of finite extension (see Fig. 2 a). It may be noted that the level of the upper surfaces does not affect the magnetic anomaly as long as this surface is not touched by the concavity of the lower surface.

The shape and situation of the model and its magnetic field is indicated in Fig. 1. A concavity from a normal Curie depth of 25 km to a depth of 10 km is depicted. The isolines of the anomaly △ T of the total intensity correspond to a magnetization of 0.3 A m^{-1} (≙ J = 30 gamma). Modifications of the model were obtained by varying the degree of concavity and the magnetization values. The inclination of the magnetization in all models was chosen to be parallel to the ambient field $I = 62°$.

The maxima and minima of the anomalies of the model variations are listed in Tab. 2. The linear dependence of the anomaly on the magnetization and depth of the model body is obvious.

For several of the model anomalies, profiles in NW-SE direction are represented in Fig. 2 a. A set of profiles of the Urach area in NW-SE direction from the magnetic map (Magnetic △ T anomalies 1976) are given for comparison in Fig. 2 b.

The comparison of the curves in Fig. 2 a and 2 b shows that, with the geometric and magnetic parameters of the concavity, only that of case (3) with the smallest amplitude might be hidden by the anomalies which form the main character of the field in that area.

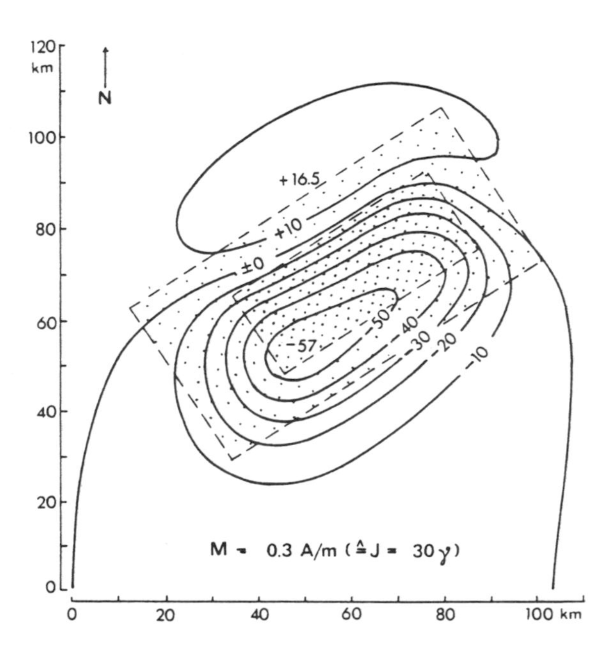

Fig. 1. Pattern of the total intensity anomaly caused by the concavity at the lower surface of a horizontal layer which is homogeneously magnetized parallel to the local magnetic field. The horizontal projection of the concavity is marked in the figure (see Fig. 2 a).

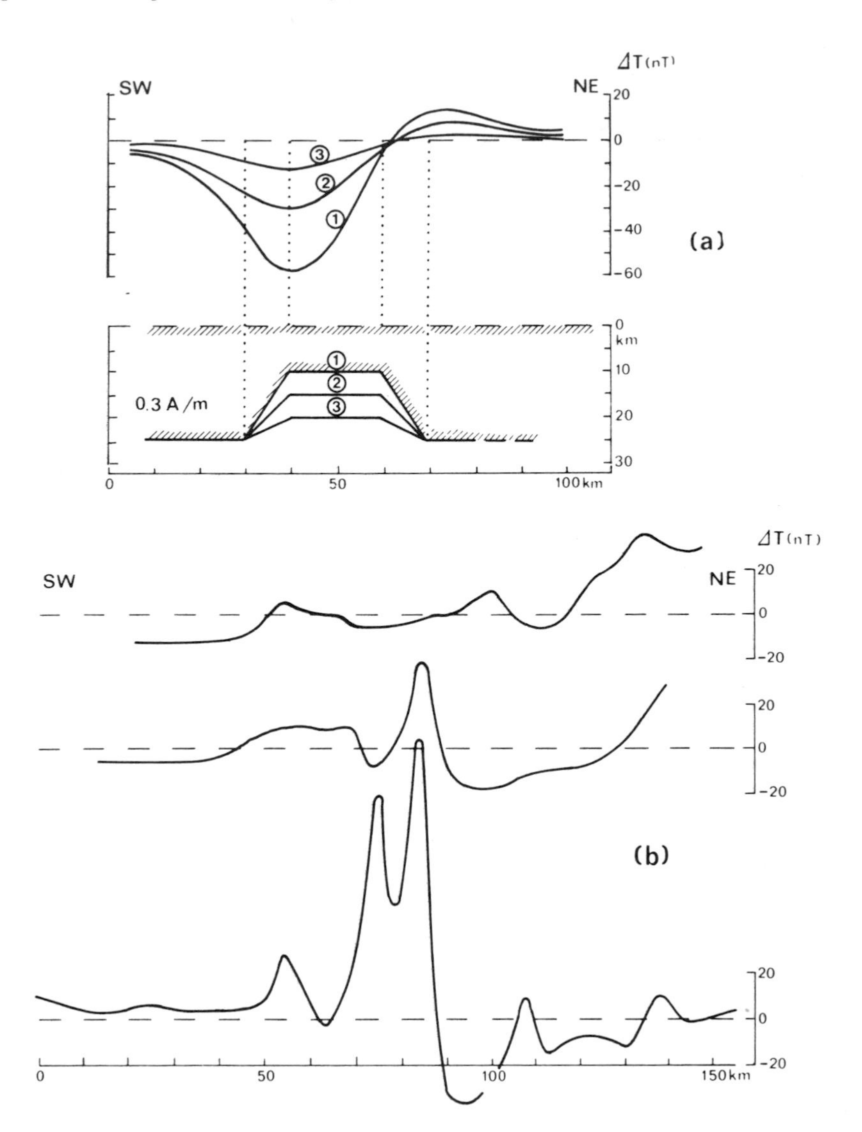

Fig. 2. Comparison of magnetic △ T-anomalies of concavities of a magnetized layer with a magnetization of M = 0.3 A m^{-1} with measured anomalies.
a) Cross section of a model and profile examples (for model modifications see Tab. 2).
b) Profiles of the Urach area of the aeromagnetic map of Germany taken in SE-NW direction.

A lower magnetization of possibly M = 0.05–0.1 A m^{-1} (≙ J = 5–10 gamma) would result in a still smaller amplitude (see Tab. 2). The same would be the case if a concavity of the Curie isotherm is extended over a larger region than just over the Urach Area. For those cases a magnetic identification is hardly possible.

Table 2. Maxima and minima of magnetic anomalies of three-dimensional model bodies (see Fig. 2 a). The lower surface is 25 km deep for all models; direction of magnetization D = 0°; I = 62°

Top of the concavity depth (km)	Magnetization in $A\ m^{-1}$ (≙ J = n · 10^2 gamma)	Minimum (in nT)	Maximum (in nT)
10	0.3	– 57	16
10	0.6	– 114	33
10	1.2	– 227	66
12	0.3	– 44	13
15	0.3	– 30	8.3
15	0.6	– 59	16.6
20	0.3	– 12	3.3
20	0.6	– 24	6.6
20	1.2	– 48	13.3

3. Conclusion

It is permissible to assume for large regions of the earth's crust that the carrier of the rock magnetization is magnetite with a Curie temperature of about 550 °C (Tab. 1, BOSUM & ULRICH 1969, LAUER 1975, PUCHER et al. 1978). The Curie temperature there is most probably not very dependent on the chemical composition of the rocks. Therefore one can expect the Curie isotherm to occur at the same depth for thermally undisturbed parts of the earth's crust.

The magnetic fields of different thermal concavities from a normal Curie depth of 25 km were computed and compared with aeromagnetic profiles of the Urach area. No indication of an uplifting of the Curie isotherm was found. If the value M = 0.3 A/m is representative for the magnetization, there might possibly be an undetectable concavity of 5 km (depth varying between 25 and 20 km). A smaller magnetization would allow a correspondingly larger one.

References

Bosum, W. & Ulrich, H.-J. (1969): Die Flugmagnetometervermessung des Oberrheingrabens und ihre Interpretation. – Geol. Rundschau, **59**, 1: 83–106, Stuttgart.

Lauer, J.-P. (1975): Contribution a l'etude de l'influence du metamorphisme thermique sur le proprietes magnetiques des roches; examples pris dans les Vosges cristallines du nord. – Thése, Universität Strasbourg.

Map of the magnetic anomalies △ T of the Bundesrepublik Deutschland, 1 : 500 000. – Bundesanstalt für Geowissenschaften und Rohstoffe (BGR), Hannover, 1976.

Skinner, N.J. & Bhogal, P.S. (1977): Total field magnetic survey at Olkaria, Kenya. – Braunschweigische Wissensch. Ges., Sonderheft 2: 10–18, Göttingen.

Pucher, R., Fromm, K., Lensch, G. & Ahrendt, H. (1978): Palaeomagnetic study of the Ivrea region, Western Alps. – Report, NLfB Hannover, Archive No. 81 414.

The Urach Geothermal Project, p. 285–288;
Schweizerbart'sche Verlagsbuchhandlung, Stuttgart, 1982

Geoelectrical Deepsoundings in the Area of the Urach Research Borehole

E.-K. BLOHM

with 5 figures

Three electrical deepsoundings have been carried out in the surroundings of the Urach 3 borehole in November 1978 by the working group Electrical Deepsounding of the Geological Survey of Lower Saxony, Hannover. The original idea was to perform a deepsounding close to the research borehole. This was not possible because of subterranean gas- water- and other pipes. Therefore, the profiles have been selected as shown in Fig. 1.

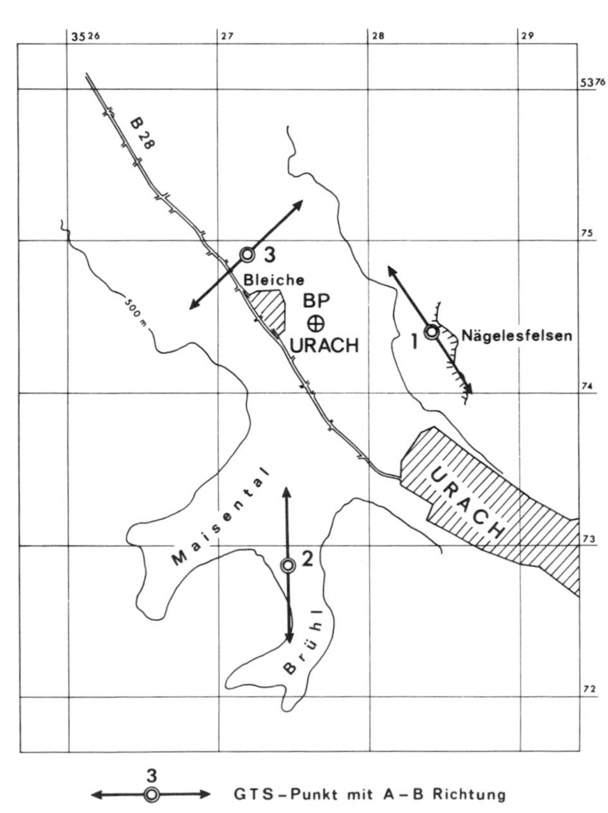

Fig. 1. The location and direction of geoelectrical deepsounding 1 Naegelesfelsen, deepsounding 2 Bruehl, and deepsounding 3 Bleiche; BP = borehole point of the research borehole Urach 3.

The results of the deepsounding 1 Naegelesfelsen are presented in Fig. 2. The distance from the research borehole is 775 m and the difference in topographic high is + 194 m. The maximal distance between the electrodes AB was limited to 4 km because of the intensively built-up area. The deepsounding 2 Breuhl (see Fig. 3) was performed at a distance of about 1625 m from the borehole, a topographic high difference of + 39 m and an electrode distance of AB = 8 km. The corresponding values for the deepsounding 3 Bleiche, for which the results are presented in Fig. 4, are 650 m, + 2 m, and AB = 8 km, respectively.

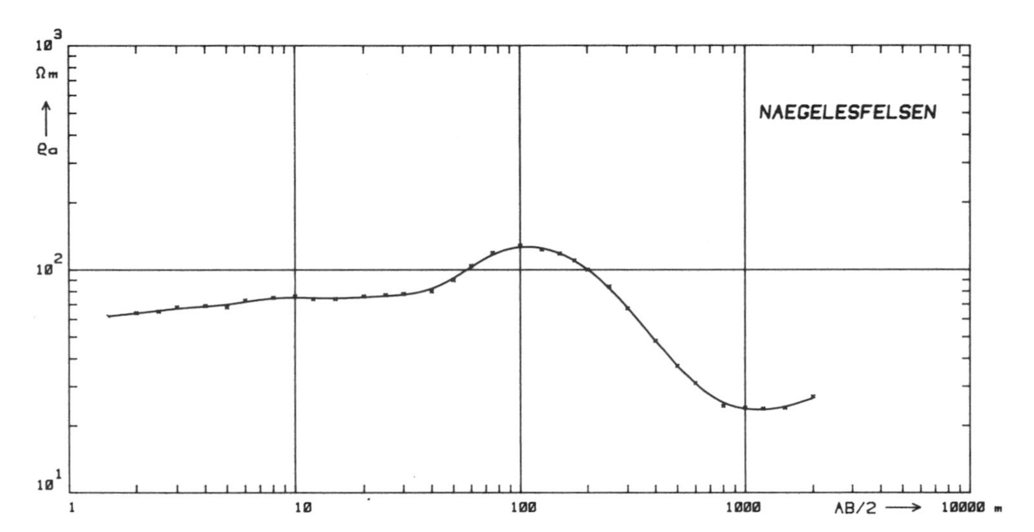

Fig. 2. The results (points) of the geoelectrical deepsounding 1 Naegelesfelsen, which are approximated by a calculated graph; AB = 4 km, δ_a = apparent resistivity.

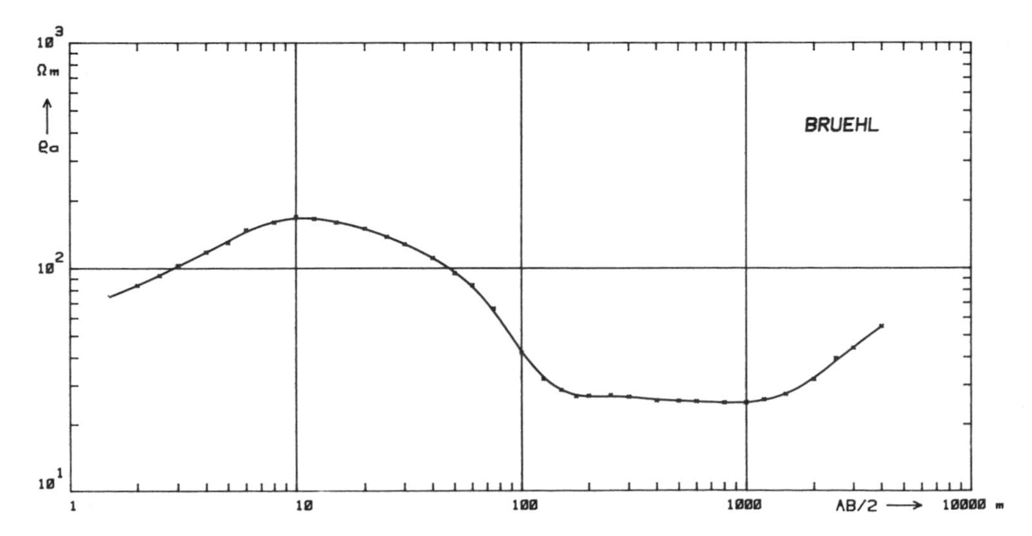

Fig. 3. The results (points) of the geoelectrical deepsounding 2 Bruelh, which are approximated by a calculated graph; AB = 8 km, δ_a = apparent resistivity.

The graph of the deepsounding 1 Naegelesfelsen differs somewhat from those of the deepsounding 2 and deepsounding 3. If a correction is carried out for eliminating the topographic high difference of + 194 m, the last part of the graph is identical with that for the deepsounding 2 and deepsounding 3. In any event, the graph of deepsounding 1 is too short for a deep reaching interpretation.

Except for the first portion down to the Malm, the graphs of deepsounding 2 Bruehl and deepsounging 3 Bleiche mutually agree very well from 283 m (Lias) down to about 3350 m (crystalline basement). Therefore, the determination of the electrical resistivity field close to the Urach 3 research borehole is based only on the deepsounding 3 Bleiche.

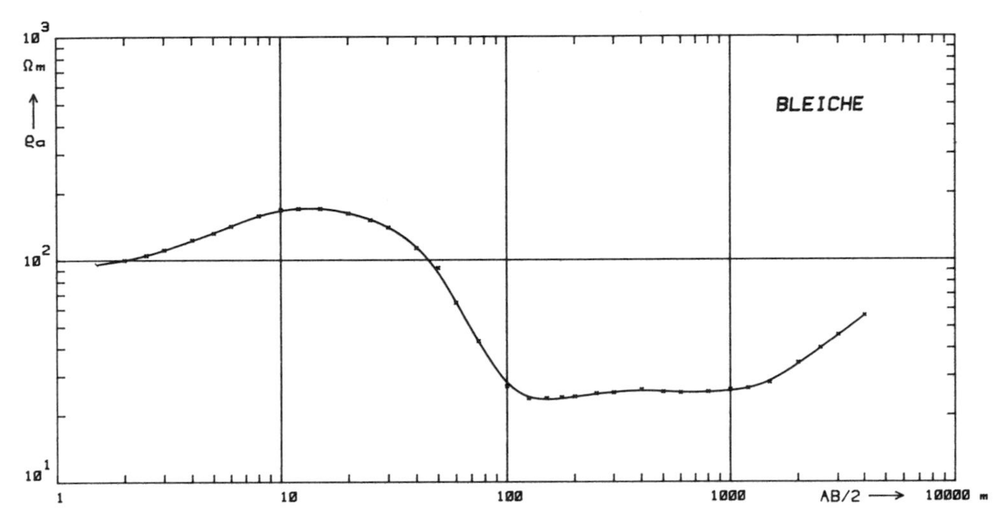

Fig. 4. The results (points) of the geoelectrical deepsounding 3 Bleiche, which are approximated by a calculated graph; AB = 8 km, δ_a = apparent resistivity.

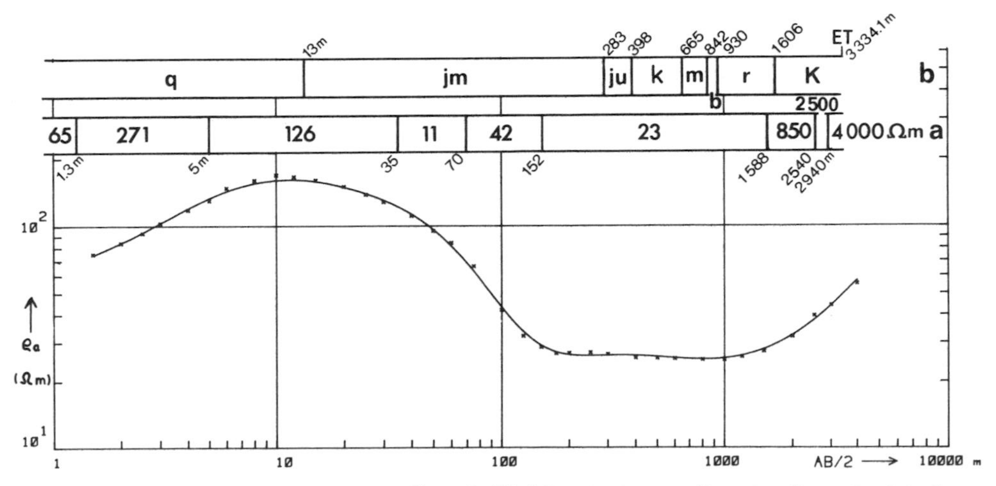

Fig. 5. The geoelectrical deepsounding 3 Bleiche corresponding to the calculated resistivity model (a) and for comparison the simplified geological profile (b); q = Quartär, jm = Dogger, ju = Lias, k = Keuper, m = Muschelkalk, r = Rotliegendes, K = Kristallines Basement.

For the interpretation of the electrical deepsounding results the Ingeso method of MUNDRY & DENNERT (1980) was employed, and the Cyber 76 of the Regional Computer Centre in Hannover was used for calculating.

For that purpose, the measured electrical deepsounding graph is preinterpreted by the "Hilfspunkt" method; the electrical resistivity values from the Urach 3 borehole after WOHLENBERG (this volume) are thereby used too, in order to have an optimal starting condition for the Ingeso method. A 9-layer model has been assumed, especially with respect to the last part of the graph. The best results are presented in Fig. 5 (horizontal column a) for a fixed deep resistivity value of δ_8 = 4000 Ohm · m and the thickness of the next-to-last layer m_7.

In Fig. 5, horizontal column b, the geological profile (DIETRICH this volume) is also presented. A comparison between geological layers and the electrical resistivity layers for the first 650 m shows no clear correlation, but good conformity with the electrical resistivity values from the borehole measurements given by WOHLENBERG (this volume) is evident.

The resistivity of 23 Ohm · m for the 6th layer, which consists of Jurassic, Triassic, Rotliegend and Permo-Carboniferous, is surprising. The low resistivity for these different geological layers can be explained by the high electrolyte content of the pore water, which has an electrical resistivity of 0.3 Ohm · m and less; further causes are the relatively high porosity and the presence of fissures.

The transition from sediment to the crystalline basement is characterized by a large increase in resistivity. Within the crystalline basement itself there is no notable decrease in resistivity, because of the mainly compact rocks. Within the basement a 3-layer system has been assumed to approximate the increasing resistivity with depth. The results are in good agreement with those measured in the borehole.

The effect of anisotropy for Mesozoic sediments has already been discussed in the paper of BLOHM & HOMILIUS (1980). From that it can be deduced that the anisotropy in such layers in the Urach region can be neglected. The low electrical resistivity is possibly caused by the water circulating within the fissured rocks, which prevents the establishment of a differentiated electrical model.

References

Mundry, E. & Dennert, U. (1980): Das Umkehrproblem in der Geoelektrik. – Geol. Jb., **E 19**: 19–38, Hannover.

Blohm, E.-K. & Homilius, J. (1980): Geoelektrische Tiefensondierungen mit großen Auslagen. – Geol. Jb., **E 19**: 39–68, Hannover.

The Urach Geothermal Project, p. 289–300;
Schweizerbart'sche Verlagsbuchhandlung, Stuttgart, 1982

Distribution of Electrical Conductivity in the Urach Geothermal Area A Magnetotelluric and Geomagnetic Depth Sounding Investigation

A. BERKTOLD and K. KEMMERLE

with 7 figures

Abstract: Magnetotelluric measurements were carried out at 29 sites and geomagnetic depth soundings at 22 sites on 2 profiles across the geothermal anomaly of Urach (SW-Germany). The measurements were performed in the period range of 6 to 1000 s. No indications were found for an anomaly of increased conductivity in the deeper parts of the crust or in the upper mantle below the geothermal anomaly. This does not, however, exclude the existence of a moderately increased temperature by about 100 °C in the deeper crust. Within the central region of Urach the lateral variation of conductivity is rather strong. First there is an area where poorly conducting basaltic magma has intruded into the highly conductive sediments of the cover. This area partially overlaps with the central part of the geothermal anomaly. Below the volcanic area the thickness of the sedimentary cover is maximal (1600–1800 m). The sedimentary thickness decreases toward the surrounding areas to an average of 300–500 m. The spatial distribution of conductivity – and as a consequence of the induced current system – is influenced by all these structures and physical parameters, but with different weighting factors. Several conductive structures were found in the central region. The most important one is a NW-SE striking structure where the strongest polarization of the induced electric field of the whole area occurs. The origin of this strong polarization seems to be located in the lower part of the crystalline basement. The integrated conductivity might be increased in this depth range because of fissures, caverns etc. filled with hot and highly conducting fluids. In the surroundings of the central region of Urach the spatial distribution of conductivity is clearly smoother than within the central region.

Magnetotelluric measurements and geomagnetic depth soundings were carried out along 2 profiles across the Urach geothermal anomaly (Fig. 1). Profile 1 ran roughly parallel to the strike of the Swabian Jura (N 62°E, stations OHN to GER) while profile 2 was roughly perpendicular to the strike of the Jura (N 44° W, stations TÜB to MAS). Magnetotelluric measurements were performed at 29 stations and geomagnetic depth soundings at 22 stations. At 4 sites Schlumberger soundings were carried out additionally. The period range of the measurements was about 6–1000 s. The magnetic variations were measured partially with fluxgate magnetometers but mainly with induction coil magnetometers. During the first field campaign the electric field variations were registered mainly on film and the magnetic variations on analog tape recorders. For these registrations the phases between electric and magnetic field variations could not be determined satisfactorily. During the second field campaign digital tape recorders were used for measuring. The digitizing rate was 2 s.

The electric and magnetic field variations were transformed to the frequency domain

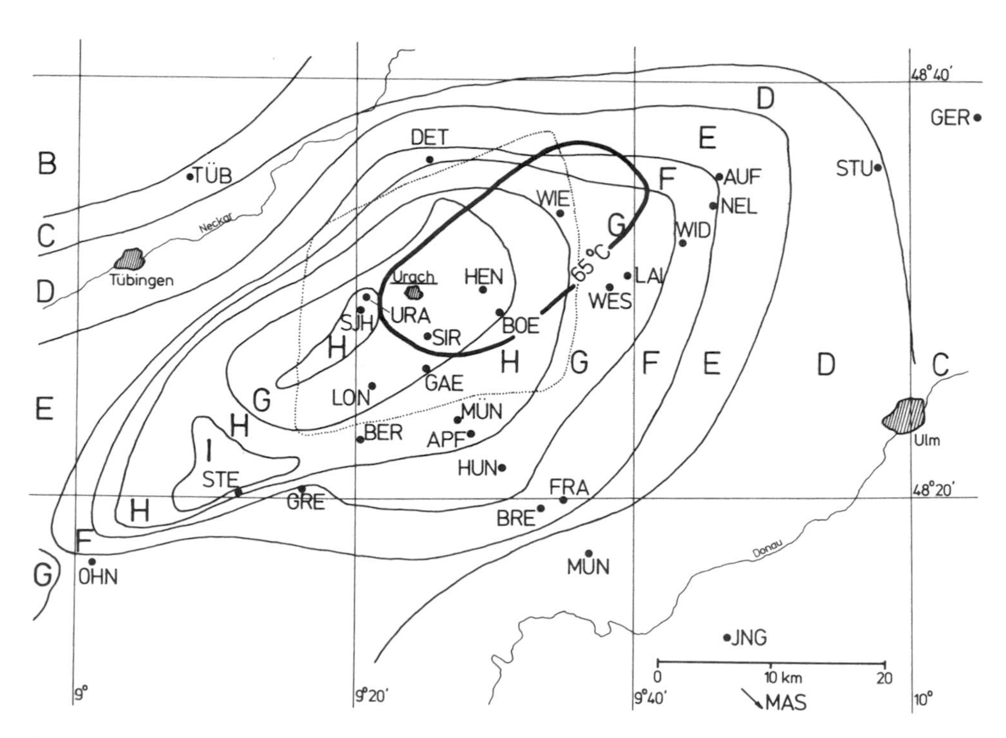

Fig. 1. Location of the measuring sites along profile 1 (stations OHN to GER) and along profile 2 (stations TÜB to MAS). – Heavy lines: Contour map of basement depths below surface after JENTSCH et al. (see this volume). – Depth intervals: B 0.2–0.4 km; C 0.4–0.6 km; D 0.6–0.8 km; E 0.8–1.0 km; F 1.0–1.2 km; G 1.2–1.4 km; H 1.4–1.6 km; I 1.6–1.8 km; K 1.8–2.0 km. – Dotted line: Approximate boundary of the area of volcanic pipes. – Boldface line: centre of the geothermal anomaly marked by the line of 65 °C at 1000 m depth.

by Fourier analysis. The coherence between perpendicular components of the induced electric field was calculated and a coordinate system was determined for each station where this coherence became minimal. In this coordinate system the average amplitude of the electric field is smallest in one coordinate direction and it is largest in the other coordinate direction. This coordinate direction (of largest average amplitude) is called the preference direction of the induced electric field. This direction was determined as a function of the period. At most of the stations the preference direction is constant for periods larger than 100 s while it tends to rotate at some sites for shorter periods. As a further parameter the degree of polarization was determined for each site as the average amplitude ratio of the electric field components parallel and perpendicular to the preference direction. At all sites the degree of polarization depends on the period T. This may be seen from Figs. 3 and 4 by the different period dependences of the curves $\rho_{a,\,par}$ (T) and $\rho_{a,\,per}$ (T). (Further results of these figures are discussed later.) For short periods (T < 20 s) there is no polarization, or only a rather weak polarization of the electric field. The polarization increases with increasing period until about 100 s. For periods larger than 100 s the polarization remains more or less constant. In Fig. 2 the preference direction of the induced electric field as well as the degree of polarization are plotted for most of the sites for periods of T = 100 s. The direction of the boldface line represents the preference direction at each site. The ratio between the lengths of the boldface line

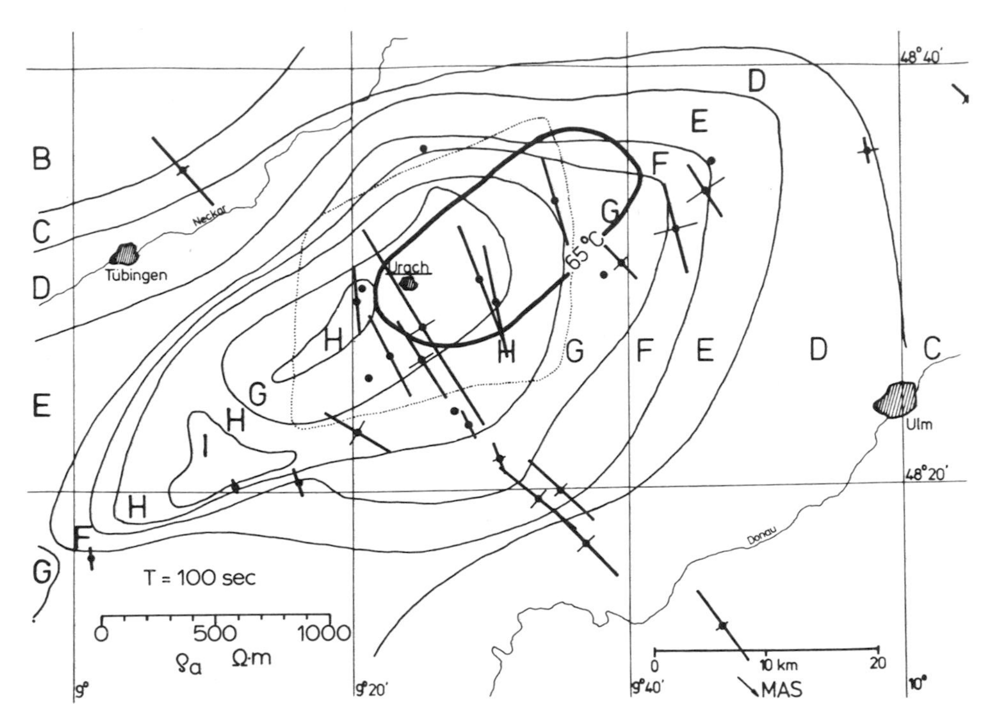

Fig. 2. Preference directions of the induced electric field for periods of T = 100 s. The heavier lines represent the preference directions. The degree of polarization is given by the ration between the lengths of the heavier line and the perpendicular finer one. The lengths of the lines are proportional to the apparent resistivities $\rho_{a,\ par}$ and $\rho_{a,\ per}$ for T = 100 s.

and the perpendicular fine line represents the degree of polarization. The length of the bars is proportional to $\rho_{a,\ par}$ and $\rho_{a,\ per}$ respectively (which are defined further below). The direction of the induced electric field is on the average N 30° W over the whole area of investigation. This direction is about perpendicular to the strike of the Swabian Alb. It is not typical for the geothermal area proper, since it was found also at other sites north of the River Donau. At some stations in the central part of the volcanic area a further increase of the polarization (for the whole period range) was observed. At these stations strongest polarization of the electric field found in the Urach area was encountered. The degree of polarization decreases to the SW and NE of this area. The region of strongest polarization seems to extend in the NW and in SE direction.

In Fig. 3 and 4 the apparent resistivity ρ_a is plotted as a function of the period T for most of the stations. $\rho_a(T)$ was analysed in the coordinate system of minimal coherence as defined above. This means that the electric field component in the preference direction was correlated with the magnetic field component of maximal coherence (which, in most cases, was the magnetic field component perpendicular to the electric component). The same correlation was carried out also for the remaining 2 field components. From these impedance elements the functions $\rho_{a,\ par}$ (T) (electric field component parallel to the preference direction) and $\rho_{a,\ per}(T)$ (electric field component perpendicular to the preference direction) were determined. The directions "parallel" and "perpendicular" thus defined are perpendicular and parallel, respectively to the average strike of the Swabian Jura.

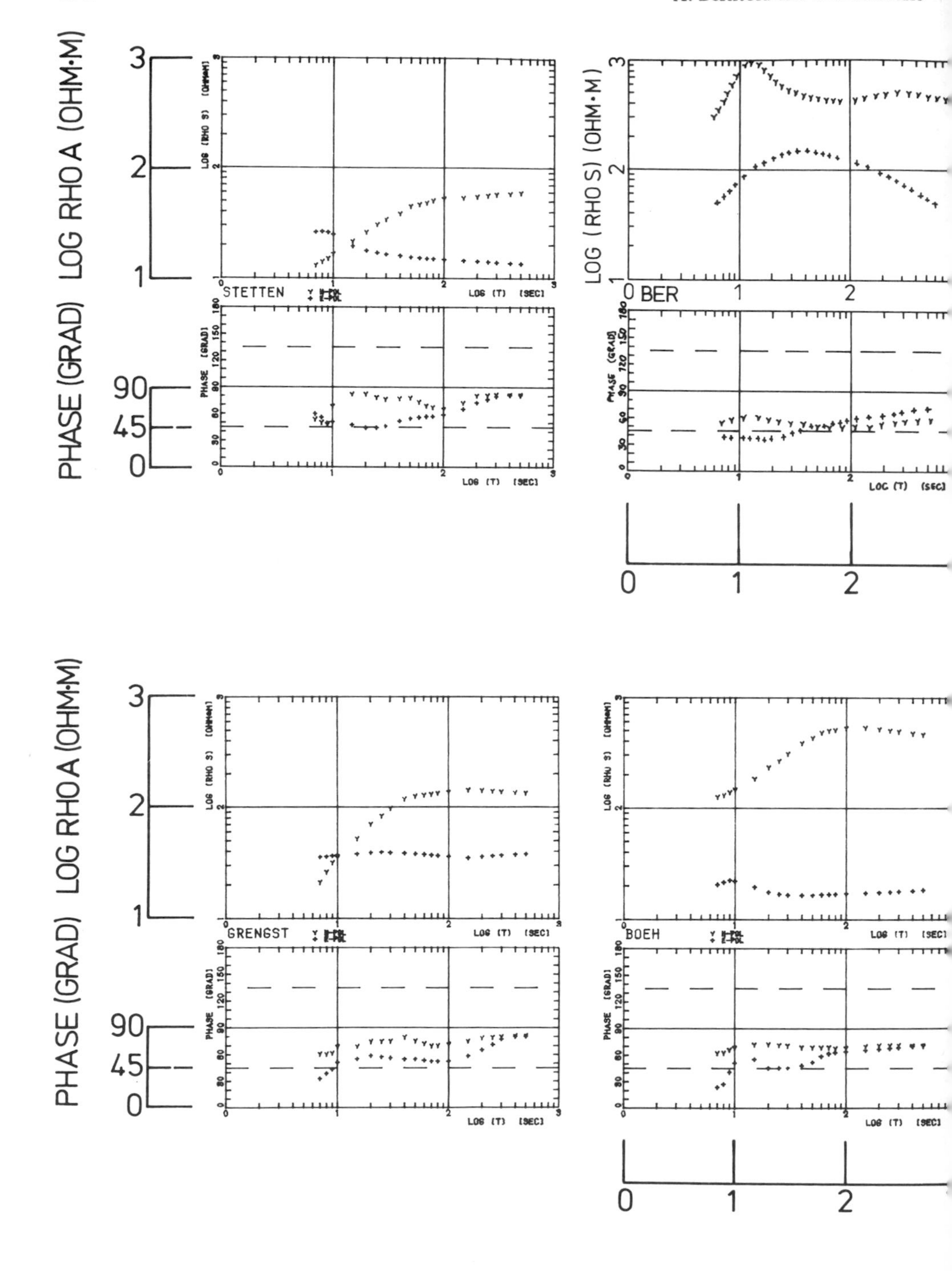

Fig. 3. The apparent resistivity RHO A and the phase retardation, as a function of the period T for some stations along profile 1. Upper part of the fig.: stations STE, BER, LAI, STU. Lower part of the fig.: stations GRE, BOE, NEL, GER.

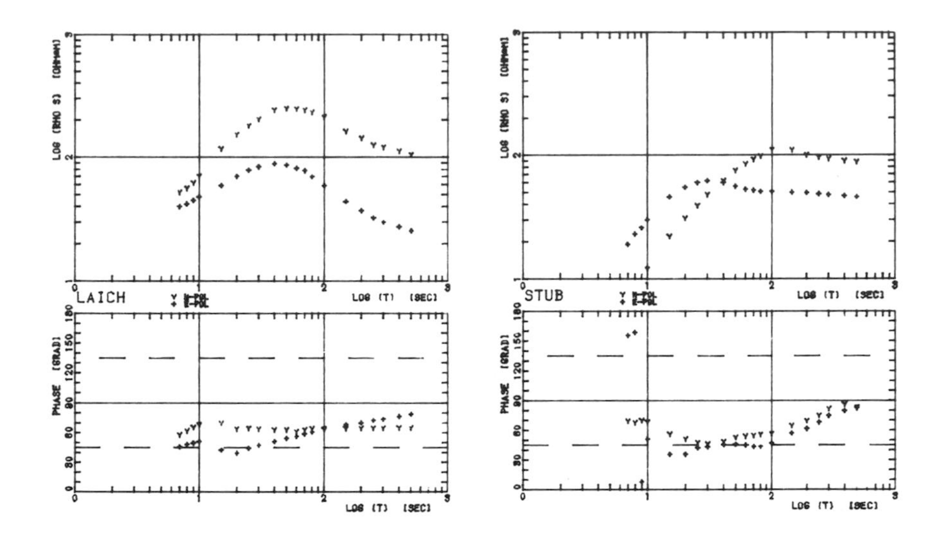

LOG T (s)

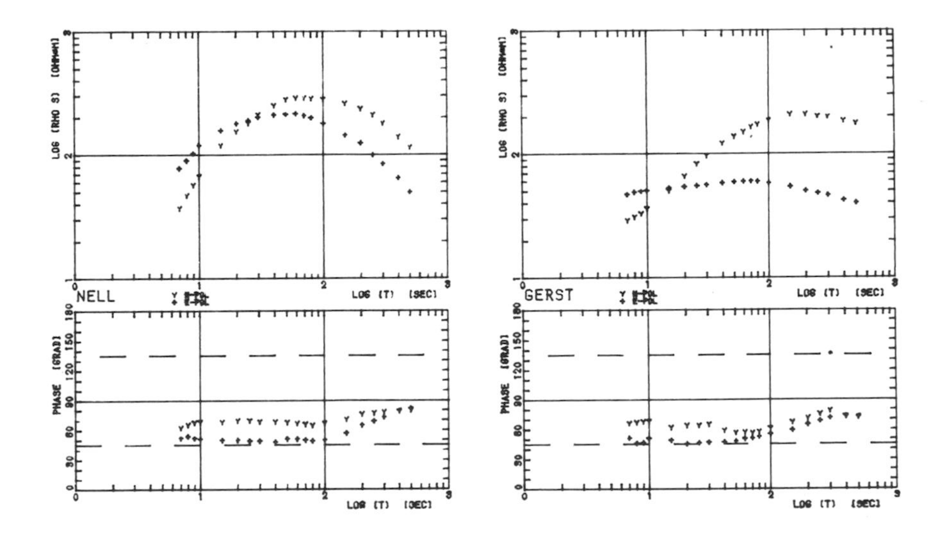

LOG T (s)

Symbol Y (curves of higher ρ_a-values: the apparent resistivity has been calculated with the use of the electric field component in the preference direction ($\rho_{a,\,par}$-curves).
Symbol + (curves of lower ρ_a-values: the apparent resistivity has been calculated from the electric field component perpendicular to the preference direction ($\rho_{a,per}$-curves).

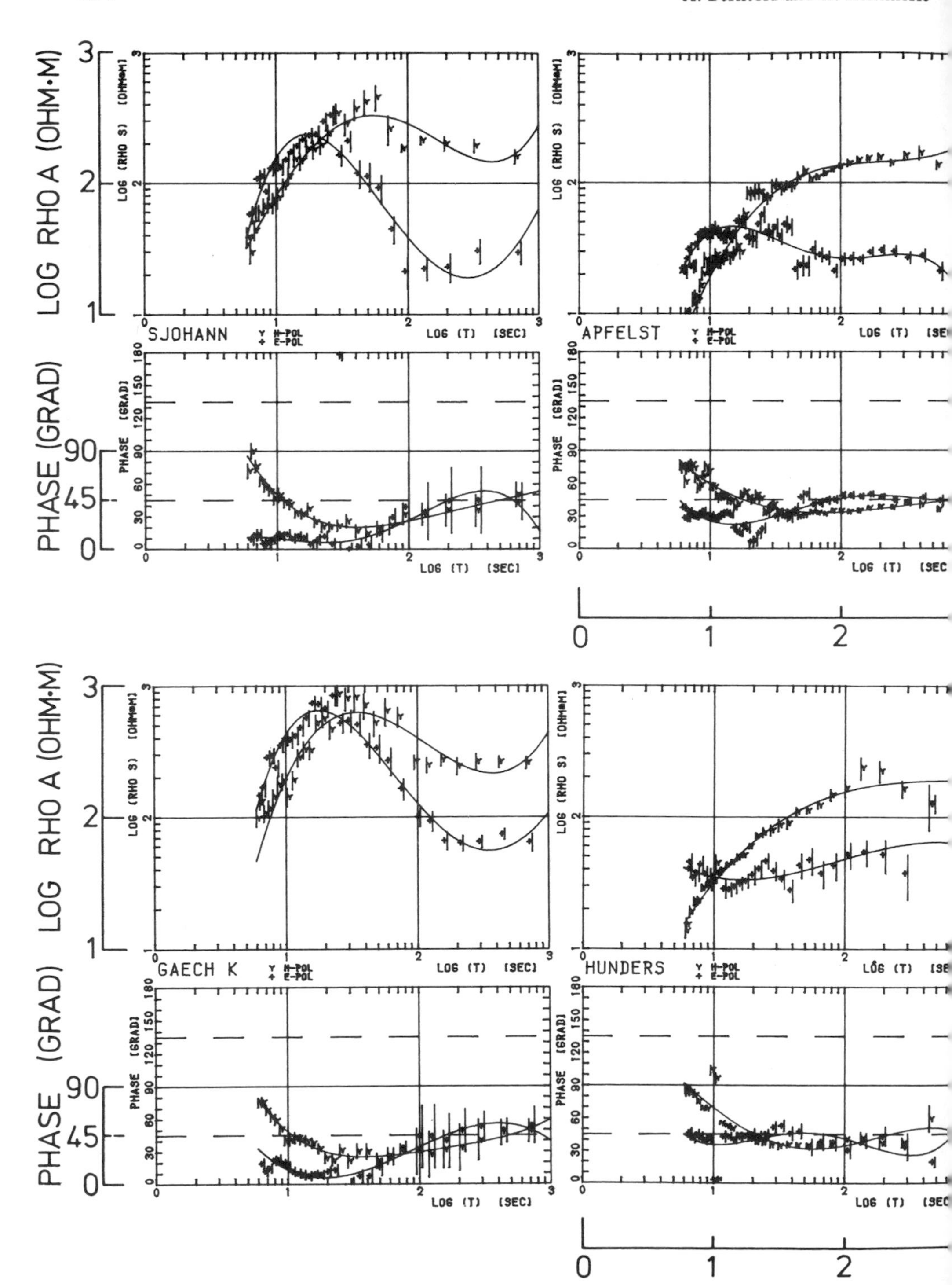

Fig. 4. See figure 3. Upper part of the fig.: stations SJH, APF, BRE, ING. Lower part of the fig.: stations GAE, HUN, MUN, MAS.

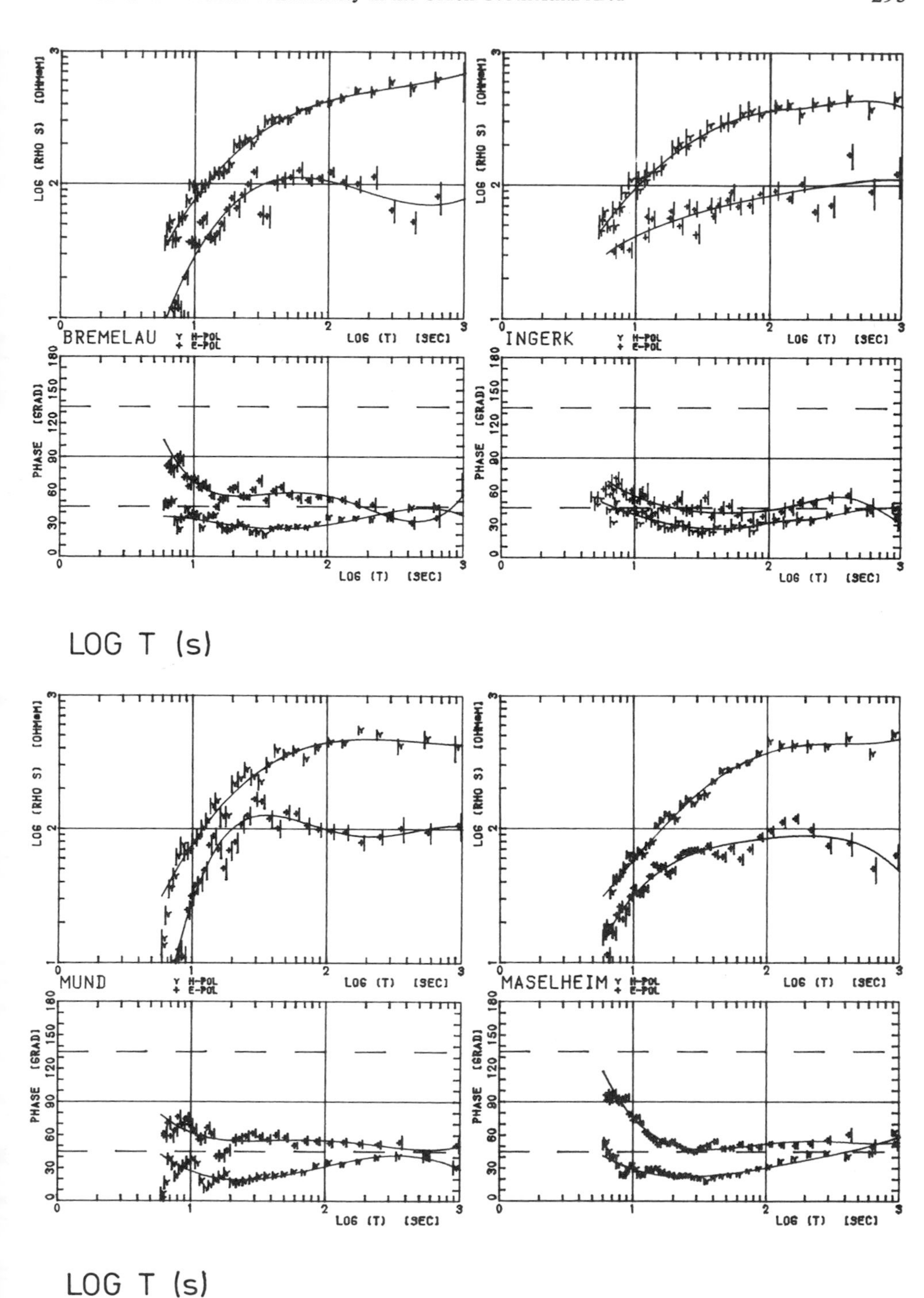

LOG (RHO S) [OHM*M]
BREMELAU
Y H-POL
+ E-POL
LOG (T) [SEC]
PHASE [GRAD]
INGERK
MUND
MASELHEIM
LOG T (s)
LOG T (s)

The dependence of $\rho_{a,\, par}$ on the period T is similar for most of the stations (e.g. STE, GRE, STU, GER on profile 1; APF–MAS on profile 2). $\rho_{a,\, par}(T)$ increases with the period from 10 to 100 s and it remains constant or decreases slightly for longer periods. At some stations SW of Urach, however, (SJH, GAE, BER) a different period dependence was found: $\rho_{a,\, par}$ is largest in the period range of about 20–70 s. It decreases toward longer as well as toward shorter periods. A somewhat similar shape of the $\rho_{a,\, par}(T)$-curves was found also for the 2 stations LAI and NEL in the NE part of profile 1. If one interpretes the $\rho_a(T)$-curves of these 5 stations by plane layered models one has to assume a zone of high resistivity in the upper part of the crust (10–30 km depth). However the shape of these curves might be caused also by the lateral conductivity distribution. The period dependence of $\rho_{a,\, per}$ is clearly weaker than that of $\rho_{a,\, par}$ at most stations. As a consequence of the different period depencences of $\rho_{a,\, par}$ and $\rho_{a,\, per}$ the degree of polarization also depends on the period, as has been discussed before. The positions of the $\rho_a(T)$-curves within the coordinate system differ from site to site. Lowest values of $\rho_a(T)$ were found for some stations in the S and SW of the volcanic area (OHN, STE, GRE, APF, HUN) and at only 1 station in the NE (STU): $\rho_{a,\, par}$ is near 100 Ohm · m for longer periods and $\rho_{a,\, per}$ about 20–30 Ohm · m. From the central volcanic area to the SE and NE the $\rho_{a,\, par}$ as well as the $\rho_{a,\, per}$ curves are shifted on the average to higher apparent resistivities (APF-HUN-BRE-MÜN-ING-MAS and STE-GRE-LAI-NEL-GER). The maximal factor of shifting is about 5. This effect of increasing apparent resistivities to the SE and NE may be caused partly by a decrease of the thickness of the sedimentary cover in these directions (see Fig. 1): The decrease of the sedimentary thickness in the lateral direction causes an increase of the current density within the sedimentary cover and therefore an increase of the apparent resistivity in the lateral direction. At some stations near the SW and NE border of the volcanic area (SJH, GAE, BER and LAI, NEL respectively) the period dependence of ρ_a clearly differs from that observed at all other stations. This effect has been discussed before. It has been interpreted as being due to the existence of zones of increased resistivity in the crust below these stations. In the area between the stations SJH, GAE, BRE in the SW and the stations LAI, NEL in the NE there are some stations (e.g. BOE, MÜN, SIR, AUE, HEN) where the strongest electric field polarization of the Urach area was observed. The $\rho_{a,\, par}(T)$-curves are shifted there to the highest ρ_a values found in this area (up to 1000 Ohm · m). This shifting of the $\rho_a(T)$-curves does not depend on the period. For station BOE this effect can be seen in Fig. 3. The most probable explanation for this strong electric field polarization may be a channeling of the induced current system in the central volcanic area which can cause an increased current density, and as a consequence an increased electric field component in the direction of channeling. This direction is – as can be seen from Fig. 2 – about N 30° W. The strong polarization of the electric field in the central volcanic area cannot be caused by the lateral variation of the sedimentary thickness. This can easily be seen from Fig. 2: If any striking direction of the sedimentary cover exists in the Urach area it is more SW-NE than NW-SE. Moreover there is no indication for an anisotropy or strong lateral variation of conductivity in the uppermost 300 m. This can be deduced from Schlumberger profiles, which have been measured at some sites in 2 perpendicular directions. The interpretation of the strong polarization effect in the volcanic area by a channelling of the induced current system may be confirmed by results of RICHARDS et al. (see this volume). They find an increased current density in the short period range in about the same area and with the same striking direction as we found. Summarizing all these results, we believe that the origin of the increased current density is in the lower part of the sedimentary cover and in the upper part of the cry-

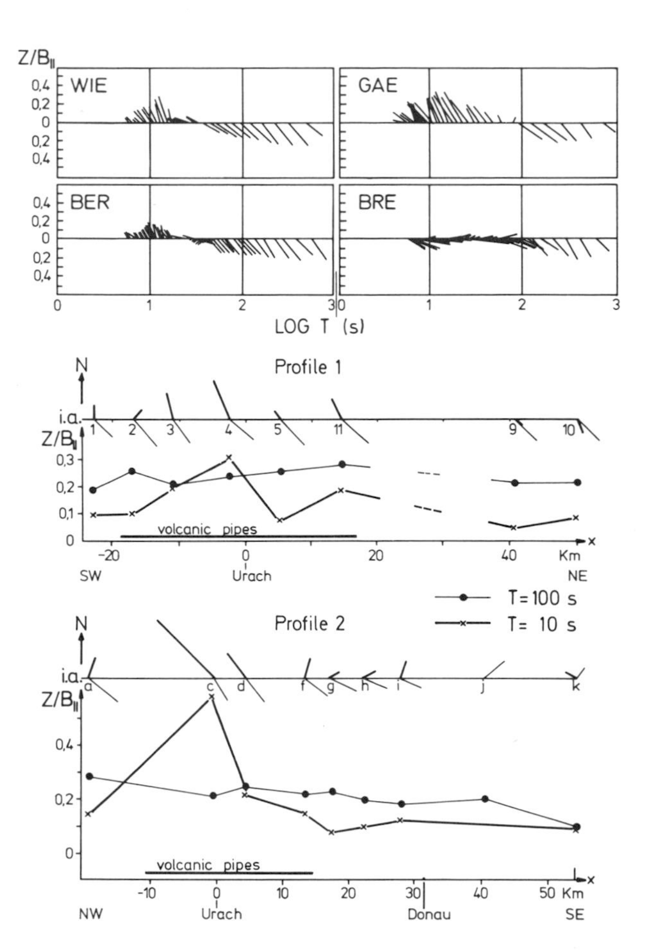

Fig. 5. Upper graphs: length and direction of the induction arrows as a function of the period T for 4 stations;
Lower graphs: length and direction of the induction arrows for T = 10 and 100 s along the 2 profiles.
I.A.: directions of the induction arrows along the profiles (N: north direction).
1: STE; 2: GRE; 3: BER; 4: GAE; 5: HEN; 11: WIE; 9: STU; 10: GER; a: TÜB; c; SJH; d: GAE; f:APF; g: HUN; h: BRE; i: MUN; j: ING; k: MAS.

stalline basement. It may be correlated with the distribution of the volcanic intrusions as well as with the distribution of deep-seated fissures etc. filled with hot and highly conducting fluids.

In Fig. 5 some results of geomagnetic depth sounding are plotted for the stations BRE, BER, GAE and WIE as well as along the 2 profiles. In the upper part of the figure the direction of the induction arrow is plotted as a function of the period for 4 stations. This direction corresponds to the direction of the horizontal component ($B_{||}$) which exhibits maximal coherence with the vertical component (Z). The length of the induction arrow represents the amplitude ratio $Z/B_{||}$. In the lower part of the figure length and direction of the induction arrows are plotted along the 2 profiles for the periods 10 s and 100 s. The symbol for the direction of the induction arrow is "i.a.". For periods longer than about 50 s the direction of the induction arrow is nearly the same for all stations. This direction is about SE. The length of the induction arrows decreases slowly from NW to SE. Both results are confirmed by former investigations at many other stations in southern Germany. This spatial distribution of direction and length of the induction arrows may be caused by a lateral increase of conductivity at greater depth below the Swabian Jura from SE to NW (BERKTOLD 1978). For periods less than about 50 s direction and length of the induction arrows are distributed more inhomogeneously.

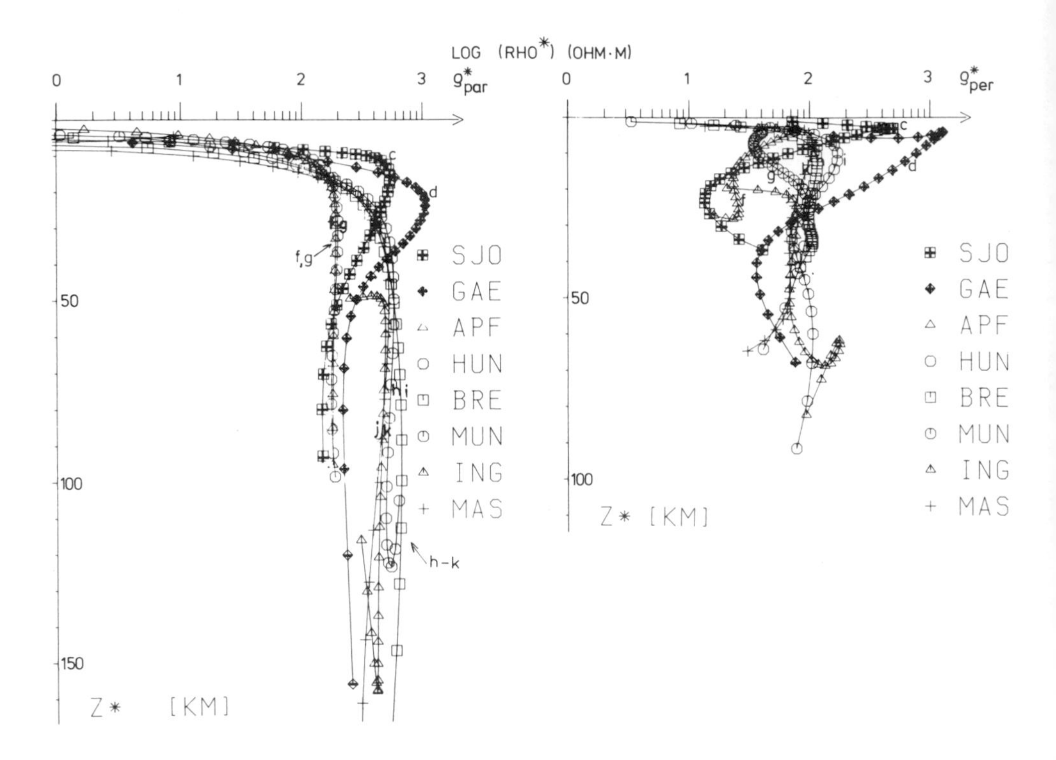

Fig. 6. $\rho^*_{par}(z^*)$ and $\rho^*_{per}(z^*)$ for most of the sites along profile 2 calculated from the apparent resistivity and phase curves of Fig. 3 and 4.

The length of the induction arrows (or the ratio $Z/B_{\parallel}$) is small and rather constant outside the volcanic area and it clearly increases toward the stations inside the volcanic area. The increased length of the induction arrows near and inside the volcanic area may be caused – at least in part – by the NW-SE striking structure which causes the strong polarization of the induced electric field. The directions of the induction arrows in the volcanic area seem to be influenced not only by the structure of strong electric field polarization, but also by other conductive structures such as the region of low apparent resistivity in the S and SW of the volcanic area.

As discussed before, the direction of the induction arrow depends strongly on the period at all stations. Therefore, no unique coordinate system can be found for a two-dimensional interpretation of the field data within the whole period range. Hence we have preferred to present only one-dimensional models of the conductivity-depth-distribution at the different sites. For this purpose the modified apparent resistivity ρ^* (SCHMUCKER 1971) was calculated as a function of the depth of penetration z* (or the depth of the "centre of mass" of the in-phase induced current system). ρ^* (z*) approximates the true conductivity-depth-distribution. ρ^* and z* were computed from the apparent resistivity and phase curves shown in Figs. 3 and 4. The results for profile 2 can be seen in Fig. 6. (The symbols "par" and "per" have the same meaning as in Figs. 3 and 4.) The depth distribution of the modified apparent resistivity ρ^* differs for the different stations as a consequence of the distortion of the induced current system by the lateral conductivity distribution. For stations with an intermediate apparent resistivity like STE, GRE, APF or HUN apparent resistivities of 200–500 Ohm · m were found for the crust, decreasing slightly with depth. In Fig. 7 $\rho^*_{par}(z^*)$ and $\rho^*_{per}(z^*)$ are plotted for some stations along the 2 profiles (mainly where good results were obtained for the phases). The graphs are shown to demonstrate the strong **lateral** conductivity distribution below the volcanic area and the smooth lateral conductivity distribution outside this area. The **vertical** distribution of conductivity, however, is not very well determined: The distortion of the induced current system near the earth's surface causes in part serious errors in the determination of the conductivity-depth-distribution.

To summarize the results and interpretations discussed up to now it is useful to subdivide the area under investigation into 2 regions, the central region and the surroundings. The central region may roughly be bordered by a circle of 30 km radius around the city of Urach (see Fig. 1). In the surroundings the spatial distribution of conductivity is rather smooth (see Figs. 5 and 7). At greater depth (deeper part of the crust, upper mantle) a lateral increase of conductivity from SE to NW seems to exist below the whole Swabian Jura. This structure is not correlated with the geothermal anomaly. No indications were found for an anomaly of increased conductivity in the deep crust or upper mantle below the geothermal anomaly. This does not exclude the existence of an increased temperature by about 50 °C in the deeper crust (as might be deduced from the temperature measurements in the deepest part of the Urach boreholes). An increased temperature by far more than 50 °C, e.g. 200 °C, cannot exist in the deeper crust. Laboratory measurements show that such an increased temperature would cause an increase of conductivity by a factor

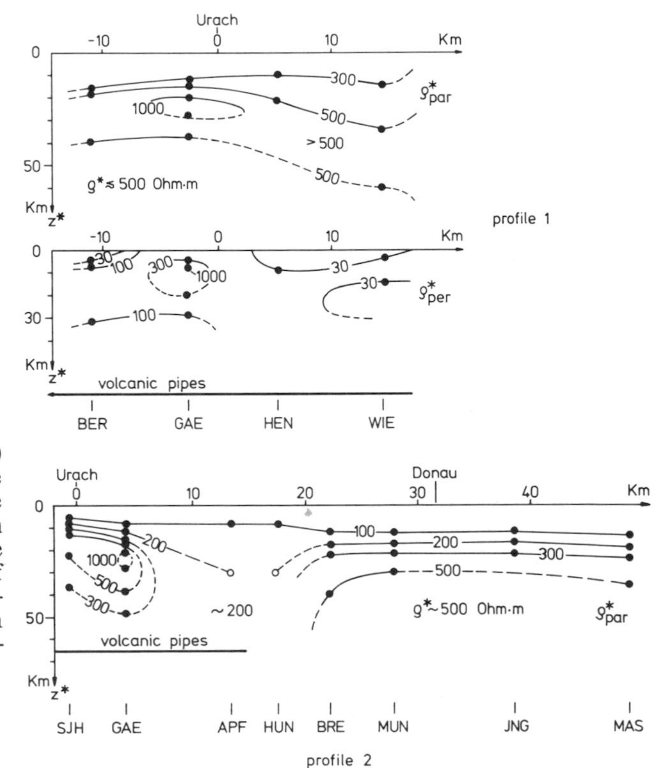

Fig. 7. $\rho^*_{par}(z^*)$ and $\rho^*_{per}(z^*)$ for some stations along profile 1 and $\rho^*_{par}(z^*)$ along profile 2. The graphs are shown mainly to demonstrate the strong lateral distribution of conductivity below the volcanic area and the smooth lateral conductivity distribution outside this area.

of about 10, which surely could be detected in this depth range. Within the central region – as defined above – the lateral distribution of conductivity is rather strong. First there is the area where poorly conducting basaltic magma has intruded into the highly conducting sediments of the cover ($\rho_{sed} \approx 20$ Ohm · m). This area partially overlaps with the central part of the geothermal anomaly (see Fig. 1). Below the volcanic area the thickness of the sedimentary cover is maximal (1600–1800 m). The sedimentary thickness decreases toward the surrounding areas to an average of 300–500 m. The spatial distribution of conductivity is influenced by all these structures and physical parameters – but with different weighting factors. Several conductive structures were found in the central region. These are the areas of low apparent resistivities in the S and SW (OHN, STE, GRE, APF, HUN), the areas of probably increased resistivity in the crust (SJH, GAE, BER and LAI, NEL) and finally the NW-SE striking structure of strong electric field polarization (BOE, MÜN, SIR, AUE, HEN). The origin of this strong polarization seems to be located in the lower part of the sedimentary cover and in the upper part of the crystalline basement. The integrated conductivity is increased in this depth range, probably because of fissures, caverns etc. filled with hot and highly conducting fluids. This zone of increased conductivity within poorly conducting basement may act as a gap for the large-scale induced current system, thus causing a channelling of this current system in a direction which is perpendicular to the strike of the uparching crystalline basement below the Swabian Jura.

Acknowledgements

The authors are indebted to Dr. M. BEBLO and Dipl.-Geophys. P. NEURIEDER and to all students who participated in the field survey. We are grateful to Prof. G. ANGENHEISTER for his interest and support during all stages of the work. Data analysis and model calculations were done at the Leibniz Rechenzentrum, München. This project has been supported by the Commission of the European Communities.

References

Beblo, M. (1977): Electrical conductivity in the geothermal anomaly of the Urach volcanic area, SW Germany – first results. – Acta Geodaet., Geophys. et Montanist. Acad. Sci. Hung. Tomus 12 (1–3), pp. 87–91.

Berktold, A. (1978): On the distribution of the electrical conductivity in the area between the Rhinegraben, the Bohemian massif and the Central Alps. – Acta Geodaet., Geophys. et Montanist. Acad. Sci. Hung. Tomus 13 (3–4), pp. 437–440.

Haak, V. (1980): Relations between electrical conductivity and petrological parameters of the crust and upper mantle. – Geophysical Surveys (4), pp. 57–69.

Haenel, R. (1979): Determination of subsurface temperatures in the Federal Republic of Germany on the basis of heat flow values. – Geol. Jb. **E 15**: 41–49, Hannover.

Schmucker, U. (1978): Auswerteverfahren Göttingen. – Protokoll, Koll. "Elektromagnetische Tiefenforschung", Neustadt/Weinstraße, p. 163–188.

Wohlenberg, J. (1980): Elektrische Untersuchungen. – Endbericht über geophysikalische Untersuchungen in der Forschungsbohrung Urach. – Report, NLfB Hannover, Archiv No. 85805.

The Urach Geothermal Project, p. 301–311;
Schweizerbart'sche Verlagsbuchhandlung, Stuttgart, 1982

Electrical Conductivity in the Urach Geothermal Area, A Geomagnetic Induction Study Using Pulsations

M.L. RICHARDS, U. SCHMUCKER and E. STEVELING

with 7 figures and 2 tables

Abstract: Magnetic deep-sounding and magnetotelluric surveys were carried out in 1978 and 1979 in the Swabian Alb of southwestern Germany to study the electrical conductivity structure in the Urach geothermal area. Magnetic field was recorded at 18 sites along the length of the Alb, and electric field at 10 of these sites. Geomagnetic pulsations have been analysed for the period range 7.5 to 2000 s. The apparent resistivity decreases continuously with depth giving little indication of anomalous deep structure in the vicinity of the Swabian volcano and Urach anomaly. An anomalous feature which is observed in the geothermal area is a shallow, 2-dimensional, conductive structure, with Urach in the center, trending approximately perpendicular to the edge of the Swabian Alb. No effect of the SW-NE trending edge itself is seen in the data, except at the shortest periods.

The electrical conductivity structure in the region of the Swabian volcano in southwestern Germany is of special importance to understanding the Urach geothermal anomaly. Electromagnetic methods of geophysics hold some promise of determining at least the broad features of the structure. Therefore, in 1978 and 1979 magnetic deep-sounding and magnetotelluric measurements were made in the Swabian Alb in southwestern Germany. Electric and magnetic pulsations have been analysed for periods of 8 s and longer. The instrumentation included 3-component, induction coil magnetometers and Filloux-Hempfling Ag-AgCl electrodes (FILLOUX 1973, HEMPFLING 1977). Data were recorded digitally at 2 s sample interval on magnetic cassettes. At any one time six magnetometers and three 2-component telluric instruments could be operated.

The survey was made along a 170 km profile, oriented E 30°N (Fig. 1). An additional short (30 km) north-south profile was established in the Urach volcanic area. Except for MOE and STH the stations were all situated on the Jura formation. The base station BOE operated continuously during both field campaigns, each about seven weeks long. At the other stations the recording time was from one to four weeks.

From each station's data set, between 40 and 170 good individual segments (effects) of 4 to 32 min length were selected for analysis of short period pulsations. A cubic least squares trend was removed and a cosine taper (hanning data window) applied before Fourier analysis. The raw spectral values, representing products of individual Fourier coefficients, were smoothed with a Parzen window and averaged over effects to obtain suitable spectra and cross spectra (SCHMUCKER 1978). For longer period variations effects of 1 to 4 h were selected, and first low-pass filtered with 80 s cutoff and decimated from 30 to 2 values/min before being analysed as for the shorter periods. For each of

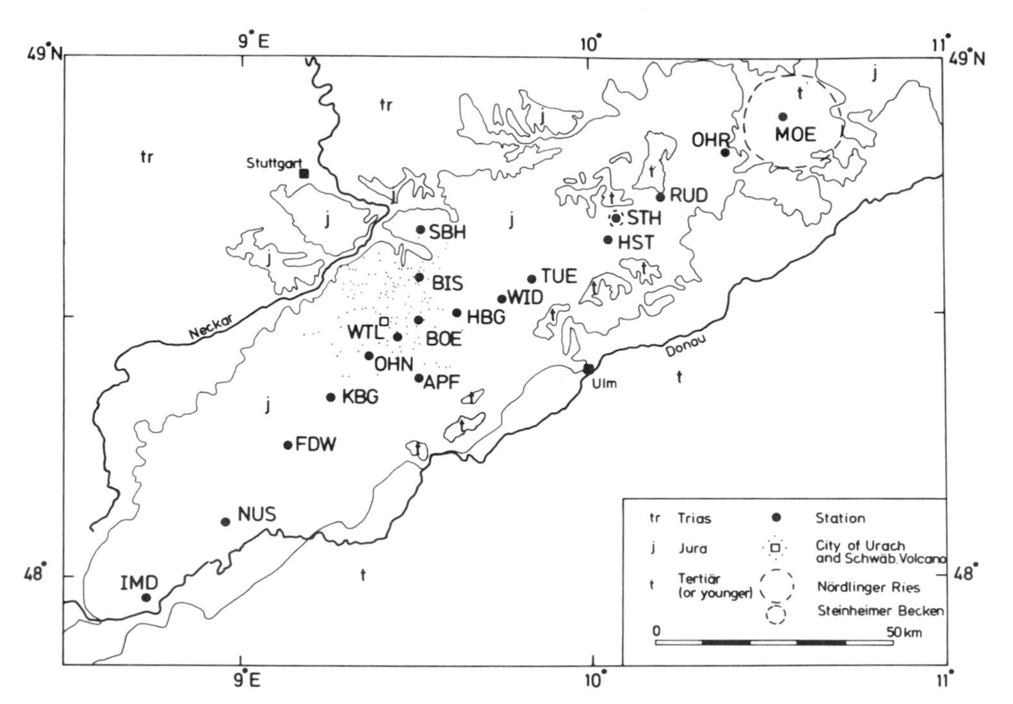

Fig. 1. Simplified geological view of the Swabian Alb (adapted from the Geologische Übersichtskarte von Südwestdeutschland 1 : 600 000) including sites of 1978 and 1979 stations.

the 18 stations in Fig. 1 the complex transfer functions h and d between the anomalous (H_a, D_a) and the normal (H_n, D_n) magnetic fields, defined by the relation

$$(1) \quad \begin{vmatrix} H_a \\ D_a \end{vmatrix} = \begin{vmatrix} h_H & h_D \\ d_H & d_D \end{vmatrix} \begin{vmatrix} H_n \\ D_n \end{vmatrix}$$

were calculated for 25 periods between 7.5 and 2000 s. In this range nine Parzen window widths were used (Tab. 1) such that for adjacent estimates there was a small window

Table 1. Periods and frequency ranges of data analysis; where ranges overlap the better choice of window varies from station to station, but few periods (brackets) are affected.

Selected Periods (s)	Frequency range (cph)	Parzen window width (cph)
2000, 1500, 1200, (1000)	1.2– 3.6	0.6
(1000), 750, 600, (500)	2.4– 7.2	1.2
(500), 375, 300, (250)	4.8– 14.4	2.4
(250), 187.5	9.6– 28.8	4.8
150, 120, (100)	12 – 36	6
(100), 75, 60, (50)	24 – 72	12
(50), 37.5, 30, (25)	48 –144	24
(25), 18.8, 15, (12.5)	96 –288	48
(12.5), 9.4, 7.5	192 –576	96

overlap and the window width increased for decreasing period. The base station BOE served as normal for both 1978 and 1979 surveys, but it was clear that this station, in the center of the conductivity anomaly, was not a suitable choice. A better reference is the arithmetic average of the transfer functions at the four stations IMD, NUS, RUD and OHR, at the SW and NE ends of the profile. All transfer functions used in the following analysis of anomalous horizontal magnetic field are refered to these average values of h_H, h_D, d_H and d_D, respectively. The transfer functions for anomalous vertical magnetic field component and the corresponding induction arrows have also been computed, but there are some inconsistencies, and no clear meaning of their spatial or frequency distribution has been found.

For the 10 stations where electric field was measured the complex transfer function Z (the impedance) between electric field (E_N, E_E) and the local horizontal magnetic field (H, D)

$$(2) \qquad \begin{vmatrix} E_N \\ E_E \end{vmatrix} = \begin{vmatrix} Z_{NN} & Z_{NE} \\ Z_{EN} & Z_{EE} \end{vmatrix} \begin{vmatrix} H \\ D \end{vmatrix}$$

was calculated for the same set of frequencies and by the same method as above. In equations 1 and 2 a noncoherent, residual term has been omitted.

Two properties of the impedance matrix, SKEWNESS and ANISOTROPY, give useful information about the nature of the conductivity structure. Small values of skewness $= |Z_{NN} + Z_{EE}| / |Z_{NE} - Z_{EN}|$ are associated with 2-dimensional structures (e.g. JONES & VOZOFF, 1978). Except for long period results at FDW the value is found to be less than 0.3 for the entire period range. Excluding FDW, KBG and IMD the result is skewness < 0.2 (Tab. 2), so with some reservations about the three southwestern stations a 2-D interpretation is expected to be valid for the profile.

Table 2. Some properties of the impedance; means and standard deviation are for the 25 periods listed in Table 1.

Station	Mean Prefered Direction (°)		Skewness Max	Skewness Mean	Anisotropy Max, T > 100 s	Anisotropy (T= 12.5 s)	1-D model τ(S)	2-D model τ(y) (S)
MOE	N28W	± 4	.16	.11	1.8	0.95	200	115
RUD	N40W	± 3	.11	.07	1.9	1.07	100	70
HST	N37W	± 6	.10	.06	1.7	1.05	60	115
TUE	N29W	± 4	.09	.06	2.2	1.20	10	55
WID	N35W	± 4	.15	.07	2.3	1.23	50	80
HBG	N38W	± 2	.11	.07	2.7	0.97	100	95
BOE 1978	N30W	± 2	.14	.08	3.4	1.45	100	
BOE 1979	N32W	± 2	.13	.08	3.4	1.45	100	150
KBG	N10W	± 4	.24	.15	3.6	1.75	120	115
FDW	N16W	± 8 1)	.23 2)	.14	2.3	1.37	120	115
IMD	N26E	± 4	.24	.15	--- 3)	--- 3)	50	50

1) This value is typical for $30 < T < 300$ s. 2) $T < 1000$ s. 3) The frequency dependence at IMD is unlike that at other stations; max = 3.3 and min = 1.9.

The skewness parameter is independent of coordinate system, but anisotropy is not. For a rotation of coordinates through angle α, reckoned positive for rotation eastward from geomagnetic north, and restricted to the range $|\alpha| \leqslant 45°$, the impedance matrix transforms to

$$(3)\quad \begin{vmatrix} Z_{xx} & Z_{xy} \\ Z_{yx} & Z_{yy} \end{vmatrix} = \begin{vmatrix} \cos\alpha & \sin\alpha \\ -\sin\alpha & \cos\alpha \end{vmatrix} \begin{vmatrix} Z_{NN} & Z_{NE} \\ Z_{EN} & Z_{EE} \end{vmatrix} \begin{vmatrix} \cos\alpha & -\sin\alpha \\ \sin\alpha & \cos\alpha \end{vmatrix}$$

Then, anisotropy = $|Z_{xy}|/|Z_{yx}|$. The average value for the 25 periods is maximum at each station for the angle shown in Tab. 2. To summarise the frequency dependence, anisotropy equals one at short period and increases with period until a relatively stable value is reached at $100 < T < 200$ s. Anisotropy is maximum at KBG and BOE, systematically decreasing away from the Swabian volcanic region.

The prefered direction α is determined by Swift's criterion (JONES & VOZOFF 1978) and for stations BOE to MOE is found to be approximately at right angles to the strike of the Swabian Alb (Tab. 2). Therefore, a rotation of $\alpha = -30^\circ$ is chosen. Taking into account magnetic declination of about 2°W, the x-coordinate has the direction N 32° W, and the y-coordinate N 58° E, or approximately perpendicular and parallel, respectively, to the survey profile.

The transfer function matrix for anomalous magnetic field transforms in the same way as the impedance. The real parts of two terms of the transformed matrix, h_y and d_y, are represented in Fig. 2 as perturbation arrows (SCHMUCKER 1970) at ten different periods for the profile stations. Particularly in the period range 25 to 150 s the arrows for stations FDW to APF are comparatively large, and are oriented in the y-direction, that is,

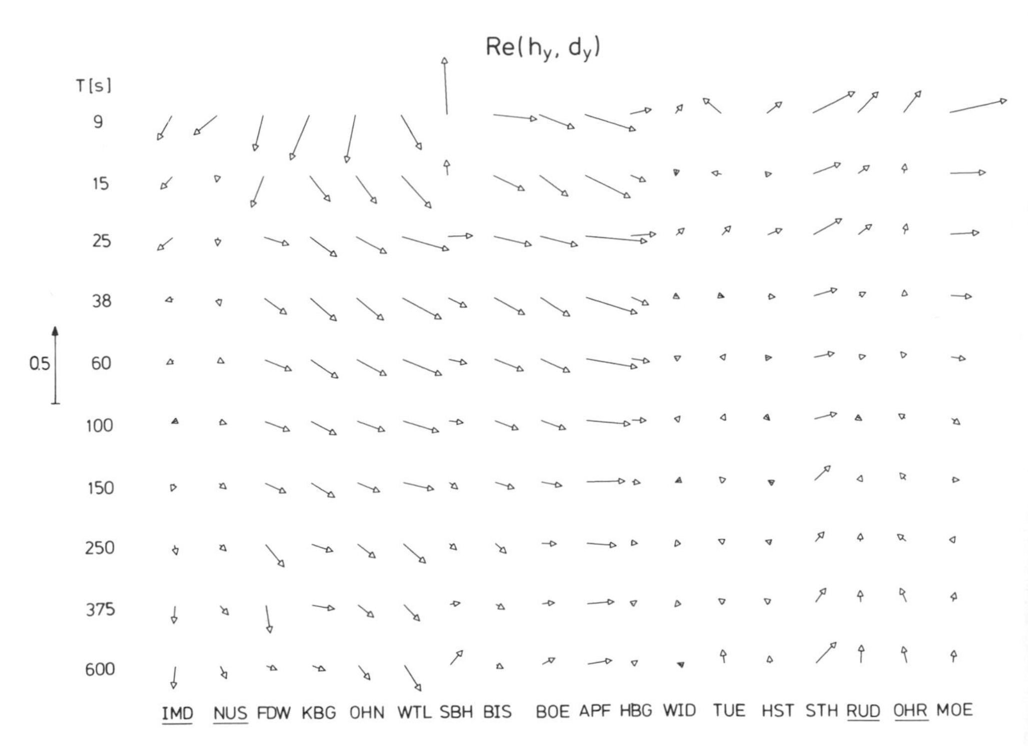

Fig. 2. Magnetic field perturbation arrows for anomalous (horizontal) magnetic field transfer function in rotated coordinates. Re(h_y) is the x-component (positive upward) and Re(d_y) the y-component (positive to the right) of the arrow. The arrows are referenced to the mean of the 4 underlined stations.

along the strike of the Alb. A comparison with Fig. 1 shows that most of these stations lie in the area of the Swabian volcano. Thus, the arrows suggest a structure with increased electrical conductivity in this region, with a trend roughly at right angles to the strike of the Alb. The period dependence argues for a superficial origin. Two exceptions to these general features are the relatively large arrows at STH and MOE, where the cause could be the highly conducting sediments filling the Steinheimer Becken and the Nördlinger Ries, respectively.

In Figs. 3 and 4 the real parts of the two pairs of impedance elements (Z_{xx}, Z_{yx}) and (Z_{xy}, Z_{yy}) are plotted as electrical analogues of the perturbation arrows. The result of the rotation is clear. The arrows (Z_{xx}, Z_{yx}) and (Z_{yx}, Z_{yy}) are nearly perpendicular, that is, Z_{xy} is large compared to Z_{yy}, and Z_{yx} is large compared to Z_{xx}. In the following we consider Z_{xy} to be the impedance for E-polarization, and Z_{yx} for H-polarization, in general agreement with the observation that the magnetic perturbation arrows are oriented in the y-direction. So, for E-polarization the large scale induced currents flow perpendicular to the trend of the Swabian Alb. The region of relatively large electric perturbation arrows, from HBG to HST, corresponds to the region of small magnetic perturbation arrows (Fig. 2). Both are indicators of relatively low conductivity. Although the electric field evidence is not complete, the opposite situation seems to occur in the region from OHN to BOE, including APF and BIS but excluding the northernmost station SBH; the magnetic perturbation arrows are large and the electric ones small suggesting higher conductivity.

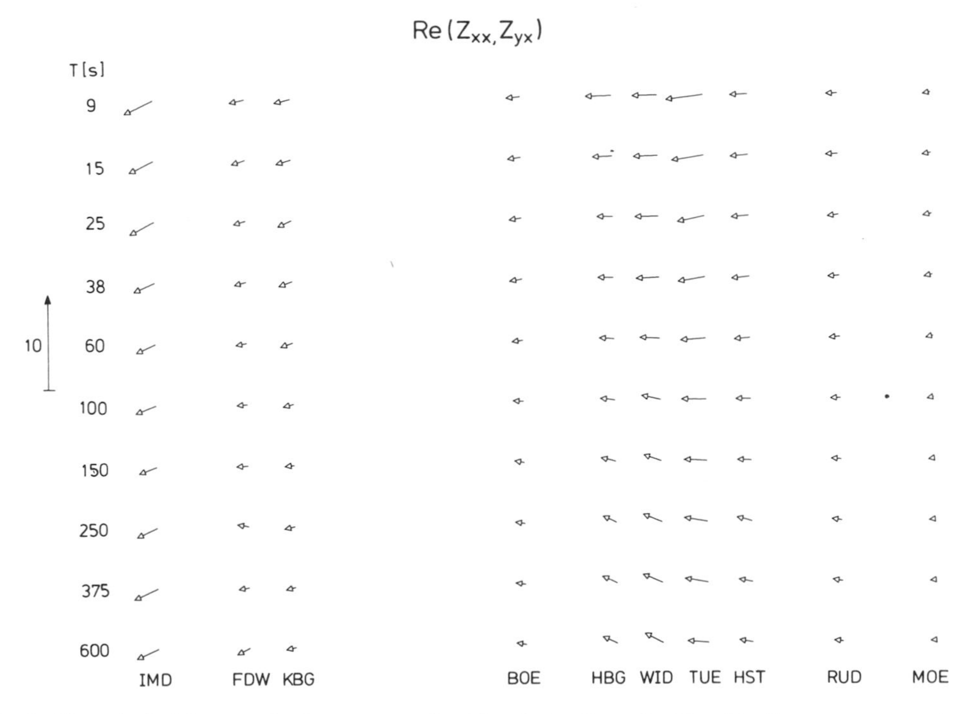

Fig. 3. Electric field perturbation arrows for H-polarization. Re (Z_{yx}) is the y-component (positive to the right) and Re(Z_{xx}) the x-component (positive upward). The scaling arrow at the left represents 10 mV/(km · nT) for the period T_o = 60 s. For other periods the scaling arrow length must be multiplied by the factor $\sqrt{T/T_o}$.

The complex penetration depth parameters for E- and H-polarization are, from SCHMUCKER (1970),

$$(4) \qquad C_E = \frac{Z_{xy}}{i\omega\mu_o} \quad \text{and} \quad C_H = \frac{-Z_{yx}}{i\omega\mu_o} \quad \text{where } \omega = \frac{2\pi}{T}.$$

For H-polarization this conversion has only a formal meaning, but estimates of apparent resistivity ρ^* and depth of penetration z^* can be derived from the E-polarization parameter C_E. In the Swabian Alb a good approximation to the conductivity structure appears to be a well conducting thin sheet with conductance τ overlaying a poorly conducting uniform substratum, in which case,

$$(5) \qquad z^* = \text{Re}(C_E),\ \rho^* = \frac{1}{2}\omega\mu_o\,(|C_E|^2 / \text{Re}(C_E))^2, \text{ and}$$

$$\tau = -(\text{Re}(C_E) + \text{Im}(C_E)) / (\omega\mu_o\,|C_E|^2).$$

The depth distribution of modified apparent resistivity ρ^* (z^*) for the E-polarization is represented in Fig. 5 for the four stations IMD, BOE, TUE and MOE. Station IMD is not typical, but the curves for all the other sites lie between and are parallel to the curves for MOE and TUE. The result at BOE is representative of the Urach area (stations KBG to HBG). The most resistive site is TUE, and there is a systematic trend to lower resistivity away from TUE in either direction along the profile. The curves are well determined. They give little indication of structure, ie, the resistivity decreases con-

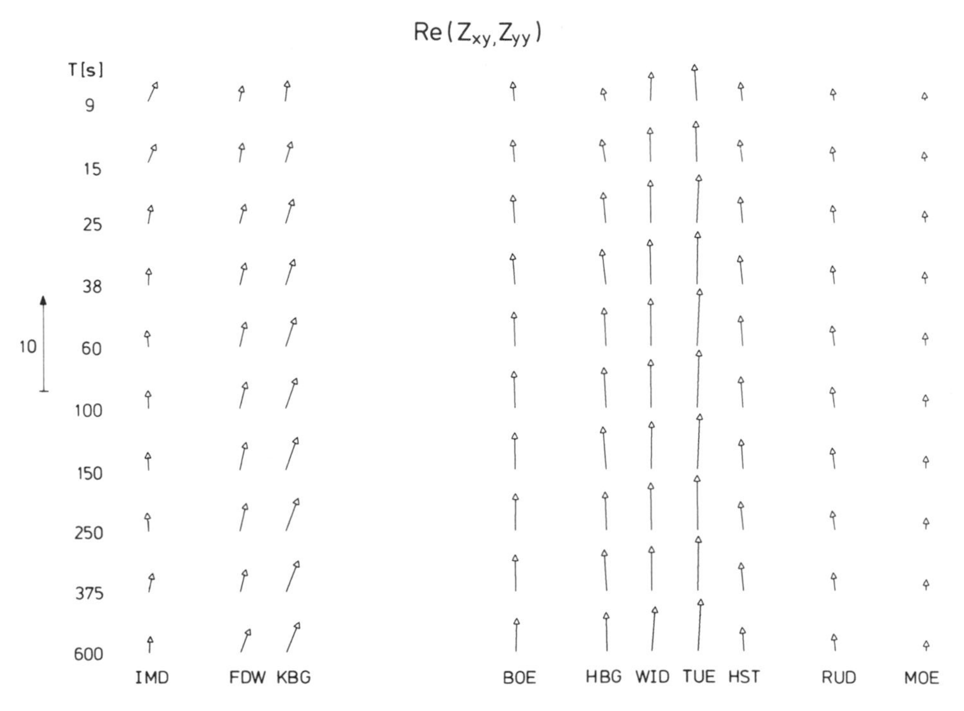

Fig. 4. Electric field perturbation arrows for E-polarization. Re(Z_{yy}) is the y-component (positive to the right) and Re(Z_{xy}) the x-component (positive upward). For remarks see Fig. 3.

tinuously through most of the depth range from the surface to 200 km. There are, for example, no clear signs of well conducting layers. In the Urach area ρ^* decreases from 1000 Ωm at $z^* = 5$ km to 200 Ωm at 200 km.

A first analysis of E-polarization C_E was made for each station with a 1-dimensional inverse model algorithm (SCHMUCKER 1974). Following the ρ^* (z^*) results, a thin sheet conductor overlying two homogeneous layers and a halfspace was assumed. The value of surface conductance (Tab. 2) was estimated from the frequency dependent τ of eqn. 5. A 3-layer model would be better for MOE and IMD, and the fit is poor for TUE, but a similar structure is observed at all the stations from FDW to RUD; the first layer is thinner and more conductive than the second, and the halfspace is a relatively good conductor. The major feature along the profile is a very resistive structure under WID and TUE. The model at WID agrees with one of BEBLO (1977).

To extend the interpretation, and to explain both E- and H-polarization observations, in particular the d_y variation along the profile, a two dimensional structure is developed. The model algorithm of SCHMUCKER (1971) is used, and for reasons explained above, the 2-D model profile is chosen parallel to the strike of the Swabian Alb. No entirely satisfactory model has been found. The simplest reasonable model is a surface thin sheet

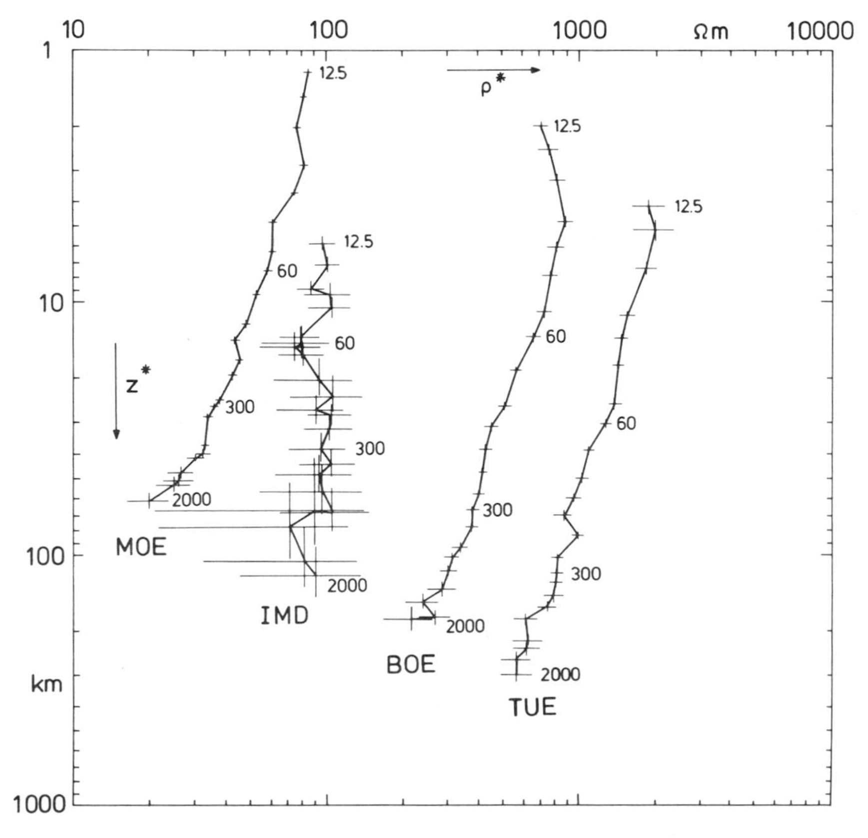

Fig. 5. Observed distribution of modified apparent resistivity ρ^* with depth z^* at 4 stations, in E-polarization. Estimates are shown with error bars, and the period in seconds is noted alongside some estimates.

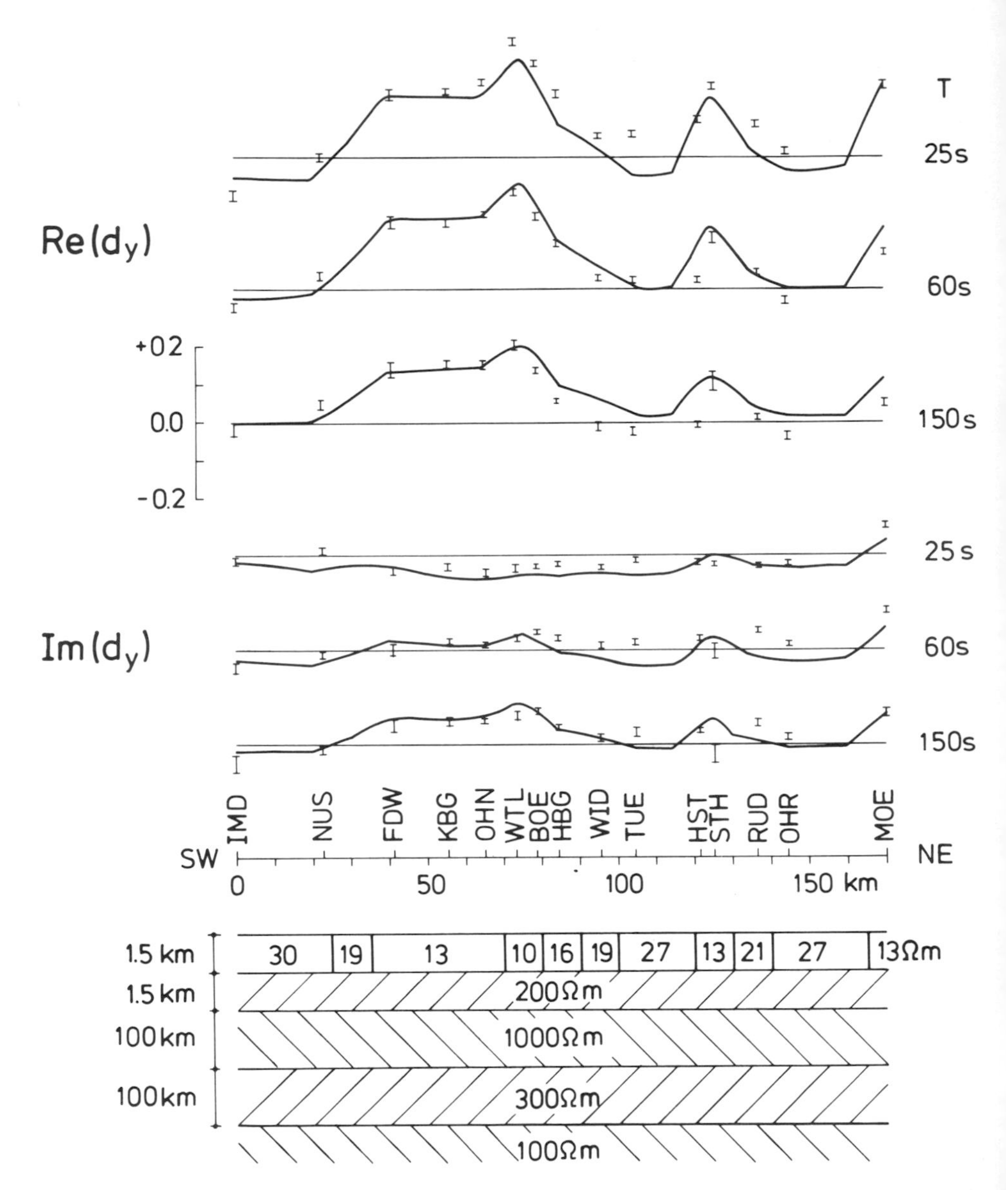

Fig. 6. 2-D resistivity model for the Swabian Alb profile, and comparison of observed and (E-polarization) calculated d_y anomaly for 3 periods T. Observed values are shown with error bars. Station locations are indicated along the y-axis of model. For each curve the horizontal line indicates $d_y = 0$.

Fig. 7. Comparison of calculated complex penetration depth (solid line) with the observed (error bars connected by dashed line) for the model of Fig. 6. In the upper half is shown the H-polarization (C_H) and in the lower the E-polarization (C_E) for the periods 25, 60 and 150 s.

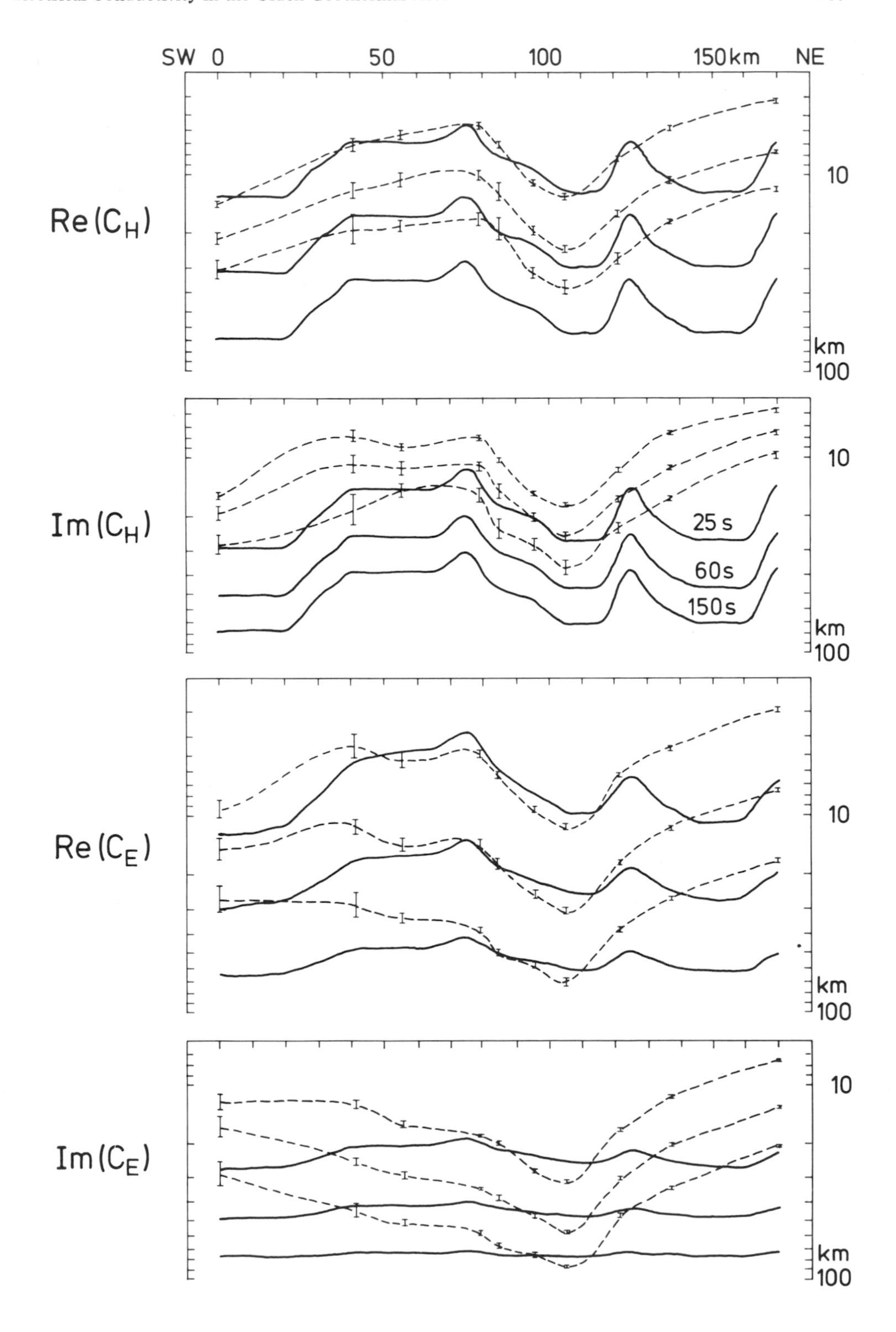
SW
0
50
100
150km
NE
Re(C_H)
10
km
100
Im(C_H)
10
25 s
60s
150s
km
100
Re(C_E)
10
km
100
Im(C_E)
10
km
100

with variable conductance τ (y) overlying a relatively resistive, layered structure. The conductance (Table 2) can be adjusted to give a very satisfactory fit to the d_y and C_E observations, but the H-polarization result C_H, though similar in shape to the observed variation along the profile, is much larger in value.

A simple modification which produces only minor changes to the E-polarization model but fits better the H-polaritzation result is to replace the thin sheet by a 1.5 km thick, inhomogeneous, surface layer. This is the model of Fig. 6. The resistivity values indicated for the upper layer of the model correspond to values of τ (y), where the thickness of this layer was chosen to agree with the Urach 3 borehole logs presented by WOHLENBERG (1980). The borehole conductivity integrated from the surface to the crystaline basement at 1.6 km depth gives a layer resistivity of between $< 12\ \Omega$m and 15 Ωm. The nearest electric field station to the borehole is BOE, where the resistivity of the upper 1.5 km in the model is 10 Ωm. For the depth range 1.6 to 3.3 km the value is 140 to 270 Ωm, so the resistivity in the second layer of the model was set at 200 Ωm. The deeper layer resistivities follow approximately the ρ^* (z^*) curve for BOE (Fig. 5).

The results of the model computations are given in Figs. 6 and 7. Above the model sketch in Fig. 6 the observed transfer function for anomalous horizontal magnetic field is plotted together with model results for E-polarization. The comparison shows that in the period range 25 to 150 s the model matches the observed anomaly rather well, that is, the variation of d_y along the profile can be explained to a large extent by lateral variation of the resistivity in the uppermost 1.5 km, corresponding to the Swabian Alb Jurassic formation, and an underlying layered resistivity very much like the ρ^* (z^*) distribution at BOE. It should be mentioned, however, that the model results for vertical magnetic field (not shown) are unlike the observed transfer functions. The calculated complex C_E and C_H are compared in Fig. 7. In E-polarization the agreement in both real and imaginary parts is good for the region between km 70 and km 110 of the profile, i. a., between BOE and TUE. Towards the ends of the profile the agreement is poorer, the model values indicating a greater depth of penetration than is observed. In H-polarization, with the excepting of the Re(C_H) for short periods, the agreement between model and observed is much poorer. The model presented gives a reasonably good explanation of the observed anomalies in the center of the profile, the region of the Swabian volcano. It appears, however, that the resistivity must be decreased in the deeper layers at both ends of the profile. Work to improve the model is not yet completed.

In conclusion, from the ρ^* (z^*) distribution there is little or no evidence of anomalous deep structure in the vicinity of the Swabian volcano and Urach geothermal anomaly. The apparent resistivity decreases smoothly from between 1000 and 2000 Ωm at a penetration depth z^* of 5 km to a value of 200 to 600 Ωm at 200 km. The strong feature which is observed in the geothermal area is a conductive structure with Urach in the center, trending approximately N 30° W, perpendicular to the Swabian Alb, and this can be explained by a 2-D conductivity anomaly in the upper 1 to 2 km. At all but the shortest periods, the NE-SW trending edge of the Swabian Alb is not visible in our data.

Acknowledgements

For their help in all phases of the field work and in the preparation of this report we acknowledge the support of the staff of the Institut für Geophysik, Göttingen University. We thank the foresters of the several state and private parks in the Swabian Alb who kindly cooperated in finding sites. Analysis and model calculations have been made at the Gesellschaft für wissenschaftliche Datenverarbeitung Göttingen. We thank the Deutsche Forschungsgemeinschaft for financial support of this research.

References

Beblo, M. (1977): Electrical conductivity in the geothermal anomaly of the Urach volcanic area, SW Germany – first results. – Acta Geodaet., Geophys. et Montanist. Acad. Sci. Hung., **12**: 87–91.

Filloux, J.H. (1973): Techniques and instrumentation for study of natural electromagnetic induction at sea. – Phys. Earth Planet. Int., **7**: 323–338.

Geologische Übersichtskarte von Südwestdeutschland, (1954), 1 : 600 000. – Geologisches Landesamt in Baden-Württemberg.

Hempfling, R. (1977): Beobachtung und Auswertung tagesperiodischer Variationen des erdelektrischen Feldes in der Umgebung von Göttingen. – Thesis, University of Göttingen, 83 pp.

Jones, F.W. & K. Vozoff, (1978): The calculation of magnetotelluric quantities for three-dimensional conductivity inhomogeneities. – Geophysics, **43**: 1167–1175.

Schmucker, U. (1970): Anomalies of Geomagnetic Variations in the south-western United States. – Bull. Scripps Inst. Oceanog., Univ. of Calif., San Diego, La Jolla, Calif., **13**: 165 pp.

– (1974): Erdmagnetische und magnetotellurische Sondierungen mit langperiodischen Variationen. – Protokoll, Koll. “Erdmagnetische Tiefensondierung”, Grafrath/Bayern, p. 313–342.

– (1974): Direkte und iterative Verfahren zur Behandlung 2-dimensionaler Leitfähigkeitsmodelle. – ibid., p. 429–441.

– (1978): Auswertungsverfahren Göttingen. – Protokoll. “Elektromagnetische Tiefenforschung”, Neustadt/Weinstraße, p. 163–188.

Wohlenberg, J. (1980): Elektrische Untersuchungen. – In Endbericht über Geophysikalische Untersuchungen in der Forschungsbohrung Urach, NLfB Hannover, Archive No. 85805, p. 26.

The Urach Geothermal Project, p. 313–321;
Schweizerbart'sche Verlagsbuchhandlung, Stuttgart, 1982

Gravity Measurements at the Geothermal Anomaly Urach

J. MAKRIS, K. MÜLLER and K.H. TÖDT

with 4 figures

Abstract: By gravity surveys the structural situation at the geothermal area of Urach was investigated. The data of the regional geophysical survey of Germany were supplemented by another 2000 stations, which were surveyed in the years 1977 and 1978. A qualitative interpretation in connection with the geothermal and seismic data showed that the gravity high of Urach and the corresponding geothermal anomaly are in good correlation. Quantitatively a 2 D-density model was established by considering drilling information and results of deep seismic soundings. The computed gravity anomaly was surplus of 0.2 g/cm^3 in the upper crust. Thus the high-density body responsible for the geothermal high is situated 6 km below the surface and its horizontal and vertical extensions are 15 km and 10 km, respectively.

1. Introduction

During 1977 and 1978 a gravity and magnetic survey was performed by the Institute of Geophysics, University of Hamburg, in the area of the geothermal anomaly of Urach.

About 2000 gravity and 1500 magnetic stations were established. Our intention was to improve the resolution of the known gravity field and to develop 2 D-density models constrained by borehole and seismic data.

The ground magnetic survey was evaluated and compared to the aeromagnetic map of the Federal Republic of Germany. It was used for a first qualitative interpretation. The results are presented elsewhere.

2. Technical Details of the Survey and Reduction of the Data

Two field crews were deployed for surveying the 2000 gravity and 1500 magnetic stations. Together with the regional data, which were made available by the Niedersächsisches Landesamt für Bodenforschung, Hannover, a field-point density of one station per 4 km^2 was achieved. The locations of the measured points are given in Fig. 1.

The following instruments were used:

Lacoste-Romberg Gravity Meters, Model G,
Elsec and Geometrics Protron Magnetometers,
Thommen Altimeters 3 B 4,
Füss aspirated Psychrometers.

The influence of the atmospheric pressure variations on the altimetric measurements was corrected by using barograph recordings. The accuracy of the altitudes is better than ± 1 m because of frequent readings taken at bench marks and triangulation points.

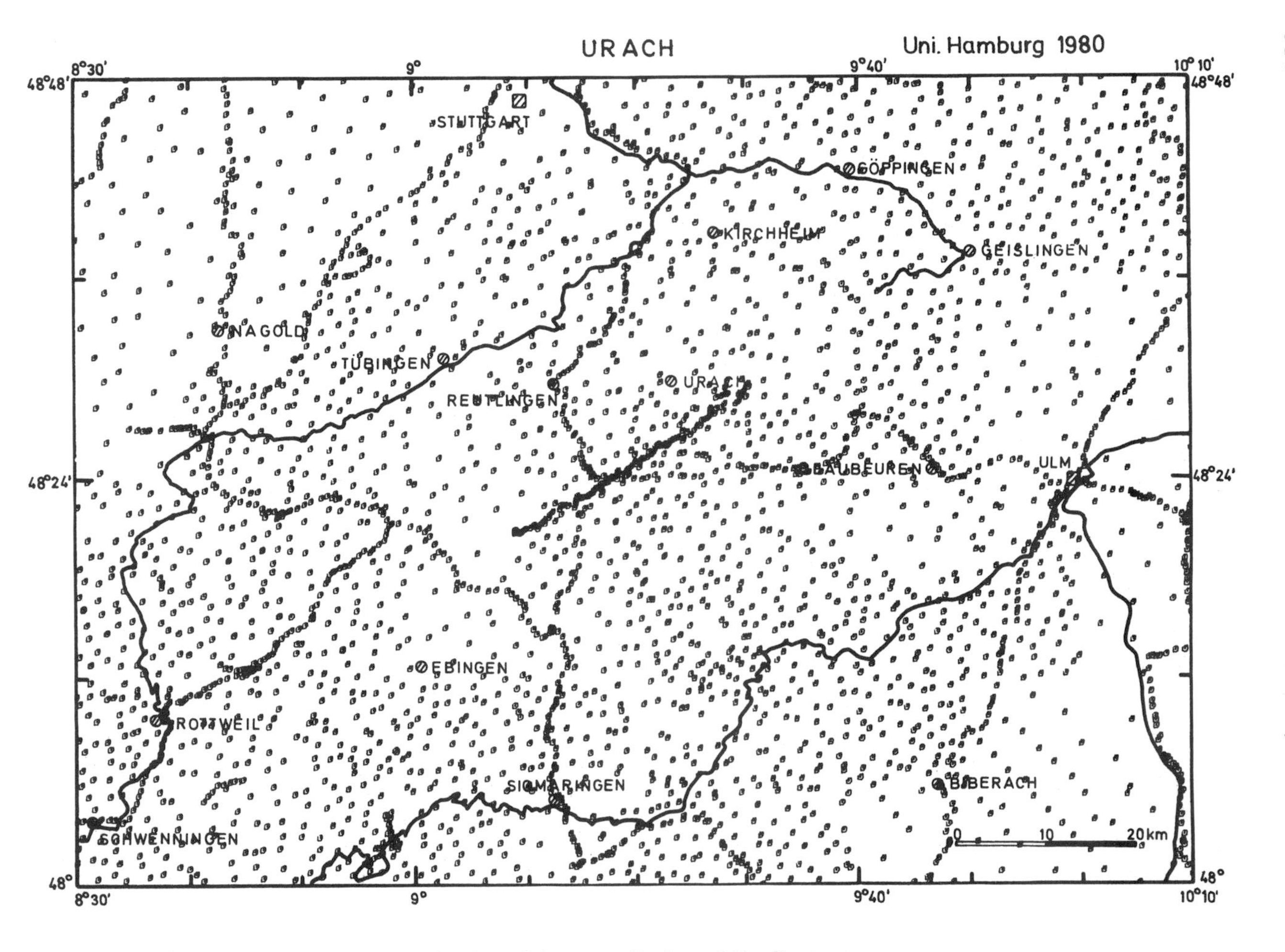

Fig. 1. Distribution of the gravity stations used for compilation of the Bouguer map.

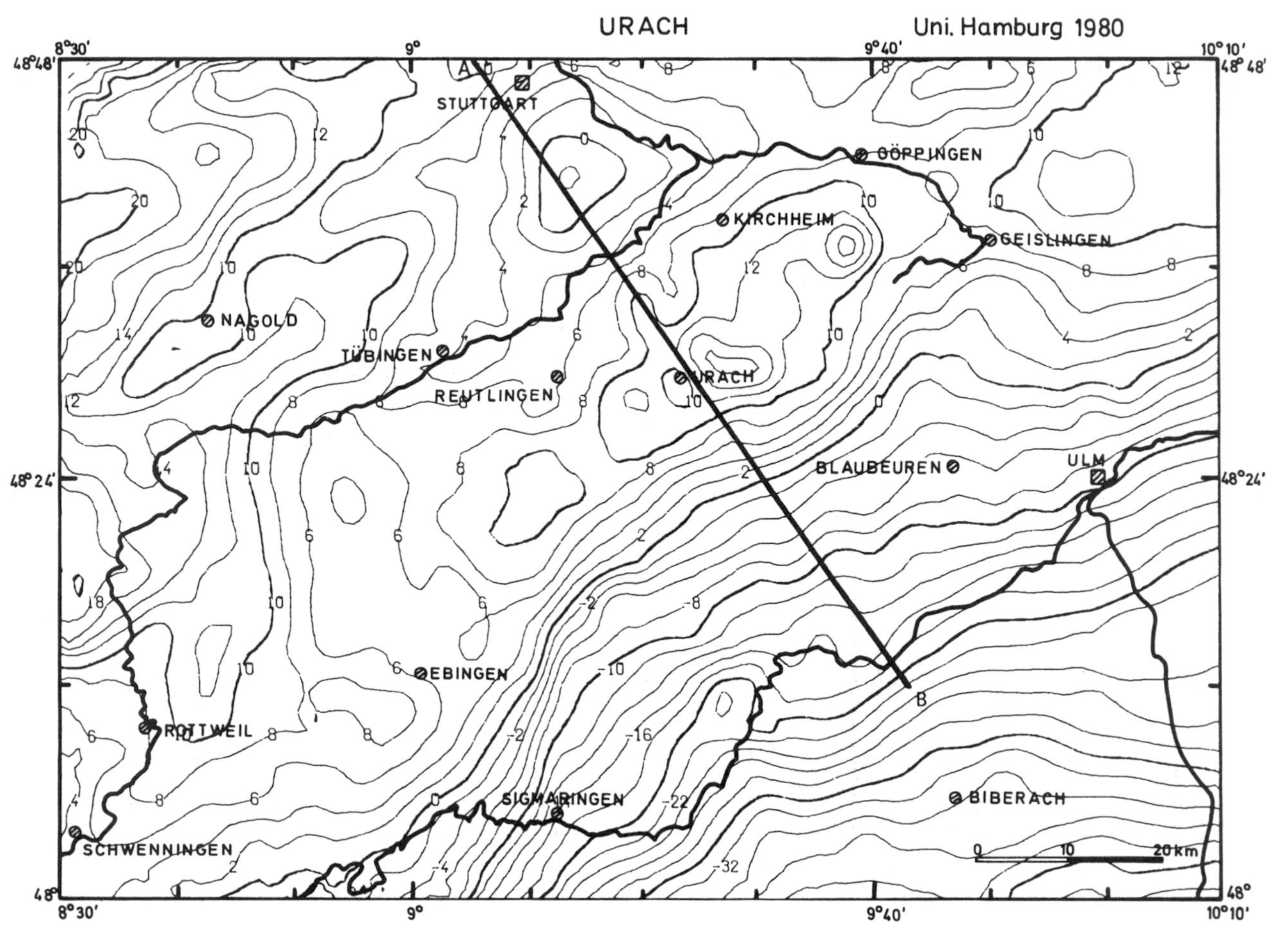

Fig. 2. Bouguer gravity map; reduction density: 2.4 g/cm^3; reference level: 300 m above sea level; contour interval: 2 mgal; the line indicates the location of the 2 D density model.

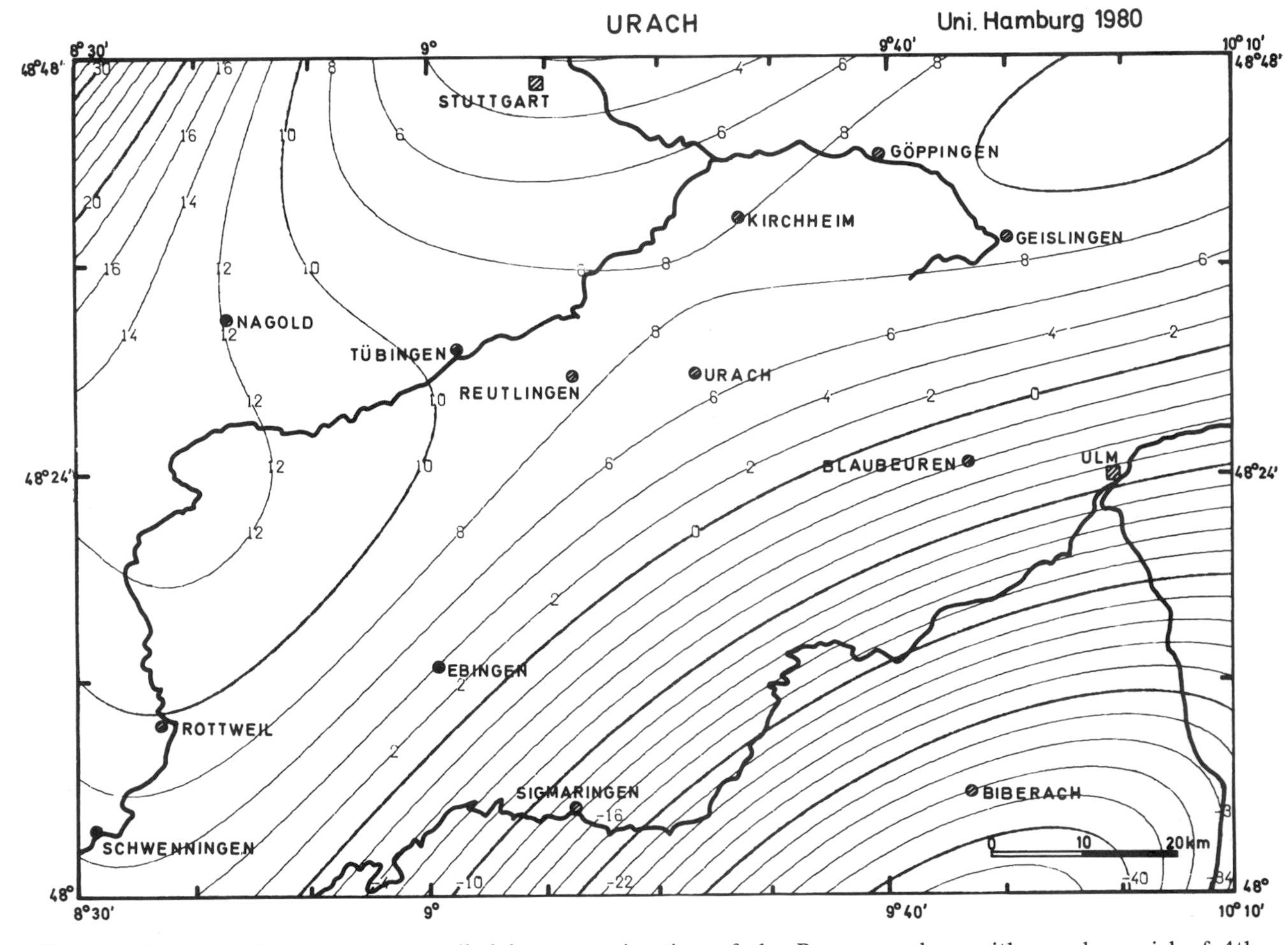

Fig. 3 a. Regional Gravity map compiled by approximation of the Bouguer values with a polynomial of 4th degree. Contour interval: 2 mgal.

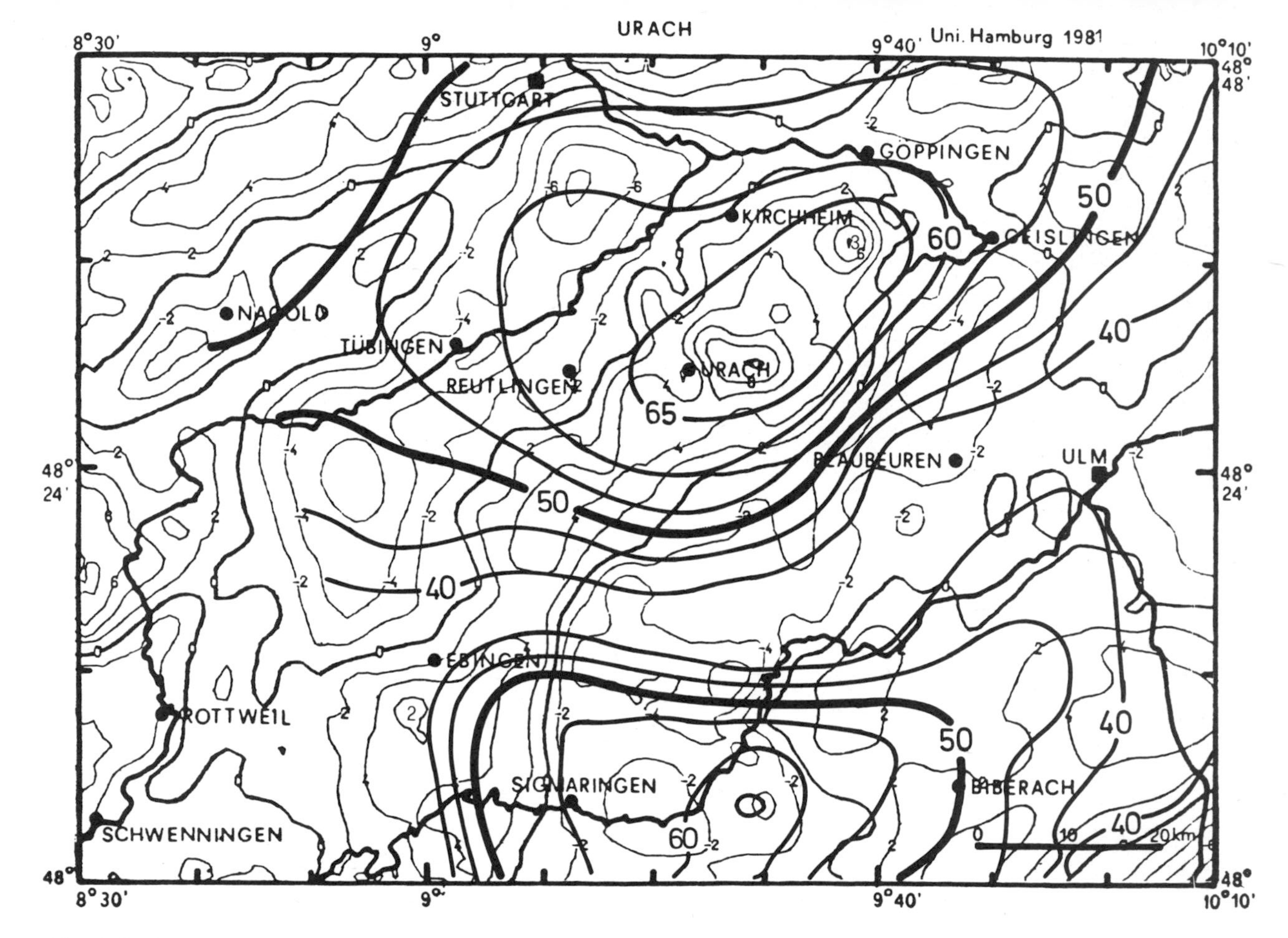

Fig. 3 b. Residual gravity map compiled by subtraction of a polynomial of 4th degree from the Bouguer map. Contour interval: 2 mgal. Isotherms at 1000 m depth. Interval of Isolines: 10 °C, ZOTH (this volume).

The international gravity formula of 1930 was used to compute the normal gravity; hence a comparison with older maps is still possible. All gravity data were reduced topographically to Hayford Zone O_2 (0–166.7 km). The topographic as well as the Bouguer reduction was computed on a spherical earth. Effects of the spherical Bouguer shell were obtained according to the formula of JUNG (1961: 75). The reduction density is 2.4 g/cm^3 and the reference level is 300 m above mean sea level. These reduction values are in agreement with those used in the regional geophysical survey of Germany. The new map produced is improved in two aspects with respect to previous ones:

- It has a greater field-point density and is therefore more reliable, with a higher resolution of the isolines than the regional survey.
- By reducing the topographic effects to Hayford Zone O_2 regional effects due to the topography have been entirely eliminated.

3. Interpretation of the Gravity Map

The Bouguer gravity map is given in Fig. 2. The anomalies are interpolated in 2-mgal isolines and show a general trend in NW-SE direction. This gravity trend corresponds to that of the major geological units. In the western part of the map the outcropping crystalline basement of the Black Forest influences the gravitational field and distrubs the NW-SE trend. The gravity level changes gradually from + 20 mgal to – 40 mgal in a NW-SE direction. A local gravity high is associated with the transition zone from the Jurassic to the Triassic formations. Its maximal values coincide with the Urach geothermal field. The southeastern part of the map covers the northern part of the Bavarian Molasse, and the decrease of the gravity level to negative values is an expression of crustal deformation and thickening of the sediments.

In order to prepare the gravity data for a qualitative interpretation we approximated the Bouguer values with orthogonal polynomials of various degrees. The regional trends are produced mainly by the Molasse in the south, the Jurassic formations in the north and northeast and the Black Forest-crystalline in the west. They are best represented by a polynomial approximation of the 4th-degree. The result of this computation is given in Fig. 3 a. In Fig. 3 b the corresponding residual anomalies are presented by interpolating the residual field in 2 mgal isolines. On this map the Urach gravity high has an amplitude of + 8 mgal. It is bordered to the north and to the south by two zones of residual negative values of – 6 and – 4 mgals respectively. In addition to the procedure described above, the interpretation of the amplitude and gradients of the residual high of Urach could be further constrained by geological information and deep seismic soundings. The distribution of the sediments was taken from Isopachen maps from GEYER & GWINNER (1968: 29, 37, 42, 45) and from the boreholes Albershausen (Alb), Buttenhausen (But), Upflamör (Upf), shown in GEYER & GWINNER (1979: 16) and Urach 3 (U III). The depth to the top of the basement was derived from explosion seismics by JENTSCH et al. (1980). The Mohorovicic discontinuity was chosen according to EMTER (1971: 50), GIESE (1976) and MEISSNER et al. (1980). The density model is presented in Fig. 4 middle and lower parts.

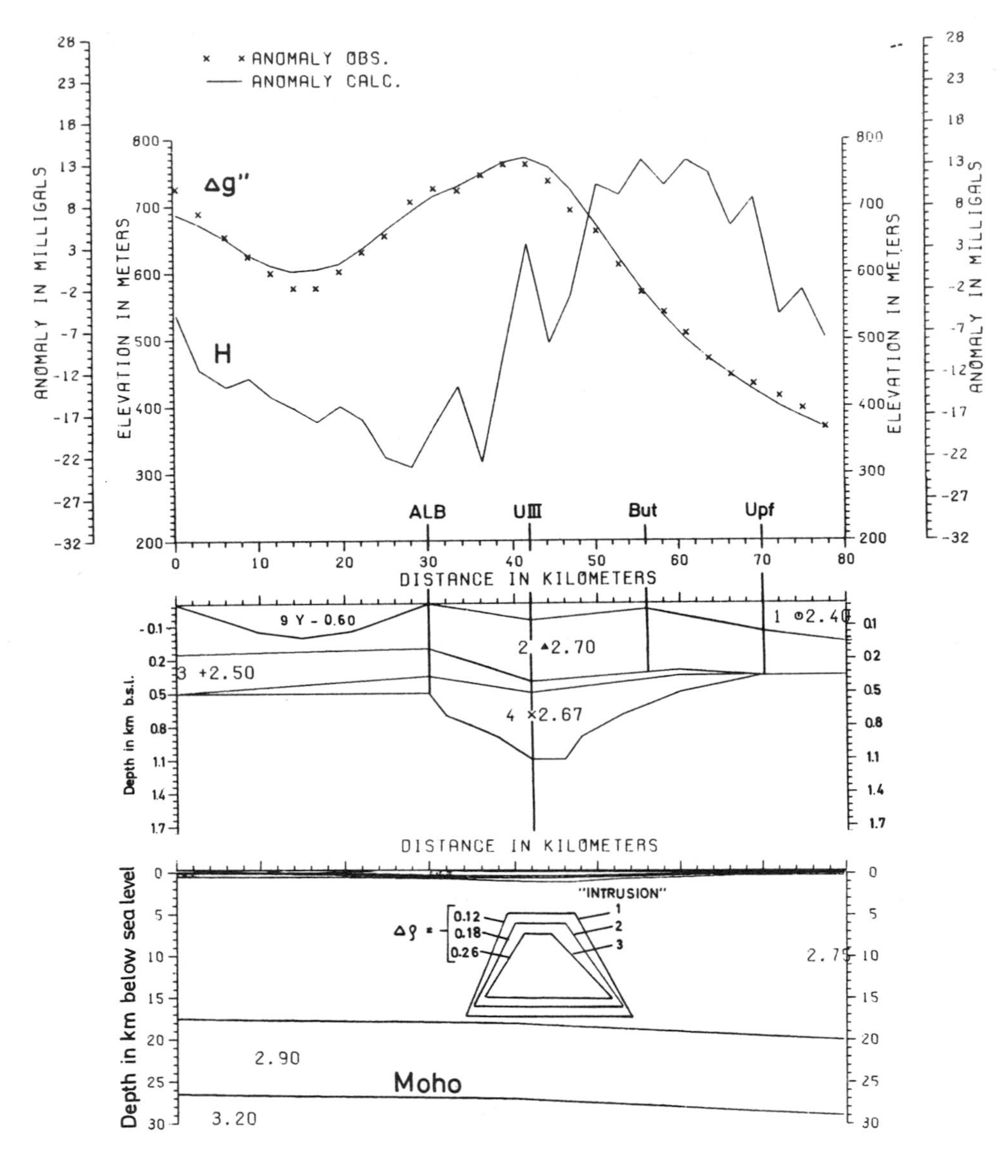

Fig. 4. Density model for the profile indicated in Fig. 2, calculated by considering borehole data and seismic information.

4. Quantitative Interpretation of the Urach Gravity High

A 2 D-density model of the Urach gravity high was developed by using a density-contrast of + 0.12 to + 0.26 g/cm^3 between the crystalline basement and the Urach intrusion. This range of density-contrast is limited by the borehole density logs (PLAUMANN & WOHLENBERG 1980), the P_g-velocity of 5.6 km/sec (WOHLENBERG 1980) and the petrology of the extruded volcanics. A further limitation of the basement density is posed by the relationship between P-wave velocity and density (BIRCH 1960). The reason for assuming that the gravity high is caused by a density increase in the crystalline basement and not in the upper mantle is that the centre of mass of the intrusion cannot be deeper than 13 km. This was estimated from a comparison of the half-width of the observed residual anomaly with that of a horizontal mass-line. The results of the computations are given in the lower part of Fig. 4. The variation of the dimensions according to different density contrasts used are presented by "Intrusion 1" to "Intrusion 3". The depth to the upper part of the anomalous mass ranges between 4 and 8 km below the surface. The most probable solution, however, is 6 km.

The algorithm used for the above computations is that of TALWINI et al. (1959).

5. Gravity Anomalies and the Urach Thermal High

By plotting the isotherms at 500 m depth (HAENEL 1980) on the residual gravity map, Fig. 3 b, one can see that the Urach gravity high and the temperature high coincide quite remarkably. From the fact that the isotherms show their maximal intensity between Ebingen und Göppingen in the NE-SW and between Stuttgart and Ulm in the NW-SE directions one is tempted to interpret the 2 D-density structure as a 3 D-density anomaly with a NW-SE axis of length nearly 3 to 4 times the NE-SW axis. The depth to the top of the 3 D-intrusion varies as a function of the observed residual gravity. In Urach the structure has reached its highest position below the surface and the observed thermal field reacts accordingly. Along the Ebingen-Urach-Göppingen line there are two further gravity highs which indicate relatively shallow depth to the intrusion top. The one south of Ebingen is quite large in extension and, if the interpretation presented is correct, it should be geothermally interesting. The anomaly east of Kirchheim has a higher value but is more restricted in its extension. We believe that these anomalies should be studied in detail and tested as possible additional low enthalpy geothermal fields.

6. Summary and Conclusions

The new Bouguer gravity map of the area of Urach shows a gravity high along the axis Ebingen-Göppingen. By separating the local effects from the regional trends the residual high associated with the Urach intrusion is resolved into two further anomalous areas of + 6 and + 7 mgal. These are located south of Ebingen and east of Kirchheim.

The residual anomaly was interpreted quantitatively by computing 2 D-density models constrained by borehole information, reflection and refraction data. The density contrast between the Urach intrusion and the igneous basement ranges between + 0.12 to 0.26 g/cm^3. The depth to the top of the intrusion, based on a density contrast of + 0.18 g/cm^3, is 6 km below surface and its horizontal and vertical extensions are 10 and 15 km, respectively. The long axis of the cylindrical intrusion strikes NE-SW and is 3 to 4 times its width. Local low enthalpy geothermal fields might be associated with the two further gravity highs described above.

7. Acknowledgements

The authors are indebted to their colleagues and students who participated in the field surveys and to the Deutsche Forschungsgemeinschaft (German Research Association) for providing the needed funds. The Niedersächsisches Landesamt für Bodenforschung, Hannover, made available data of the regional survey of Germany. This study is a part of the Geothermal Programme of the Federal Republic of Germany, which is a multiinstitutional effort in geology, geochemistry and geophysics.

References:

Birch, F. (1960): The velocity of compressional waves in rocks up to kbar. – J. Geophys. Research **65**: 1083–1102.

Emter, D. (1971): Ergebnisse seismischer Untersuchungen der Erdkruste und des oberen Erdmantels in Südwestdeutschland. – Dissertation Universität Stuttgart.

Giese, P. (1976): Results in the Generalized Interpretation of the Deep-Seismic Sounding Data. – In: Exposion Seismology in Central Europe, ed. by P. Giese, C. Prodehl, A. Stein, Springer Verlag Berlin Heidelberg New York.

Geyer, O.F. & Gwinner, M.P. (1968); Einführung in die Geologie von Baden-Württemberg, Schweizerbart, Stuttgart.

– (1979): Die Schwäbische Alb und ihr Vorland. – Samlg. Geol. Führer, **67**, Gebr. Bornträger, Berlin Stuttgart.

Jentsch, M., Bamford, D., Emter, D., Prodehl, C. (1980): A Seismic Refraction Investigation of the Basement Structure in the Urach Geothermal Anomaly, Southern Germany. – this volume.

Jung, K. (1961): Schwerkraftverfahren in der angewandten Geophysik. – Akad. Verlagsges. Geest & Portig KG, Leipzig.

Meissner, R., Bartelsen, H., Krey, T., Schmoll, J. (1980): Combined Reflection and Refraction Measurements for Investigating the Geothermal Anomaly of Urach. – this volume.

Plaumann, S. and Wohlenberg J. (1980): Gravimetrische Untersuchungen. – In: Endbericht über Untersuchungen in der Forschungsbohrung Urach, by R. Haenel, Report NLfB Hannover, Archive Nr. 85805.

Talwani, M., Worzel, J.L., and Landismann, M. (1959): Rapid Gravity Computations for Two-dimensional Bodies with application to the Mendocino Submarine Fracture Zones. – Journal of Geophys. Research, V. **64**: 49–59.

Wohlenberg, J. (1980): Seismische Untersuchungen. – In: Endbericht über Untersuchungen in der Forschungsbohrung Urach, by R. Haenel, Report NLfB-Hannover, Archive No. 85 805.

The Urach Geothermal Project, p. 323–343;
Schweizerbart'sche Verlagsbuchhandlung, Stuttgart, 1982

III. THE POSSIBLE UTILIZATION OF GEOTHERMAL ENERGY

Results of the Fracture Experiments at the Geothermal Research Borehole Urach 3

K. SCHÄDEL and H.-G. DIETRICH

with 11 figures and 2 tables

Abstract: In the research well Urach 3 the diatexite of crystalline basement was fractured both in open hole (3320–3334 m) and through perforations in the cased hole (P 1: 3259–3264, P 2: 3271–3276 m, P 3: 3293–3298 m). The break down pressure for frac creation was a few hundred bar lower than expected. This points to a reopening of natural fissures. Propping material could be pumped into the fracture systems. The pressure field initiated through one fractured system interacted with the stress field in such a way, that the neighbouring fissures system above and below reduced their width. Although, only one single borehole is available at Urach and a circulation-loop was created between open-hole (fracture 1) and P 3 (fracture 4). During water injection through the hole a simultaneous water flow was measured into all four fractures, but more than 86 % of flow rate occured through the open hole fracture, which has the lowest impedance.

1. Introduction

The purpose of this paper is to consider the following questions:
- Can the basement rock at Urach 3, a diatexite in the lower part (mica syenite) be fractured by means of high hydraulic pressures?
- If the rock is fractured at several places, is it possible to establish a water circulation loop as a basis for a hot-dry-rock system in one borehole?

2. Concepts in Planning (SCHÄDEL)

Since practical experience in the fracturing of crystalline basement rock did not exist in Germany, we started with two different concepts in the planning of the fracture experiments:
- On the one hand, the crude oil technique: This represents the concept of a one-borehole system in which several fractures were to be accomplished one above the other

along the wall of a borehole (ERNST 1977, 1978). Since the rocks are under minimal horizontal stress, the circulation should vertically take place in a fractured space.
– On the other hand, the model of geologists: Creation of a circulation by means of the system of natural open joints (= tensile fractures in the rock) necessarily having to occur without necessarily requiring hydraulic fracturing.

In fact, the resulting concept was a compromise between both. For that purpose, a frac section of 14 m length in the open hole and above it three five-metre-long perforated frac sections in the 7"-casing have been prepared.

The frac project was directed by the "Frac-Crew of Urach", which represented the partner of the two companies, Preussag and Halliburton. The Frac-Crew of Urach consists of: ALTHAUS, DIETRICH, HAENEL, HEINLE, NETH, RUMMEL, SCHÄDEL, SCHWEIZER, WOHLENBERG, and ZOTH. The state of the borehole previous to the fracture experiments is reported by DIETRICH & SCHÄDEL (1978) and DIETRICH (this volume).

3. **Procedure** (DIETRICH)

According to the plan, the diatexite encountered in the lower borehole section below about 3000 m depth has been hydraulically fractured in four different zones. The attempt to accomplish a hydraulic connection between the different frac zones has thereby been undertaken in order to establish a circulation system. In the available single-borehole system, consisting of a cased borehole with a drill pipe, the circulation is possible if the annulus and drill pipe are sealed with a special packer. The RTTS-packer (Retrievable, Test, Treat, Squeeze-packer) provided by the Company Halliburton is very adequate for all of the experiments in the borehole Urach 3.

In order to avoid a possible solidification of the RTTS-packer caused by caved-in rock fragments and/or proppings, the circulation experiments were always carried out from the upper to the lower fracture: Injection of water via annulus into the fractured rock and from there back into the borehole and through the drill pipe up to surface. For financial reasons, these experiments were performed without insulation of the tubing.

The first fracture experiments were performed in the non-cased borehole section (Frac zone 1) between 3320 m (casing shoe 7"-terminal casing) and 3334 m (final depth), thereafter consecutively in descending sequence (Frac zones 2–4) through perforated 7"-cased sections, each about five metres in length (P 1: 3259–3264 m, P 2: 3271–3276 m and P 3: 3293–3298 m). The perforations were accomplished by using the (Hyper) Jet bazooka-type system with 4"-casing guns from Preussag Company (HK 4" in the perforation section P 1 and P 2) and Schlumberger Company (4" CG-EL in P 3). The shot density amount to about 13 shots per metre.

For the technical tasks in the 7"-terminal casing, such as driving a 5 7/8"-roller bit or a 7"-scraper, the incorporation and extraction of a 7"-RTTS-packer, or unexpected recovery tasks as well as for the drilling of the well itself, a drill rig of type Cabot-Franks-Explorer 900 was employed. The technical data have been already mentioned in the Technical Details of Geothermal Well Urach 3 (DIETRICH this volume). For the fracture experiments a 3 1/2" °E-drill pipe in combination with 30 drill collars (about 270 m in length) of 4 3/4" in diameter was employed. A double Shaffer-Preventer 7 11/16" x 1000 psi with pressure reserve system from URACA Company has been installed at the well head.

The Halliburton Company was commissioned to execute the frac treatments and the circulation tests. This company also supplied the necessary equipment, materials and chemicals, such as a 36 m^3 water tank, a blender, a suction-and-pressure manifold, as well as various pump units (e. g. HT 400) with the necessary connections for the 3 1/2" IF-drill pipe (frac-head S 15) for pressures up to 700 bar. A combination of such pumps yielded up to about 1470 kW (2000 PS) and produced more than 20 l/s at surface pressures up to 700 bar. A logging truck (Frac-Van) was at our disposal for the measurement of the frac-pressures, pump rates and the total volume.

The measurement of the pressures was performed by means of the Martin-Decker-Recorder and the Automatic Digital-Fracrecorder. The pump rates were indicated by the flowmeter and printed by the recorder. The accuracy of the flow rate instruments is about ± 0.5 % and greater deviations occur during the pumping of gel.

During the circulation experiments, a logging truck with recorder, flowmeter and pressure gauges was employed instead of the Frac-Van.

The bauxite "Superprop-Norton-Propsand 20/40 E 362" was used as propping material during the fracture experiments. The propping material was mixed with "Versa-gel" as carrier medium and pumped through the perforations into the hydraulically fractured rock.

The Versa-gel has been prepared by mixing several chemicals: gelation medium and fraction reducer, gel-stiffening medium, pH-value-reducer and -raiser, gel-breaker and defoamer with the type specifications WG-11, CL-11. HYG-3, K-34, AP-Breaker and NF-1.

For generating the pressure in the annulus (maximum 350 bar) during the fracture and circulation experiments, an additional vehicle (HT-600) with a maximal power of about 440 kW (600 PS) was employed.

The Bridge Plugs (Baker) necessary for the fracturing were set by the Preussag Wire-Line-Service. This Service has also measured the 7" casing with a 2.5"-gyro compass (Humphrey Gyro) for a second directional survey.

The temperature measurements in the borehole were carried out prior to the fracture experiments and during the circulation experiments also by Preussag with electronic devices (Gearhart Owen- and Hewlett-Packard-Instruments). On of the latter was applied during the circulation experiment also in combination with mechanical measuring equipment, the Kuster gauge from the Niedersächsisches Landesamt für Bodenforschung (NLfB). A skid unit was thereby available as measuring hoist.

Unfortunately, the temperature measurements in the annulus with a Maihak gauge near the perforations (= injection temperatures) could not be realized. The electric cable was damaged at a depth between 800 and 1000 m.

A gas separator with a volume of 3 m^3 from the same Preussag-Service made it possible to separate and measure the gases during the circulation. For this analysis a chromatograph from the Geo-Data-Company stood at our disposal.

For the chemical field analysis of water samples an atomic absorption spectroscope from the Mineralogical Institute of the University Karlsruhe was available.

The total volume of injected and extracted water and its distribution relative to the perforated cased sections and the open borehole have been determined by temperature and flowmeter measurements carried out by Schlumberger (see chapter 11).

The fracture and circulation experiments were performed from 7th May to 2nd August, 1979. In order to limit the noise disturbance in the neighbourhood, most of the work was usually performed only on workdays between 6 a.m. and 10 p.m. on a 2-shift basis. A concise summary of the course of the work is given in Tab. 1.

Table 1. Chronological summary of the frac and circulation tests performed at Urach 3 in 1979.

Date executed 1979		Frac/circulation tests Current exp. no.	Frac section/ test no.	Principal operations
May	7– 8	–	–	Delivery of drilling rig; Leak-Off-Test in open hole by University of Bochum
	8– 9	–	–	Temperature measurements down to final depth (FD) with Kuster-gauge by NLfB
	9–21	–	–	Rig not in use
	22–24	–	–	Rig up; final preparation for the frac tests
	25	–	–	Removal of 2 7/8'-Killstring; placement of Kelly pipe
	26–30	–	–	Running the borehole with 5 7/8" bit and with 7" scraper down to 3334 m (FD); drill pipe contents for geochemical depth profile (33 samples) extracted; circulation of fresh water; bumping the 3 1/2" drill pipe and above ground conduits with 350 bar
	30	–	–	Electrical borehole measurements: CBL, VDL, CCL
	30	1	1	Leak-off-Test in open hole (3320–3334 m)
	30–31	–	–	Incorporation of frac equipment; packer placed at 3303 m
	31	2 3	1. 1 1. 2	2 fracs in open hole (11 + 25 m^3 water resp.): pump rates up to about 750 l/min, well head pressures up to 250 + 430 bar
June	31– 1	–	1	Measurements of pressure reduction and back-flow
	1	–	–	Packer removal; GR/CCL measurement; Bridge plug placed at 3306.3 m
	2	–	–	7"/13 3/8" annulus bumped with a pressure of 200 bar
	2	–	2	1st Perforation of 7" casing (3259–3264 m = P 1)
	3– 4	–	–	Interruption of work
	5	–	–	7"-RTTS-packer placed at 3250 m
	6	4	2. 1	Frac at P 1: injection of about 35.2 m^3 water with WG–11 max. flow rates about 1000 l/min, max. well head pressure about 550 bar
	6– 7	–	2	Measurement of pressure reduction and back-flow
	7– 8	–	–	Removal of frac equipment, repair of the draw works
	8	–	3	2nd perforation of 7" casing (3271–3276 m = P 2)
	9–11	–	–	Running the 7" scraper, pressure tests in the annulus; 7"-RTTS packer placed at 3267 m
	11	5	3. 1	Frac at P 2: injection of about 87 m^3 water and 16 m^3 gel; flow rates 350 to about 1000 l/min, well head pressures 100–350 bar
	11–12	–	3	Measurement of pressure reduction and back-flow
	12	6	2. 2	Injection of water through P 1 via 3 1/2"/7" annulus. Flow rates 140–250 l/min, max. well head pressure about 350 bar. During repeated injections and decompressions, no hydraulic

Table 1. cont'd

Date executed 1979	Frac/circulation tests Current exp. no.	Frac section/ test no.	Principal operations
			connection but pressure connections to P 2 are noticeable.
12–13	–	2	Measurement of backflow from P 1
13	7	3. 2	Chiksan pipework tested with 685 bar pressure; frac at P 2: injection of about 30 m^3 water, max. about 1200 l/min, max. well head pressure 670–690 bar; no hydraulic connection between P 1 and P 2 observed.
13	8	3. 3	Frac at P 2: injection of 10 m^3 water, 10 m^3 gel and 20 m^3 gel + proppings (30–120 g/l); 550–870 l/min; max. well head pressure about 590 bar; pressure connections but no hydraulic connection between P 2 and P 1 observed
13–15	–	2 + 1	Measurement of pressure reduction and back-backflow from P 2 and P 1
15	–	–	Removal of frac equipment
June 16	–	–	Gyro compass directional survey; 2nd Bridge plug placed at 3269 m
17	–	–	Interruption of work
18	9	2. 3	7"-RTTS packer placed at 3216 m; Frac at P 1: injection of 25 m^3 gel and 25 m^3 gel + proppings (90–240 g/l) at 350–800 l/min and max. well head pressure about 650 bar; "Sandout" at injection 240 g/l ?
18–19	–	2	Measurement of pressure reduction and back-flow from P 1
19	10	–	3 Leak-off-Tests at P 1
19–20	–	–	Pressure tests in 7" casing with RTTS-packer: 2-stage cementer reveals slight leak at 2520 m
20–22	–	–	Insertion of 5 7/8" bit with fishing pipe; during drilling, 2nd Bridge plug pushed to 3304 m onto 1st plug: bit removed, RTTS packer placed at 3267 m
23	11	2. 4/ 3. 4	1st circulation test with alternating injections through P 1 and P 2: pressure equilibrium proven
24	–	–	Interruption of work
25	–	–	Removal of 7"-RTTS-packer; 2-stage cementer bumps with 200 bar: no pressure reduction
25–26	–	4	3rd Perforation of 7" casing (3293–3298 m = P 3); 7"-RTTS-packer placed at 3280 m
26	12	4. 1	Frac at P 3: injection of 11.2 m^3 water, 5 m^3 gel and 25 m^3 gel + proppings (90–240 g/l) at 50–800 l/min and max. well head pressure about 670 bar: "Sandout" at injection 240 g/l!
26–27	–	4	Measurement of pressure reduction
27–30	–	4	Measurement on top proppings and washout of Superprop thru 1"-endless tubing (Nowsco Co.) with injection of Tylose mud; RTTS packer extracted and incorporated – placed at 3267 m
30	13	2.5/ 3. 5	2 nd circulation test thru P 1 (via annulus), P 2 + P 3 (drill pipe): injection of 4.3 m^3 thru P 1 and 22.2 m^3 thru P 2 + P 3 at small flow rates and max. well head pressure of 350 bar:

Table 1. cont'd

Date executed 1979		Frac/circulation tests Current exp. no.	Frac section/ test no.	Principal operations
				pressure connection but no circulation observed
	30–	–	2 + 3	Measurement of pressure reduction; interruption
July	1			of work
	2– 6	–	–	Removal of 7"-RTTS packer; washout of prop material; run with 5 7/8" bit; 7"-Bridge plug milled with Economill in combination with Navi-Drill; 7"-RTTS packer placed at 3302 m
	6	14	1. 3/ 4. 2	3rd circulation test (CT): injection of 32.3 m^3 water via annulus (P 1, P 2, P 3) at 100 l/min and max. well head pressure of about 350 bar: backflow via drill pipe – circulation in the rock probable.
	6– 7	–	1. 3/ 4. 2	Backflow measurements
	7	15	1. 4/ 4. 3	4th CT: injection of 48.8 m^3 water via annulus (P 1, P 2, P 3) with flow rates of 75–130 l/min and max. well head pressure about 350 bar; circulation proven with tracer (uranin)
	7– 9	–	1 + 4	Backflow measurements
	9	16	1. 5/ 4. 4	5th CT: injection of 63.3 m^3 at 100–63 l/min and max. well head pressure 350 bar via annulus (P 1, P 2, P 3); circulation successful
	9–10	–	1 + 4	Backflow measurements
	10	17	1. 6/ 4. 5	6th CT: injection of 21.5 m^3 water via annulus (P 1, P 2, P 3), at 110–65 l/min and max. well head pressure 350 bar: circulation successful
	10	18	2. 6 + 3. 6/ 1. 7 + 4. 6	7th CT (7"-RTTS-packer placed at 3284 m): injection of 3.8 m^3 water at 90–35 l/min and max. well head pressure 350 bar via annulus (P 1, P 2): circulation not proven
	10–12	–	1 + 4	Backflow measurements; removal of RTTS packer; failure in attempt to insert temperature gauge into the annulus; 7"-RTTS packer placed at 3302 m
	13	19	1. 8/ 4. 7	8th CT: injection of 39.4 m^3 water via annulus (P 1, P 2, P 3) with 140–50 l/min and max. well head pressure 350 bar; temperature measurements: circulation successful
	13–14	–	1 + 4	Backflow measurements
	14	20	1. 9/ 4. 8	9 th CT: injection of 5 m^3 hydrochloric acid 5 % via annulus (P 1, P 2, P 3) with 124–60 l/min: circulation reduced from 60 to 37.5 l/min, partial stop due to increased well head pressure beyond 350 bar
	14–16	–	1 + 4	Backflow measurements
	16	21	1.10/ 4. 9.	10th CT: injection of 31 m^3 water via annulus (P 1, P 2, P 3) at 58–36 l/min and max. well head pressure 350 bar: circulation successful
	16–17	–	1 + 4	Backflow measurements
	17	22	1.11/ 4.10	11th CT: injection of 19.6 m^3 water via annulus (P 1, P 2, P 3) at 43–30 l/min and max. well head pressure 350 bar: circulation successful
	17	–	1 + 4	Backflow measurements

Table 1. cont'd

Date executed 1979	Frac/circulation tests Current exp. no.	Frac section/ test no.	Principal operations
17	23	2. 7 + 3. 7/ 1.12 + 4.12.	7"-RTTS packer placed at 3284 m; 12th CT: injection of 3.5 m^3 water via annulus (P 1, P 2) at 45–35 l/min, partial stop due to increase beyond 350 bar well head pressure: circulation not proven
17–18	–	1 + 4	Backflow measurements
18–19	–	1 + 4	Flowmeter measurements (Schlumberger) during injection of 57.1 m^3 water at rates of 400–300 l/min and max. well head pressure 350 bar as well as during backflow out of the rock
19–20	–	–	Combined temperature measurements (Hewlett-Packard device with Kuster gauge), gyro compass directional survey, backflow measurements
20–21	–	–	1st insert test with Baker production packer for 3 1/2" tubing; 7" scraper run
22	–	–	Temperature measurements with Kuster gauge, gyro compass directional survey with Humphrey instrument
23	–	–	Attempt to insert production packer stopped at 3276 m
23–24	–	–	Run with Taper Mill: resistance at 3276–78 m and 3283–84 m
23–24	–	–	Attempts to insert production packer without success: packer, packer-setting tool and CCl remain in the borehole
25–27	–	–	Running the 7" casing with Economill, Overshot and impression block: fishing jobs postponed to the planned longtime circulation test
28–30	–	–	Preparation for rig down; assembly of christmas tree with stuffing box
July/ Aug. 31– 2	–	–	Rig down; transport of the rig.

4. The Rock (SCHÄDEL)

The rock in the frac section consists of diatexite (mica syenite), a micaceous variety of granite. The rock is jointed; on the average 9 m of core have 7 joints. The joint planes are slanted at an angle of 40–70°. Hidden vertical joints having often been observed in plutons which were so important for the project in Fenton Hill (Los Alamos), could not be found. In part, the joints have incrustations of calcite and pyrite; however they are open, which means that the incrustations are not cohesive. Others are closed, compact with quartz or other fillings. Fig. 1, left, illustrates a core section and Fig. 1, rigth, the deformation of the same core which resulted from strain relief during drilling and the removal from the depth. Thus, the great horizontal stress which prevails deep in the rock can be estimated. In this core section, the maximal stress is directed along the vertical strike slips of the relaxation joint planes which tend to N 80 °E. The most important lithophysical parameters were determined by RUMMEL et al. (this volume). Their average values for this frac section are as follows:

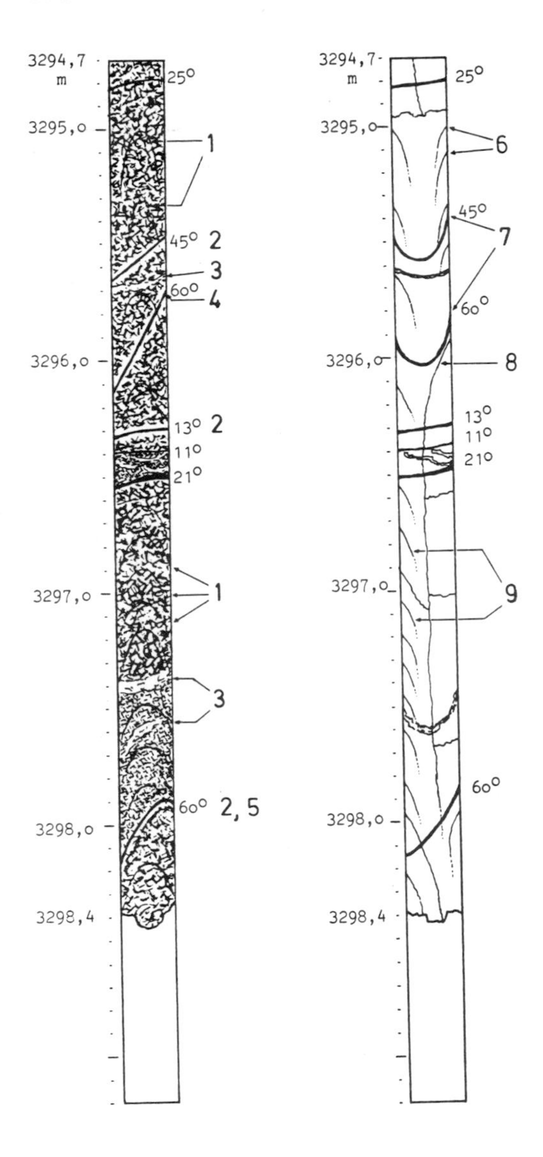

Fig. 1. The basement rock cored from section 3294.7–3298.4 m (compiled from detailed core descriptions by NETH); core recovered: 3.7 m; at the top of the core about 0.15 m sloughing: greenish-light gray rock (bleached "mica syenite"?).
1 – Release joints
2 – Bleached zone
3 – Aplitic streaks
4 – Bleached zone; flat-lying joint with clayish fillings and pyrite up to 2 cm thick
5 – Flat-lying joint compact with calcite filling
6 – Release joints; about 26 in number were found in the overall core length
7 – Flat-lying joints, with streaked joint filling
8 – Release joint, more than 2 m long; divides the core into two halves starting at 3296.0 m
9 – Release joints lying opposite; to be found about 35 times in the overall core length

tensile strength = 180 bar
elastic moduli $E_{dyn} = 6.5 \cdot 10^6$ N/cm^2
and $E_{stat} = 6.0 \cdot 10^6$ N/cm^2
angle of internal friction = 38° (at 610 par pressure).

In order to have an idea of the three-dimensional structure of the rock before the fracturing work started, a valumetric joint model was constructed as far as cores were available. This model was very helpful in explaining the circulation which took place later.

5. Natural Permeability, Leak-off Test and Opening Pressure (SCHÄDEL)

The mean natural apparent permeability of the open borehole section at 3320–3334 m depth was tested in a preliminary test by RUMMEL. The resulting value is 3 μD. The actual permeability of non-jointed rock is about zero. If the permeability is related to the joints alone, then it is one to two powers of ten higher. This is an estimated value. According to the definition, the specification in darcy is not suited for describing the permeability in joints, especially because, as will later be shown, the width of joints is pressure-dependent (see chapter 9).

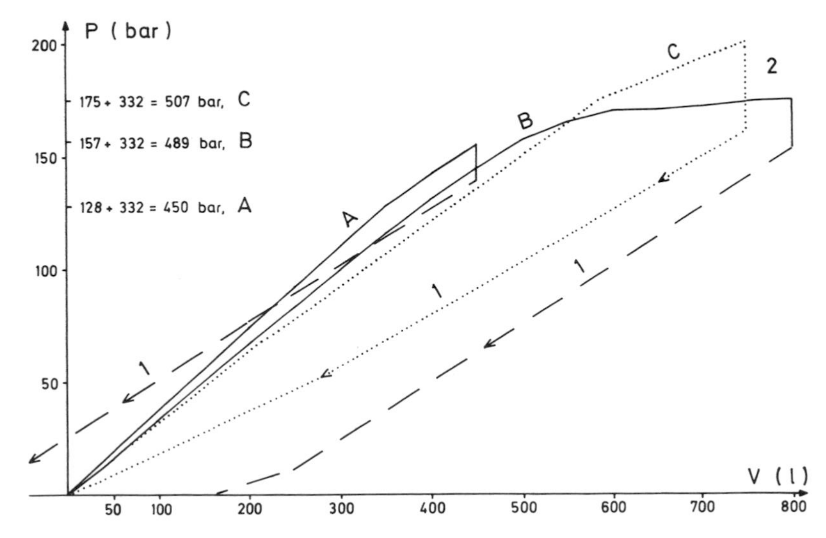

Fig. 2. Leak-off test; rock pressure about 580 bar.
1 – Backflow; 2– Pressure decline; V – Accumulated volume; P – Well-head pressure; A – 1. test, about 38 l/min; B – 2. test, about 35 l/min; C – 3. test, about 180 l/min.

At the beginning of the actual fracturing procedure, a leak-off test was undertaken in the open hole at a depth at 3320–3334 m in order to determine the pressure limit, above which the rock begins to assimilate water. Fig. 2 shows the three successive results. Accordingly, the rock opens clearly at a well-head pressure of 170–200 bar.

The well-head pressure of 170–220 bar corresponds in reality to a bottom hole pressure of 500–530 bar. At this pressure, either new fracture planes are produced or joint planes in the rock near the borehole are – because of elastic behaviour – so deformed that considerable amounts of water can suddenly penetrate into the rock at pressures above this limit. One can assume that the in situ stress due to the rock pressure, which effects a normal tension in the joint planes, is probably overcome at this pressure. In other words, the opening pressure and the rock pressure are approximately in equilibrium. If the dip of the planes and the normal rock pressure are known, the magnitude of the least horizontal stress can be estimated; this amounts here to about 500 bar. If the uplift pressure is still fully effective – which is not certain – the vertical pressure amounts to 580 bar, which means that hydrostatic pressure conditions already prevail in the rock.

If one interprets the opening as a tension failure, that is, as a genuine fracture, then an effective horizontal stress cannot prevail, because the tensile strength of the rock is approximately equal to the applied pressure.

6. Fracture Experiments in the Open Hole (SCHÄDEL)

Both of the following experimental fractures in the open hole show that the rock can assimilate greater amounts of water without difficulty at pressures above that of the opening pressure. Apparently a simple linear relationship exists between the applied pressure and the injection rate (Fig. 3). However, the pressure level in the second test is higher than that in the first (which is an argument against a genuine hydraulic fracture). The same observation was already made during the three successive leak-off tests which followed. An increasing tendency in the opening pressure was found here as well. The injected water partly returns to the surface after pressure release in opening the borehole.

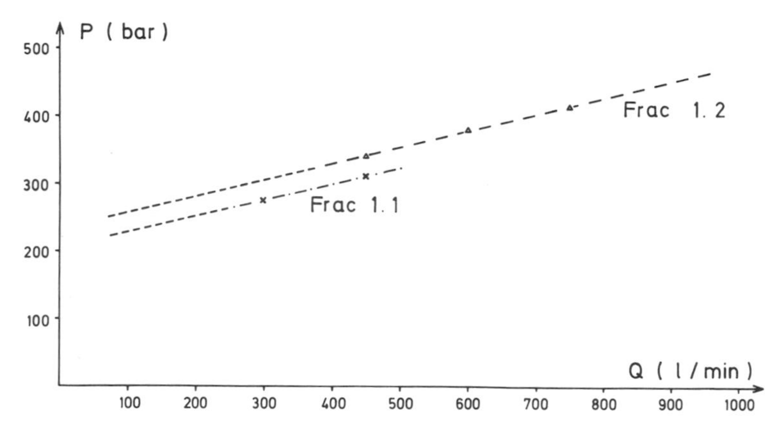

Fig. 3. Pressure at the mean injection rate Q; P – Well-head pressure.

7. Experimental Fractures Induced by Perforation (SCHÄDEL)

As to the experimental fractures induced by perforating the casing of the borehole, the following questions were posed: Is a normal perforation, as employed in oil-well drilling, satisfactory in rendering rock without intergranular porosity accessible to the water from the borehole (see SCHÄDEL 1980, Fig. 6). Does the cementation hold if a circulation between two neighbouring perforations is attempted at an interval of only seven metres?

Both questions were answered affirmatively in various experiments; the first, however, with limitations. High pressure drops probably occur between the perforation holes and the rock. In future experiments, the opening of the casing should by all means be attempted by sandblasting in order to guarantee a better communication between the borehole and the rock with its joints.

All of the fracture experiments undertaken in the perforated casing demonstrated that the rock can also assimilate large amounts of water (up to 20 l/s) through the perforation. However, a connective flow between two perforations, the one situated at the short distance of seven metres (perforation P 1 and P 2) above the other, was probably not attained. The rock closes after the dropping of the pressure. During repeated opening, higher pressures were often observed, as compared to the previous test.

Nevertheless, a connection could be accomplished in one case between Frac 2 (P 1) and Frac 3 (P 2). This remained limited, however: first the connection occured only in a downward direction; second, it was observed only within the pressure interval of 310–350 bar.

Discontinuous pumping was performed in Frac 2 (about 0.5–0.8 l/s), whereas Frac 3 below it was held at a constant pressure level by controlled pumping. The fluctuations in pressure in Frac 2 were recorded with reduced amplitudes in Frac 3 (see Fig. 4). Pressures above 350 bar in both fracs interrupted the connection. A possible explanation is demonstrated in Fig. 4 on the left. It must be borne in mind here that it

– functions essentially in a single downward direction and
– only in the pressure interval of 310–350 bar.

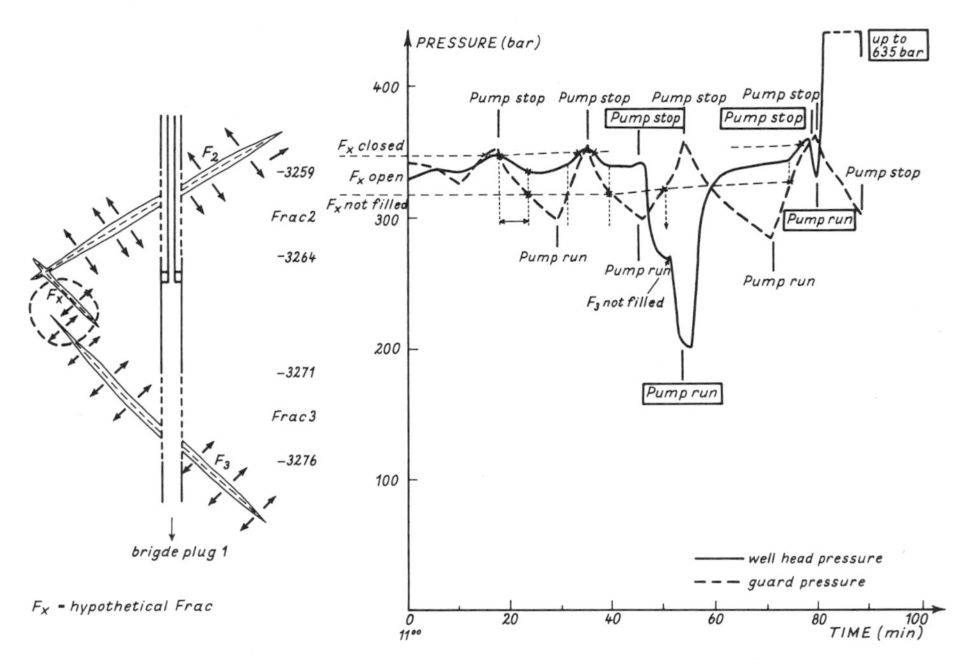

Fig. 4. The pressure graph of fracture 2 (guard pressure) and the fracture 3 (weel-head pressure) on the right (after NETH & ZOTH, see DIETRICH et al. in press) and the possible explanation on the left.

The connection is conceivable as being accomplished via a narrow fault which communicates directly with Frac 2, from which it can be filled and enlarged during the pressure interval. This in turn is involved in a field of stress at a higher pressure, however, and is compressed by the transverse expansion. The spatial position is so favourable with respect to Frac 3 that it is subject to the influence of its own field of stress while in the state of expansion without, however, being able to be filled and expanded via Frac 3. At pressures below 300 bar during "shut in", a pressure equilibrium between Frac 2 and 3 results. With increasing duration of the experiment, a moderate increase in the pressure level was observed.

8. Use of Proppings (SCHÄDEL)

In order to hold the rock open as much as possible, the incorporation of propping material in the joints was attempted. Because of the high rock pressures, there was the danger that the proppings would be compressed. Therefore, a sintered bauxite was used with a grain size of 20/40 mesh (the greatest diameter being about 1.8 mm). The incorporation was performed with the aid of a gel which was mixed with progressively increasing concentrations of the propping material. The incorporation was accomplished without difficulty in two of three tests; in the third test (Fig. 5), a "sand out" occurred at the highest concentration (240 g/l). The transition between the perforations in the casing and the joints was possibly obstructed. Fig. 6 shows a scheme of the "propping columns" used for the incorporation and distribution of the propping material in a roughly circular, expanded fracture plane.

9. Fracture Size (SCHÄDEL)

The incorporation of proppings yields a lower empirical value for the width of the fracture. Since the largest grain diameter was 1.8 mm, a width of at least 2 mm can be assumed; otherwise, an incorporation would not have been possible. A calculation of the fracture width was performed by SCHWEIZER (in SCHÄDEL & SCHWEIZER 1979). Nearly static

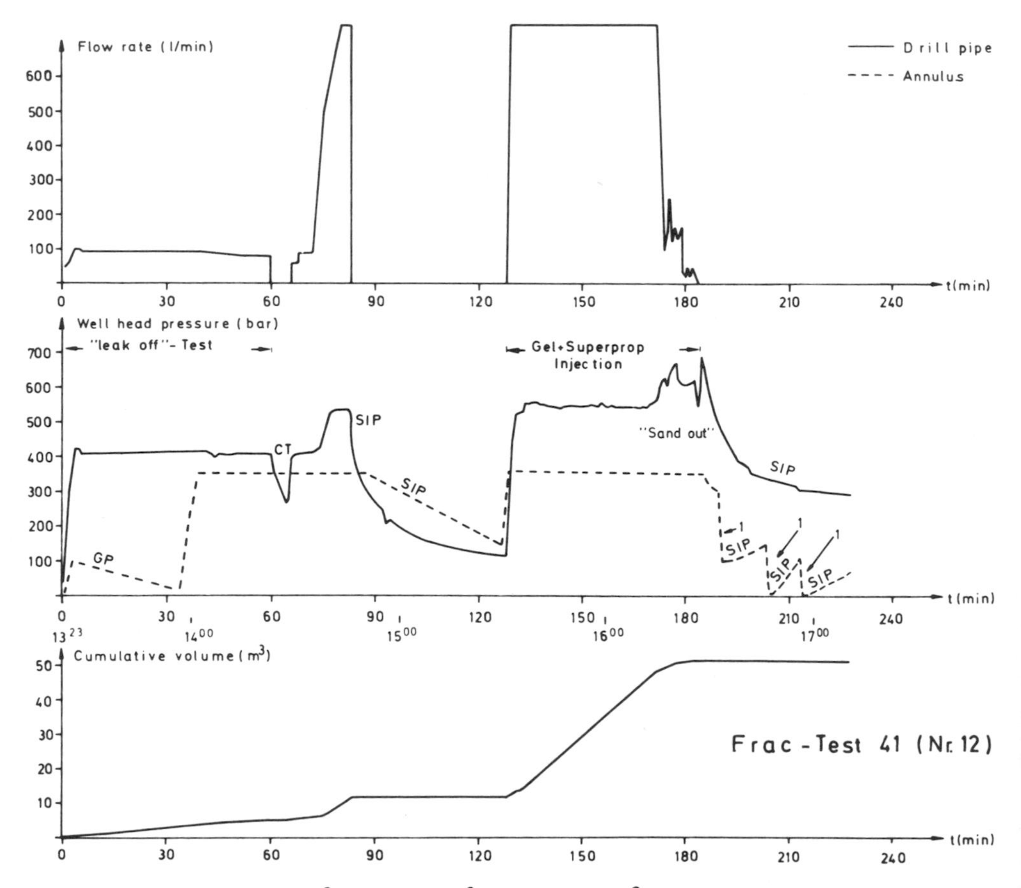

Fig. 5. Injection of 11.2 m^3 water, 5 m^3 gel and 25 m^3 gel with proppings (90–240 g/l) in frac test 4.1 (no. 12). "Sandout" during incorporation of 240 g/l. (CT = Circulation test; GP = Guard pressure; SIP = Shut in pressure; 1 = Pressure decreased) compiled from NETH & ZOTH.

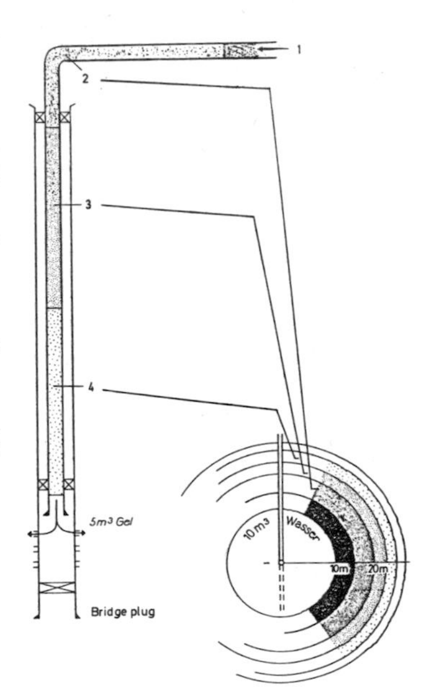

Fig. 6. Incorporation of proppings in the Urach 3 borehole in a penny-shaped fracture with an assumed fracture width of 2.5 cm.
1 – 240 g propping per 1000 cm^3 gel, in total 5 m^3, then 10 m^3 water
2 – 150 g propping per 1000 cm^3 gel, in total 10 m^3
3 – 120 g propping per 1000 cm^3 gel, in total 5 m^3
4 – 90 g propping per 1000 cm^3 gel, in total 5 m^3, corresponding to a length of 1470 m in the drill pipe
The used propping material: Superprop 20–40 mesh, density 3.63 g/cm^3, bulk density – wet: 2.2, bulk density – dry: 2.14, density of gel: 1.0 ± 0.05 g/cm^3.

conditions were thereby assumed, since the flow rate in the rock at a distance of several metres from the borehole must be assumed to be relatively low (order of magnitude: cm/s–dm/s). The amount of the deformations is affected by the applied high pressures (effective injection pressure in the rock at a depth of about 3300 m in general: 530–930 bar). It is also responsible for the expansion of the injection surfaces basically depends on the frac radius and on the E-modulus and have an order of magnitude of centimetre (SCHWEIZER). The expansion is elastic; the fracture closes with a decline in pressure.

This large fracture width causes difficulties in understanding sufficient space in the rock for the water pressed into it. Fissures which are merely a few millimetres in width, or fractions thereof, can assimilate only a limited quantity of fluid per square metre of surface area and, therefore, very large surfaces, and thus a deep penetration into the rock are required. This is not possible without great pressure drops because of friction. Below the pressure limit of 530–700 bar (bottom hole pressure), however, the absorptivity of the rock is too small to make the observed injection rates possible. Probably, the bulk of the injected volume remains in a relatively small region near the borehole, in which the pressure is higher than the opening pressure. Beyond the vicinity of the borehole, the pressure is lower than the opening pressure whereby water is scarcely absorbed. The reason for this initially hypothetical assumption is the spontaneous variability the absorptivity at pressures higher than the opening pressure, which is in equilibrium with the rock pressure that varies in the surroundings of the injection locality. This results in a permeability with a negative skin factor (Fig. 7) between – 2.0 and – 5.4 (SCHWEIZER this volume). Following the injection, of course, the injected fluid disperses in the rock and thus a distribution of pressure results, but this requires a long time.

In the summer of 1981, two years following the conclusion of the project, a significant pressure increase due to the injections into the borehole was still noticeable, as the comparison of the standing water levels in the borehole previous to and following the fracture experiments demonstrates (Fig. 8).

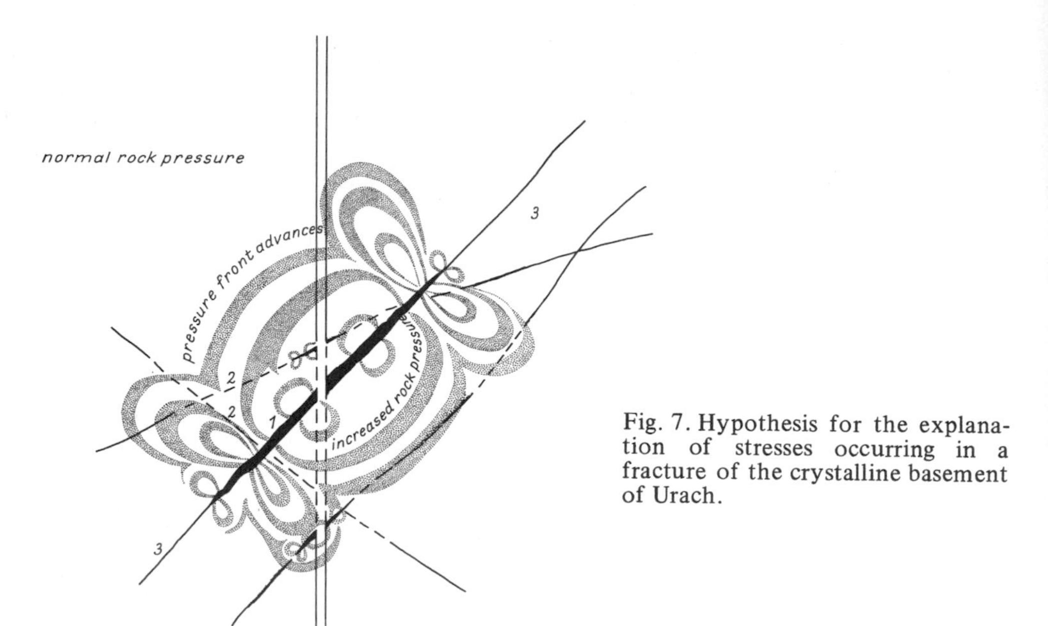

Fig. 7. Hypothesis for the explanation of stresses occurring in a fracture of the crystalline basement of Urach.

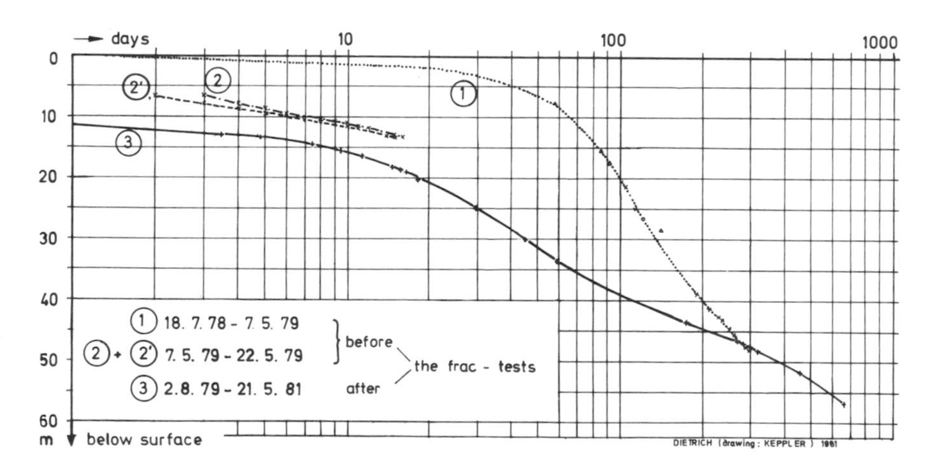

Fig. 8. Development of water level before and after frac tests in well Urach 3.

10. Circulation Experiments (SCHÄDEL)

A circulation was carried out successfully several times from the propped Frac 4 (3296–3301 m) through the rock down to the open hole (below 3320 m depth); see Fig. 9. Because joints were observed in oriented cores, a likely passage of 35–40 m in length through the rock can be assumed as a reasonable explanation; see Fig. 10, KAPPELMEYER, HAENEL & RUMMEL 1979, DIETRICH et al. 1980. Evidence of the connection was established with uranin (sodium-fluorescein) as tracer.

The volume of the circulation system could be calculated during the uranin test from the difference between the injected fluid and the borehole volume and amounted to about 3.4 m^3. This determination is inexact since the resulting volume amounts to only 4 % of the total amount of fluid injected into the borehole and the rock. A second possibility of determination is given by the flow curves of the ionic concentrations, especially that of the calcium ion (ALTHAUS, this volume). A volume of 4 m^3 can therby be inferred.

The water circulated from the annulus, through the rock into the open hole and via drill pipe back to the surface. The temperature in the drill pipe was therby measured (see Fig. 10). The maximal amount of water that flowed at the beginning of the experiments, was 1.1–1.3 l/s (DIETRICH et al. 1980). In the course of the experiments, which lasted from 6 to 12 hours respectively, the rate diminished to about 0.5 l/s.

During each of the circulation experiments, the flow restistance increased with time. The reason is most likely the following: The amount that had to be pumped in was always somewhat more than the amount that flowed back. In other words, in each time unit of the experiment during pumping, a portion of the fluid was retained in the rock. This absorbes a certain volume and thereby induces additional tension, which presumably affects their lowest parts and reduces their diameters via the flow passages.

After the termination of pumping, water still flowed out of the strained rock for a long time. If one provides for free backflow in the annulus and in the drill pipe as well, the rock relaxes after a while, and the degree of permeability during the subsequent test is equivalent to the initial value. Another explanation, which was expressed from the mineralogical-chemical point of view (personal communication by ALTHAUS), interprets this phenomenon only as an obstruction of the passageways due to chemical fillings. Afterwards, these would be pressed aside upon renewed pumping into the rock. This view is not shared by the authors.

In further circulation experiments, it is planned to investigate this and especially to improve the permeability in the lower part of the passageway by accurate acidification.

11. Flowmeter Tests (DIETRICH)

As demonstrated schematically in the circulation model of Fig. 10, the circulation in the rock results from fractures which form a triangle with the borehole. In this model, it is assumed that the water which is injected into the rock through the perforation 3 circulates almost exclusively by way of individual, hydraulically opened fractures. This assumption was to be further investigated in flowmeter tests at the end of the entire project.

As a result of the particular test sequence, however, only a partial success could be expected; for technical reasons the 7"-RTTS-packer had to be set at a depth of 408 m. When the packer was set at greater depth the spinner of the 1 11/16"-fullbore-flowmeter had always been obstructed repeatedly in passing the 3 1/2" drill pipe. Therefore, it was only possible to investigate all of the perforations at once. The open borehole below the casing shoe could not be reached because of a cave-in. Moreover, the surface pressure for injection was not to exceed about 340 bar as a safety measure for the protecting casing.

The values of the flowmeter tests are summarized in Tab. 2. This results in:

– During the injection of water at least about 86 % of the total volume were pumped

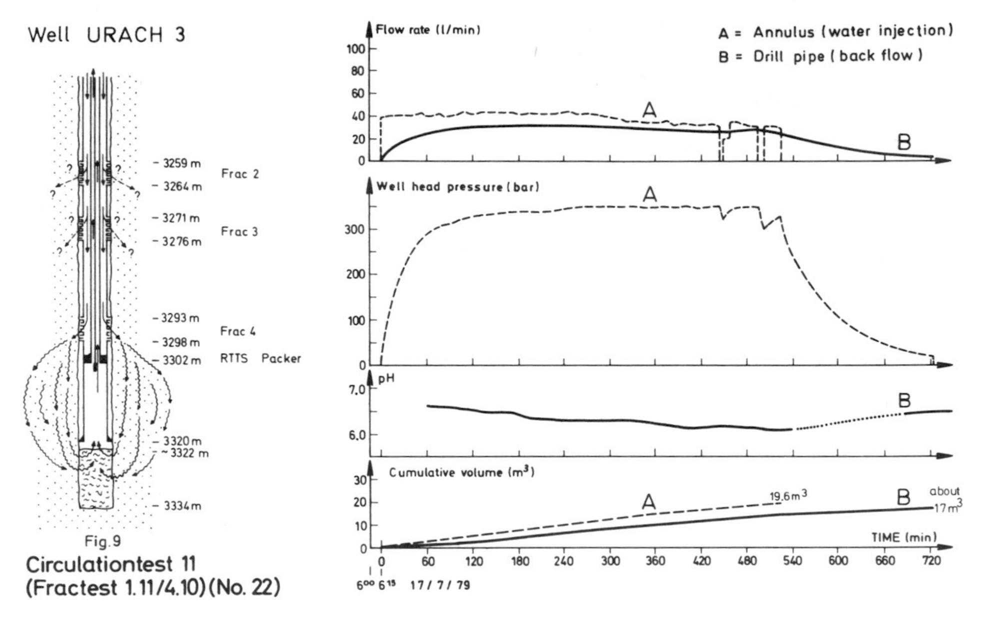

Fig. 9. Circulation test 11 (Frac-Test 1.11/4.10 (No. 22) after NETH & ZOTH, see DIETRICH et al., in press.

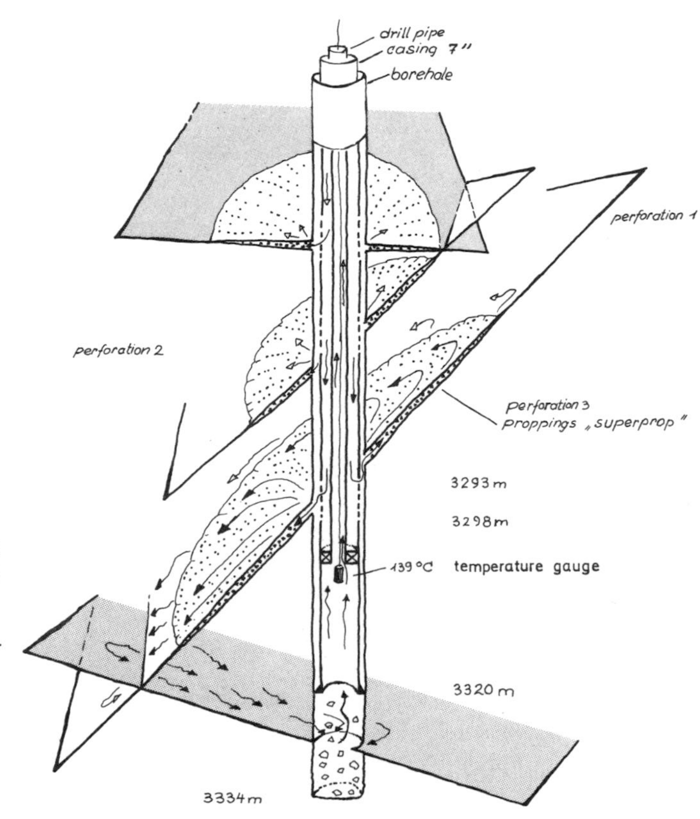

Fig. 10. The assumed circulation loop within the Urach 3 borehole.
1 – Superprop gel
2 – Probe for temperature measurement
3 – 7" casing shoe

Table 2. Injection and Backflow (Production) Measurements in Urach on 18./19.7.1979 (n. m. = not measurable, H = Halliburton Service, S = Schlumberger Service, Pressure = Well Head Pressure).

Injection and backflow measurements with flowmeter and tank levels

Test-Type Test-No. Flow rates:	Evaluation Schlumberger-Measurements	Perforation 1 3259–3264 m	Perforation 2 3271–3276 m	Perforation 3 3293–3298 m	Open Hole 3320–3334 m
Injection-Test 1 (18.7.79) Pressure: about 60–120 bar	l/min	–	–	–	about 216
H = about 207 l/min S = about 216 l/min	%	–	–	–	100 %
Injection-Test 2.1 (19.7.79) Pressure: about 285–355 bar	l/min		about 9.6	about 20.8	about 289.6
H = about 320 l/min S = about 320 l/min	%		about 3 %	about 6.5 %	about 90.5 %
Injection-Test 2.2 (19.7.79) Pressure: about 340 bar	l/min	–	–	4.6	about 225.4
H = about 210 l/min S = about 230 l/min	%	–	–	2 %	about 98 %
Backflow (Production) Test 1.1 (19.7.79) Pressure: about 255–240 bar	l/min		about 3.5	about 11	about 88
H = about 120/130 l/min S = about 102.5 l/min	%		about 3.4 %	about 10.7 %	about 85.9 %
Backflow (Production) Test 1.2 Pressure: about 235–215 bar	l/min	–	n.m.	about 5.1 (?)	about 97.4 (?)
H = about 110/120 l/min S = about 102.5 l/min (?)	%	–	n.m.	about 5 %	about 95 %

into the crystalline rock through the open hole. The same results were measured during the backflow.
- At low flow rates and at small or decreasing surface pressures perforation 3 also shows a measurable portion of the total flow rate.
- At high pressures and high flow rates an injection is detectable also in perforations 1 and 2 (Fig. 11).

The continuous temperature curves, which were also recorded during these flowmeter tests, confirm these conclusions and shows small temperature changes or slight gradient deviations. The perception of individual fractures was feasibly only near perforation 3, since the open borehole absorbed most of the injected water. The extracted core (No. 42, DIETRICH this volume) from approximately the same depth also reveals a very brittle and hydrothermally decomposed rock. Unfortunately, the perforation sections 1 and 2 are not identical with the cored interval.

These results are also of interest for a two-borehole system. These various tests have demonstrated that it is in principle possible to simultaneously inject water into several fractured and reactivated fissure zones lying one above the other. According to the preceding circulation tests, these fissure zones are in part not connected hydraulically (cf. Fig. 4).

However, the great differences in the fractional distribution of the flow rates over the various zones do not appear at all promising for a multiple injection or circulation system according to the hot dry rock concepts for one or more wells.

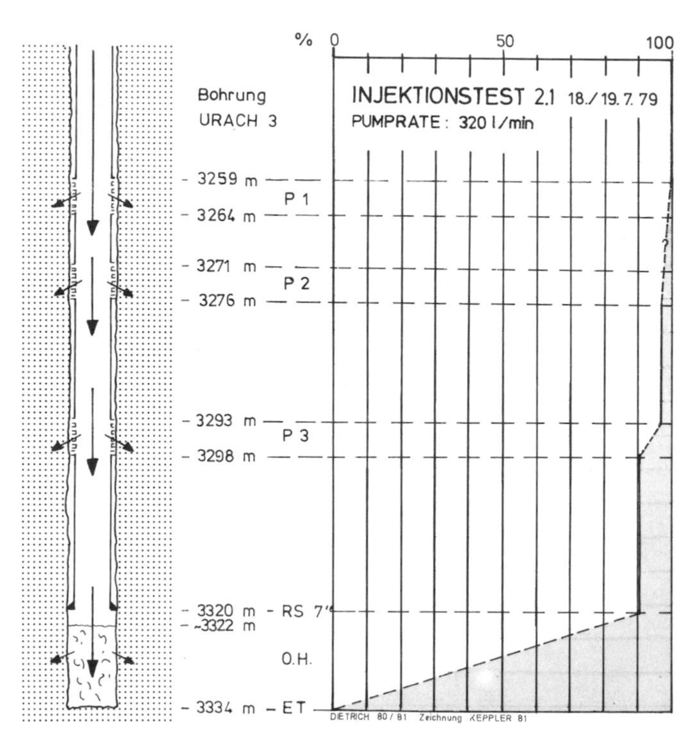

Fig. 11. Flowmeter test after completion of the circulation experiments at a pressure build-up of 285–355 bar and a percentage distribution of the total flow rate of 320 l/min respective to the three perforations of the 7" terminal casing and the open hole (8 1/2").

In appraising the results obtained (Tab. 2, Fig. 11), however, it should be considered that the highest injection rates, and thus the least resistance to entry, coincide with the zone of the open hole. This uncased borehole section is 14 m long and includes several pronounced, partially open, natural fissure zones. In contrast, the three perforated intervals, each about 5 m in length, are only punctual links between the borehole and the rock (about 13 shots per metre). Furthermore, it is not certain everywhere whether similarly pronounced fissure zones are located immediately behind the perforated casing, as in the case of the open hole, or whether those fissure zones which may actually be present coincide exactly with the perforations. Hence a higher frictional resistance during the injection through these perforations cannot be excluded.

The question of whether the frictional closses could in fact be decisively decreased by means of slotted casing sections, but under otherwise identical test conditions in the crystalline, must be clarified by additional experiments. If it were possible to reduce the frictional loss, a larger fraction of the quantity of water pumped could be injected into the zones of the formation situated behind the casing. In this case, an enhancement of the energy extraction according to the hot dry rock concept would result. This will be all the more the case if the most suitable borehole sections with sufficiently pronounced, natural fissure zones are fractured, on the basis of geological and geophysical borehole data, and if in addition further experience with frac treatments is applied (e.g. BRINKMANN et al. 1980, FUHRBERG et al. 1980, JOHN 1980).

The problem of the additional influence which may be exerted by propagating pressure fronts (cf. chapt. 7, Fig. 4 and Fig. 7) on the fractional distribution of the injection rates must be taken into account, moreover, especially with closely adjacent fissure zones which have been subjected to frac treatment (for instance between P 1 and P 2); this question must be considered separately.

12. Summary

Through the open borehole as well as through perforation of the cased borehole a large amount of water, gel or proppings with the aid of gel can be injected when a certain pressure limit is surpassed. The fracture pressures (opening pressures) are higher than those in Los Alamos. Probably, the primary horizontal stress is greater in Urach than in Los Alamos because of the different topographic and tectonic situation. In addition to the primary stress conditions, a secondary field of stress results from the injection of water – i.e. through its spatial consumption; this effect is superimposed upon the primary stress and is manifested in the changed opening pressures.

The permeability increases near the borehole during the injection, but is low with regard to spatial consideration of all the joint passages – with the exception of one or a few principal directions in which the flow moves in advance of the induced field of stress. In general, therefore, a negative skin effect results. Over long time periods the injected water disperses within the joint network of the rock; this process is not yet comleted two years thereafter. The rock can be held open to a limited extent by the use of proppings. A circulation could be accomplished from a propped joint through the rock to the deepest borehole section.

The technical equipment (casing, cementation, packer, drill pipe, gel and proppings) for conventional oil-well drilling satisfied the requirements. The measuring instruments must be improved.

According to the existing results, the one-borehole system is poorly suited for hot-dry-rock heat exploitation for two reasons:

– In the borehole, a nearly complete heat transfer between the drill pipe and the annular occurs (as expected). Insulation is possible, but it consumes space and increases flow resistance.
– In regions with a greater horizontal stress, the induced field of stress, which necessarily occurs during the injection of fluid, will impede the return flow to the borehole because the joints which lead back to the borehole will thereby be compressed.

These two drawbacks do not apply to the two-borehole system. According to the experience gained in Urach, the principle of energy exploitation by means of the hot-dry-rock system is found to be absolutely sound. The means of realizing it depends – as in crude oil recovery – on the respective geological and tectonic situation.

13. Perspectives (SCHÄDEL)

The work in Urach represents an example of hot-dry-rock experiments with a horizontal stress greater than in Los Alamos. Since a depth of more than five kilometres is necessary for the actual energy exploitation, a higher horizontal stress and a more complicated field of tension must be expected. In future experiments, therefore, more consideration should be given to:
– elasticity and plasticity of the rock, cleavage, deformation of joints ("balooning effect"), development of secondary stress;
– water in fractures at high pressures and high temperatures, uplift pressure at great depths;
– drilling technique and fracturing technique under great horizontal stress; and the
– development of theoretical concepts for the field of stress and a suitable measuring technique.

References

Aamodt, R.L., Albright, J.N. Batchelor, A.E., Fisher, H.N., Murphy, H.D., Potter, R.M., Spence, R.W. & Tester J.W. (1978): Reservoir Geometry Modeling. – LA – 7109 PR, 193–221, Los Alamos Scientific Laboratory.

Aamodt, R.L., Blatz, L.A., Brown, D.W., Counsil, J., Fisher, H.N., Grigsby, C.O., Holley, C.E., jr., Lawton, R.G., Murphy, H.D., Potter, R.M. & Tester, J.W. (1978): Field Studies of Reservoir Properties. – LA – 7109 PR, 165–192, Los Alamos Scientific Laboratory.

Brinkmann, F., Fuhrberg, H.-D. & Schöber, K. (1980): Frac-Behandlungen in tiefen, geringpermeablen Gaslagerstätten – derzeitiger Stand und weitere Aussichten. – Erdöl-Erdgas., **96**: 37–44.

Camponuovo, G.F., Freddi, A. & Borsetto, M. (1980): Hydraulic Fracturing of Hot Dry Rocks. Tridimensional Studies of Cracks Propagation and Interaction by Photoelastic Methods. – In: Strub, A.S. & Ungemach, P. (Eds.): Advances in European Geothermal Research. International Seminar on the Results of EC Geothermal Energy Research, 2 d, Strasbourg, 1980, p. 947–956, D. Reidel Publ. Comp., Dordrecht/Boston/London.

Dietrich, H.-G., Haenel, R., Neth, G., Schädel, K. & Zoth, G. (1980): Deep Investigation of the Geothermal Anomaly of Urach. – In: Strub, A.S. & Ungemach, P. (Eds.): Advances in European Geothermal Research, 2nd, Strasbourg 1980, p. 253–266, D. Reidel Publ. Comp., Dordrecht/Boston/London.

Dietrich, H.-G. & Schädel, K. (1978): Untersuchung der geothermischen Anomalie in Urach auf eine mögliche Nutzung durch eine Untersuchungsbohrung bis tief ins Kristallin (ET 4023 B). – Progr. Energieforsch. u. Energietechn. 1977–1980, Statusrep. 1978 – Geotechnik und Lagerstätten, 1: 79–85, 3 Abb.; Jülich (Projektleitung Energieforsch. (PLE), KFA Jülich).
Ernst, P.L. (1977): A Hydraulic Fracturing Technique for Dry Hot Rock Experiments in a Single Borehole. – SPE 6897, 52nd Annual Fall Conf. SPE, Colorado.
– (1978): Frac-Studie zur Gewinnung geothermischer Energie aus einem Bohrloch. – Progr. Energieforsch. u. Energietech. 1977–1980, Statusrep. 1978 – Geotechnik und Lagerstätten, 1: 101–109, Jülich (Projektleitung Energieforsch. (PLE), KFA Jülich).
Fuhrberg, H.-D., Brinkmann, F. & Schöber, K. (1980): Erfahrungen mit Fracbehandlungen in tiefen, geringpermeablen Gaslagerstätten. – In: Strub, A.S. & Ungemach, P. (Eds.): Advances in European Geothermal Research. International Seminar on the Results of EC Geothermal Energy Research, 2 d, Strasbourg, 1980, 994–1008, D. Reidel Publ. Comp., Dordrecht/Boston/London.
John, P.-W. (1980): Stimulation geringpermeabler tiefer Gaslagerstättten, – Programm Energieforsch. und Energietech. 1977–1980. Statusrep. 1980 – Geotechnik und Lagerstätten, 2, 919–942, Jülich (Projektleitung Energieforsch. (PLE), KFA Jülich).
Perkins, T.K. & Kern, L.R. (1970): Width of Hydraulic Fractures, Well Completions. – Petroleum Transactions Reprint S. No 5.
Kappelmeyer, O., Haenel, R. & Rummel, F. (1979): HDR-activities in the F.R. Germany. – Hot Dry Rock Geothermal Information Conference, Santa Fe, Sept. 18, 1979.
King Hubert, M. & Willis, D.G. (1970): Mechanics of Hydraulic Fracturing, Well Completions. – Petroleum Transactions Reprint S. No 5.
Raleigh, C.B., Healy, J.H. & Bredhoeft, J.D. (1972): Faulting and Crustal Stress at Rangely. – Colorado geophysical monograph 16, Washington, D.C. 1972.
Rummel, F., Alheid, H., Winter, R. & Wöhrl, Th. (1979): Gesteinsphysikalische Daten, Bohrung Urach 3. – Zwischenbericht 1979, RUB, Bochum.
Schädel, K. (1980): Ergebnisse der Frac-Versuche in der Erdwärme-Forschungsbohrung Urach III (ET 4023 C). – Programm Energieforsch. und Energietechn. 1977–1980. Statusrep. 1980 – Geotechnik und Lagerstätten, 1, 171–189, Jülich (Projektleitung Energieforsch. (PLE), KFA Jülich).
Schädel, K. & Schweizer, R. (1979): Ingenieurgeologisches Gutachten zum Herstellen von Wärmeaustauschflächen in der Bohrung Urach 3 (Landkreis Reutlingen, TK 25, Bl. 7422 Dettingen/Erms). – Gutachten, III – 1099/79, Geologisches Landesamt Baden-Württemberg, Freiburg.

The Urach Geothermal Project, p. 345–350;
Schweizerbart'sche Verlagsbuchhandlung, Stuttgart, 1982

Pressure Build-up Curves of the Hydraulic Fracturing Experiments in the Borehole Urach 3

R. SCHWEIZER

with 7 tables

Abstract: The hydraulic fracturing tests of the borehole Urach 3 are evaluated with a modification of the straight-line method of HORNER (1951). This allows the determination of transmissibility, skin-factor and flow capacity. The results were interpreted accordingly for the several perforations.

1. Introduction

In the following, analytical equations which are valid for a porous water-bearing layer (aquifer) will be presented. These equations will then be applied to an artificially created fracture as used in the hot dry rock technology.

2. Analytical Methods

2.1 Determination of Transmissibility

Beginning with the equation for the pressure build-up in a borehole (HORNER 1951), we have the following boundary conditions:

- infinite extent of the reservoir
- homogenity and isotropy
- constant fluid compressibility
- constant reservoir parameters

The equation for the case of fluid injection ($q > 0$) is:

$$p_{wf} = p_i - \frac{q \cdot \mu}{4 \pi k h} \cdot E_i \left(- \frac{\phi \mu c r_w^2}{4 k t} \right), \tag{1}$$

where:
p_{wf} = flow pressure at the borehole bottom during the production phase
p_i = initial pressure, reservoir pressure
ϕ = porosity
μ = fluid viscosity
c = fluid compressibility

r_w = effective radius
k = permeability, 1 cm² △ 100 μD
h = reservoir thickness
q = flow rate
Ei = exponential integral
γ = fluid density

Upon injection of a fluid, the pressure relation is always $p_{wf} \geqslant p_i$.

In the case of long time intervals, i.e. $t \geqslant \frac{12.5\, \phi\, \mu\, c\, r_w^2}{k}$ equation (1) can be simplified (linearized) as follows:

$$(2) \qquad p_{wf} = p_i - \frac{9 \cdot \mu}{4\, \pi\, k\, h} \cdot \ln \left(\frac{\gamma\, \phi\, \mu\, c\, v_w^2}{4\, kt} \right),$$

$$\triangle p = p_{wf} - p_i.$$

In the relationship of $\triangle p$ to $\log_{10} t$, (2) constitutes a linear equation.

With the slope of these straight lines, one can determine the transmissibility k · h:

$$(3) \qquad kh = \frac{0.1832\, \mu\, q}{m},$$

whereby m is the slope of the straight lines. If △ p is related to $\log_{10}$ (t = △ t)/ △t for the shut-in periods, the transmissibility k · h can likewise be determined with equation (3), where:

t = mean pressure build-up time, and
△t = time following the shut-in.

2.2 Determination of the Skin-Factor

According to VAN EVERDINGEN (1953) the skin effect is defined as a constant s which relates the pressure drop $\triangle p_{skin}$ in the skin zone to a dimensionless flow rate. The skin zone is a region near the borehole which has a decreased or an increased permeability in comparison to that of the rocks outside this zone. The associated pressure drop is given by:

$$(4) \qquad \triangle p_{skin} = s \left(\frac{q\, \mu}{2\, \pi\, k\, h} \right).$$

When the skin effect is present, a decrease in the pressure in equation (2) results, thus:

$$(5) \qquad p_{wf} = p_i - \frac{q\, \mu}{4\, \pi\, k\, h} \left(\ln \frac{\gamma\, \phi\, \mu\, c\, r_w^2}{4\, k\, t} - 2\, s \right).$$

In the ideal case, a pressure decrease of $\triangle p_{skin}$ should result immediately following the shut-in.

After subtraction of equation (5) from:

$$p_{ws} = p_i + \frac{q\, \mu}{4\, \pi\, k\, h} \ln \left(\frac{t + \delta t}{\triangle t} \right)$$

and after several transformations, the following equation results for the skin effect:

(6) $$s = 1.151293 \left[-\log_{10} \frac{k}{\phi \mu c r_w^2} - \frac{p_{ws} - p_{wf}}{m} - \log_{10} \delta t - 0.351378 \right]$$

where p_{ws} = pressure following the shut-in at time $\triangle t$.

2.3 Determination of the flow Capacity

The flow capacity is defined as the relation of the actual output index to the output index without skin effect (s = 0) (MATHEWS & RUSSEL 1967):

(7) $$Fl = \frac{p^* - p_{wf} - \delta p_{skin}}{p^* - p_{wf}}$$

where p* = pressure following infinitely long shut-in time:

$$\frac{t + \triangle t}{\triangle t} \rightarrow 1 \text{ (extrapolated)}$$

$$\triangle p_{skin} = 0.87 \text{ s m.}$$

3. Results

For the determination of the transmissibility of a fracture, the permeability must first be from pumping tests with the use of equation (3). From both values the reservoir thickness h, which now represents the "theoretical lenght" of a fracture can be calculated.

The tests and perforations mentioned in the following are listed by DIETRICH (this volume). A separate treatment of each fracture interval is necessary for the interpretation of the results.

3.1 Open-Hole Zone

During the frac procedure the transmissibility $k \cdot h$ increased to 80 μD m (frac 1.1) and to 549 μD m (frac 1.2), respectively (Tab. 1).

When one compares both values of the shut-in in Tab. 5 (40 μD m and 42 μD m) with their respective skin-factors (– 3.02 and – 2.97), the following conclusion can be deduced: The permeability hardly increased further as a result of the second frac operation. The rock reacts elastically that is, after the frac tests the initially increased permeability returns again to its original value. The initial value of the transmissibility $k \cdot h$ is estimated to be 35 μD m; during pumping tests k was determined to be 3 μD, which results in about 45 μD m for $k \cdot h$, since $h \approx 15$ m.

3.2 Perforation 1

The determination of $k \cdot h$ was not possible here in frac 2.31. The maximal value for $k \cdot h$ was attained in frac 2.1 (1st frac) with 216 μD m (Tab. 2).

The values of the shut-in phases are between 7 and 15 μD m for $k \cdot h$, presumably with one aberration at 35 μD m (frac 2.1) (Tab. 6). The behaviour of $k \cdot h$ is contrary to that of s and FL in both fracs. This contradiction can be explained only by a possible increase in h (together with considerably increased flow rates).

3.3 Perforation 2

As in perforation 1, k · h could not be determined in all of the pressure build-up curves (fluctuating flow rates). A maximum of 366 μD m was reached for k · h (Tab. 3). During the first three stages of test 3.1, a reduction in the k · h-values (Tab. 7) from 34 μD m to 16 μD m was observed; subsequently an increase to 40 μD m after two further stages occurred. This behaviour was confirmed by that of s and FL. In the tests 3.2 and 3.3, an increasing trend of k · h from 7 to 13 μD m is present, and is also portially confirmed by s and FL. The peculiarity is indeed the strong reduction of k · h at the end of test 3.1 from 40 μD m to value of 7 μD m. Because of the large amount of water which was previously injected, additional resistance could have occurred, or the rock sealed off the flow channels again after a while, as a result of elasticity.

3.4 Perforation 3

The highest value for k · h during the pressure buildup was attained in perforation 3 with 747 μD m (Tab. 4). Values for k · h, however, could not be determined in half of the pressure build-up stages. In general, one finds an increase in the k · h-values (shut-in phase), which is not confirmed, however, by the behaviour of s and FL.

4. Summary

In practically all the frac tests, the k · h-values observed during the shut-in phase returned again to the order of magnitude of the initial vlues. Thus, one can assume that the rock is elastic, that is, the rock recloses, and thus the flow resistance is increased. A real improvement in the permeability of the rock due to the hydraulic fracs can be presumed only in perforation 2, although a strong elastic reaction of the rock was observed here as well.

Table 1.
Values of k · h in the open-hole zone.

	k · h (μD m)	Flow-rate (m^3/s)
frac 1.1	80	0.0050
frac 1.1	52	0.0075
shut in	40	–
frac 1.2	353	0.0075
frac 1.2	549	0.0100
shut in	42	–

Table 2.
Values of k · h in perforation 1.

	k · h (μD m)	Flow-rate (m^3/s)
frac 2.1	216	0.0098
shut in	7	–
frac 2.1	56	0.0110
shut in	35	–
frac 2.31	?	0.0123
shut in	8	–
frac 2.32	37	0.0123
shut in	10	–
frac 2.32	75	0.0145 (prop)
shut in	10	–
frac 2.32	69	0.0166
shut in	15	–

Table 3.
Values of k · h in perforation 2.

	k · h (μD m)	Flow-rate (m^3/s)
frac 3.1	30	0.0068
shut in	33	–
frac 3.1	123	0.0120
shut in	28	–
frac 3.1	67	0.0150
shut in	16	–
frac 3.1	?	0.0173
shut in	32	–
frac 3.1	42	0.0124
shut in	40	–
frac 3.2	?	0.0167
shut in	7	–
frac 3.2	?	0.0183
shut in	7	–
frac 3.3	47	0.00975 (superprop)
frac 3.3	366	0.0111 (superprop)
shut in	13	–

Table 4.
Values of k · h in perforation 3.

	k · h (μD m)	Flow-rate (m^3/s)
frac 4.1	64	0.0015
shut in	3	–
frac 4.1	64	0.0015
frac 4.1	747	0.0113
shut in	21	–
frac 4.1	?	0.0125
frac 4.1	586	0.0133
frac 4.1	?	0.0138
frac 4.1	?	0.0133
frac 4.1	?	0.0106
shut in	44	–

Table 5. Shut in period in the open-hole zone.

	Pumped volume (m^3)	k · h (μD m)	Skin-factor	Flow-capacity
frac 1.1	10.8	42	– 3.02	2.11
frac 1.2	25.4	40	– 2.97	1.97

Table 6. Shut in period in the perforation 1.

	Pumped volume (m^3)	k · h (μD m)	Skin-factor	Flow-capacity
frac 2.1	11.4	7	– 4.6	7.0
frac 2.1	59.8	35	– 2.0	1.5
frac 2.3	12.1	8	– 4.4	5.1
frac 2.3	10.0	9	– 4.3	4.9
frac 2.3	13.3	10	– 4.1	4.0
frac 2.3	146.2	15	– 3.7	2.2

Table 7. Shut in period in the perforation 2.

	Pumped volume (m^3)	k · h (μD m)	Skin-factor	Flow-capacity
frac 3.1	19.9	34	– 4.4	2.9
frac 3.1	5.1	28	– 4.1	2.8
frac 3.1	13.7	16	– 3.9	2.9
frac 3.1	20.3	32	– 3.2	2.1
frac 3.1	103.2	40	– 5.4	3.0
frac 3.2	4.2	7	– 3.5	5.3
frac 3.3	13.8	7	– 4.7	7.2
frac 3.3	37.9	13	– 5.0	4.8

References

Horner, D.R. (1951): Pressure Build-Up in Wells. – Proc., Third World Pet. Cong., E.J. Brill, Leiden II, 503 p.

Matthews, O.S. & Russel, D.G. (1967): Pressure Buildup and Flow Texts in Wells. – Society of Petroleum Engineers of AIME Monograph Series Dallas Texas **1**, 4: 103.

Van Everdingen A.F. (1953): The Skin effect and its Influence on the Productive Capacity of a well. – Trans., AIME **198**: 171–176.

The Urach Geothermal Project, p. 351–353;
Schweizerbart'sche Verlagsbuchhandlung, Stuttgart, 1982

Analytical Model Calculation on Heat Exchange in a Fracture

H. RODEMANN

with 2 figures

The exploitation of geothermal energy by means of fractures in hot dry rock can be simulated by model calculations.

Various types of numerical model calculations have already been carried out by the finite difference method (LAWTON et al. 1976) as well as by the finite element method (LAWTON 1974). The numerical modelling allows the inclusion of most of the important physical processes, such as heat conduction within the rock, heat exchange between the fluid and the rock, water motion including buoyancy, thermal variation of viscosity, and thermal cracking. On the other hand complicated programs are necessary; thus simplifications (e. g. two-dimensional instead of three-dimensional calculations) are required in order to reduce the amount of computer storage space needed.

Analytical calculations offer the advantage of simpler algorithms which can be programmed on desk computers and allow a quick overview of the influence of the different parameters, but they require more simplifications than numerical calculations and are thus less realistic. The need for such calculations is demonstrated by the use of the rectangle-fracture model (MILORA & TESTER 1976) considering heat exchange rates in fractures and using the solutions of LAUWERIER (1955) and BODVARSSON (1969).

The analytical calculations used up to now can be adapted to a more realistic geometry as follows: A fracture with the shape of a flat circular cylinder and a free choice of the positions of the injection and extraction boreholes is assumed. The equations used are (RODEMANN 1978, 1979, 1981).

– continuity equation $$\operatorname{div}(\rho_w \vec{v}) = \frac{\partial \rho_w}{\partial t}$$

– Darcy equation $$\vec{v} = -\frac{k}{\eta}\operatorname{grad} p$$

– heat conduction equation $$\lambda_r \operatorname{div}\operatorname{grad} T = c_r \rho_r \frac{\partial T}{\partial t}$$

– heat conduction equation including a velocity term $$\lambda_r \operatorname{div}\operatorname{grad} T = c_w \rho_w \left(\frac{\partial T}{\partial t} + \vec{v} \cdot \operatorname{grad} T\right)$$

with ρ = density, $\vec{v}$ = velocity, t = time, k = permeability, η = viscosity, p = pressure, λ = thermal conductivity, T = temperature, c = specific heat capacity, w = water, and r = rock.

The potential theory approach to treating the fluid flow in the fracture as one-dimensional resembles that of (GRINGARTEN & SAUTY 1975 a, 1975 b) in the case of an aquifer with a borehole doublet or multiplet. The main advantage of the calculation is the inclusion of the influence of borehole positions on the performance of the fracture, especially the reduction of the thermal lifetime by reduction of the distance between the boreholes.

A comparison with calculations carried out with a finite difference computer program furnished by the LASL-HDR group (McFARLAND 1975) showed a good agreement when an identical geometrical and physical situation was considered. Thus the algorithms used were mutually confirmed.

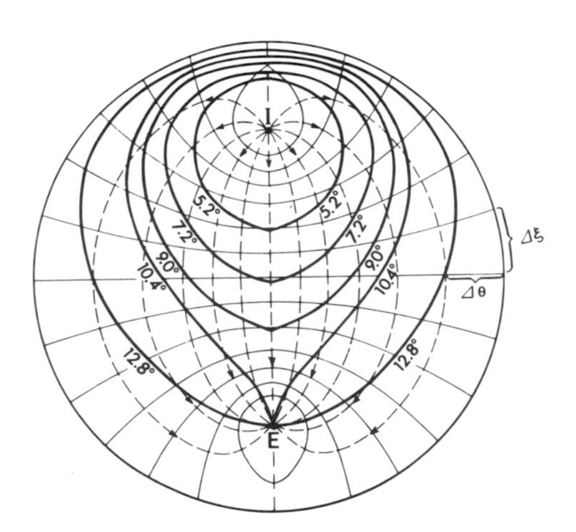

Fig. 1. The distribution of pressure (–), fluid flow (- - -) and temperature (–) for conditions as planned for the Falkenberg hot dry rock project after one week of injection. The following model parameters have been used (I = point of injection, E = point of extraction):

injection temperature	: 3 °C
rock temperature	: 13 °C
fracture radius	: 50 m
half distance between boreholes	: 31.25 m
flow rate	: 3 l s^{-1}
water viscosity	: 1.4 · 10^{-3} Pa · s (8 °C)
rock density (granite)	: 2500 kg m^{-3}
specific heat capacity of rock	: 836 J kg^{-1} K^{-1}
thermal conductivity	: 2.93 W m^{-1} K^{-1}
specific heat capacity of water	: 4180 J kg^{-1} K^{-1}
density of water	: 10^3 kg m^{-3}
fracture width	: 10^{-3} m.
$\triangle \xi$	: 0.04 bar
$\triangle \theta$	: 0.25 l s^{-1}

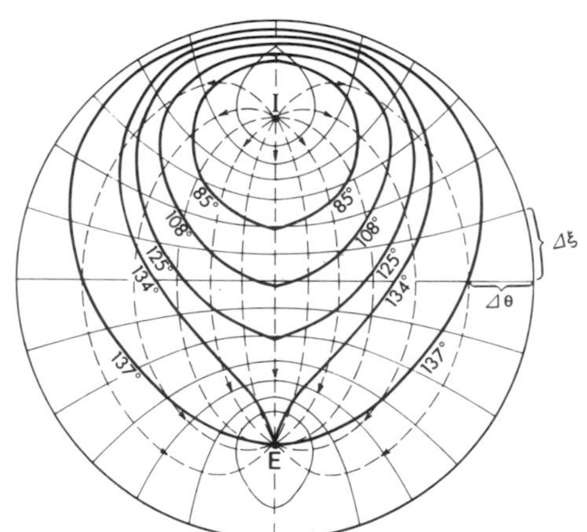

Fig. 2. The distribution of pressure (–), fluid flow (- - -) and temperature (–) for conditions as given in the Urach hot dry rock project after three days of injection. The following model parameters have been used (I = point of injection, E = point of extraction):

injection temperature	: 60 °C
rock temperature	: 140 °C
fracture radius	: 50 m
half distance between boreholes	: 31.25 m
flow rate	: 3 l s^{-1}
water viscosity	: 0.3 · 10^{-3} Pa · s (100 °C
rock density (gneiss)	: 2700 kg m^{-3}
specific heat capacity of rock	: 840 J kg^{-1} K^{-1}
thermal conductivity	: 2.7 W m^{-1} K^{-1}
specific heat capacity of water	: 4180 J kg^{-1} K^{-1}
density of water	: 10^3 kg m^{-3}
fracture width	: 10^{-3} m
$\triangle \xi$	: 0.01 bar
$\triangle \theta$	: 0.25 l s^{-1}

The main deviations between finite difference and analytical calculations occurred when:

- the more realistic elliptical dependence of the fracture width on distance from the fracture centre was introduced, and instead of a constant width
- allowance was made for buoyancy in the finite difference model.

In the models considered the neglect of the fracture width variation results in cooling rates (thermal drawdown) of the extracted water which are too slow by a factor 1.5 to 2. This factor decreases with growing ratio of borehole distance of radius. This behaviour results from the neglected concentration of the water flow to the fracture regions with the maximal width.

If strong in relation to the drag forces buoyancy forces can have a dominating influence on the flow pattern in the fracture and consequently on the extraction temperature drawdown. The effect may reduce the drawdown, especially in the case of a thermal short circuit due to a small distance between the boreholes, but the inverse effect is also possible. This cannot be included in the analytical calculations.

In Figs. 1 and 2 the isobars, the flowlines and the isotherms for the Falkenberg hot dry rock project as well as for the Urach hot dry rock project are shown.

References

Bodvarsson, G. (1969): On the temperature of water flowing through fractures. – J. G.R., **74**: 1987–1992.

Gringarten, A. & Sauty, J.P. (1975 a): Simulation des transferts de chaleur dans les aquifères. – Bull. B.R.G.M. (2) **III**: 25–34.

– (1975 b): A theoretical study of heat extraction from aquifers with uniform regional flow. – J.G.R., **80**: 4956–4962.

Lawton, R.G. (1974): The AYER heat conduction computer program. – Report LA–5613–MS, LASl, Los Alamos, USA.

Lawton, R.G., Murphy, H.D., Tester, J.W. & McFarland, R.D. (1976): Geothermal reservoir modelling. – In: Blair, A.G., Tester, J.W. & Mortensen J.J. (Ed.) – LASL, Hot Dry Rock Geothermal Project, progress report LA–6525–PR, LASL, Los Alamos, USA.

Lawerier, H.A. (1955): The transport of heat in an oil layer caused by the injection of hot fluid. – Appl. Sci. Res., **A 5**: 145–150.

McFarland, R.D. (1975): Geothermal reservoir models – crack plane model. – Report LA–5947–MS, LASL, Los Alamos, USA.

Milora, S.L. & Tester, J.W. (1976): Geothermal energy as a source of electric power – Thermodynamic and economic design criteria. – MIT Press.

Rodemann, H. (1978): Ein Besuch bei der Hot-Dry-Rock Arbeitsgruppe Los Alamos, USA, 5.–6.10.1978. – Report, NLfB Hannover, Archive No. 81653.

– (1979): Modellrechnungen zum Wärmeaustausch in einem Frac. – Report, NLfB Hannover, Archive No. 81990.

– (1981): Theoretische und experimentelle thermische Untersuchungen an einem künstlich erzeugten Frac in Falkenberg. – Report, NLfB Hannover, Archive No. 85725.

The Urach Geothermal Project, p. 355–360;
Schweizerbart'sche Verlagsbuchhandlung, Stuttgart, 1982

Energy Extraction Study for the Urach Borehole

R. HAENEL

with 6 figures

Abstract: The increase in temperature of water pumped through the frac in the Urach borehole was measured at a depth of 3300 m, where the water was extracted from the frac. The measured values were simulated by computer. The computer simulation shows that, in spite of uninsulated tubing, a temperature of 50–60 °C is to be expected at the well head after three weeks. For a borehole with insulated tubing, a thermal power output of 350–450 kW over a 30-year period has been estimated. No indication of fracture size can be expected from a huff-puff experiment.

Introduction

Water was circulated for less than a day through a small frac at a depth of 3300 m, with injection and extraction from a single borehole.

The objective was to simulate one of these circulation tests. A threeweek circulation test is planned for the future. The temperature expected at the well head during this experiment is also to be calculated in advance. The possibilities for utilizing the Urach borehole for heating purposes are being considered. Therefore, the maximum heat extraction that can be maintained over a long period of time was estimated.

Finally, the question of whether information can be obtained from a huff-puff experiment about the size of an artificial fracture was investigated.

1. Simulation of a Short Circulation Test in the Urach Borehole

1.1 Basic Equations

The amount of heat that can be extracted can be calculated by the differential equation for heat conduction for non-steady-state conditions. This equation includes a term for water flow:

$$\rho\, c \frac{\partial T}{\partial t} = \operatorname{div}(k \operatorname{grad} T) + \vec{v} \operatorname{grad} T.$$

Rewritten in cylindrical coordinates, the equation takes the following form:

$$\frac{\partial T}{\partial t} = \kappa \left(\frac{\partial^2 T}{\partial r^2} + \frac{1}{r}\frac{\partial T}{\partial r} + \frac{\partial^2 T}{\partial z^2} \right) + \frac{1}{\rho c}\left(\frac{\partial k}{\partial r}\frac{\partial T}{\partial r} + \frac{\partial k}{\partial z}\frac{\partial T}{\partial z} \right) + \frac{v_z}{\rho c}\frac{\partial T}{\partial z}\,.$$

So the finite difference method can be used, the differential quotients have to be replaced by difference quotients:

$$\frac{T_{i,j}^{n+1} - T_{i,j}}{\Delta t} = \kappa \left(\frac{T_{i+1,j}^{n+1} + T_{i-1,j}^{n+1} - 2T_{i,j}^{n+1}}{\Delta r^2} + \frac{1}{r_o} \frac{T_{i+1,j}^{n+1} - T_{i-1,j}^{n+1}}{2\,\Delta r} + \frac{T_{i,j+1}^{n} + T_{i,j-1}^{n} - 2T_{i,j}^{n}}{\Delta z^2} \right)$$

$$+ \frac{1}{\rho c} \left(\frac{k_{i+1,j} - k_{i-1,j}}{2\,\Delta r} \right) \left(\frac{T_{i+1,j}^{n+1} - T_{i-1,j}^{n+1}}{2\,\Delta r} \right) + \frac{1}{\rho c} \left(\frac{k_{i,j+1} - k_{i,j-1}}{2\,\Delta z} \right) \left(\frac{T_{i,j+1}^{n} - T_{i,j-1}^{n}}{2\,\Delta z} \right) + \frac{v_z}{\rho c}$$

Where T = temperature, t = time, ρ = density, c = specific heat capacity, k = thermal conductivity, κ = thermal diffusivity, z = depth, r = radius, grad T = temperature gradient, v = velocity, and n = time increment.

The equation can be solved by the method of MINEAR & TOKSOEZ (1970), as well as that of PEACEMAN & RACHFORD (1955).

For the calculations described below, a borehore model is used as given in Fig. 1. It should be noticed that the cross-sectional area of the tubing is one quarter that of the annulus. Thus, the water velocity in the tubing is four times greater than in the annulus.

A sketch of the artificial fracture at the bottom of the borehole is shown in Fig. 2.

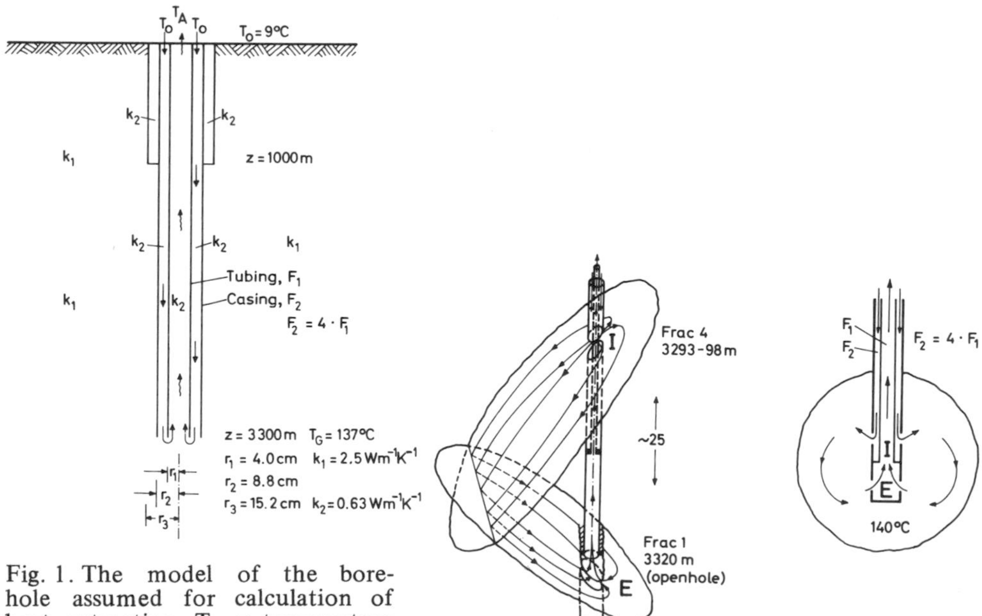

Fig. 1. The model of the borehole assumed for calculation of heat extraction. T_o = temperature of injected water, T_A = temperature of the rock at the borehole bottom, k_1 = thermal conductivity of the rock, k_2 = thermal conductivity of the water, r = radius, F = cross-section area of the tubing and casing.

Fig. 2. The artificial fracture at the bottom of the Urach 3 borehole (left) and the model assumed for calculation (right). In both cases the distance IE ≈ 25 m, I = injection point, E = extraction point. The radius of the fracture is assumed to be r = 20 m.

1.2 Field Test

Water was circulated on July 13, 1979, through a fracture near the bottom of the borehole as shown in Fig. 2.

The temperatures of injected and extracted water were measured at the well head and at the bottom of the tubing. A graph of the bottom hole temperatures is shown in Fig. 3. The drop in temperature is due to the circulation of cold water through the well itself before the circulation test was begun. The water injection and extraction rates are also given in the figure. Both values varied with time and only a part of the injected water could be recovered during the circulation test. It should be further pointed out that before this test, there was another set of tests that disturbed the temperature field. Therefore, the temperature at 3305 m depth was only 139.8 °C on the day of the test instead of the normal 142.2 °C.

The computer model allows only constant flow rates. The best approximations for the injection and extraction rates were determined to be 1 and 0.5 l/s, respectively. The calculation was carried out for the circulation system described above, using the theory of heat extraction from a fracture presented by RODEMANN in this volume. The calculated temperatures are higher than the measured ones (see Fig. 3). It must be kept in mind that part of the injected water is adsorbed by the rock. This influence on the heat balance means that the simulation method must yield values that are too high.

A second calculation was made in which the influence of the fracture was not included. The best fit was for a flow rate of about 0.8 l/s. The temperatures are lower than the measured temperatures due to the lack of heat from the fracture.

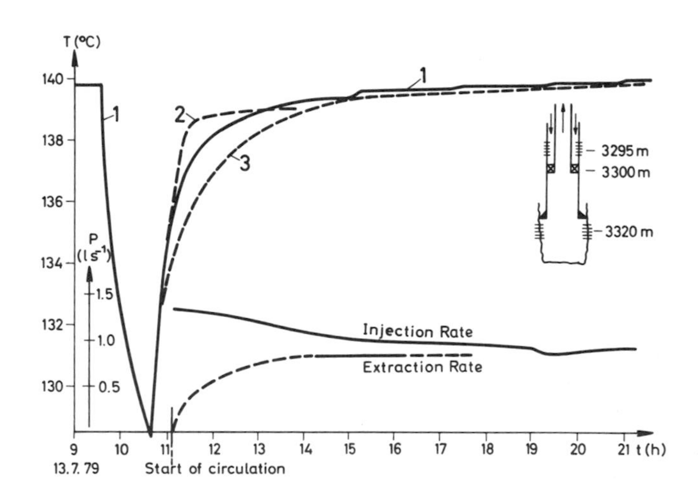

Fig. 3. The temperature at about 3300 m depth measured at the end of the tubing (curve 1). The injection and extraction rates are also presented in $l\,s^{-1}$. The calculated temperature at 3300 m depth is given by curve 2 for a mean extraction rate of $0.5\,l\,s^{-1}$ inclusive the fracture, and by curve 3 $0.8\,l\,s^{-1}$ without the fracture; in both cases the water is injected through the annulus and extracted through the tubing.

2. Expected Temperature at Well Head: Short-term Circulation

A three-week duration is planned for the next circulation test. The results of model calculations for this test are presented in Figs. 4 and 5. Well head temperatures were calculated also for opposite flow directions and different flow rates (1 l/s, 5 l/s, and 10 l/s). Uninsulated tubing was assumed. The calculated temperatures are relatively high when injection is through the annulus. The reason for this is the greater flow rate in the tubing, resulting from its smaller cross-sectional area relative to that in the annulus. Heat loss into the annulus from the tubing is, therefore, small. The results show that the influence of the small frac (see Fig. 2) on the extracted water is less than 1 °C. Therefore, heat extraction is almost exclusively from the borehole.

The measured well-head temperatures from the circulation test on July 13, 1979, are also shown in Fig. 4, but a comparison is not possible due to the disturbed temperature field at that time.

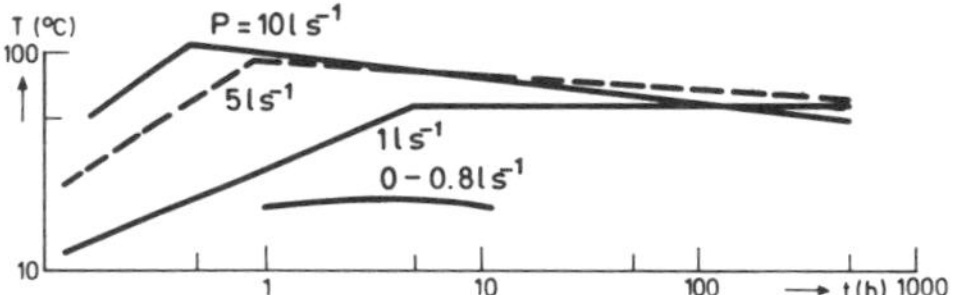

Fig. 4. The temperature at the well head with injection through annulus and fracture and extraction through the tubing.

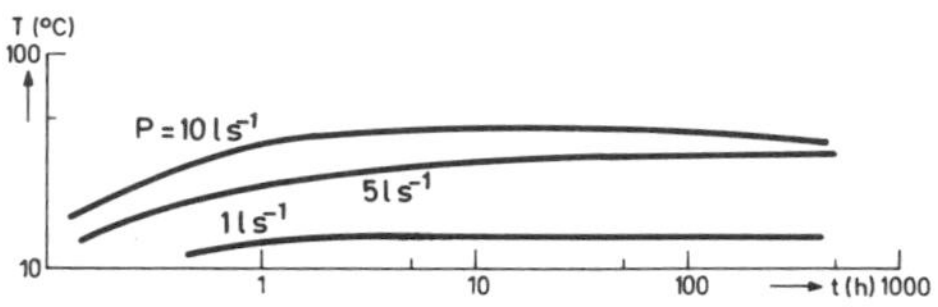

Fig. 5. The temperature at the well head with injection through the tubing and fracture and extraction through the annulus.

3. Expected Temperature at Well Head: Long-term Circulation

The possibility of extracting heat through insulated tubing over a long period of time has been under consideration. Because the influence of the small fracture used in the first circulation test can be neglected, only heat extraction from the borehole will be considered in the following calculation.

The method used in the computer program for the calculations discussed in the previous section does not allow long-term calculations due to instabilities resulting from the iterations. On the other hand, heat loss through the insulated tubing into the annulus can be expected to be small. Therefore, the calculation can be done with an equation given by CARSLAW & JAEGER (1959). This equation assumes a constant temperature difference between the rock and the water in the borehole:

$$q = \frac{2k \Delta T}{r} \left(\frac{1}{\ln(4S) - 1.15} - \frac{0.577}{(\ln(4S) - 1.115)^2} \right)$$

where the dimensions of q are watt per sq. meter. Rewriting this equation to obtain a value for the total Urach borehole surface area ($A = 2\pi rz$):

$$Q = 4\pi k z \Delta T \left(\frac{1}{\ln(4S) - 1.15} - \frac{0.577}{(\ln(4S) - 1.15)^2} \right)$$

where $S = \kappa t/r$, $\kappa = k/\rho c$ = thermal diffusivity, k = thermal conductivity, ρ = density, c = specific heat capacity, t = time, r = borehole radius, z = depth of borehole, $\triangle T$ = rock temperature – water temperature.

The amount of water necessary to transport this amount of heat is given by:

$$P = \frac{Q}{c_w \triangle T_w}$$

where P = l/s, c = specific heat capacity of water, and $\triangle T_w$ = temperature of extracted water – temperature of injected water.

As can be seen in Fig. 4, the maximum temperature that can be expected for the extracted water is about 60 °C. More reasonable temperatures would be 40–50 °C. The following values were assumed for the calculation of P:

$k = 2.7\ Wm^{-1}k^{-1}$, $\kappa = 1.2 \cdot 10^{-6}\ m^2\ s^{-1}$, r = 0.089 m (for a 7" casing), z = 3330 m, $c = 4.184 \cdot 10^3\ J\ kg^{-1}K^{-1}$, T_o = 9 °C (injected water temperature),
and T = 60, 50, 40 °C
$\triangle T_w$ = 51, 41, 31 °C
$\triangle T$ = 40, 45, 50 °C.

The results are shown in Fig. 6. With heat extraction only from the borehole, between 355 and 444 kW can be expected at the end of 30 years of continuous extraction. These are maximum values, of course, because no heat loss is assumed through the insulation of the tubing.

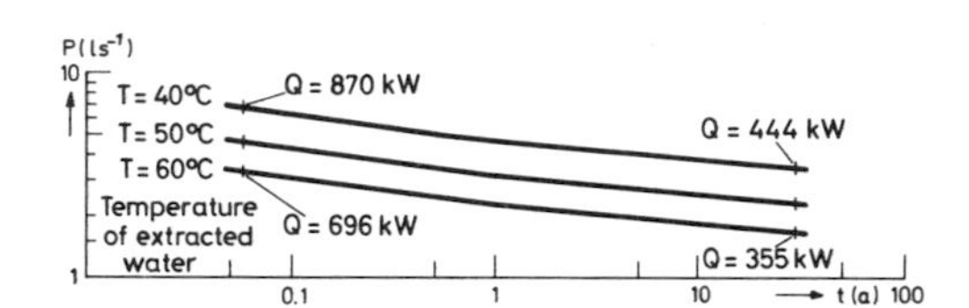

Fig. 6. The necessary flow rate in $l\ s^{-1}$ to keep the temperature of extracted water at well head constant; Q is the corresponding power output.

A constant temperature at the well head was assumed for the extracted water because this is important for its use for house heating, heat pumps, etc. If the borehole cost is not considered and the consumer is nearby, so the above-mentioned energy output can be utilized economically. Only a water circulation pump and a short pipe to the consumer would be necessary.

4. Huff-Puff Experiment

A huff-puff experiment was simulated to determine whether information can be obtained from this kind of test on the size of a fracture if the thickness is assumed as given by SNEDDON (1964) or DELISLE (1978). CARSLAW and JAEGER (1959, p. 100, Fig. 11) derived an equation for the length of time needed to uniformly heat a layer of infinite extension. Accordingly, the temperature in the middle of a fracture is given by:

$$T(0, t) = T_w + (T_r - T_w) \cdot x(t)$$

assuming a thickness (width of the fracture) 2 L = 0.5 cm and a thermal diffusivity κ = 1.72 10^{-3} cm^2 s^{-1}, and where T_r = rock temperature, T_w = water temperature, and x (t) is a factor for the convergence of the water temperature to the rock temperature. The results are as follows:

t	= 1,	5,	10,	50,	100 sec
x (t)	= 0,	0.12,	0.35,	0.95,	0.99

This table shows that the water will adapt the rock temperature within about 100 seconds. Because it is impossible to vary the length of time the water remains in the fracture within the time span, no indication of the fracture can be expected from a huff-puff experiment.

References

Carslaw, H.S. & Jaeger, J.C. (1959): Conduction of heat in solids. – 2nd Edit., Clarendon, Oxford.
Delisle, G. (1978): Berechnungen zur raum-zeitlichen Entwicklung eines nach dem Hydraulic Fracturing – Verfahren erzeugten Risses unter Berücksichtigung der geologischen Situation bei Falkenberg/Oberpfalz. – BGR-Archive, Nr. 80328, Hannover.
Minear, J.W. & Toksöz, M.N. (1970): Thermal regime of a downgoing slab and a new global tectonics. – J. Geophys. Res. **75**: 1397–1419.
Peaceman, D.W. & Rachford, H.H.Jr. (1955): The numerical solution of parabolic and elliptic differential equations. – J. Soc. Appl. Math. **3**: 28–41.
Sneddon, I.N. (1946): The distribution of stress in the neighbourhood of a crack in an elastic solid. – Proc. Roy., London, A 187, 229.

The Urach Geothermal Project, p. 361–365;
Schweizerbart'sche Verlagsbuchhandlung, Stuttgart, 1982

Recovery of Chemical Energy in Geothermal Systems

E. ALTHAUS

with 2 figures and 1 table

Abstract: Mineral reactions between rocks and thermal exchange fluids usually are exothermic. They produce, therefore, chemical energy that can be exploited in addition to the physically extractible heat. The total amount represents a vast energy reservoir. Reaction rates have been shown to be high enough in order to gain access to this latent energy. This is coupled, however, with the dissolution of rather large amounts of matter in the heat exchange solutions.

Introduction

Mineral reactions between rocks and aqueous geothermal fluids bring about not only compositional changes but also energetic effects. They can be endothermic or exothermic depending on the chemical conditions. Endothermic reactions consume part of the heat to be exploited, whereas exothermic reactions liberate additional energy. If this could be extracted along with the physically generated heat from cooling of the reservoir, a substantial contribution could be made to the overall energy gain. The question is, however, whether the chemical energy potential due to mineral reactions is large enough to be of interest for technical purposes and, if so, whether it is possible to recover it under reasonable conditions. If so, a way of obtaining maximal heat production from the hydrothermal, mineralogical processes must be sought. Endothermal reactions should be avoided if possible. Most reactions that can be expected in hydrous or non-hydrous rocks are exothermal (Tab. 1); a comprehensive compilation can be found in ALTHAUS (1979).

A rock most likely as reservoir rock for hydrothermal fluids is of more or less granitic composition. A typical granite composition is close to 30 % potash feldspar (Ksp), 30 % albite (Ab), 30 % quartz (Qz), 5 % anorthite as component in plagioclase (An), and 5 % biotite (Biot). If a rock of this composition is cooled, a certain amount of heat is extracted. From the data given in the literature (ROBIE & WALDBAUM 1968; Handbook of Chemistry and Physics, 1975; NBS Technical Note, 1968) this energy can be calculated if the extent of cooling is known. Table 2 shows the results; even if a (perhaps unrealistic) cooling of 300 K is assumed, the extracted energy is not greater than 27900 J (6675 cal) per 100 g of rock. Compared to this, the chemical (i. e. the reaction enthalpy) contribution is not negligible. If it is assumed that reactions no. 2, 9, 21, and 23 of Tab. 1 occur in our model rock, the amount of heat set free by chemical energy conversion is equal to 14330 J (3429 cal) per 100 g, i. e. of the same order as the heat extractible physically under the most favourable conditions. Chemical reaction enthalpy is therefore a vast energy potential if it is possible to gain access to it.

This depends on two factors: reacitvity and reaction rates. It was shown elsewhere (ALTHAUS this volume) that reactivity depends mostly on deviation from chemical equilibirum composition of fluids coexisting with mineral assemblages. The reaction rate determines the energy liberation per unit time. i. e. the power generation of a reacting system. Only if reaction rates are high enough, can the large potential of chemical energy be tapped and furnish a significant contribution to the total energy budget.

From Reaction rate studies by HELGESON (1971) on the basis of experimental data of other authors it can be concluded that this effect is possible even to an economically interesting extent.

Table 1. Reaction enthalpies.

No.	Reactants	Products	Enthalpy
a)	K-feldspar (microcline)		
1)	$3\ Kfsp + 2\ H^+$	$= 1\ Musc. + 2\ K^+ + 6\ Qz$	– 9200 J
2)	$2\ Kfsp + 2\ H^+ + H_2O$	$= 1\ Kaol + 2\ K^+ + 4\ Qz$	– 37700 J
3)	$1\ Kfsp + 4\ H^+ + 4\ H_2O$	$= Al^{3+} + K^+ + 3\ H_4SiO_4$	– 266000 J
4)	$1\ Kfsp + H^+ + 7\ H_2O$	$= 1\ Gibb + K^+ + 3\ H_4SiO_4$	– 271000 J
5)	$3\ Kfsp + 2\ H^+ + 12\ H_2O$	$= 1\ Musc + 2\ K^+\ 6\ H_4SiO_4$	– 66600 J
6)	$1\ Kfsp + 4\ HCl + 4\ H_2O$	$= AlCl_3 + KCl + 4\ H_4SiO_4$	– 244000 J
7)	$2\ Kfsp + 2\ HCl + 9\ H_2O$	$= 1\ Kaol + 2\ KCl + 4\ H_4SiO_4$	– 418000 J
b)	Albite (low)		
8)	$3\ Ab + 2\ H^+$	$= 1\ Par + 2\ Na^+ + 6\ Qz$	– 24700 J
9)	$2\ Ab + 2\ H^+ + H_2O$	$= 1\ Kaol + 2\ Na^+\ 4\ Qz$	– 62700 J
10)	$1\ Ab + 4\ H^+ + 4\ H_2O$	$= Al^{3+} + Na^+ + 3\ H_4SiO_4$	– 292000 J
11)	$1\ Ab + H^+ + 7\ H_2O$	$= 1\ Gibb + Na + 3\ H_4SiO_4$	– 296000 J
12)	$3\ Ab + 2\ H^+ + 12\ H_2O$	$= 1\ Par + 2\ Na^+ + 6\ H_4SiO_4$	– 165000 J
13)	$1\ Ab + 4\ HCl + 4\ H_2O$	$= AlCl_3 + NaCl + 3\ H_4SiO_4$	– 276000 J
14)	$2\ Ab + 2\ HCl + 9\ H_2O$	$= 1\ Kaol + 2\ NaCl + 4\ H_4SiO_4$	– 205000 J
c)	Intermediate reactions, decay of muscovite and paragonite		
15)	$1\ Musc + 10\ H^+$	$= 3\ Al^{3+} + K^+ + 3\ H_4SiO_4$	– 348000 J
16)	$1\ Par + 10\ H^+$	$= 3\ Al^{3+} + Na^+ + 3\ H_4SiO_4$	– 380000 J
d)	Kaolinite		
17)	$1\ Kaol + 5\ H_2O$	$= 2\ Gibb + 2\ H_4SiO_4$	– 186000 J
e)	Quartz		
18)	$1\ SiO_2 + 2\ H_2O$	$= 1\ H_4SiO_4$	– 70600 J
f)	Biotite (F-phlogopite)		
19)	$1\ Biot + 10\ H^+ + 2\ H_2O$	$= Al^{3+} + 3\ Mg^{2+} + 1\ K^+ + 2\ HF + 3\ H_4SiO_4$	– 389000 J
20)	$2\ Biot + 2\ H^+ + 7\ H_2O$	$= 2\ Chry + 1\ Kaol + 2\ K^+ + 4\ HF$	– 100000 J
21)	$2\ Biot + 14\ H^+$	$= 1\ Kaol + 6\ Mg^{2+} + 2\ K^+ + 4\ HF + 4\ Qz + 3\ H_2O$	– 160000 J
22)	$2\ Biot + 2\ H^+ + 9\ H_2O$	$= 2\ Kaol + 6\ Mg(OH)_2 + 2\ K^+ + 4\ HF + Qz$	– 86100 J
	Anorthite		
23)	$1\ An + 2\ H^+ + H_2O)$	$= 1\ Kaol + Ca^{2+}$	– 174000 J
24)	$1\ An + 8\ H^+$	$= 2\ Al^{3+} + Ca^{2+} + 2\ H_4SiO_4$	– 350000 J

Kfsp = potash feldspar
Musc = muscovite
Kaol = kaolinite
Gibb = gibbsite
Ab = albite
Par = paragonite
Chry = chrysotile
Biot = biotite
An = anorthite
Qz = quartz

Reaction rate studies

From the data given elsewhere (ALTHAUS this volume) reaction rates can be calculated. They depend on reactivity of solution, temperature, and, to a lesser extent, pressure. In The case of feldspars it also depends on the area of the contact surface. This must not be so in all cases; it was shown by KRONIMUS (1979) that for mafic minerals (olivine, diopside) the amount of matter leached from the solids (and, hence, the rate at which energy is set free) does not depend on the surface area of the solids but only on the concentration of solutions (which are saturated even after short contact periods). The former reactions can be called area controlled, the latter concentration controlled. Feldspar reactions are typically area controlled.

From the mass transfer determined in leaching experiments the amounts of matter passing through unit areas can be calculated (ALTHAUS this volume, Figs. 1 to 5). For a single square millimetre only rather small numbers result, but if this is extrapolated to square kilometres (i. e. the areas necessary for economic energy extraction) fairly large amounts of matter enter the heat exchange fluids. The energy production is, as shown above, directly proportional to the mass transfer. Since this depends on chemical conditions, the same must hold for the liberation of chemical energy.

The type of reaction is not always the same, but depends strongly on the metal-to-hydrogen ion ratio in the solution. It was shown by microprobe investigations that kaolinite + mica was produced with 0.001 n HCl (starting concentration), gibbsite with 0.01 n HCl and total stoichiometric dissolution with 0.1 n HCl. According to the different reaction enthalpies coupled with these solutions, energy production is subject to chemical parameters. It can be stated, however, that in principle the more energy is produced the higher the amount of leached matter and the higher the hydrogen ion concentration in the starting solution.

If a high rate of energy production is wanted, this is necessarily coupled with a high rate of production of dissolved material. This interdependence is shown in Figs. 1 and 2. In Fig. 1, the strong influence of the pH value (as a measure for the deviation from equi-

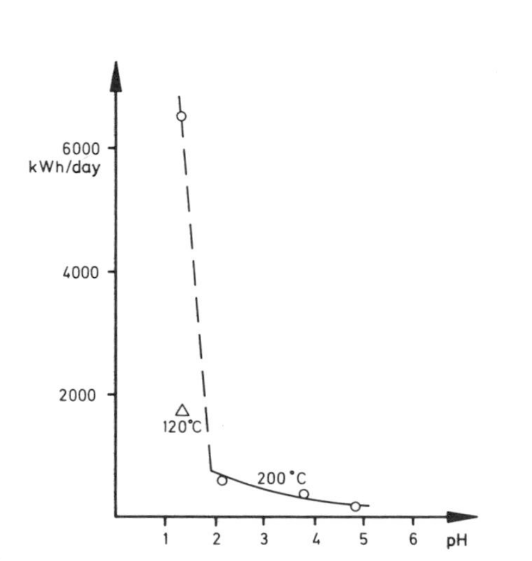

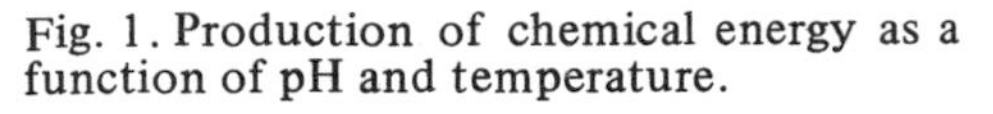
Fig. 1. Production of chemical energy as a function of pH and temperature.

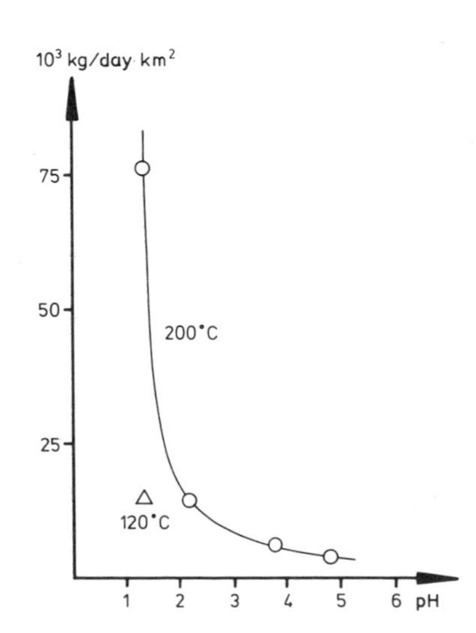

Fig. 2. Production of dissolved matter as a function of pH.

librium) is shown. High energy production results from low pH values (i. e. small metal cation-to-hydrogen ion ratios). The highest value corresponds to a power density of about 250 kW/km^2. The strong influence of temperature is evident, too.

The energy production necessarily is coupled with solute production. The absolute contents even in the most reactive solutions are not extreme (several 1000 to several 10 000 ppm), but if reasonable flow rates are assumed (e.g. 500 $l/s\ km^2$) these sum up to vast amounts of dissolved matter. Fig. 2 shows that, in the case of the most reactive fluids employed in the experiments, about 75 metric tons of matter would be dissolved in the exchange fluids per square kilometre of contact area. It is not desirable, therefore, for a geothermal system to be operated at maximum energy production rate. That is, serious problems will probably be connected with the large amounts of dissolved matter. It is necessary, therefore, to choose optimal and maximal conditions. These can be different, however, for different rocks. Basic rocks react in a different way than felsic rocks, and these in turn behave other than carbonates. The tool for directing the reactions is the composition of the heat exchange fluid. Injection (or re-injection of used) liquids can stimulate reactions of very different kinds, either dissolution of rocks (at a corrosion rate of up to several 100 micrometres per day) or formation of precipitates. If a high energy production rate is wanted, compositions of exchange fluids should deviate strongly from equilibrium. If, however, high concentrations of dissolved material are undesirable, solutions should be composed more closely to equilibrium conditions.

One additional effect has to be considered: Heat exchange operates most effectively if a fresh, uncontaminated and hot mineral surface is in contact with the exchange fluid. If a secondary mineral layer is formed on the original surfaces, only diffusive energy transport along channels through this layer will occur. If the layer has a certain thickness, heat exchange will be slowed down considerably since diffusion is a slow process in porous material with low permeability. Such a layer will act, therefore, as a thermal insulator. The reactivity of the fluids must be strong enough to prevent excessively thick layers from being formed, but weak enough not to produce solutions too concentrated for handling at the surface.

Under alkaline conditions, dissolution of silica is the dominant process, all other reactions being of subordinate importance only. Reactions in alkaline solutions do not contribute markedly to energy production; in many cases, they probably even cool down the geothermal systems. It seems to be reasonable, therefore, to think of acid solutions in the first place when looking for heat exchange fluids in geothermal systems.

The statements that have been made so far were derived from experimental and theoretical considerations. An attempt was made to test them in a field experiment in the Urach borehole. In the circulation experiments described elsewhere (ALTHAUS this volume) not only dissolved material was produced, but, judging from the kind of reaction inferred from the observed changes in liquid composition, energy was produced, too. Unfortunately, no temperature sensor had been installed in the borehole during these circulation tests and hence no direct proof of energy production is possible. However since large amounts of cold water had been pumped down into the hole and circulated over a rather small area around its bottom, a definite physical cooling effect was expected. When, however, after completion of the experiments, the temperature was measured, no cooling was observed with respect to previous measurements performed when the borehole was at rest and probably in thermal equilibrium. It is tempting to ascribe this effect to the liberation of chemical energy by hydrothermal mineral reactions.

References

Althaus, E. (1979): Mineralogical and chemical effects in a MAGES. – Final Report, Op. Agent Kernforschungsanlage Jülich GmbH, im Auftrag der IEA. Jülich.

Handbook of Chemistry and Physics (1975). – CRC Press Juc., Boca Raton Fla.

Helgeson, H.C. (1971): Kinetics of mass transfer among silicates and aqueous solution. – Geochim. Cosmochim. Acta **35**: 421–469.

Kronimus, B. (1979): Mineralogische Experimente zur Gewinnung geothermischer Energie. – Reaktionen zwischen einer hydrothermalen, fluiden Phase und den Mineralen Olivin und Diopsid. – Diplomarbeit Karlsruhe.

NBS Technical Note (1968) **270–3**.

Robie R.A. & Waldbaum, D.R. (1968): Thermodynamic properties of minerals and related substances at 298.15 K (25.0 °C) and one atmosphere (1.013 bar) pressure and at higher temperatures. – Geol. Surv. Bull. **1259**.

The Urach Geothermal Project, p. 367–380;
Schweizerbart'sche Verlagsbuchhandlung, Stuttgart, 1982

The Large-Scale Project in Oberschwaben – Swabian Alb for the Utilizaion of Geothermal Energy

J. WERNER, G. STRAYLE and E. VILLINGER

with 3 figures

Abstract: On the basis of the geothermal demonstration project in Saulgau, a large-scale project involving several deep boreholes is proposed for the area of Oberschwaben – Swabian Alb. The properties of the aquifers and the regional geothermal conditions, especially in the Malm and the Upper Muschelkalk, the most important thermal groundwater storeys, as well as the possibility of utilizing them as sources of energy, are thereby to be investigated. The energy resource of the thermal water alone (not including the rock) in this region is nearly as large as the energy of the oil reserves in the entire northern Alpine foreland.

1. Einleitung

Abgesehen von den zunächst rein wissenschaftlichen Grundlagenversuchen bei Urach wird sich die Nutzung geothermischer Energie in Baden-Württemberg zunächst auf die Grundwasserstockwerke des Buntsandsteins, des Muschelkalks und des Malms (Weißjuras) beschränken. Eines der höffigsten Gebiete ist das oberschwäbische Molassebecken mit der Möglichkeit der Erschließung thermaler Wässer aus dem oberen Malm und dem Oberen Muschelkalk. Im Bereich des Oberrheingrabens ist zwar eine deutliche positive Temperaturanomalie nachgewiesen, doch sind hier infolge der intensiven Bruchtektonik keine sehr ausgedehnten Grundwasserleiter vorhanden. Im Gegensatz dazu fehlen zwar im Molassebecken derart hohe Wärmeanomalien, dafür ist aber der geologische Bau viel einheitlicher, sodaß hier ausgedehnte Grundwasserstockwerke genutzt werden können.

2. Konzipierung des Großprojektes

Im Jahre 1979 häuften sich die Absichten baden-württembergischer Gemeinden geothermische Energie zu nutzen. Da es wenig sinnvoll erschien, eine größere Anzahl von Einzelprojekten ohne Beziehung zueinander aufzustellen, kam man überein, die hydrogeothermischen Einzelprojekte in Oberschwaben und im Gebiet der Schwäbischen Alb zu einem Großprojekt zusammenzufassen. Damit soll erforscht werden, wie weit sich die thermalen Stockwerke des verkarsteten Malms und des Oberen Muschelkalks großräumig zur Heizenergiegewinnung eignen und welche hydraulischen und hydrothermischen Wechselwirkungen sich zwischen verschiedenen Einzelobjekten ergeben (Wassermengen- und Wärmebilanzierung).

Auf Anregung des Projektträgers Energieforschung bei der Kernforschungsanlage Jülich sollte zur Ausarbeitung des Großprojekts eine Arbeitsgruppe gebildet werden, bestehend aus Vertretern des Geologischen Landesamtes Baden-Württemberg und des Niedersächsischen Landesamtes für Bodenforschung (Dr. HAENEL) sowie dem Ing.-Büro FRITZ (Urach).

3. Geologische und geothermische Übersicht

3.1 Geologische Abgrenzung

Das Großprojekt konzentriert sich auf das Molassebecken, da das für eine Erdwärmenutzung wichtigste Grundwasserstockwerk des Malms (oberer Weißer Jura) nur hier in den erforderlichen thermalen Tiefenbereich abtaucht.

Das zweite bedeutende, jedoch nur im westlichen Molassebecken vorhandene Grundwasserstockwerk ist der Obere Muschelkalk (Fig. 1). Dieser sollte ebenfalls untersucht werden, und zwar schon deshalb, weil im Falle des Fehlschlagens einer Bohrung auf das

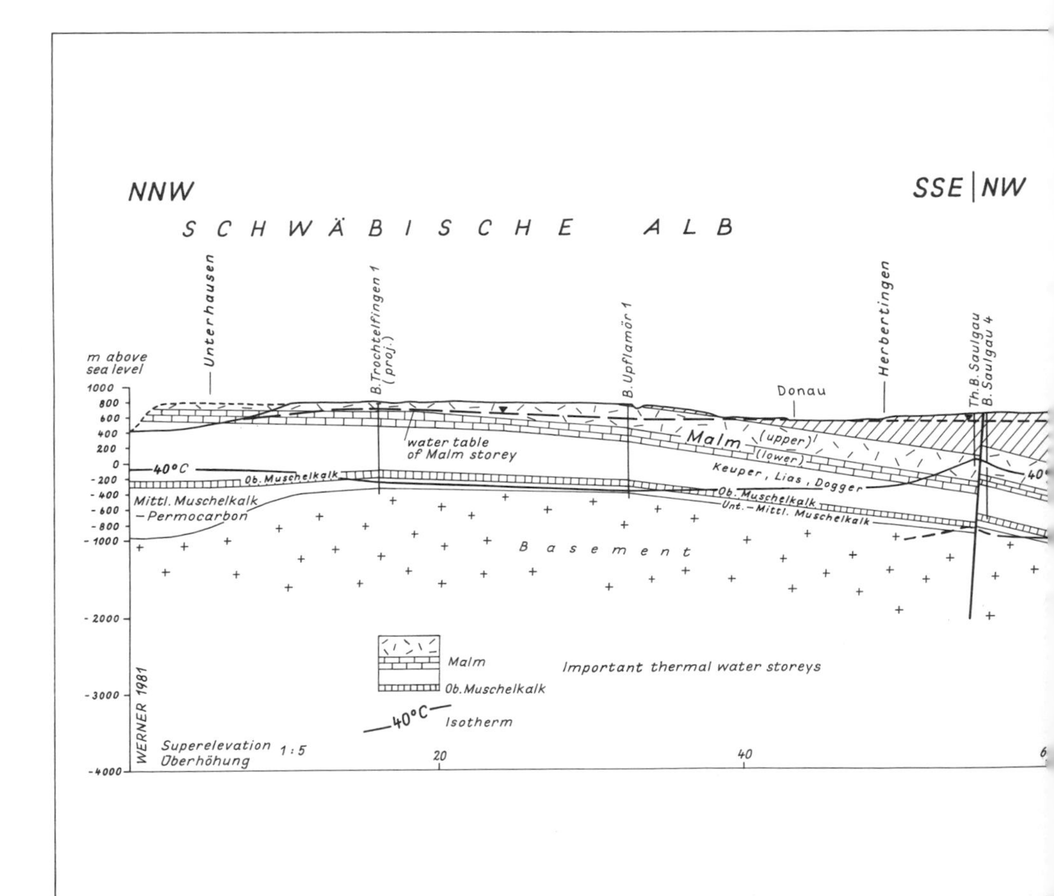

Fig. 1. Hydrogeologischer und geothermischer Schnitt durch das Molassebecken. Schnittlinie s. Abb. 2.

Malm-Stockwerk die Möglichkeit des Ausweichens in dieses tiefere, im ganzen geringer ergiebige, aber verläßlichere Stockwerk gegeben ist.

Die Abgrenzung des Untersuchungsgebietes nach Süden ist durch eine Faziesgrenze im Malm festgelegt. Sie scheidet die teilweise verkarstete und durchlässigere schwäbische Ausbildung des Malms im Norden von der nach bisherigen Kenntnissen weniger durchlässigen helvetischen Ausbildung (Quintnerkalke) im Süden. Der Übergangsbereich beider Faziestypen verläuft etwa entlang der Linie SE Konstanz – Ravensburg – Leutkirch (vgl. Fig. 2).

Die Abgrenzung im Osten ist durch das rasche Ausdünnen des Oberen Muschelkalks in dieser Richtung etwa ab der Iller gegeben.

Die Untersuchung des Malm-Stockwerkes könnte in nordwestlicher Richtung etwa an der 40°-Isotherme seiner Grundwassertemperatur enden. Diese verläuft ungefähr entlang der Linie Biberach – Saulgau – NW Überlingen – Radolfzell. Da aber der Obere Muschelkalk mit untersucht werden soll, empfiehlt es sich, die Schwäbische Alb miteinzubeziehen.

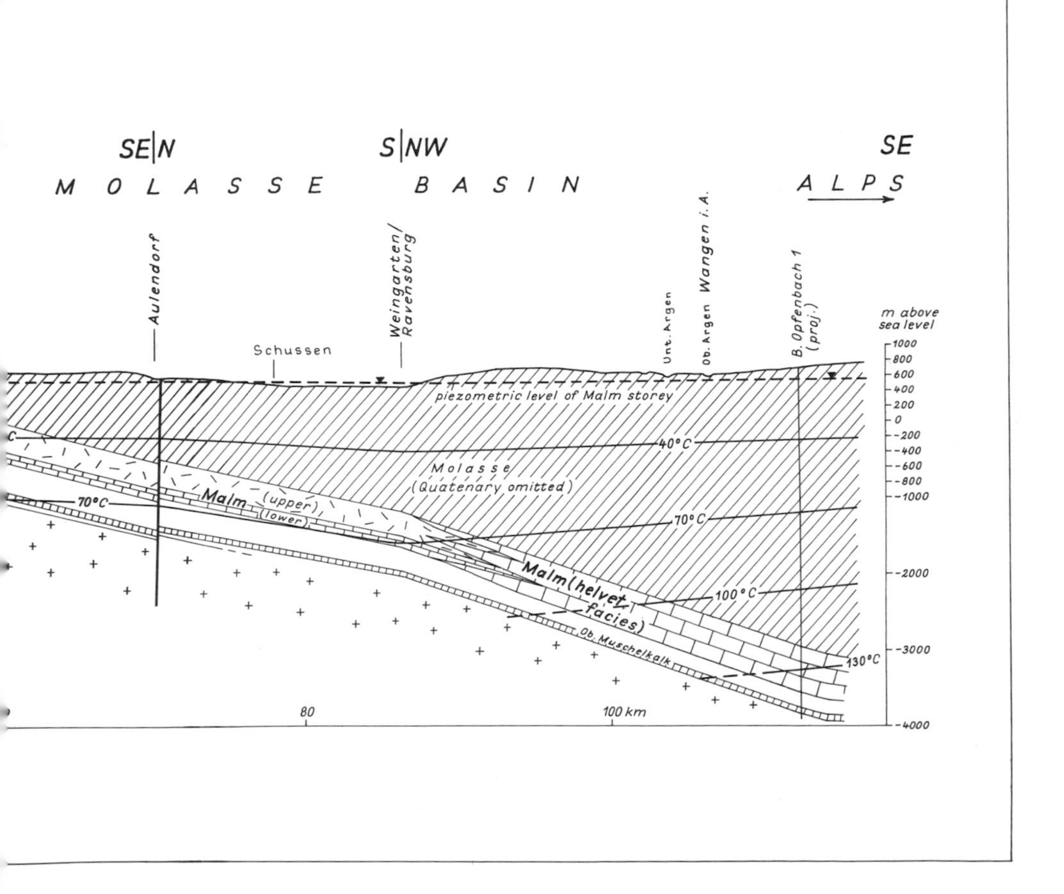

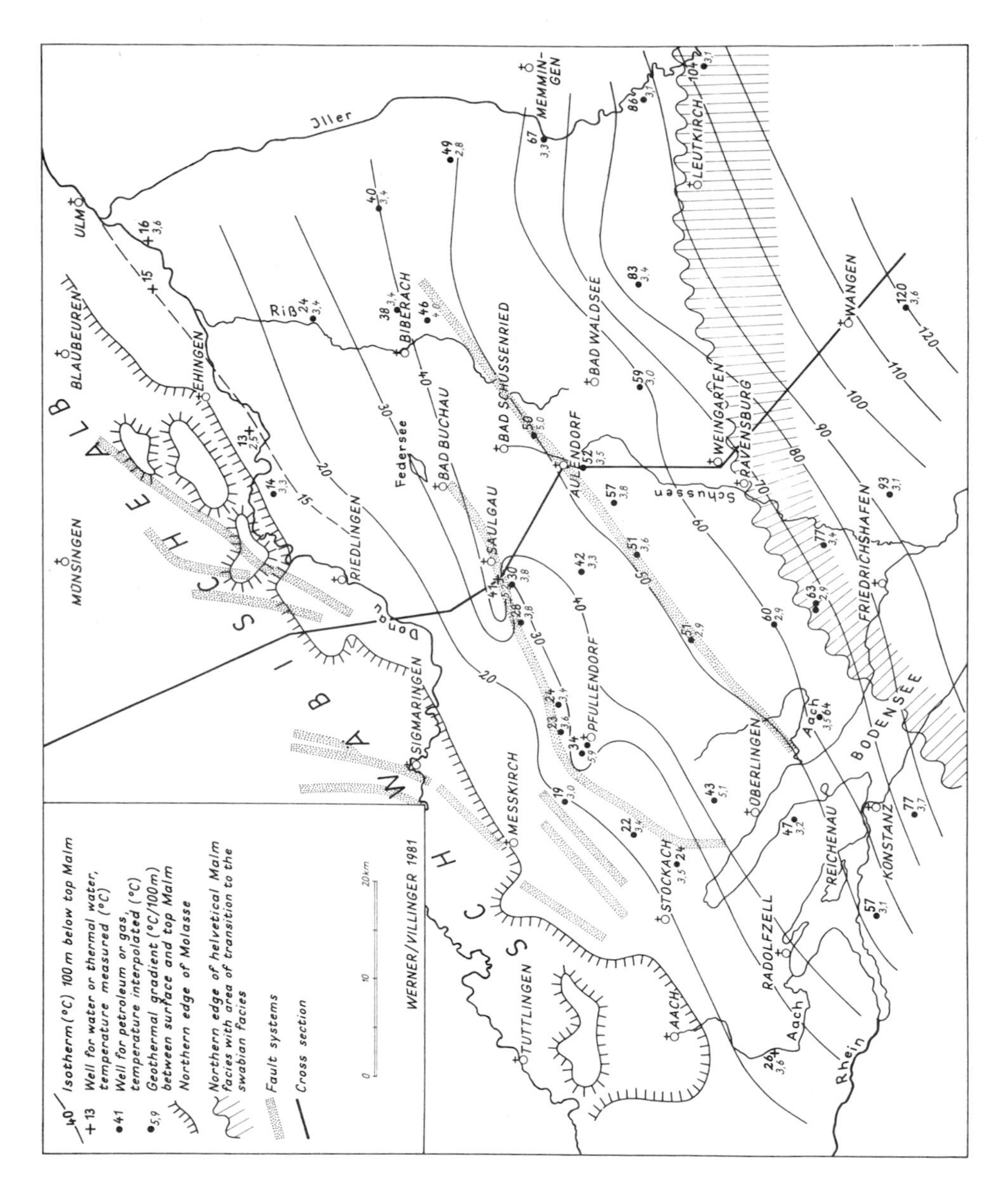

Fig. 2. Temperatur im überdeckten Karstaquifer des Weißjuras 100 m unter seiner Obergrenze. Daten der beiden Bohrungen SW Konstanz errechnet nach RYBACH et al. (1980; 288). Temperaturdaten der Erdöl-/Erdgasbohrungen interpoliert aus Einzelmessungen der Maximaltemperatur.

Nach Südwesten endet das Untersuchungsgebiet am Bodensee bzw. an der Schweizer Grenze.

3.2 Geologischer Überblick

Das süddeutsche Molassebecken ist zwischen der Schwäbisch-Fränkischen Alb und den Alpen tief eingesunken und mit tertiären Ablagerungen, der Molasse, verfüllt. Nach Südosten wird die Mächtigkeit dieser Beckenfüllung immer größer und erreicht vor dem Alpennordrand mit ungefähr 4000 m ihren Höchstwert. Den Boden dieses Beckens bildet in Oberschwaben die mächtige Kalksteinfolge des Malms (Weißer Jura), also dasselbe Gestein, aus dem die Schwäbische Alb besteht. Darunter folgen Gesteine, die in ähnlicher Ausbildung das schwäbisch-fränkische Schichtstufenland nördlich der Schwäbischen Alb aufbauen. Von oben nach unten sind dies: Dogger (Brauner Jura), Lias (Schwarzer Jura), Keuper und Muschelkalk. Nur im äußersten Westen von Oberschwaben folgt darunter noch etwas Buntsandstein und – in einer relativ schmalen Zone etwa auf der Linie Konstanz-Bad Waldsee-Memmingen Rotliegendes (Perm). Im übrigen Bereich von Oberschwaben liegt der Muschelkalk unmittelbar dem kristallinen Grundgebirge aus Graniten und Gneisen auf.

Im gesamten oberschwäbischen Teil des Molassebeckens fallen die Schichten im allgemeinen nach Südosten ein. Das gleichmäßige, beckeneinwärts zunehmende Schichtfallen wird von überwiegend durch Zerrung verursachten, tektonischen Lagerungsstörungen (Verwerfungen) unterbrochen, die an der Oberfläche wegen der quartären Überdeckung meist nicht erkennbar sind. In Fig. 1 sind zwei der wichtigsten bekannten, antithetischen Störungen eingezeichnet. Die Sprunghöhen der Verwerfungen im Molassebecken bleiben unter 150 m.

Die Schwäbische Alb schließt im Nordwesten an das Molassebecken an. In ihr tritt in breitem Ausstrich der Malm zutage. Da die Schwäbische Alb im Zusammenhang mit dem Einsinken des Molassebeckens schräggestellt wurde, überwiegt auch hier südöstliches Schichtenfallen, allerdings mit geringerem, nach NW zum Albtrauf hin stark abnehmendem Betrag (vgl. E. VILLINGER, Fig. 2, in diesem Band).

3.3 Temperaturverhältnisse

Unabhängig vom Schichtenaufbau der Molasse ist im Molassebecken durchweg ab 400–600 m Tiefe ein deutlich verringertes Ansteigen der Temperatur mit der Tiefe festzustellen. Dies äußert sich in einem Knick in den Temperaturkurven. Im östlichen Oberschwaben entsprechen die Temperaturgradienten weitgehend den mitteleuropäischen Normalwerten (0–500 m 4,1 °C/100 m, bis 1000 m: 3,3 °C/100 m, bis 2000 m: 2,9 °C/100 m nach HAENEL 1979) oder liegen nur wenig darüber.

Im Raum Saulgau-Pfullendorf gibt es positive Wärmeanomalien (Fig. 2; HAENEL 1980, vgl. auch Fig. 1 u. 4 von E. VILLINGER und ZOTH in diesem Band). Vermutlich besteht ein kausaler Zusammenhang zwischen diesen Anomalien und den großen antithetischen Verwerfungen von Pfullendorf-Saulgau und Fronhofen (Fig. 1). Die Wärmeanomalie von Saulgau (bis zum Malm-Stockwerk in 550 m Tiefe grad T = 5,2°/100 m) machte diese Lokalität für die Durchführung eines geothermischen Demonstrationsprojekts geeignet (vgl. WERNER 1978 und FRITZ & WERNER, dieser Band).

In Fig. 2 sind die 100 m unter Obergrenze des Malms zu erwartenden Temperaturen im Molassebecken wiedergegeben. Der Schnitt (Fig. 1) zeigt die Tiefenlage der 40°-, 70°-, 100°- und 130 °C-Isolinien der Gebirgstemperaturen. Die günstigsten Temperaturverhältnisse herrschen demnach einerseits im nordwestlichen Randbereich des Molassebeckens bei Saulgau-Pfullendorf, andererseits im alpenrandnahen Beckeninnern.

Im Gebiet der mittleren Schwäbischen Alb steigt der geothermische Gradient mit Annäherung an die Uracher Wärmeanomalie nach Norden hin stark an: In der Thermalwasserbohrung Urach 1 wurde an der Obergrenze des Muschelkalks mit 63 °C eine Temperatur gemessen, wie sie auch im Raum Saulgau-Pfullendorf zu erwarten ist (vgl. E. VILLINGER, Fig. 4, in diesem Band).

4. Für Erdwärmenutzung in Frage kommende Grundwasserstockwerke

4.1 Allgemeines

Aus dem Schnitt (Fig. 1) geht die unter der Schwäbischen Alb und im Molassebecken zu erwartende Schichtenfolge hervor. Die ungefähre Tiefenlage der Schichtgrenzen und die Schichtmächtigkeiten wurden unter Verwendung der Ergebnisse von Erdöl- und Erdgasbohrungen ermittelt. Naturgemäß geben jedoch Erdöl- und Erdgasbohrungen keine direkten Auskünfte über die Grundwasserführung in den verschiedenen Stockwerken.

Von vornherein sind Kalksteine, Dolomite, gröbere Sandsteine als teilweise durchlässig und damit als mehr oder weniger grundwasserleitend anzusehen. Erfahrungen über die Grundwasserführung dieser Schichten aus Gebieten, in denen sie in geringer Tiefe oder zutage anstehen, die Untersuchung von Bohrkernen aus Tiefbohrungen, das Auftreten von Spülungsverlusten sowie die Ergebnisse geophysikalischer Bohrlochmessungen und Tests in Erdöl- und Erdgasbohrungen geben Anhaltspunkte über die zu erwartenden Grundwasserergiebigkeiten. Hinzu kommen die Ergebnisse einiger Thermalwasserbohrungen im Molassebecken und im Schichtstufenland, die präzisere Aussagen über die Aquifer-Eigenschaften der Oberen Meeresmolasse, des oberen Malms und des Oberen Muschelkalks ermöglichen.

In Betracht gezogen werden hier nur die tieferen Grundwasserstockwerke, die unterhalb der für eine Energienutzung interessanten Temperaturgrenze liegen (vorläufig 40 °C). Ihre Tiefenlage geht aus Fig. 1, ihr ungefährer geografischer Verlauf im Malm-Stockwerk aus Fig. 2 hervor.

4.2 Oberer Malm (Weißjura)

4.2.1 Strömungs- und Druckverhältnisse

Das bedeutendste thermale Grundwasserstockwerk des Molassebeckens wird durch den oberen, etwa 350 m mächtigen Abschnitt des Malms (Mittel-Kimmeridge bis Tithon) gebildet. Seine günstigen Aquifereigenschaften beruhen auf der Verkarstung und Klüftung der in dieser Folge überwiegenden Massekalke.

Da die Verwerfungen im westlichen Molassebecken einen Sprungbetrag von der Größenordnung 150 m kaum überschreiten, können sie das etwa doppelt so mächtige Grundwasserstockwerk nirgends unterbrechen bzw. in Schollen zerlegen. Trotz Störungen ist somit der Malm ein großräumig zusammenhängender Grundwasserleiter.

Nach den bisherigen Kenntnissen (VILLINGER 1977, 1979) gehört das Malm-Grundwasser zwei durch eine Druckwasserscheide getrennten Strömungssystemen an (Fig. 3): Nördlich und östlich der Wasserscheide hat es die Donau zum Vorfluter. Dabei ist zu unterscheiden zwischen dem randnahen Bereich, in dem eine kräftige Neubildung durch Niederschläge vom Ausstrichgebiet des Malms (Meßkirch – Sigmaringen – Riedlingen) her stattfindet, und dem Beckeninnern, in dem eine ganz geringe Grundwasserneubildung nur durch die überdeckende Molasse hindurch erfolgen kann. Südlich und westlich der Grundwasserscheide wirken möglicherweise der Überlinger See, vor allem aber der Rhein abwärts Schaffhausen und die Aare als Vorfluter. Die hier abfließenden Grundwassermengen dürften wesentlich geringer sein als nördlich der Druckwasserscheide.

Das Druckniveau des Malm-Grundwassers liegt im Molassebecken großenteils tief unter Gelände (z.B. Saulgau: 103 m u. Gel., WERNER 1978). Im Bereich größerer Depressionen (nördliches Bodenseeufer, Schussenbecken) liegt es jedoch wahrscheinlich über Gelände (vgl. Fig. 1), so daß hier beim Anfahren des Malms mit artesischem Grundwasseraustritt zu rechnen ist. Die Darstellung der Potentiallinien in Fig. 3 beruht im östlichen Oberschwaben auf nur wenigen, recht ungenauen Wasserspiegelangaben von Erdölbohrungen, die aus Druckmessungen errechnet worden sind. Im Westen sind die Daten genauer, zudem liegen einige Wasserspiegelmessungen vor. An den Vorstellungen über Druckverhältnisse und Strömungsdynamik des Malm-Grundwassers können sich mit zunehmender Zahl genauer Messungen deshalb noch erhebliche Änderungen ergeben. Fig. 3 ist daher nur als erste Arbeitsgrundlage zu betrachten.

4.2.2 Ergiebigkeit

Die Massenkalke des oberen Malms sind von der Schwäbischen Alb her als stark verkarstet bekannt. Unter der Molasse-Überdeckung ist der Malm-Massenkalk in geringerem Maße verkarstet. Die Thermalwasserbohrung Saulgau hat gezeigt, daß zumindest in der randnahen Zone des Molassebeckens dem Stockwerk des oberen Malms große Grundwassermengen entnommen werden können (Größenordnung 50–100 l/s). Derart hohe Ergiebigkeiten sind jedoch nicht als „normal“ anzusehen. Die Ergiebigkeit jeder Malm-Bohrung hängt davon ab, in welchem Ausmaß Karsthohlräume angetroffen werden. Da es kein Verfahren gibt, Klüfte und Hohlräume von der Oberfläche her direkt zu orten, ist dies weitgehend dem Zufall anheimgegeben. Für tektonische Störzonen besteht die Wahrscheinlichkeit einer erhöhten Klüftigkeit. Daher sind nach Möglichkeit beim Ansatz einer Bohrung solche Zonen aufzusuchen.

Wie die geographische Verteilung der in Erdöl- und Erdgasbohrungen im Malm aufgetretenen Spülungsverluste zeigt, nimmt die Verkarstung vom Nordrand des Molassebeckens zum Alpenrand hin ab. Entsprechend verringern sich in gleicher Richtung die Aussichten, durch eine Bohrung im Malm hohlraumreiche Partien anzutreffen. Bereiche, in denen sich die Potentiallinien scharen (Fig. 3), sind als weniger durchlässig anzusehen. Dies dürfte vor allem für das Gebiet zwischen Karstwasserscheide und NW-Ende des Bodensees zutreffen.

Nach allen vorliegenden Beobachtungen und Erfahrungen ist die Wahrscheinlichkeit, daß eine Bohrung in den Massenkalken des Malms völlig trocken bleibt, auch für das Beckeninnere nicht sehr groß. Bei zunächst kleiner Ergiebigkeit (z.B. Thermalwasserbohrung Singen: ca. 1 l/s) besteht die Möglichkeit, Bohrungen durch Drucksäuerungen an benachbarte Kluft- und Hohlraumsysteme hydraulisch besser anzuschließen.

Auf die Abgrenzung gegen die wahrscheinlich geringdurchlässige helvetische Fazies des Malms (Quintnerkalke) wurde bereits im Abschnitt 3.1 eingegangen (vgl. Fig. 2 u. 3).

4.2.3 Temperaturen

Die im Malm-Stockwerk zu erwartenden Temperaturen sind Fig. 2 zu entnehmen. Die eine Erdwärmenutzung nach NW begrenzende 40 °C-Isotherme verläuft ungefähr entlang

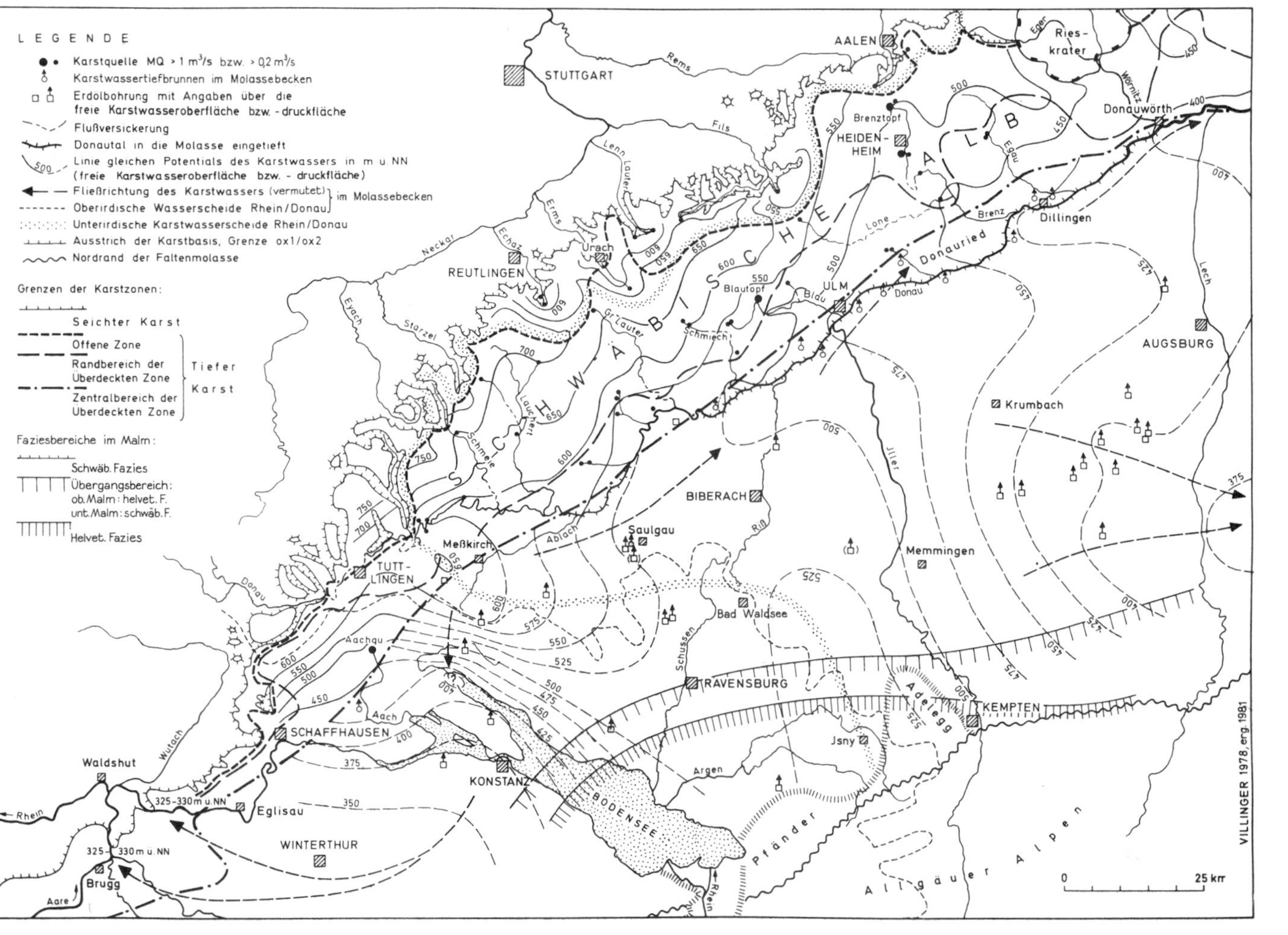

Fig. 3. Linien gleichen Potentials im Karstaquifer des Weißjuras (aus VILLINGER 1979, ergänzt).

der Linie Radolfzell – NW Überlingen – Saulgau – Biberach – Kirchberg. Bei Leutkirch können, wahrscheinlich im Grenzbereich zwischen schwäbischer und helvetischer Malm-Fazies, 90–100 °C erreicht werden (Tiefenlage der Untergrenze des Malm-Stockwerks: etwa 3000 m).

4.2.4 Mineralisation

Die Mineralisation des Grundwassers steigt im allgemeinen mit zunehmender Tiefe und Temperatur an. Eine Ausnahme von dieser Regel macht das Stockwerk des Malms unter der Molasse-Überdeckung. Aus Erdöl- und Erdgasbohrungen ist nach LEMCKE (1976) bekannt, daß der Malm Süßwasser führt, während die hangenden Molasseschichten – wie auch alle unter dem Malm folgenden Grundwasserstockwerke – Salzwasser enthalten. Diese Erscheinung ist auf die erhöhte Durchströmung des insgesamt vergleichsweise gut durchlässigen Malm-Stockwerks zurückzuführen. Die Thermalwasserbohrung Saulgau traf im Malm ein Thermalwasser an, das, abgesehen von geringfügig erhöhtem Schwefel- und Eisengehalt, Trinkwasserqualität hatte (WERNER 1978), dessen Mineralisation also weit unter der Mineralwassergrenze (1 g/kg gelöste Feststoffe) lag und typologisch zu den Ca-HCO_3-Wässern gehört. Beckeneinwärts steigt die Mineralisation des Malm-Grundwassers jedoch etwas an. So wurde in einer Bohrung des Erdölfeldes Fronhofen ein NaCl-Gehalt von 0,34 g/kg bei einer Gesamthärt von 6° d festgestellt. Es ist anzunehmen, daß weiter beckeneinwärts der NaCl-Gehalt des Malm-Grundwassers infolge zunehmender Tiefenlage und abnehmender Durchströmung kräftig ansteigt.

4.3 Oberer Muschelkalk

4.3.1 Aquifereigenschaften

Die Kalksteinserie des Oberen Muschelkalks ist vom Schwäbischen Schichtstufenland und von einigen Thermalwasserbohrungen am Nordrand der Alb wegen ihrer gleichmäßigen Ausbildung und ihrer auch bei größerer Tiefenlage vorhandenen Klüftung als guter und zuverlässiger Grundwasserleiter bekannt. Seine Obergrenze liegt im nördlichen Molassebecken etwa 850 m, in Alpenrandnähe etwa 750 m unter der Obergrenze des Malms (Fig. 1). Wegen der im Bereich der Schwäbischen Alb nach NW anschwellenden Mächtigkeit von Keuper, Lias und Dogger liegt er nahe dem Albrand in Bezug auf das Malm-Stockwerk erheblich tiefer. Seine Gesamtmächtigkeit beträgt in Oberschwaben etwa 60 m. Sie steigt zum Albrand hin auf etwa 80 m an. Nach Osten dünnt der Obere Muschelkalk rasch aus, so daß er etwa ab der Iller als Grundwasserleiter nicht mehr in Betracht zu ziehen ist.

Durch die in Kap. 4.2.1 erwähnten antithetischen Störungen wird der Muschelkalk-Aquifer in Oberschwaben in mehrere, hydraulisch vermutlich voneinander weitgehend unabhängige Schollen zerlegt.

Insbesondere in seinem obersten Abschnitt, dem in Oberschwaben rund 15 m mächtigen Trigonodus-Dolomit, wurden in einer Vielzahl von Erdöl- und Erdgasbohrungen starke Spülungsverluste beobachtet. Der Obere Muschelkalk ist demnach auch im Molassebecken so geklüftet und durchlässig, daß er als guter Grundwasserleiter wirkt. Aus einer Bohrung erzielbare Ergiebigkeiten von 5–10 l/s erscheinen auch für das Molassebecken möglich. Das Verfahren der Drucksäuerung zur Verbesserung des Anschlusses einer Bohrung an benachbarte Kluftzonen ist anwendbar.

Im Gegensatz zum Malm-Stockwerk und trotz seiner wahrscheinlich guten Grundwasserleitfähigkeit ist das Muschelkalk-Stockwerk im Bereich des Molassebeckens wohl nicht oder kaum in die meteorisch-tellurische Zirkulation einbezogen. Daß dieses System weitgehend geschlossen ist, wird belegt durch das Fehlen einer Regenerationsmöglichkeit, die hohe Mineralisation des Muschelkalk-Grundwassers (bis zu 85 g/kg NaCl) und den sehr niedrigen Aquiferdruck (z.B. in der Erdölbohrung Owingen 1 Druckspiegel etwa 300 m u. Gel. = etwa 230 m + NN).

Dies gilt nicht für den Bereich der Schwäbischen Alb, wo eine geringe Regenerationsmöglichkeit durch die überdeckenden Schichten hindurch wahrscheinlich noch besteht, wo das Muschelkalk-Grundwasser eine bedeutend geringere Mineralisation aufweist und wo auch die Druckverhältnisse ein Abströmen zum Vorfluter Neckar erkennen lassen (vgl. E. VILLINGER, Fig. 3, in diesem Band).

Das gebietsweise sehr niedrige Druckniveau des Muschelkalk-Aquifers bedingt entsprechend große Förderhöhen bei der Erschließung.

4.3.2 Temperaturen

Aus dem Schnitt (Fig. 1) und aus E. VILLINGER (Fig. 4 in diesem Band) geht hervor, daß der Obere Muschelkalk im Molassebecken in Gebirgstemperaturbereichen von wenig über 40 °C am Nordwestrand bis über 140 °C am Alpenrand (Wangen i.A.: etwa 130 °C) lagert. Im Bereich der Albhochfläche sind die Gebirgstemperaturen im Oberen Muschelkalk vermutlich etwas kleiner als 40 °C. Zum Albrand hin steigen sie im Raum Urach bis auf über 60 °C an.

4.4 Sonstige Grundwasserstockwerke

4.4.1 Molasse

Das bezüglich der Ergiebigkeit zuverlässigste und beste Grundwasserstockwerk der Molasse ist der Baltringer Horizont der Oberen Meeresmolasse, ein mittelkörniger Sandstein. Seine Gebirgstemperatur liegt nur im alpenrandnahen Teil des Molassebeckens über der in Abschnitt 3.1 genannten Grenze von 40 °C bei sicher erzielbaren Ergiebigkeiten von nur 5–10 l/s, so daß er für ein geothermisches Energienutzungsprojekt vorläufig kaum in Frage kommt.

In der Unteren Süßwassermolasse auftretende Grundwasser führende Horizonte (sog. Hauptsande; Süßwasserkalke) sind bezüglich ihrer Ergiebigkeit unzuverlässig und von geringer Bedeutung. Obwohl ihre Temperaturen nur wenige °C unter denen des Malms liegen, sollten sie daher ebenfalls zur Vermeidung bohrtechnischer Komplikationen überbohrt werden.

Nur im tiefen Beckenbereich ist die Untere Meeresmolasse ausgebildet. Der ihr angehörende Bausandstein besitzt möglicherweise ähnliche Aquifereigenschaften wie der Baltringer Horizont. Er sollte bei Bohrungen auf den Malm ebenfalls getestet werden.

4.4.2 Jura und Keuper

Die Schichten des unteren Malms, Doggers, Lias und Keupers (vgl. Fig. 1) enthalten mehrere Grundwasser-Stockwerke von geringer Ergiebigkeit. Das relativ interessanteste von ihnen ist der Stubensandstein im Mittleren Keuper. Diese fein- bis grobkörnige, teilweise auch tonige Sandsteinserie liegt etwa 400 m unter der Untergrenze des Malm-Stockwerks und besitzt im mittleren Trogbereich eine Mächtigkeit von etwa 10 m. Ihre Ergiebigkeit dürfte höchstens eine Größenordnung von 1 l/s erreichen und damit für eine Erschließung ausscheiden.

4.5 Einbohrloch-Verfahren oder Dublette?

Bei der Nutzung von Wärmeenergie tiefer Grundwasser-Stockwerke sind grundsätzlich zwei Verfahren möglich:

Beim **Einbohrloch-Verfahren** wird heißes Grundwasser aus einer Bohrung entnommen, entwärmt und als Abwasser oberirdisch abgeleitet. Dieses Verfahren setzt somit einen großen abwirtschaftbaren Grundwasservorrat bzw. ausreichendes Nachströmen von Grundwasser aus dem Einzugsgebiet und Aufheizen voraus. Damit stellt sich auch die Frage nach einer Beeinflussung anderer Entnahmestellen aus dem gleichen Stockwerk. Ferner muß sichergestellt sein, daß die im entwärmten Grundwasser (Abwasser) enthaltene Mineralstoffracht schadlos in den Vorfluter eingeleitet werden kann.

Bei der sog. **Dublette** wird heißes Grundwasser aus einer Bohrung entnommen und nach Erwärmung über eine zweite Bohrung dem gleichen Grundwasserleiter wieder zugeführt. Dieses Verfahren beruht somit im Gegensatz zum Einbohrloch-Verfahren auf dem Wärmeaustausch, der zwischen dem von der Versenkbohrung zur Entnahmebohrung zirkulierenden Wasser und dem Grundwasserleiter stattfindet.

Es wird also der im Grundwasserleiter und seiner Gesteinsumgebung zwischen zwei Bohrungen vorhandene Wärmevorrat abgewirtschaftet. Dieser erneuert sich nur zu einem kleinen Teil aus dem aufsteigenden Erdwärmestrom. Im Zusammenhang mit der Forschungsbohrung Urach angestellte überschlägige Berechnungen zeigen, daß der durch eine Dublette abwirtschaftbare Wärmevorrat für einige Jahrzehnte ausreichen dürfte.

Wegen der hohen Salzfracht des Muschelkalk-Grundwassers im Molassebecken scheidet bei seiner Verwendung zur Energienutzung das Einbohrlochverfahren aus. Dort ist also von vornherein eine Dublette einzuplanen.

Dagegen dürfte die Wärmenutzung des Malm-Stockwerks nach dem Einbohrlochverfahren nicht nur im randnahen Bereich des Molassebeckens, sondern auch weiter beckeneinwärts möglich sein.

5. Abschätzung der Energievorräte

Grundlage für eine Bilanzierung des Wasserinhaltes und des Energiepotentials der beiden Thermalwasserstockwerke bildet eine Fläche, die wie folgt abgegrenzt wird:

- Die Ostgrenze bildet die Iller, da östlich dieser Linie kein Muschelkalk mehr verbreitet ist.
- Die Südgrenze bildet etwa die Linie Leutkirch – Ravensburg – SE Konstanz, da der Malm südlich dieser Grenze in die helvetischer Fazies übergeht (vgl. Fig. 2 und 3).
- Die Nordwestgrenze ist durch die 40 °C-Isotherme im Malm gegeben: Sie verläuft etwa von der Insel Reichenau, vorbei an Pfullendorf, Saulgau und Bad Buchau zur Iller.

Nur innerhalb dieser Grenzen liegen beide Stockwerke in einer Tiefenstufe, die für die Nutzung der Erdwärme ausreicht.

5.1 Malm

Die so abgegrenzte Fläche beträgt rund 2300 km^2. Bei einer mehrfach durch Bohrungen nachgewiesenen Mächtigkeit von 300 m ergibt sich ein Gesteinsvolumen von 690 Mrd. m^3, das über 40 °C temperiert ist. Bei einem angenommenen Kluftvolumen von 0,1 % ergibt sich ein Thermalwasservolumen von 690 Mio m^3. Wegen der Elastizität des gespannten Aquifers ist der Speicherkoeffizient etwas größer, sodaß sich rund 750 Mio. m^3 Thermalwasser im Malm befinden (ohne Berücksichtigung des Zustroms aus dem Einzugsgebiet). Dieses Wasser hat bei einer möglichen Temperaturausnutzung von 25 °C einen Energiegehalt von 21,6 Mrd. kWh. Dies entspräche einer Heizölmenge von 1,8 Mio. Tonnen bei einem unteren Heizwert von 42 000 kJ/kg. Diese Zahlen sind Mindestwerte, weil das Hohlraumvolumen sehr niedrig angesetzt ist.

5.2 Oberer Muschelkalk

Die Fläche des nur westlich der Iller verbreiteten, nutzbaren Muschelkalk-Aquifers reicht viel weiter nach Norden, wie zahlreiche Thermalwasserbohrungen nördlich der Schwäbischen Alb beweisen. Seine Mächtigkeit ist mit 20 m relativ gering. Legt man die Nordgrenze für die Bilanzierung auf die geografische Breite von Ulm, so erhält man rund 60 Mrd. m^3 Gesteinsvolumen, das über 55 °C temperiert ist. Das Wasservolumen darin liegt etwa bei 60 Mio. m^3 (angesetztes Kluftvolumen 0,1 %). Mit der etwas höheren Temperaturausnutzung von 40 °C ergibt das einen Energieinhalt von 2,7 Mrd. kWh, was ca. 0,2 Mio. Tonnen Heizöl entspricht.

5.3 Vergleich mit Erdölvorräten

Das Energiepotential des in den beiden Thermalwasser-Stockwerken enthaltenen Wassers beträgt rund 24,3 Mrd. kWh. Der gesamte Energieinhalt der beiden Aquifere ist wesentlich höher, da der Wärmeinhalt des Gesteins noch nicht berücksichtigt ist. Da eine wirtschaftliche Nutzung bisher nicht nachgewiesen wurde, sind die hier genannten geothermischen Energievorräte als Resourcen zu bezeichnen. Die Größe des Energiepotentials läßt sich am anschaulichsten mit den Erdölvorräten im Alpenvorland vergleichen.

Die amtliche Reserveschätzung gibt für das gesamte Alpenvorland an gesicherten und wahrscheinlichen Erdölvorräten 2,25 Mio. Tonnen an. Umgerechnet ergibt dies eine Wärme von rd. 26,2 Mrd. kWh. Die geothermischen Resourcen der beiden Stockwerke allein innerhalb Oberschwabens entsprechen somit etwa den gesamten Reserven der Erdölvorräte.

6. Skizzierung eines Großprojekts

6.1 Fragestellungen

Ein Großprojekt Oberschwaben – Schwäbische Alb ist geowissenschaftlich nach folgenden Fragestellungen auszurichten:

6.1.1 Malm im Molassebecken

- Erkundung und Abgrenzung strömungsdynamisch unterschiedlicher Zonen (Ergiebigkeit, Bohrrisiko, regionale Aquifer-Parameter, Mineralisation);
- Bilanzierung dieser Zonen (Grundwassermengenbilanz, Wärmebilanz, mögliche Belegungsdichte ohne gegenseitige Beeinflussung usw.).
- Wie weit ist es notwendig, beim Ansetzen von Bohrungen tektonische Strukturen zu berücksichtigen?
- In welchem Maße sind Ergiebigkeitssteigerungen in Bohrungen mit unbefriedigendem Ergebnis durch Drucksäuerungen und Hydrofrac-Verfahren möglich?
- Ist gebietsweise eine Versenkung von entwärmter Muschelkalk-Sole in das Malm-Stockwerk vertretbar?
- Ermittlung von Zonen unterschiedlicher Eignung für Erdwärmenutzung.

6.1.2 Muschelkalk

- Abgrenzung des hochmineralisierten gegen den niedrigmineralisierten Bereich (Schwäbische Alb). Erkundung der Druckverhältnisse und der Strömungdynamik;
- Erkundung der Ergiebigkeiten und des Bohrrisikos; Ermittlung der Aquifer-Parameter;
- Bilanzierung (Grundwassermenge, Wärme);

– Ermittlung der erforderlichen Abstände und der Lebensdauer von Dubletten;
– Ermittlung von Zonen unterschiedlicher Eignung für Erdwärmenutzung.

6.2 Projektvorschlag zur Lösung der aufgezeigten Fragen

6.2.1 Ergänzung und Erweiterung des Projekts Saulgau (nordwestlicher Beckenrandbereich des Malms)

Im Rahmen des Projekts Saulgau wird der Strömungsmechanismus des Malm-Karstwassers im nordwestlichen Randstreifen des Molassebeckens untersucht, mit dem Ziel einer Wassermengen- und Wärmebilanzierung für dieses System. Um gesicherte Ergebnisse zu erhalten, sind zur Ergänzung des Projekts Saulgau zumindest zwei weitere Tiefbohrungen in den Malm erforderlich, und zwar eine in dem 11 km grundwasserstromab im Schichtsteichen gelegenen **Bad Buchau** sowie eine entweder bei **Aulendorf** (Entfernung 13 km ungefähr in Richtung des Schichtfallens und quer zur mutmaßlichen Hauptfließrichtung) oder im gleichweit entfernten, aber aus der Richtung des Schichtfallens nach NE abweichend gelegenen **Bad Schussenried**.

Die 3-Punkte-Anordnung (mit Saulgau) böte u.a. die Möglichkeit, richtungsabhängige unterschiedliche Druckreaktionen auf Interferenz-Pumpversuche (zeitliche Verschiebung, Amplitude) zu messen und vielleicht zur Gewinnung regionaler Aquiferparameter auszuwerten. Ferner lassen sich aus Temperatur- und radiohydrometrischen Messungen (s. Projekt Saulgau, FRITZ & WERNER, dieser Band) möglicherweise genauere Daten über die Strömungsdynamik und den Wärmetransport des Systems gewinnen.

Eine Erweiterung der Untersuchungen nach Osten wäre durch ein Einzelprojekt **Biberach** möglich.

6.2.2 Westlicher Beckenrandbereich des Malms

Dieser gehört dem zum Hochrhein gerichteten Strömungssystem an (vgl. Abschnitt 4.2.1, Fig. 3). Als erster Untersuchungspunkt bietet sich die Insel **Reichenau** an, da sie hydrogeologisch-geothermisch günstig liegt und bezüglich der Wärmeverwendung sehr gut geeignete Objekte aufweist (Gärtnereien).

6.2.3 Mittlerer Beckenbereich des Malms

Bohrungen in diesem Bereich, in dem die Temperaturen des Malms über 60 °C erreichen, haben zunächst die Frage nach ausreichender Durchlässigkeit und damit Ergiebigkeit des Grundwasserleiters zu beantworten. Am besten geeignet erscheinen die dicht beieinanderliegenden und sich hydrogeologisch nur wenig unterscheidenden Einzelobjekte **Weingarten** und **Ravensburg**. Ihre tiefe Lage im Schussental läßt zumindest für Weingarten artesischen Wasseraustritt erhoffen. Nicht nur hierdurch, sondern auch durch ihre Entfernung von der Grundwasserscheide sind sie gegenüber **Bad Waldsee** bevorzugt, durch die Randlage zur Quintner Fazies jedoch möglicherweise benachteiligt.

6.2.4 Tiefer Beckenbereich des Malms

Da die Ausläufer der helvetischen Malm-Fazies maximal bis etwa zur Linie SE Konstanz – Ravensburg – Leutkirch reichen, wäre ein Einzelobjekt **Leutkirch** von Interesse (Temperatur annähernd 100 °C). Wegen des hohen Risikos sollte man diesem Objekt aber erst nach einem Erfolg im mittleren Beckenbereich nähertreten.

6.2.5 Regionale Untersuchungen des Malms

Entsprechend der Ergänzung des Projektes Saulgau durch ein regionales hydrogeologisches, geothermisches und radiohydrometrisches Untersuchungsprogramm sollten auch in einem Großprojekt Oberschwaben-Schwäbische Alb die Tiefbohrergebnisse durch regionale Untersuchungen ergänzt werden. Hierzu gehören Untersuchungen zur Auffindung möglicher thermaler Grundwasseraustritte im Überlinger See und am Hochrhein.

6.2.6 Oberer Muschelkalk

Als zweiter Untersuchungspunkt des Oberen Muschelkalks im Molassebecken ist **Bad Buchau** vorgesehen. Dort soll auch der Malm getestet werden (Ergänzung des Projektes Saulgau). Ausschließlich auf den Muschelkalk zielende Objekte sind **Unterhausen** (nördlich außerhalb von Fig. 2) und **Blaubeuren**, beide noch im Bereich der Schwäbischen Alb gelegen. Beide könnten im Verein mit Bad Buchau wichtige Ergebnisse über die Eignung des Oberen Muschelkalks zur Erdwärmenutzung im Gebiet Schwäbische Alb – nordwestliches Molassebecken erbringen (Druckverhältnisse, Ergiebigkeit, Mineralisation). Andere, weiter beckeneinwärts gelegene Objekte ergeben sich möglicherweise bei Mißerfolgen von Bohrungen, die zunächst auf den Malm angesetzt wurden.

Literatur

Haenel, R. (1979): Determination of Subsurface Temperatures in the Federal Republic of Germany on the Basis of heat flow Values. – Geol. Jb., **E 15**: 41–49, Hannover.

– (1980): Atlas of Subsurface Temperatures in the European Community. – Schäfer Druckerei GmbH, Hannover.

Lemcke, K. (1976): Übertiefe Grundwässer im Alpenvorland. – Bull. Ver. schweiz. Petroleum-Geol. u. -Ing., **42**: 6–18, Basel.

Rybach, L., Büchi, U.P., Bodmer, Ph. & Krüsi, H.R. (1980): Die Tiefengrundwässer des schweizerischen Mittellandes aus geothermischer Sicht. – Eclogae geol. Helv., **73**: 293–310, Basel.

Villinger, E. (1977): Über Potentialverteilung und Strömungssysteme im Karstwasser der Schwäbischen Alb (Oberer Jura, SW-Deutschland). – Geol. Jb., **C 18**: 3–93, Hannover.

– (1979): Aspekte der Grundwassernutzung im Karstaquifer der Schwäbischen Alb. – Clausth. Geol. Abh., **30**, Schönenberg-Festschrift: 318–335, Clausthal-Zellerfeld.

Werner, J. (1978): Die Thermalwasserbohrung Saulgau (Württemberg). – Abh. Geol. Landesamt Baden-Württ., **8**: 129–164, Freiburg i. Br.

The Urach Geothermal Project, p. 381–393;
Schweizerbart'sche Verlagsbuchhandlung, Stuttgart, 1982

Geothermal Demonstration Project Saulgau

J. FRITZ and J. WERNER

with 9 figures and 1 table

Abstract: The geothermal energy below the pre-Alpine Tertiary will be used for heating purposes (energy cascading). The energy price amounts to about 9 Pfennige per kilowatt-hour. The hydrogeological and geothermal data of interest are:
Karst aquifer in Upper Malm, 350 m thick, covered by 550 m of Tertiary sediments, low salinity, well-head temperature 42 °C, temperature gradient down to 650 m: 4.8 °C/100 m, expected production rate 50–100 l/s. The extractable heat amounts to 13 MW or 24 GWh/a; if heat pumps are used: 33 GWh/a.

1. Introduction

The Geothermal Demonstration Project Saulgau was approved at the end of 1979. As a result of a German Federal Government grant of about 8.2 million DM, plus financial assistance from the regional government and the town of Saulgau itself, the financial basis for the exploration work, excluding the technical equipment, is assured.

In the region of the Molasse basin in the northern pre-Alpine region a huge reservoir of thermal groundwater in the Upper Malm, the enthalpy of which can be utilized for extracting energy is to be explored. The Saulgau pilot project is designed to demonstrate the efficiency of using this resource for the combined purposes of heating and of supplying drinking and industrial water for a mixed consumer structure. The hydrogeological, geothermal, technical and economic research is designed to serve both a scientific and a practical basis for projects in areas with corresponding or similar conditions. This is to be combined with the balancing of the quantity of heat available in the northern pre-Alpine region with regard to the partial saving of fossil energy and the development of a technical prototype. The project manager is Dipl.-Ing. J. FRITZ, Ingenieurbüro Fritz in Urach, the geological and hydrogeological consultant is Dr. J. WERNER, Geologisches Landesamt Baden-Württemberg, and the geothermal consultant is Dr. R. HAENEL, Niedersächsisches Landesamt für Bodenforschung. Dipl.-Geol. B. BERTLEFF (hydrogeology), Dipl.-Geophys. W. MICHEL (geothermics) and Dipl.-Ing. F. CAMMERER (thermodynamics) are charged with the performance of the project.

2. Geoscientific Basis and Results of Former Exploration

2.1 Geology

Saulgau is situated south of the Swabian Jura near the northwestern margin of the South German Molasse basin. A local positive heat anomaly, which has no connection with the wellknown Urach anomaly (see ZOTH this volume), occurs here. The latter is located north of Saulgau in the Swabian Jura. The origin of the former is not yet clear, but there is possible connection with the Saulgau fault. This fault runs along the strike of the Molasse basin (see WERNER et al. this volume). It is an antithetic fault with a throw of about 170 metres.

On the deeper block of the fault, the thermal water borehole Saulgau 1, has been drilled to the thermal aquifer of the massive Massenkalk of the Upper Malm in 1977. It is the most important aquifer in the southwest German stratum step region, forming the Swabian Jura.

As a result of the general dipping to the SE at the southern margin, the Malm submerges under the Molasse basin.

Around Saulgau the Upper Malm is covered by Tertiary Molasse sediments more than 500 metres thick, and its Massenkalke are strongly karstic, as shown in the Saulgau 1 borehole. In correspondence to this fact, this aquifer has a high efficiency of 50–100 l/s.

The water from the karstic limestones explored has a measured maximum temperature of 42 °C in the borehole. The Malm groundwater has the quality of drinking water (apart from minor quantities of iron and sulphur).

2.2 Results from the Thermal Water Borehole Saulgau 1

The first data concerning the Saulgau field were obtained from the thermal water borehole Saulgau 1 in 1977, and are as follows: final depth: 650 m; tested aquifer: karstic limestones of the Upper Malm, 610 to 650 m in depth; water level at hydrostatic pressure: 103.7 to 104.6 m below surface; pumping test: 36 l/s (= maximum power of pump); reduction about 12 m; measured maximum temperature of water in the borehole: 41.9°; volume of gas: approximately 2 % (97 % of this was nitrogen).

The borehole has been drilled down to the deep block of the Saulgau fault.

2.3 Hydrogeology

The pre-Tertiary strata in the southwest German section of the Molasse basin contain two productive aquifers: the Massenkalk of the Upper Malm with a thickness of up to approximately 350 metres, and the Upper Muschelkalk which is situated about 800 metres deeper and has a thickness of about 65 metres. The two aquifers are completely seperate; each has its own circulation system.

2.3.1 The Flow System of the Thermal Malm Karst

The Probable Recharging Area: The water of the thermal Malm karst flow parallel to the basin, whereby the Danube is the receiving body of water. This follows from the distribution of the potential lines (see VILLINGER this volume). Hence the recharge area for the Malm aquifer is the outcrop of the Malm in the region of Meßkirch-Sigmaringen.

Probable Emergence Area: The emergence area of the thermal karst is the region where the Danube again crosses the Malm while flowing through the Molasse basin. Also probably are regions where only a thin permeable layer seperates the Malm and the Danube. The first possible emergence area is in the area of Munderkingen. Other areas are located further downstream. Springs with increased temperatures are known.

2.3.2 Deeper Aquifers

In the Molasse basin, several sandstone layers are located beneath the Malm. These strata have a good porosity, as found in geological prospecting for petroleum. However, these strata are not appropriate for extracting water. The second possible aquifer in the Saulgau region is the Upper Muschelkalk. The depth of the Muschelkalk is about 1500 m and its thickness about 65 m.

Tests for productivity have not been conducted.

2.4 Geophysical Investigations

Within the scope of the "Geothermal Demonstration Project Saulgau" the geophysical problem of exchange of fluids and heat in a karstic aquifer has to be solved. This requires the determination of the hydraulic and geothermal properties of the fluids. Hence the dynamics and temperature of fluids in an aquifer have to be determined. With the results of the geophysical investigations it is possible to estimate the productivity and the life time of one well or of a system of wells (doublet). Another problem is the mutual interference among several wells. This problem is a further geophysical task.

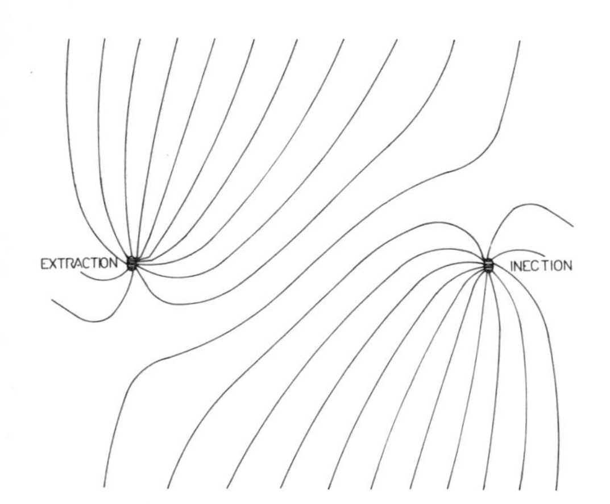

Fig. 1.

Fig. 2.

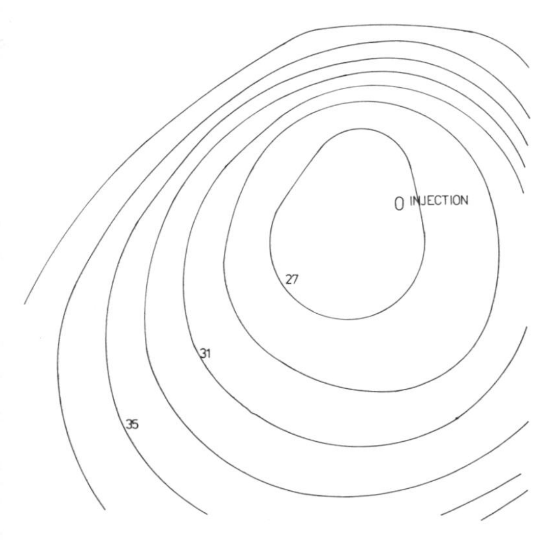

Fig. 1. The stream lines caused by a water extraction borehole and a water injection borehole. Both wells are of equal strength. The uniform flow takes place at a constant specific discharge Qo in a direction with an angle $p = -90^{\circ}$ with the x axis.

Fig. 2. The potential isolines caused as by the example given in Fig. 1.

Fig. 3. The temperature isolines around the injection borehole caused as by the example given in Fig. 1 (in 30 years).

2.4.1 Geothermics

The geothermal investigations include the developing of a computer-based model to determine the temperature in flowing fluids in aquifers. Heat transport in flowing fluids is more than heat conduction. It also includes heat transfer caused by convection. Heat conduction causes heating of the fluids by the surrounding rocks. A further effect of conduction is cooling of the confining rocks. For the differential equation describing the problem, appropriate methods of solution are to be found. These methods are analytical or numerical, and have different advantages and disadvantages.

The advantages of the analytical method is simple handling; the programming is not as difficult as it is with the numerical method. Another advantage is that the solution can be applied to different values of parameters and input data and that it clearly shows the influence of each parameter. If the shape of aquifer boundaries is irregular the analytical solution is often impossible. The types of boundary condition may vary along these boundaries. Inhomogeneous storativity and transmissivity cannot be represented in the form of analytical expressions. Hence the aquifer model has to be idealized. For example, the horizontal extension of the aquifer is assumed to be infinite and the aquifer is assumed to be isotropic.

The disadvantages of the analytical method are identical with the advantages of the numerical method and vice versa. It is possible to investigate different systems of recharging and pumping wells in uniform flow with fixed boundary and initial conditions.

The analytical example is a doublet with pumping and recharge wells. Fig. 1 shows the stream function, Fig. 2 the potential distribution and Fig. 3 the temperature around the injection well at a certain time after starting the system.

2.4.2 Geophysical Well Logging

Geophysical methods of investigation are applied on boreholes in direct exploration, and for the identification of geological formations and formation fluids encountered in boreholes.

Most important purposes of well logging in the Saulgau Project is the location and identification of fissures in the karstic aquifers (Malm and Muschelkalk). These require other surveys than porous aquifers. Several electrical, sonic and nuclear methods shall be tried with respect to this problem.

2.5 Isotopic-Hydrological Investigations

The isotropic hydrological investigations, chiefly of stable hydrogen isotopes, are suitable for examining the flow system of the thermal aquifer. These data are important as input for the mathematical models. Isotopic investigation of cores completes this test programme.

2.6 Research Boreholes Saulgau TB 2 and 3

The Saulgau TB 2 borehole will first explore the formation of the Upper Malm. The diameter is large enough to insert a pump with the high output of 100 l/s into the well. The Saulgau TB 2 well will be deepened into the deep block of the Saulgau fault. The depth is estimated at 900 m. After a positive result of TB 2, the third borehole (TB 3) will be drilled simultaneously. The main aim is to investigate the Muschelkalk aquifer. The depth is estimated at about 1600 m.

3. Utilization of the Geothermal Energy for Heating

On the basis of a presumed production rate of 100 l/s, corresponding to 360 m^3/h, and a temperature difference of 32 °C, the heat content amounts to 13 MW. If 1800 hours of peak heating consumption are assumed, the result is an output of 24 GWh/a.

Central Heating Station: In prepareation for the distribution of geothermal energy, a central station for transferring heat to an external distribution system is planned.

The central heating station will be realized in different sections, thus facilitating a step-by-step connection to several consumers.

The first section provides one heat pump system with two electrically driven rotary screw compressors. The installed heating power of this first block amounts to 4.4 MW. The district hospital, vocational school and a thermal bath can thus be served. When natural gas becomes available to Saulgau through the Fronhofen connection, it will be possible to drive the subsequent two sections – block 2 and block 3 – with gas engine heat pumps.

The total heating power of the three blocks amounts to 18 MW. If 1800 hours of peak heating consumption are assumed, the heating station produces 32.4 GWh/a.

The necessary heating power increases with falling out door temperature; see Fig. 4.

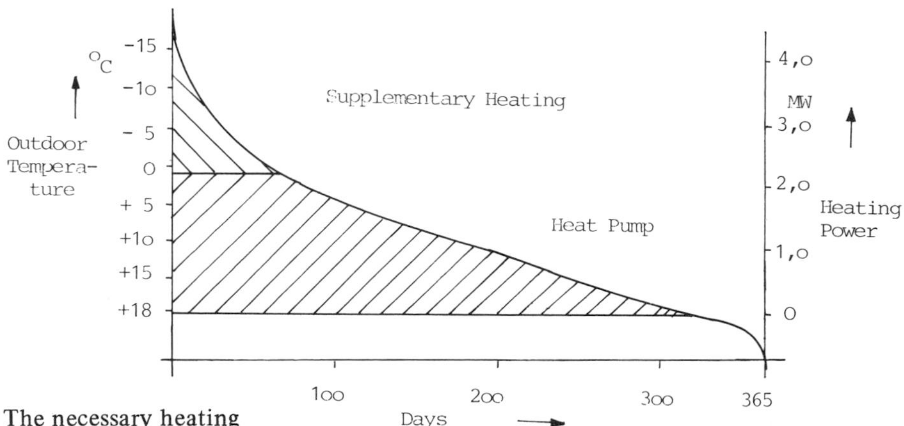

Fig. 4. The necessary heating capacity during a year.

At an outdoor temperature higher than about 0 °C, the heat pumps are able to provide the consumers with sufficient energy for heating by monovalent operation. Below this temperature the conventional heating boiler must provide for supplementary heating (bivalent operation).

If the heating system is designed for low feeding and return temperatures monovalent operation by heat pumps is possible (see Fig. 5).

The low temperature heating system requires larger radiators and heating surfaces; however, the higher investment is compensated by more economical operation.

Provided that a geothermal well produces hot water at more than 60 °C, direct heat exchange with the external heating circuit is possible. In France there are projects operating according to this principle (Melun, Creil etc.).

With this supposition a bivalent operation with both direct heat exchange and heat pumps, or a trivalent operation (Fig. 6) with conventional heating boiler and high feeding temperatures above 70 °C is economical.

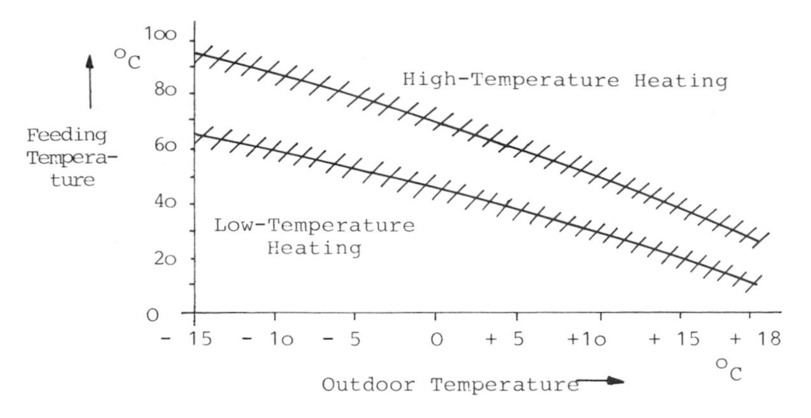

Fig. 5. The necessary heat for a consumer depending on the air temperature.

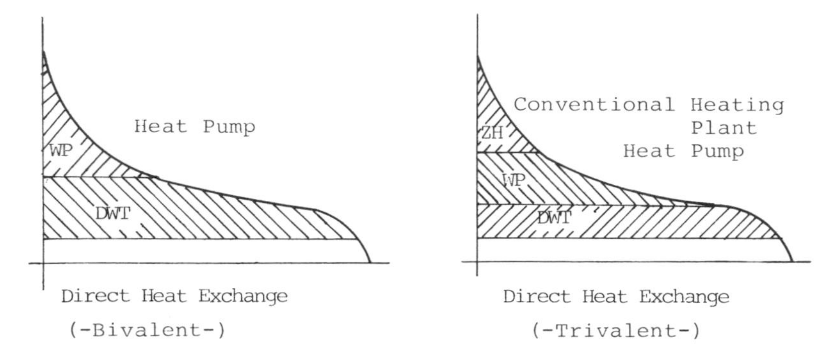

Fig. 6. The schematical heat support by a bivalent (a) or trivalent (b) working system.

3.1 Heating Station – Block 1

An underwater pump situated in the deep Saulgau 1 well (TB 1) delivers the thermal water through a thermally insulated pipe into a storage tank; see Fig. 7. A trickling installation effects the methane gas evolution. After heat exchange in the heat pumps, the water is injected into the ground at 12 °C. After this recharge, the water can be used for drinking puposes. Drain water and waste water from the thermal therapy bath undergoes heat exchange in the heat pump system too, and then flows into the urban sewage disposal system.

Two transformers (1600 kVA and 400 kVA) supply electric power. Thus the heating station supplies the thermal therapeutic bath with low and intermediate voltage too.

3.2 Key Plan

The extraction of geothermal energy from water produced at 115.2 m^3/h and 40 °C proceeds in two parallel streams of 57.6 m^3/h; see Fig. 8. One part will be utilized in the thermal therapeutic bath. The temperature of 40 °C is too high for the thermal swimming pool; hence the water will be cooled in the primary water cycle from 40 °C to 37 °C. The reduction of quantity in the swimming pool, caused by evaporation and splashing,

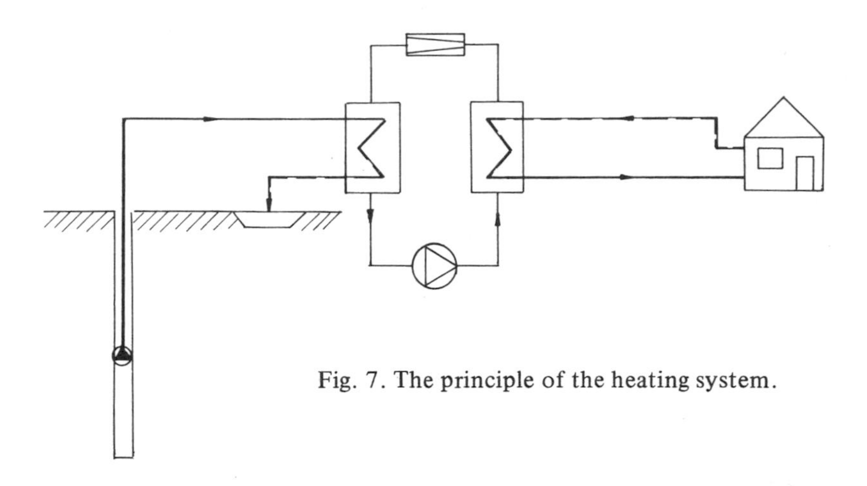

Fig. 7. The principle of the heating system.

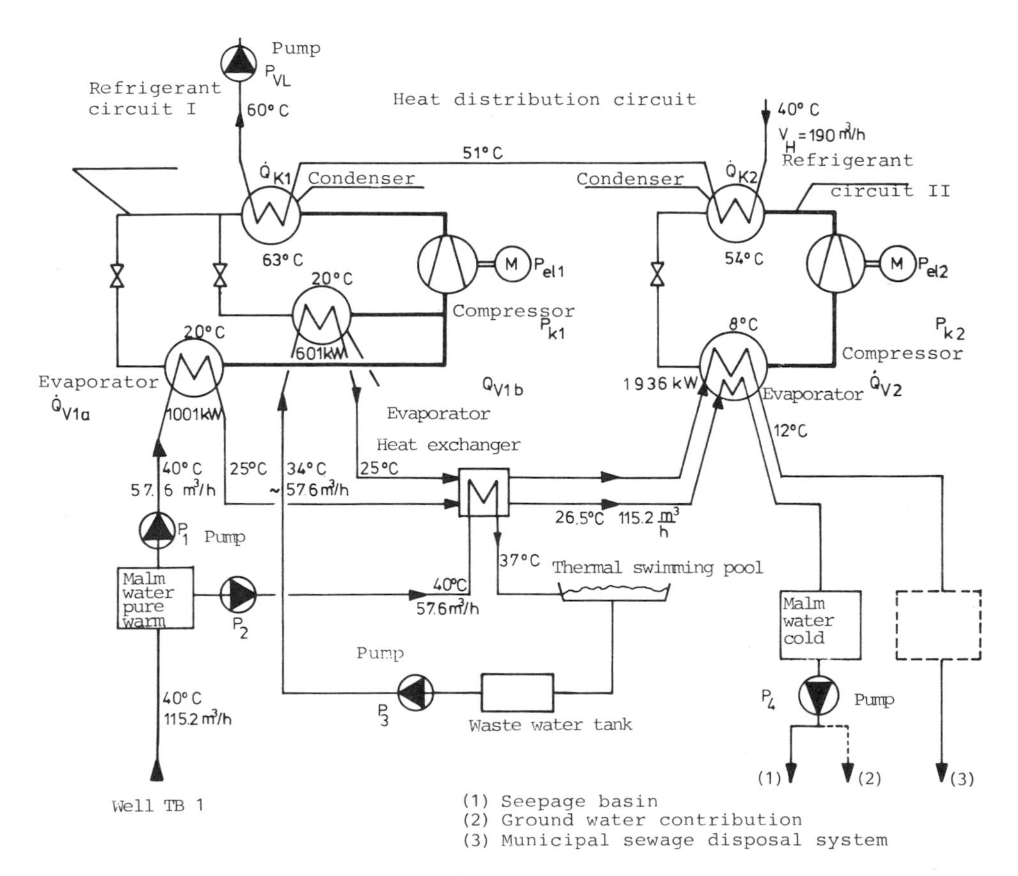

Fig. 8. The detailed flow chart for utilizing the geothermal energy in the Saulgau project.

is not considered. The loss by surface evaporation and cooling in the swimming pool causes a further decrease of temperature from 37 °C to 34 °C.

Both streams (40 °C and 34 °C) are introduced separately to the evaporators Q_v 1 a and Q_v 1 b of the first coolant circuit. After a heat exchange of 1001 kW and 601 kW in the evaporators, both warm water streams are cooled down to 25 °C and are supplied separately to the heat regenerator described for tempering the swimming pool. The warm water streams (26.5 °C) coming out of the heat regenerators transfer a thermal output of 1936 kW in the evaporator Q_v 2 to the coolant circuit II. This water is discharged at an output temperature of 12 °C to the groundwater and the urban sewage disposal system.

The evaporation temperature in the coolant circuit I is 20 °C, that in the coolant circuit II, 8 °C.

The rotary screw compressor PK 1 in the coolant circuit I increases the temperature from 20 °C to 63 °C. The rotary screw compressor PK 2 connected in parallel in the coolant circuit II increases the temperature from 8 °C to 54 °C.

The electric power consumption of the motors which drive the cimpressors amounts to 453 kW and 599 kW.

In the consumer circuit about 190 m³/h will be circulated. Water from the feedback will be warmed in the condenser QK 2 of the coolant circuit II from 40 °C to 51 °C and in the condenser QK 1 of the coolant circuit I from 51 °C to 60 °C. In this process, heat will be exchanged at rates of 2422 kW and 1970 kW.

The ratio of the heat to the electric input power of the propelling engines yields the coefficient of performance:

$$\epsilon_{\Sigma} = \frac{QK}{Pel} = \frac{2422 + 1970}{453 + 599} = \frac{4392}{1052}$$

$$\epsilon_{\Sigma} = 4.17$$

Three water pumps (3 kW, 2 x 4 kW) drive the volume of water through the evaporators and heat exchangers. The fourth pump delivers the cold water to the drain well as ground water.

3.3.1 Compressor

To compress the refrigerating medium one rotary screw compressor will be used for each circuit. Each compressor has one pair of rotors with an asymmetric profile.

The seal of the motor shaft is a cooled, pressure and vacuum sealed axial face seal. The compressor I is directly coupled with the electric motor but a flanged spur gear is interconnected.

A distributing regulator makes possible the output regulation and the start without load.

Technical Data:

	Compressor I		Compressor II
Refrigerating capacity	1602 kW		1936 kW
Power consumption	453 kW	about	599 kW
Rotational speed	2950 min^{-1}		3115 min^{-1}

3.3.2 Electric Motor Drive

For driving both compressors standard 3-phase H.T. motors with medium capacity are employed.

Driving for Compressor $P_{K\,1}$:

Power consumption	453 kW
Rating	600 kW
Rotational speed	2950 min^{-1}
Form	B 3
Protective system	IP 54
Voltage (50 Hz)	6000 V

Driving for compressor $P_{K\,2}$:

Power consumption	599 kW
Rating	710 kW
Rotational speed	2950 min^{-1}
Form	B 3
Protective system	IP 54
Voltage (50 Hz)	6000 V

3.3.3 Evaporator

The maximal heat exchange with compact construction is achieved with an inside aluminium star profile. The refrigerating medium evaporates inside the pipes. The casing tubes are weldless steel tubes, the covers and reversing baffle are also fabricated of steel.

WP 1	Evaporator WV 1 a	Evaporator QV 1 b
Refrigerating power	1001 kW	601 kW
Quantitiy of water	57.6 m^3/h	57.6 m^3/h

WP II	Evaporator WV 2
Refrigerating power	1936 kW
Quantity of water	115.2 m^3/h

3.3.4 Condenser

The compressed refrigerant will be condensed in ribbed tube condensers.

The ribbed tubes are made of copper and the casing, pipe bottoms and covers are fabricated of steel.

	WP I	WP II
Refrigerating power	1970 kW	2422 kW
Quantity of water	190 m^3/h	190 m^3/h
Input temperature of water	51 °C	40 °C
Output temperature of water	60 °C	51 °C

3.3.5 Heat exchanger

The heat exchanger which is used to cool the water for the swimming pool is made of steel.

3.3.6 Control- and Monitoring System

The controlling and monitoring functions of the heat pump plant, as well as storage in case of disorder or failure, are automated. The plant will be controlled and checked by means of an indicator panel.

The control system permits automatic operation of the plant. The start and running up to speed of the motors, compressors and servo units are easily effected by pressing an activator button. Should the occasion arise, the refrigerant circuit and therefore the heating capacity of the heat pump plant can be adapted to a lower heat consumption by the consumers by means of a slide valve gear of the rotary screw compressors. An optimal load distribution for both heat pump units is thereby guaranteed. With the frequency-controlled underwater pump situated in the deep well, the primary heat flow can be controlled too, thus ensuring the correct input to the evaporator.

The monitoring system measures values for all important parts of the plant: – bearings and stator temperatures of the motors, oil pressure and oil temperature of the compressors, temperature and pressure values of the refrigerating circuits, process messages of servo units –. The instantaneous load can be viewed on a schematic flow diagramme on the indicator panel. Disorders and failures are recorded and can be displayed on the indicator panel after a machine break-down too.

3.4 Endurance – Working Life

The heat pump plant is designed to operate without servicing for about 20 000 hours. After that time, inspection and servicing is necessary; this takes about 14 days. The working life with a duty period of 6000 to 8000 hours per year is between 20 and 25 years. The plant is free of maintenance, except for changing of the lubrication oil in the journal bearings of the electric motors.

3.5 Thermal Water Pipeline from the Well to the Heat Station

From the deep well (TB 1) to the heat station a buried, insulated pipeline has been installed. Because the Geothermal Project is to be enlarged with another deep well (TB 2) later, the pipeline was planned with two diameters: DN 150/280 from TB 1 to TB 2 and DN 300/500 from TB 2 to the heat station; see Fig. 9. The length of the DN 150/280 section is 320 metres, and that of the DN 300/500 section is 500 metres. The deep wells are coupled in order to test the doublet. The distance of 400 m between the two wells is a good assumption for the geotherm model "Extraction – Injection".

3.6 Energy consumption – energy costs

The data for energy consumption and energy costs are given in Tab. 1.

The chart compares the consumer's heating costs incurred with the use of a conventional heating system and with the use of the heating station.

The first section of the heating station will save energy cost at the rate of:

Table 1. The data are referred to 1 year.

Consumer	Heating power (kW)	Service life (h)	Energy (kWh)	Oil volume (1 EL)	Natural gas volume (m^3)	Oil price (DM/a)	Gas prise (DM/a)	Elektric power consumption price
District hospital	1976.7	2900	5.731	$\cdot 10^6$	878850	947865	527310	473930
Vocational School	348.83	250	8.72032	$\cdot 10^5$	133721	144222	80233	72111
Thermal swimming pool	1395.3	3600 (300 days x 12 h)	5.023	$\cdot 10^6$	770214	830670	462130	415340
Heat pump (electric motor)	(4400.0) 1120.0	3600	4.032	$\cdot 10^6$				483840
					Σ	$1.07 \cdot 10^6$	$1.32 \cdot 10^6$	$0.48 \cdot 10^6$

Oil price: 0.60 DM/1 EL
Gas price: 0.50 DM/m^3
Electric power consumption price: 0.12 DM/kWh
(1 EL = fuel oil No. 2)

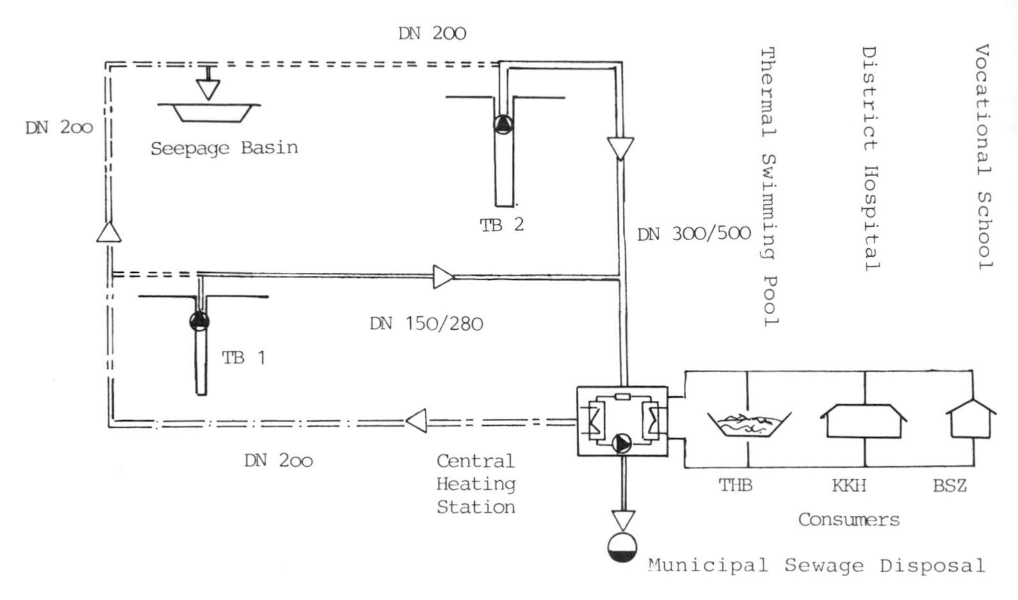

Fig. 9. The utilization of geothermal energy by energy cascading and the diameter of the pipes.

	Gas	Oil
Heating station with electric motor	478 000,- -	586 000,- - DM/a
Heating station with gas motor	607 000,- -	715 000,- - DM/a

The operation of a heat pump plant with a gas engine is more economical than the electrically driven heat pump plant, but the urban gas supply capability is not sufficient.

3.7 Cost of the Project

Building	DM 1 200 000.- -
Trickling installation	DM 50 000.- -
Heat pump plant	DM 1 300 000.- -
Transformer and switchboard	DM 160 000.- -
Installation	DM 110 000.- -
Pumps	DM 90 000.- -
Thermal water pipeline	DM 320 000.- -
Water conduit	DM 240 000.- -
Total	DM 3 470 000.- -

3.8 Overall Operating Cost per Year

The yearly operating costs of the project due to depreciation interest and maintenance amount to DM 353 000.- -; the energy cost is DM 484 000.- -. Hence the expenditures are DM 837 000.- - per year.

As compared with the heating requirement of 11.626 GWh/a, for the district hospital, vocational school and thermal therapy bath a marginal energy prince of 7.2 Pfennige per kWh is incurred.

If the urban public utilities sell the energy for 8.3 Pfennige per kWh the resulting energy price corresponds to the annual gas cost of DM 962 000. - -. In comparison with the annual oil cost of DM 1 070 000. - - a selling price of 9.2 Pfennige per kWh results. With the annual profit of 1.1 to 2.0 Pf/kWh, Saulgau will be able to build a remote heating facility. An annual investment of DM 125 000. - - to DM 233 000. - - is feasible. This provides a joint capital of DM 1 562 500. - - to DM 2 912 500. - - respectively at a rate of 8 % for interest, depreciation and maintenance. With this capital, the remote heating system can be extended over three kilometres.

Because the rise in prices for fossil fuels for the coming years cannot be estimated, an economic calculation is not feasible.

The "Geothermal Demonstration Projekt Saulgau" shall furnish proof that it is possible to utilize geothermal energy efficiently, to slow down the consumption of fossil fuel and to invest the capital thus liberated economically and operationally.

The Urach Geothermal Project, p. 395–399;
Schweizerbart'sche Verlagsbuchhandlung, Stuttgart, 1982

Energy from Geopressured Zones in the Molasse Basin

H.K. MEIDL

with 3 figures

Abstract: In southern Germany there exists a geopressured reservoir. This reservoir is quite well known from hydrocarbon boreholes. Additional tests are necessary to prove the economic feasibility of utilization of the geothermal energy and of the methane. A set of tests, to be carried out in the Endorf 2 borehole is recommended.

1. Introduction

Subsurface regions having abnormally high formation pressures are known to exist around the world. They can be formed by a number of different geological mechanisms.

In the Federal Republic of Germany one of the geopressured zones of primary interest for geothermal energy is located in Bavaria, and a lot of information is available from the hydrocarbon exploration.

The most important and objective data have been obtained from an area called "Wasserburger Trog". Geographically the Wasserburger Trog is the western part of the east Bavarian Tafelmolasse which is bounded in the north by the River Donau, in the east by the Rivers Inn and Salzach, and in the south by the Alps. The western border crosses München; see Fig. 1.

The Rupel clay marl layers are probably the initial layer of a number of hydrocarbons and they normally show large hydrostatic overpressures increasing to the south near the geostatic pressure (p_{geost} = 0.2 bar/m). Recently drilled hydrocarbon production wells and exploration wells like Darching 1, 2, 3 and 4, Höhenrain 3, 4 and 6, Teisenham 1 and Miesbach 1 confirm this assumption; see Fig. 2.

An interesting observation comes from the interpretation of the reservoir temperature, measured in 1964 in the course of pressure tests. The temperature gradient thereby decreases nearly independently of the measuring depth from the north (well Pauluszell 1 about 4.5 °C/100 m) with increasing thickness of the Molasse sediments to the south (Hohenrain 1 and 2 about 2 °C/100 m); see Fig. 3.

In 1963 the hydrocarbon wildcat Endorf 2 was drilled down to a depth of 4849 m but did not discover any commercial hydrocarbons. During the drilling operation the well spudded through the permeable and overpressured Rupel sands (4229–4264 m) which are impregnated with brine (20 g/l) and gas (95 % methane) with a gas-water ratio GWR of 6 : 1 m^3 and an initial pressure of 837 bar at 4200 m and a reservoir temperature of 115 °C.

Since 1973 the jodide brine is used for medical baths by the Jod-Thermalsole AG Endorf, Oberbayern.

Considerable interest has recently been expressed by the management of the Jod-Thermalsole AG Endorf in the so-called non-conventional sources of natural gas and geothermal energy, since this offers the possibility of a major substitute for energy obtained by conventional means.

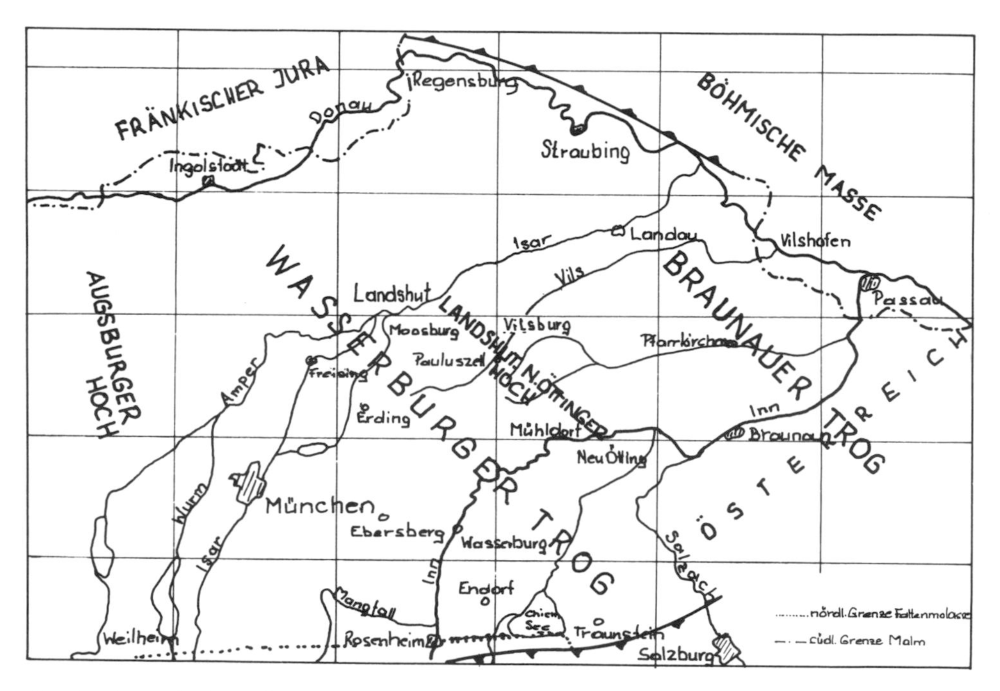

Fig. 1. General view of the east Bavarian Tafelmolasse.

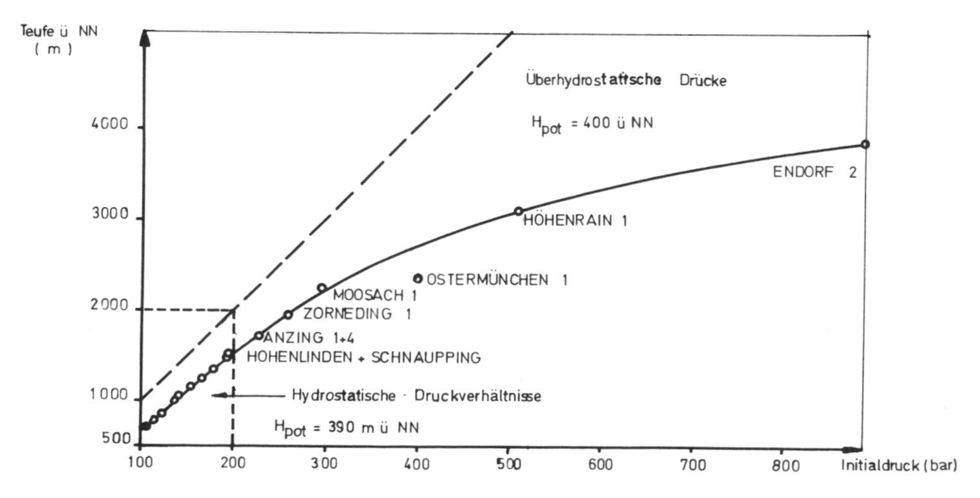

Fig. 2. Pressure-depth conditions for a number of boreholes in the Wasserburg Trog area.

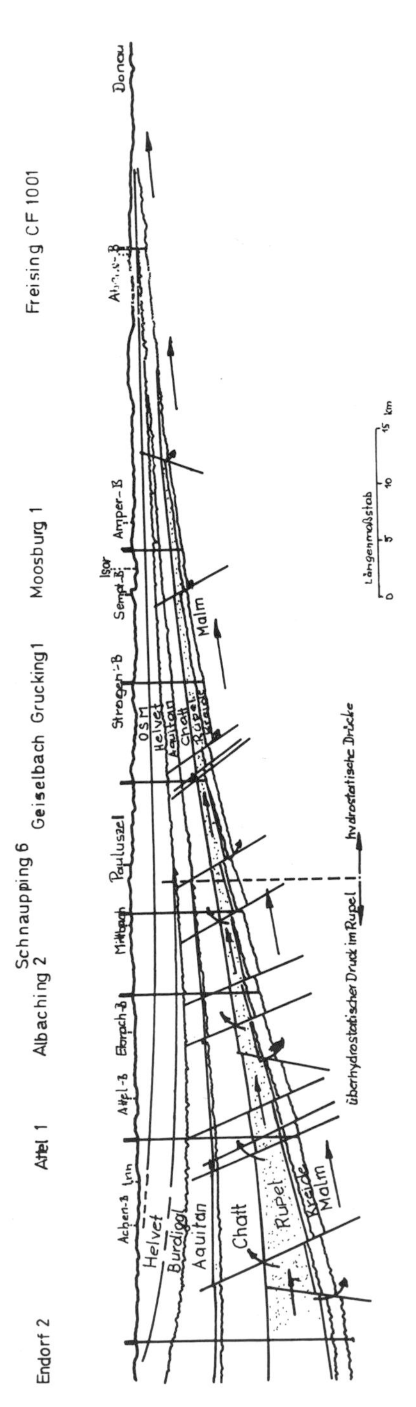

Fig. 3. North-south-cross-sectional view of the Wasserburger Trog.

2. Commercial and Technical Aspects

One of the major commercial facts – drilling a test well – has been solved in this case, and from tests conducted in 1964 a rough reservoir estimate is possible.

Thus, we know that the amount of water in place (WIP) is at least one million m^3:

$$\mathrm{WIP} = 10^6 \ \mathrm{m}^3,$$

and the gas in place (GIP) is about:

$$\mathrm{GIP} = 6 \cdot 10^6 \ \mathrm{m}^3.$$

That means that the geothermal energy is $1.2 \cdot 10^6$ kWh per 1°C from the WIP and $58 \cdot 10^6$ kWh from the GIP. However, for an economical utilization of a geopressured aquifer pilot project and for making exact cost estimates for ancillary equipment, reinjection well cost and power requirements for operation, two critical factors must be accurately known.

The first of these is the amount of methane per unit of brine that can be recovered. There is still a great uncertainty as to whether the formation water is saturated, and whether free gas is present.

The second critical factor is the flow rate and magnitude of brine production.

However this information can be provided at low cost. These primary parameters needed to project geopressured brine production rates and the additional parameters like:

- areal extent of the reservoir;
- average pressure and temperature at the well head and downhole;
- pay thickness, porosity and permeability;
- drive mechanisms; etc.

can be easily supplied.

The realisation of a measuring programme could decisively clarify the question of commercial feasibility not only for the Endorf reservoir, but also more generally for geopressured aquifers in Germany. If it can be shown that the energy recovery from the Endorf reservoir is economically feasible, there are good prospects for a lot of similar reservoirs in southern Germany.

3. Recommendation for Data Sampling

3.1 Tests

To delimit the areal extent of the reservoir, reservoir life time and pressure display, production tests have to be run. Therefore, three pressure build-up tests are necessary; for each test the Endorf 2 well be produced for 3 days and than shut in for seven days. With a bottom hole pressure instrument the pressure is continuously measured and the data automatically recorded on surface.

The next test is an isochronal test in which the well is produced for 4 hours and shut in for 10 hours. This test will be carried out three times at three different production rates. Measuring and data recording are similar to those for the pressure build-up procedure.

Before and after the first production test and after the last isochronal test a temperature log has to be run. The production will be separated into brine and gas and the volumes will be continuously measured and recorded. Between the tests up to three bottom hole samples are to be withdrawn.

3.2 Data Evaluation

The results will be field checked while testing and if necessary flow rates and shut in time can be varied. From the detailed evaluation after testing, from the pressure build-up and material balance calculation, the reservoir parameters can be determined.

From the laboratory analysis of the bottom hole samples results about phase changes of the brine under in situ conditions and surface conditions can be expected. All facts together, including temperature measurements, give the basic information for the energy balance, and the total energy content of the reservoir can be determined and alternative production concepts developed.

After a principal check of economic feasibility the alternative production concept will be proven in detail. Thus, the investment and operating cost of each concept can be determined, and the reservoir life time and energy content indicates the specific energy cost. A comparison with the cost of conventional energy sources provides guidelines for further acitvities.

The Urach Geothermal Project, p. 401–411;
Schweizerbart'sche Verlagsbuchhandlung, Stuttgart, 1982

IV. SUMMARIZED RESULTS

A Model of the Urach Geothermal Anomaly as Derived from Geophysical Investigations

A. BERKTOLD, R. HAENEL and J. WOHLENBERG

with 9 figures

Abstract: Between 1977 and 1980 the low enthalpy geothermal anomaly of Urach in SW Germany was investigated using a variaty of geoscientific methods. An attempt has been made to set up one single model of the subsurface of Urach from the results of the different methods. This clearly demonstrates the advantages of using several methods in spite of the fact that not all the results could be interpreted by the one model.
No major anomaly of any of the measured physical parameters was found below the Urach area. However, small but clear changes were detected in some of these physical parameters in the Urach region indicating a low velocity body in the region around Urach at a depth range of 5 to 25 km. The temperature of this body at a depth of 5 km is about 200 °C which is approximately 50 °C above the average temperature for the region. At a depth of 25 km the temperature is about 200 °C above the average. The primary heat source is most likely to be located in the upper mantle. The heat is probably transported through the crust mainly by large scale hydrothermal convection systems.

1. Introduction

The growing interest in geothermal resources and reserves, as part of the available alternative energy sources, has stimulated geoscientists to test and develop different geophysical methods for the exploration and exploitation of heat. A thermal reservoir may be defined as an enrichment of heat in an exploitable amount and at a workable depth. Such an enrichment may occur in areas of normal temperature-depth-distribution e.g. in the deeper part of a sedimentary basin. It may, however, also occur as an anomaly of increased temperature caused by an increased heat transport (conduction or convection) to the earth's surface. The spatial distribution of temperature in the area of a thermal reservoir may be investigated by determining the spatial variation of physical parameters of rocks (e.g. elastic properties, density, magnetization, electrical conductivity). If the temperature dependence of these physical parameters is known, the spatial distribution of temperature may be deduced from the lateral variation of these parameters in an area of interest. Anomalies of increased temperature are often correlated with young subsurface structures, recent volcanoes or active fractures. This renders a second way to investigate a thermal reservoir by determining the spatial distribution of rock units and structural elements in the subsurface to find out how they are correlated with the distribution of hot water and steam.

Up to now only a few thermal reservoirs have been investigated world-wide by methods of geoscience. One of these is the geothermal anomaly of Urach.

Two main problems had to be solved:

- to compare different methods of geoscientific exploration and to test their effectiveness in the investigation of a geothermal anomaly, and
- to deduce a model of the Urach heat anomaly by using the results of the different field measurements.

The geothermal anomaly of Urach is probably not an ideal test site for the application of geophysical exploration methods. For the Federal Republic of Germany, however, where stronger geothermal anomalies are quite rare, it may be a typical test site.

Geoscientific methods are used also during the exploitation of a thermal reservoir, e.g. for the estimation of reserves, the investigation of extension and depth range of the reservoir and the observation of temperature during the extraction of heat. The genesis of the thermal reservoir has to be investigated as an important factor for understanding the regeneration of the heat resource and predicting the useful lifetime of the reservoir. Last but not least, possible negative influences on the environment have to be observed carefully.

These few examples may give an idea of the many problems to be solved during exploration and exploitation of geothermal reservoirs by geoscientific methods.

2. Dependence of Rock Parameters on Temperature

The success of the geophysical methods in the exploration of geothermal anomalies depends considerably on the variation of the rock parameters with temperature. Therefore, these relations will be considered in more detail.

In Fig. 1 the v_p-velocity in a granite is plotted as a function of temperature. Similar graphs are known also for other rocks, e.g. from FIELITZ (1971) and KERN (1978). The v_p-velocity increases with pressure but decreases with temperature. As a result of the subsurface temperature increase to 50 °C at about 5 km depth and up to 200 °C at about 25 km depth (see Fig. 7) a velocity decrease of not more than 50 to 200 m s^{-1} can be expected. To observe such small differences in the v_p-velocity, a high resolution of the reflection and/or refraction method is required.

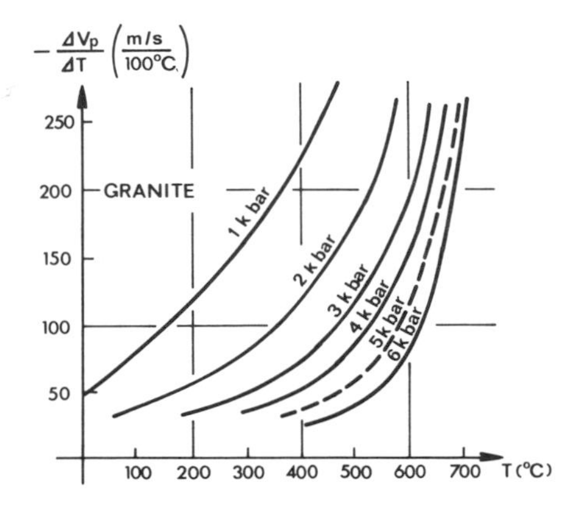

Fig. 1. The temperature dependence of the v_p-velocity per 100 °C at selected pressures (BARTELSEN et al.).

The temperature dependence of the seismic quality factor Q can also be used to analyse the geometry of the geothermal anomaly. The seismic quality factor Q is the ratio of the peak energy in a cycle of a seismic wave to the energy dissipated, and the absorption is about 268/Q db per wavelength. To recognize a change in Q a minimum size of the body crossed by the seismic waves is necessary with regard to the absorption. The peak energy decreases with increasing temperature. Therefore, the ratio of Q (T_o) for a region with normal temperature conditions T_o to Q (T_a) for a region at a higher temperature T_a is greater than one. This is shown in Fig. 2. A possibly detectable ratio of 1.5 would be caused by a temperature increase of about 100 °C.

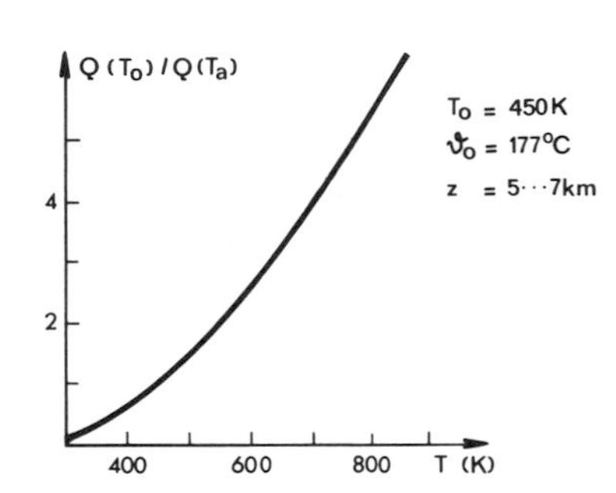

Fig. 2. The temperature dependence of the seismic quality factor Q (T_o) related to Q (T_a) for the depth interval z = 5 to 7 km and the reference temperature T_o and ϑ_o (SCHNEIDER).

The electrical conductivity of the rocks forming the sedimentary cover and the upper part of the crystalline basement depends strongly on the conductivity of the electrolytes in the rocks. The conductivity of the rock minerals is rather low at temperatures below about 400 °C. The conductivity of rocks saturated with electrolytes may be described by the following equation:

$$\sigma_o = \frac{1}{F}\sigma_w + \sigma_s$$

with σ_o = total electrical conductivity of the rock, σ_w = conductivity of the electrolyte, σ_s = surface conductivity within the saturated rock, and F = formation factor. The formation factor F depends on the porosity and the tortuosity of the rock.

If one calculates the electrolytic conductivity of rock as a function of temperature it is sufficient to calculate only the temperature dependence of the conductivity of the electrolyte. The conductivity of an electrolyte depends on several factors, such as the concentration of the electrolyte, the degree of dissociation, the electrostatic interaction

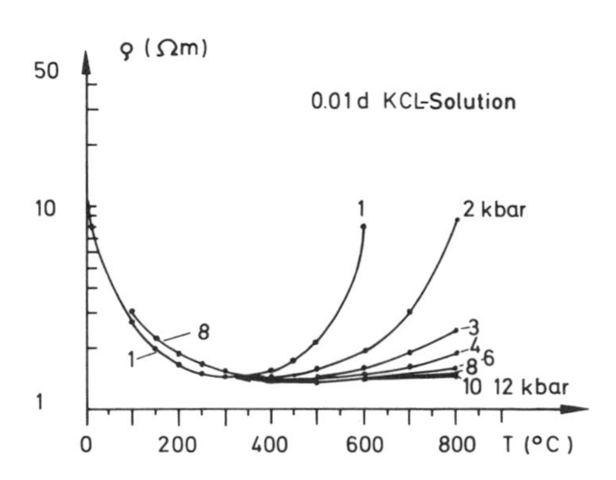

Fig. 3. The temperature dependence of the resistivity of a KCl-electrolyte at selected pressures (BERKTOLD 1980).

of the ions (mainly at high concentrations), the charge number and the mobility of the ions. The degree of dissociation, the interaction of the ions and the mobility of the ions depend – as a function of density, viscosity and dielectric constant of the solvent water – on pressure and temperature of the electrolyte. Density, viscosity and dielectric constant of the water depend in a different way on the temperature. As a result the conductivity of electrolytes (and of rocks filled with electrolyte too) increases strongly with temperature starting from room temperature to 300–400 °C. For temperatures > 400 °C the conductivity of electrolytes decreases with further increase in temperature (see Fig. 3).

In the middle parts of the crust where the electrolytic conduction mechanism is negligible and where the temperature normally is below 400 °C the conductivity of rocks is rather low (resistivity ρ roughly 10^4 ohm · m). For temperatures > 500 °C the conductivity of rocks increases with temperature as a result of the semiconductivity of rocks at high temperatures. The conductivity of rocks at high temperature can be described by:

$$\sigma(T) = \sum_i \sigma_{o,i} \cdot \exp(-E_i/kT),$$

with $\sigma_{o,i}$ = preexponential factor, E_i = activation energy for conduction mechanism i, k = Boltzmann constant, and T = absolute temperature.

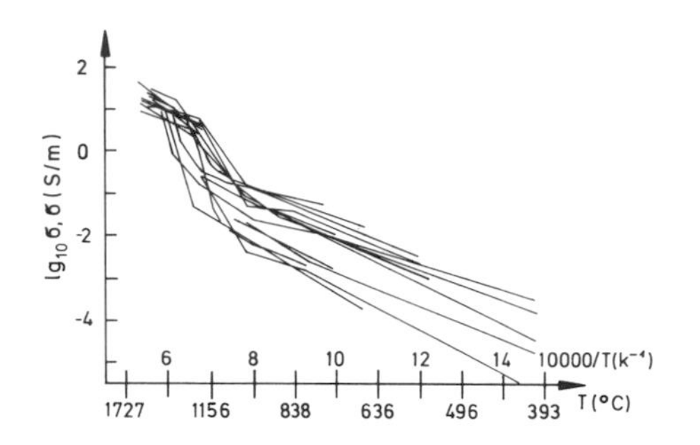

Fig. 4. The temperature dependence of the electrical conductivity of basaltic rocks (HAAK 1980).

As an example, the conductivity of basalts (a rock type which probably occurs at greater depth below the Urach area) is given in Fig. 4 as a function of the temperature.

Under normal conditions the temperature is about 100 °C at a depth of 3000 m. Within the Urach 3 borehole the temperature is about 140 °C in the same depth. The increase in temperature by about 40 °C in the same depth causes – as can be seen from Fig. 3 – an increase of conductivity of the pore electrolyte and of the rock itself by a factor of 1.4. This increased conductivity might be detected by magnetotelluric methods if the spatial distribution of conductivity around the area of increased temperature were rather homogeneous. It is known, however, that strong lateral variations of conductivity, caused mainly by the changes of rock types, occur in the area of the geothermal anomaly; these variations conceal the effect of increased temperature on the conductivity.

If a temperature increase of about 150 °C is assumed in the middle and deeper part of the crust this may cause a decrease of resistivity from some thousands of ohm · m to some hundreds of ohm · m. This slight decrease of resistivity might be detected by magnetotelluric methods if no further anomalous conductivity exists. There exist, however, several known conductivity anomalies in the uppermost part of the geothermal anomaly; these probably conceal the effect of a weak conductivity anomaly in the middle and deeper crust.

The density ρ decreases only slightly with temperature for all rocks and its temperature dependence is well known. The calculated change of the density ρ with temperature for a spatial temperature coefficient $\beta = (1-3) \cdot 10^{-5}\ K^{-1}$ using the relation $\rho = \rho_0 (1 + \beta \triangle T)$ is given in Fig. 5. The chosen temperature coefficient is typical for basaltic rocks (SKINNER 1966) under isobaric conditions, as considered for the Urach anomaly and the surroundings. A temperature increase of about 200 °C causes a density decrease of about 0.015 g cm^{-3}.

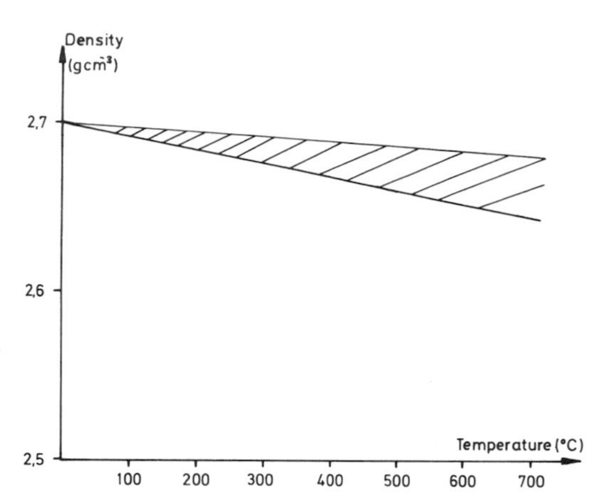

Fig. 5. The temperature dependence of the density ρ calculated with the formula $\rho = \rho_0 (1 + \beta \triangle T)$, with $\beta = (1-3) \cdot 10^{-5}\ K^{-1}$ for basaltic rocks.

The saturation magnetization shows no simple relation with temperature change. To use magnetic properties as a tool for geothermal prospection it is always necessary to determine new relations between magnetization and temperature of the different rock types. In principle, the saturation magnetization decreases with increasing temperature.

3. Interpretation of Geophysical Results and the Resulting Model

Before the Urach project was started, not much was known of the geophysical and geological situation in the area. From different boreholes in the vicinity of the Urach test site, temperature-depth functions were derived. In comparison to the regional temperature-depth function (WOHLENBERG 1979), a temperature increase of approximately 50 °C in a depth range between 2000 and 4000 m was expected from these data (Fig. 6). Considering this trend and using heat flow density data to calculate temperatures at greater depths, a temperature difference $\triangle T$ of about 150 °C could be expected between 10 km and 25 km, as compared to the regional temperature curve. By plotting the patterns of isotherms into a sketch of the geology after SCHÄDEL (Fig. 7), one gets an idea of the setting of the geothermal anomaly of Urach:

- a rather good estimate was available for the geological situation down to the crystalline basement, and
- the increase of the temperature $\triangle T$ compared to that in the undisturbed surroundings was expected to amount to 50 °C at a depth of 3000 m and approximately 150 °C between 10 km depth and the Mohorovičić discontinuity.

In Fig. 8 the location of different geophysical measurements, carried out in the Urach area, is given.

The geological situation as given in Fig. 7 was confirmed by refraction seismology after JENTSCH et al. The topography of the top basement is in agreement with the top crystalline as found in all available boreholes. The velocity of the compressional wave within

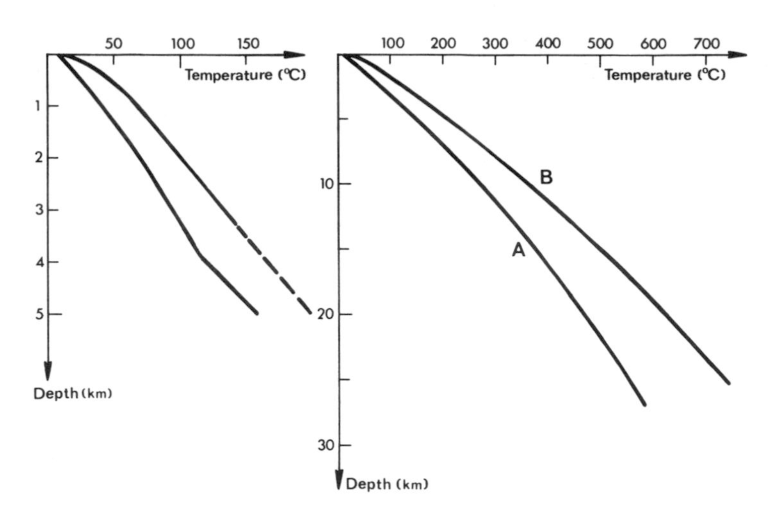

Fig. 6. The temperature versus depth. A – The reference temperature after WOHLENBERG (1979) and calculated from a heat flow density (70 mW m^{-2}), both assumed to be typical for this region. B – The measured and calculated temperatures from the heat flow density (86 mW m^{-2}) determined in the Urach 3 borehole after HAENEL & ZOTH.

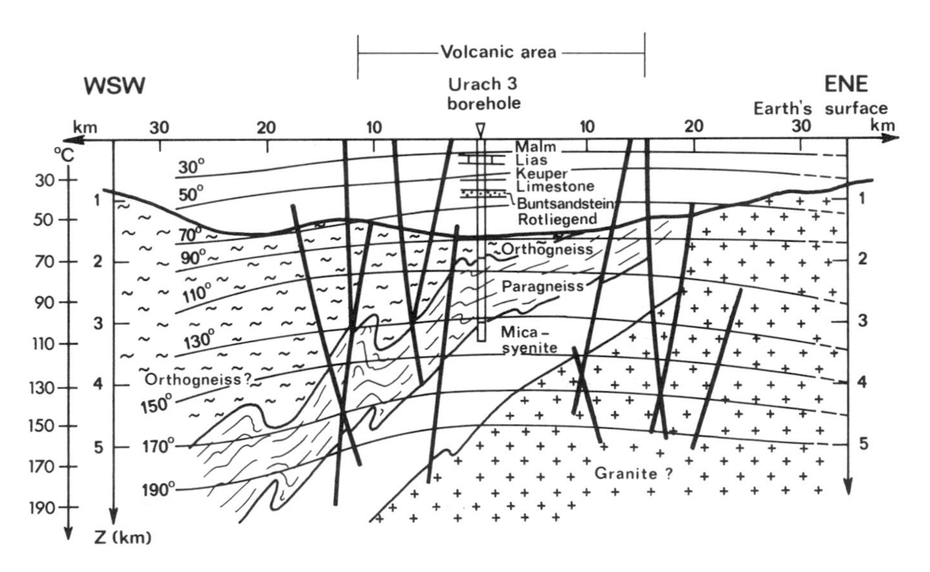

Fig. 7. The geological cross section (DIETRICH, SCHÄDEL, JENTSCH et al.) including the isotherms (ZOTH) down to 5 km; profile AB (see Fig. 8).

Fig. 8. The location of the 60 °C isotherm at 1 km depth (ZOTH), the volcanic area (MÄUSSNEST 1978), the reflection-refraction profiles U 1 and U 2 (BARTELSEN et al.), the magnetotelluric profiles as a hatched area (BERKTOLD & KEMMERLE, RICHARDS et al.). For the location of refraction profiles see JENTSCH et al.

the upper 5 km of the crust was 5.66 km s^{-1}. Variations of this velocity due to anisotropy and/or higher temperatures were not observed. According to these results, JENTSCH et al. concluded that a possible origin of the thermal anomaly of Urach must be situated deeper than 4–5 km. Changes of the velocities v_p to be expected because of temperature anomalies of 50 to 150 °C amount to approximately 50 to 150 m s^{-1}. In view of the accuracy needed and the single-coverage technique used, refraction seismology certainly has reached limits of its effectivity.

A more detailed analysis of v_p velocities using refraction seismic techniques obviously is possible if sufficient coverage of observation is applied along the profiles (BARTELSEN et al.). Thus by using a 16-fold coverage along the profile U 1 crossing the Urach anomaly SW-NE, the velocities down to a depth of 3 km could be determined with an accuracy better than 1 %. This accuracy was sufficient to observe the influence of temperature on the velocity. A low-velocity anomaly of about 2 % was detected below Urach. According to the laboratory data this corresponds to a temperature increase of about 100 °C. Velocity changes within the deeper part of the geothermal anomaly have been analysed using the stacking velocities from near angle reflection investigations. For reflection intervals 8.5 s $\leqslant t_o \leqslant$ 10 s (t_o = travel time) for example, a deviation from the average stacking velocities of about 4 % was observed. As a result of the mapping of various stacking velocities along the profile U 1 BARTELSEN et al. postulated the existence of a body with lower velocities between the upper 5 km of the earth's crust and the Moho discontinuity.

As boundary conditions for a model of the geothermal anomaly of Urach, the results of BARTELSEN et al. can be summarized as follows (see also Fig. 9):

- The uppermost 5 km of the earth's crust within the anomaly of Urach do show only a slight decrease of the seismic velocities.
- Between a depth of 5 km and the Moho discontinuity an extended region, characterized by a strong velocity decrease of up to 10 % could be detected.

Results from investigations of traveltime and absorption of seismic signals which have travelled through the temperature anomaly of Urach from different focal regions (SCHNEIDER) also support the model. The analysis of traveltimes also provides evidence of lower v_p-velocities in the uppermost crystalline formations. The studies of the frequency- and

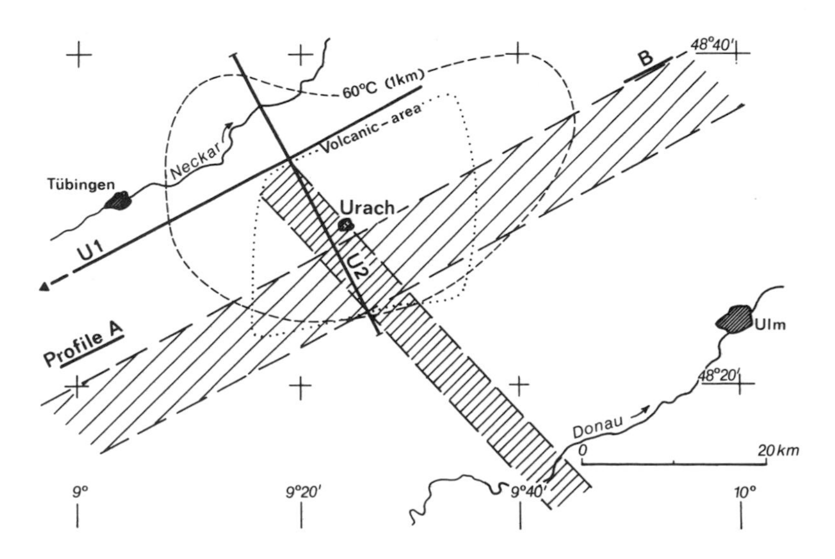

temperature-dependent absorption coefficient also indicate a temperature increase of about 100 °C in the crustal region.

The main result of magnetotelluric and geomagnetic depth sounding measurements (BERKTOLD et al., RICHARDS et al.) is the existence of a structure of increased electrical conductivity below the volcanic area and the central part of the geothermal anomaly. Model calculations show that the depth range of maximum increase of conductivity is the middle and lower part of the sedimentary cover and the upper part of the crystalline basement (about 150–4000 m depth). This result corresponds quite well with the results of measurements performed in the Urach 3 borehole (WOHLENBERG) and with the results of 3 Schlumberger soundings carried out near the Urach 3 borehole (BLOHM). Both results indicate a low resistivity of about 20 ohm · m averaged over the sedimentary cover and a resistivity of some hundreds of ohm · m in the upper part of the basement. These rather low resistivities might be caused by an increased tortuosity in this depth range, whereby the pores, fissures and caverns might be filled with hot and highly conducting fluids. In the same depth range BARTELSEN et al. found a slight decrease of the P-wave velocity by about 1 %. The structure of increased conductivity strikes about NW-SE, a preferred direction, which is not found in the lateral distribution of the sediments. This direction is probably caused by a channelling of the largescale NW-SE directed induced current system within the zone of increased conductivity. Several smaller structures of increased and decreased conductivity were found within the region of the geothermal anomaly; these may be partially correlated with structures found by seismic and gravitational measurements. The lateral variation of conductivity in the uppermost few kilometres is clearly stronger inside the volcanic area and the central geothermal anomaly than in their surroundings.

In the middle and deeper part of the crust and in the upper mantle below the geothermal anomaly no clear conductivity anomaly was found. A temperature increase of 50–150 °C in this depth range would reduce the resistivity from several thousands of ohm · m to several hundreds of ohm · m. Such a small increase of conductivity is in

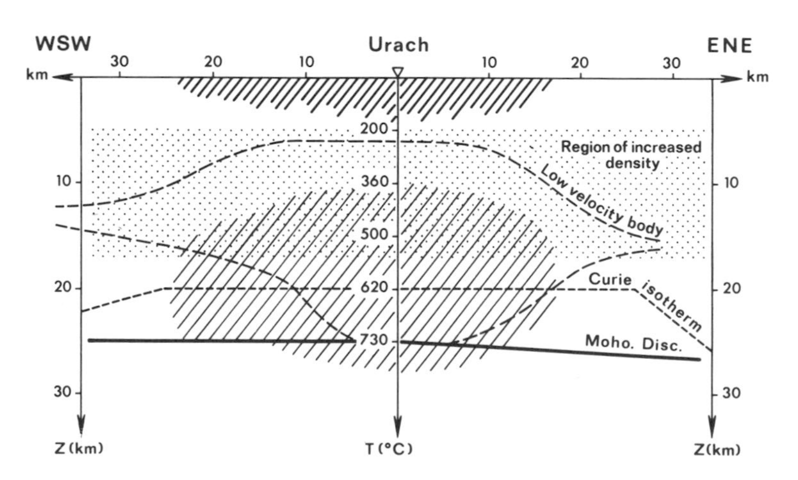

Fig. 9. The profile AB (see Fig. 8) representing a crustal model derived from the geophysical exploration.

– – – – – – – low velocity body
/ / / / / / / / / / / decreased resistivity region
- - - - - - - - - - - - Curie isotherm
. . . . increased density

agreement with the results of magnetotelluric and geomagnetic depth sounding. The location of this anomaly of slightly increased conductivity would coincide with the location of the large body of decreased P-wave velocity found by BARTELSEN et al. (see Fig. 9).

The results of the aeromagnetic survey (PUCHER & HAHN) are not in disagreement with the model derived for the Urach geothermal anomaly from results of seismic and electromagnetic prospecting experiments. The geothermal anomaly is marked neither by the existence nor by the non-existence of magnetic anomalies: there is no clear evidence for the existence of a high temperature body within the crust. If the Curie isotherm is supposed to be located at a depth of approximately 25 km under undistrubed conditions, a temperature increase of 100 °C at this depth would cause an up-doming of the Curie isotherm by about 5 km; see Fig. 9. In view of the observed magnetic field, an anomaly caused by such an up-doming of the Curie isotherm cannot be excluded. An up-doming of the Curie isotherm to a depth of 20 km would be in agreement with the model of the Urach geothermal anomaly.

The integration of results from the interpretation of gravitational measurements (MAKRIS et al.) into the model of the geological structure of Urach is difficult. The geothermal anomaly of Urach is accompanied by a positive Bouguer anomaly of about 8 mgal. As origin of the Bouguer anomaly, an intrusive body with a positive density contrast is postulated in a depth range between 6 and 18 km. The positive density contrast obviously contradicts the observed higher temperatures which lead to a small density decrease. Size and position of the intrusion, however, agree with the observed low velocity body. The attempts hitherto made to overcome this contradiction are not satisfying. The existence of a dense body which shows at the same time low seismic velocities in the depth range discussed may be explained by a dense intrusion, reducing seismic velocities because of partial melting. Partial melting, however, does not seem very plausible in view of the observed temperatures and the age (11 to 16 million years) of the volcanic events around Urach. There is also no reasonable and sufficient evidence for the existence of extended deep-seated hot intrusive bodies. The existence of a larger amount of dense basic rocks in connection with the volcanic pipes in the Urach region also does not seem very likely. Most of the observed pipes in the area are filled with tuffs. To explain the decrease of seismic velocities, finally, deep-reaching fissured or otherwise altered rocks may be presumed in addition to the temperature effect.

4. Possible Origin of the Geothermal Anomaly

As shown in Fig. 6 the temperature increase ΔT amounts to about 40 °C at shallow depth and to about 200 °C at great depth. Following HAENEL & ZOTH, the temperature in the subsurface might be somewhat higher if the effect of the ascending water is taken into account.

As a possible origin of the temperature anomaly, heat transport by water or gas ascending from great depth may be assumed. Paths may be provided by the numerous fracture and fissure systems within the crystalline basement and the sediments as well as numerous pipes and pipe systems. Diffusion of water also cannot be excluded; it may proceed through a quasi-impermeable aquiclude of clay. The hydraulic conductivity k_f of clay or similar layers is low but not negligible under these conditions; $k_f = 10^{-9} - 10^{-10}$ m s^{-1}. For geological time spans this diffusion causes detectable geothermal anomalies, as shown by HAENEL & ZOTH.

WERNER refers to the fact that ascending water can be regarded as a vehicle for heat transport since about 100 000 years, but not be considered as a postvolcanic phenomenon. In contrast, ascending gas possibly can be assumed to be a postvolcanic exhalation from the Urach volcano, which was active about 15 Ma ago.

After BUNTEBARTH & TEICHMÜLLER the geothermal anomaly must already have existed for several million years, as indicated by the observed palaeogeothermal gradient, and must have been caused by an extensive heat source at great depth. Whether the heat is transported by water or gas has not yet been discussed by the authors.

Water ascending from great depth is also indicated by the chemical and isotopic investigations carried out by KOLLER & FRIEDRICHSEN, and BALKE et al. as well as from the geotectonical point of view by DIETRICH.

STEINWACHS explains the increased microseismic activity within the Urach anomaly as being due to percolating water or gas (steam), probably in the sediments. Temperature maps also indicate near-surface water movements, as shown by ZOTH.

Fluids with a high electrical conductivity in the upper crystalline basement are indicated by magnetotelluric measurements (BERKTOLD & KEMMERLE).

Another possibility for explaining the origin of the geothermal anomaly is given by E. VILLINGER & SCHÄDEL. They assume a heat source at shallow depth within the Lias and Malm caused by oxydation of hydrocarbons and pyrite. However, they have not yet studied whether such an assumption is realistic or not.

On the basis of the above-mentioned integrated exploration the geothermal anomaly of Urach seems to be caused by heat transport by water or gas from depth; see also chapter 3.

Since a large volcanic body is not to be expected within the crust (D. WERNER, WIMMENAUER), the heat source must be assumed to be situated in the upper mantle. A similar conclusion resulted from an investigation in the Eifel area (SCHMINCKE et al. 1980).

5. Final Remarks

More than 6 years have elapsed since the first meeting on "exploration and exploitation of geothermal energy" took place in Mayschoß/Eifel in November, 1974. During this meeting a working group of geophysicists was established as part of the FKPE (Research Committee, Physics of the Earth's Interior). This group was joined in the following years by scientists of neighbouring disciplines, in order to study the Urach geothermal anomaly in a multidisciplinary programme. The main field activity was carried out in the years between 1977 and 1980. When starting the project most of the scientists had only limited experience in the investigation of a geothermal anomaly, the geological and structural setting and its probable correlation with the Tertiary volcanism of the area. The transport of heat to the earth's surface is still a matter open to discussion; it probably is an effect of deep water circulation into the hydrothermally altered crystalline basement or, at least partly, a heating by postvolcanic gas exhalations from great depth.

The Urach geothermal anomaly may not have been the ideal object for testing our methods, since it is only a low-enthalpy anomaly. It may, however, be one of the best test sites available in the Federal Republic of Germany. On the other hand, the small magnitude of the spatial variation of physical parameters as a function of the moderately increased temperature stimulated us to develop our methods to yield higher resolution and precision. From this point of view it may have been an advantage to test and develop the methods in an area of only low enthalpy.

We are convinced that geophysical exploration techniques have contributed essential information towards the understanding of the Urach geothermal anomaly. The experiment carried out has demonstrated that no single geophysical method was powerful enough to elucidate origin, geometry and structure of this thermal reservoir. Only the integrated efforts of several geophysical teams cooperating closely with other geoscientific disciplines lead to a successful exploration.

It is to be hoped that the scientific investigations may initiate an increased commercial exploitation of geothermal energy in the future.

We wish to express our gratitude to Prof. KERTZ, Braunschweig, for his stimulating activities when initiating the programme.

The project has been supported essentially by the German Ministry of Research and Technology, the Commission of the European Community, the German Research Foundation, and the community of Urach. We wish to thank the administrative members of these groups who were responsible for the performance of the Urach geothermal project.

References

(References cited without year of publication are taken from this volume)

Berktold, A. (1980): Electrical conductivity of pure salt solutions and natural waters. – Landolt-Börnstein V/1, Kap. 5.3 (in press).

Fielitz, K. (1971): Untersuchungen zur Temperaturabhängigkeit von Kompressions- und Scherwellengeschwindigkeiten in Gesteinen unter erhöhtem Druck. – Diss. Techn. Univ. Clausthal-Zellerfeld, S. 1–113.

Haak, V. (1980): Relations between electrical conductivity and petrological parameters of the crust and upper mantle. – Geophysical Surveys, (**4**): 57–69.

Kern, H. (1978): The effect of high temperature and high confining pressure on compressional wave velocities in Quartz-bearing and Quartz-free igneous and metamorphic rocks. – Tectonophysics, **44**: 185–203.

Mäussnest, O. (1978): Karte der vulkanischen Vorkommen der Mittleren Schwäbischen Alb und ihres Vorlandes (Schwäbischer Vulkan) 1 : 100 000. – Landesvermessungsamt Baden-Württemberg, Stuttgart.

Schmincke, H.U., Risse, R., Wörner, G., Boogard, P.v.d. & Viereck, L. (1980): Geothermal potential of late Quaternary East Eifel Volcanic Field. – Proceed. 2. Int. Sem. on Results of EC Geothermal Energy Research, Strasbourg, 4–6 March 1980, ed. A.S. Strub & P. Ungemach. – Reidel Publ. Comp. Dordrecht/Holland, London/ England, p. 99–108.

Skinner, B.J. (1966): Thermal expansion. – In: Handbook of Physical Constants, The Geol. Soc. Am. Mem. 97.

Wohlenberg, J. (1979): The subsurface temperature field of the Federal Republic of Germany. – Geol. Jb., **E 15**: 3–29.

The Urach Geothermal Project, p. 413–419;
Schweizerbart'sche Verlagsbuchhandlung, Stuttgart, 1982

Aspects of Utilizing Energy within the Urach Temperature Anomaly Urach and the Surroundings

R. HAENEL

with 2 figures and 2 tables

Abstract: In the Federal Republic of Germany geothermal energy is used for heating purposes in a few cases; this is also the case in Urach.
The results presented in this volume by the geothermal working group indicate the possibility of an economical utilization of hot water from the Upper Malm (Jurassic) and probably to a limited extent also from the Middle Triassic (Tertiary) within the Swabian Alb and the western part of the Molasse basin. The water can be used for heating, provided that the consumer is located nearby and that the heat can be offered as basic heat with a high load factor. The chance of economical utilization increases if the water temperature of the utilized water can be lowered down to the mean annual air temperature, by energy cascading and/or application of heat pumps.
About 15 communities around Urach have been encouraged to think about the application of geothermal energy for heating purposes. Three of them have already started with the exploitation.
Economical heat extraction from dry rocks, from a dry borehole or a fracture, is possible only under limited conditions. The extraction of heat from shallow boreholes (about 100 m in depth) by means of heat pumps is already offered by a few companies. Heat extraction from deep boreholes is recommendable only if the borehole has been drilled for other purposes and can be subsequently reused at low cost, and if favourable conditions with regard to the consumer are given. Economical heat extraction from hot dry rocks is not yet possible.

1. Introduction

At present geothermal energy is utilized only on a small scale in the Federal Republic of Germany. The hot water is utilized for heating the bath houses and the water for showers by means of heat exchangers in Urach at a power up to 0.4 MW (Urach 1977) and in Wiesbaden up to 0.6 MW (KLOTZ 1959). The heat is also used for swimming pools such as in Wallensen (Hameln/Bad Pyrmont) and in Oberbergen (Kaiserstuhl) (HAENEL & ZOTH 1972, CREUTZBURG 1963) and has been used for a long time for medical baths as in Baden-Baden, Wiesbaden Bad Teinach, etc. Furthermore, increasing interest is being shown for utilizing shallow boreholes down to about 100 m depth as heat exchanger for heat pumps serving small units.

A small industrial application is given by utilizing geothermal water in oil fields (e.g. Landau, Upper Rhine Graben). The objective is to increase the oil production by increasing the oil viscosity, which depends on the temperature.

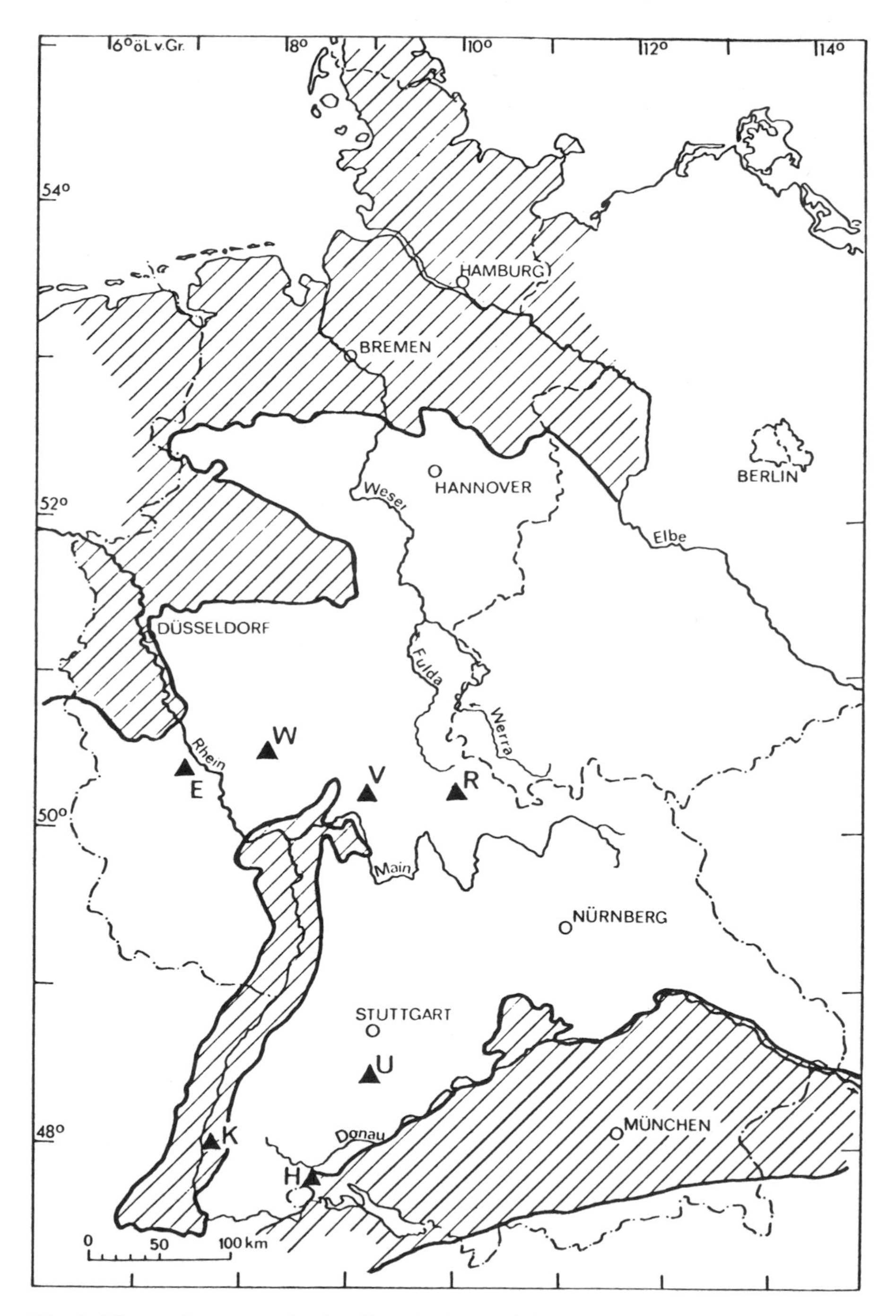

Fig. 1. The sedimentary basins (hatched areas) in which hot water can be expected for heating purposes, and the volcanic areas (▲). E = Eifel, W = Westerwald, V = Vogelsberg, U = Urach, K = Kaiserstuhl, H = Hegau.

No natural steam is available in Germany, but hot water can be obtained, especially in the large basin structures of the Northern German Basin, the Upper Rhine Graben and the Molasse Basin in the south of Germany (see Fig. 1). Moreover, favourable conditions can be expected also in the Eifel area, in view of the last volcanic activity about 10 000 years ago (SCHMINCKE et al. 1980), and in the Swabian Alb within the carstic formation of the Upper Malm (Jurassic).

The objective of the Urach project with regard to geothermal energy is twofold:

- to check the possibility of utilizing ground water for heating purposes; and
- to study the hot rock fracturing technique with regard to extracting heat in the near future.

2. Heat Extraction from Water-Bearing Zones

Four water-bearing zones are known from the Urach 3 research borehole, as well as the boreholes Urach 1 and Urach 2 already used (DIETRICH); see Tab. 1.

Table 1. The water-bearing zones from the Urach 3 borehole.

Formation	Depth range (m)	Hydrostatic water level below surface (m)	Temperature (°C)	Salinity ($g\,l^{-1}$)	Yield ($l\,s^{-1}$)
Upper Malm	2 – 18			1	
Middle Triassic (Upper Muschelkalk)	542.2– 871.3	160	58		12.6+)
Transition zone Lower Muschelkalk/ Bunter	834 – 871.3	220	61	26.6	~ 1.3
Fissured zone in the basement	1759.6–1779.9	220	102	60.9	~ 3.3

+) limitation given by pump capacity

The two deep water-bearing zones are of less interest because of the low yield. The more suitable water-bearing zones are the carstic Upper Malm and the Middle Triassic. This is more or less valid for the Swabian Alb as a whole as well as for the western part of the Molasse Basin (FRITZ, WERNER et al.). The Middle Triassic water from the Urach 3 borehole cannot be extracted, because a reciprocal effect exists with the Urach borehole 1 and 2, but outside this area there are no objections.

The relatively high mineral content of the water from the Middle Triassic requires a reinjection into the subsurface after heat extraction, preferably into the same horizon. That means, that a heat station also requires an injection well, besides the extraction well. The system is then called a dublette. The cost of such a dublette is so high that the system will not work economically for satisfying the total heat demand, as is shown in Fig. 2.

In the case that the heat is used only as basic heat (and if heat pumps are included where necessarly) during the entire year and the peak demand is satisfied by a conventional heating system (see Fig. 2), the dublette can be operated economically. The economical conditions can be improved if a multiple application, called energy cascading is possible. For example, in the first step the heat is used for house heating with the inclusion of a heat pump, in the second step for swimming pools and medical baths, and finally for soil heating in a greenhouse. An additional possibility of improvement is given by the utilization of radiators with low input temperatures, as constructed in France, and/or floor heating systems.

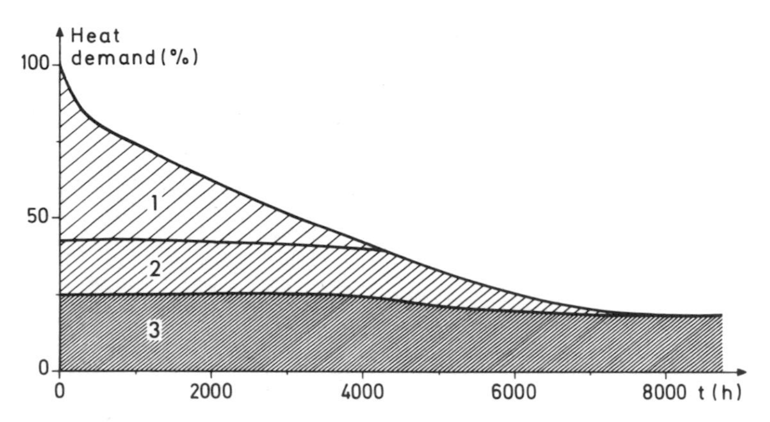

Fig. 2. The total hatched area shows the heat demand for house heating during one year (= 8760 hours).
1 – peak demand covered by conventional heating system, 2 – heat demand for the main heating period covered mainly by heat pumps, and 3 – geothermal heat used by heat exchangers as basic heat.

In the past the limits for economical use of water for heating purposes have been given by aquifer transmissivity, yield and temperature (DELISLE et al. 1974, HAENEL 1977). The data are related to the total heat demand of consumer, as shown in Fig. 2. With regard to the utilization of basic heat, the given limitations are not valid. The lowest limitations cannot be presented, because it is a function of several local parameters, such as water quality, temperature, yield, aquifer depth, consumer and local market price. In any case, the consumer must be close to the borehole. An idea of economical utilization may be obtained from the characteristic data of dublettes working economically in France; see Tab. 2. From this table the minimal flow rate seems to be about 25 l s^{-1} or somewhat less for a temperature range of 30 to 70 °C.

Within the Swabian Alb and the Molasse Basin the expected yield form the carstic Middle Triassic is about 5 to 10 l s^{-1} (WERNER 1980). The flow rate can be increased by a factor of up to three by a chemical treatment. Therefore, under favourable conditions an economical utilization of the water from the Middle Triassic seems to be feasible.

The most promising water-bearing zone is the carstic Upper Malm, which may yield up to 100 l s^{-1} of water with low mineral content. Therefore, an injection borehole is not necessary. In spite of the low temperature at shallow depth an economical utilization can be expected if heat pumps and energy cascading are applied (FRITZ). This must be true even for greater depths.

Exact data about economic heat extraction from the subsurface will be prepared by the Commission of the European Community and presented as an Atlas of Geothermal Energy Potential indicating the geothermal resources and geothermal reserves.

Table 2. Economically working heating plants in France.

Location	Formation	Depth (m)	Temperature (°C) Extraction	Injection	Yield ($l\,s^{-1}$)	Salinity ($g\,l^{-1}$)	Remarks
Blagnac-Ritouret	Inframolasse	1600	60	?	12	1.2	no reinjection
Creil	Dogger Layer	1650	57	10	30–40	29	2 dublettes, 3 heat pumps
Luxeuil-les-Bains	?	?	62	33	3.5		
Maison de la Radio	Albian	600	27	7	?		6 heat pumps
Mee-Sur-Seine	Dogger	1690	72	~18	55	13	3 heat pumps, artesian water
Melleray	Trias	1620	68	28	42–55	36	no heat pump
Melun l' Almont	Dogger	1800	70	?	25	13	heat pumps, artisian water
Mont-de-Marsan	Senoian-Turonian	1850	59	20	30–80	1	heat pump, no reinjection
Villeneuve-La-Garenne	Dogger	1790	54	?	15	25	artesian water

In connection with the Urach project about 15 communities around Urach have been encouraged to check the possibilities of utilizing geothermal energy for heating purposes (WERNER 1980, WERNER et al. and MEIDL). Three communities (Saulgau, Bad Buchau and Aulendorf) started in the meantime to exploit the water from the Malm and Middle Triassic, respectively.

3. Heat Extraction from Hot Dry Rocks

3.1 Dry Borehole

The extraction of energy from the rock wall of a borehole is in principle possible, of course. However, it is not economical for deep boreholes because of the high drilling cost. For shallow boreholes down to 50 m/100 m depth, the system can be operated economically in conjunction with a heat pump, and is already being offered by some companies and being applied at a few places. The higher the ground water table is, the better the system works.

From a theoretical point of view, the already (free of charge) existing Urach 3 research borehole can be used economically for hot water circulation, as shown by HAENEL. Hence, if a borehole already drilled for other purposes can be reused free of charge, the possibility of economical utilization should be checked.

3.2 Fracture

In the test field of Urach it has been shown that it is possible to create a single-borehole system similar to the two-borehole system in Los Alamos (USA) or other places (BATCHELOR et al. 1980, CORNET 1980, KAPPELMEYER & RUMMEL 1980). With regard to the required high flow rate (requires large diameter of tubing and annulus) and the large fracture size, however, a two-borehole system seems to be more realistic.

The heat extraction from the fracture created in the Urach 3 research borehole was low. The reason is that the temperature difference between the injected and extracted water at the fracture inlet and outlet, respectively, is small because of the low flow rate. The amount of extracted heat is higher for a sufficiently high flow rate and size of fracture. For a two-borehole system it is estimated to be of the order of $P = 30$ to $200 \, l \, s^{-1}$ and the fracture extension of the order of $F = 0.5$ to $6 \, km^2$. After a 30-year period this leads to $N = 10$ to 60 MW (BLAIR et al. 1978).

In the case of a single-borehole system it is estimated to have an influence of the order of 10 °C if the flow rate is about $25 \, l \, s^{-1}$ and the fracture size about $0.002 \, km^2$. Of course, with increasing heat extraction from the fracture the percentage of borehole heat extraction decreases.

A cost estimate for a two-borehole system has already been presented by LASL HDR Project Staff (1978). In any event a lot of problems, such as the possibility of creating large fractures, determination of location and direction of fracture by geophysical means, decreasing flow impedance and lowering drilling costs, still have to be solved for both systems. At present, therefore, energy extraction from a hot dry rock system is not economically feasible in West Germany.

References

(References cited without year of publication are taken from this volume)

Batchelor, A.S., Pearson, C.M. & Halladay, N.P. (1980): The enhancement of the permeability of granite by explosive and hydraulic fracturing. – Advances in European Geothermal Research, D. Reidel Publ. Comp., Dordrecht/Boston, London, p. 1009–1031.

Blair, A.G., Tester, J.W. & Mortensen, J.J. (1976): LASL hot dry rock geothermal project, July 1, 1975–June 30, 1976. – Progress Report LA – 6525 – PR, Los Alamos Scientific Laboratory, Los Alamos New Mexico, USA.

Cornet, F.H., (1980): Analysis of hydraulic fracture propagation, a field experimentation. – Advances in European Geothermal Research, D. Reidel Publ. Comp., Dordrecht/Boston/London, p. 1032–1043.

Creutzburg, H. (1963): Geothermische Messungen in Oberbergen. – Report, NLfB Hannover, Archive No. 5614.

Delisle, G., Kappelmeyer, O. & Haenel, R. (1975): Prospects for geothermal energy for space heating in low-enthalpy areas. – Proceed. 2. UN Symposium on the Development and Use of Geothermal Resources, U.S. Government Printing Office, Washington, 2283–2289.

Haenel, R. (1977): Betrachtungen zum Temperaturfeld im Oberrheingraben. – Report, NLfB Hannover, Archive No. 78 941.

Haenel, R. & Zoth, G. (1972): Messungen der Bodentemperatur in Wallensen (Landkreis Hameln – Bad Pyrmont). – Report, NLfB Hannover, Archive No. 8652.

Kappelmeyer, O. & Rummel, F. (1980): Investigations on an artificially created frac in a shallow and low permeable environment. – Advances in European Geothermal Research, D. Reidel Publ. Comp., Dordrecht/Boston/London, 1044–1053.

Klotz, E. (1959): Wiesbadener Thermalquellen. – Heilbad und Kurort, p. 178–180, Flöttmann, Gütersloh.

LASL HDR Project Staff (1978): Hot Dry Rock geothermal energy development project, annual report, fiscal year 1977. – LA – 7109 – PR, Los Alamos Scientific Laboratory, Los Alamos, New Mexico, USA.

Schmincke, H.-U., Risse, R., Wörner, G., Bogaard, P.v.d., & Viereck, L. (1980); Geothermal potential of late quaternary east Eifel volcanic field. Advances in European Geothermal Research, D. Reidel Publ. Comp., Dordrecht/Boston/London, p. 109–112.

Urach (1977): Presseinformation. – Kultur und Verkehrsamt der Stadt Urach vom 20.3. 1977.

Werner, J. (1980): Großprojekt hydrogeothermische Energienutzung Oberschwaben – Schwäbische Alb. – Arbeitsvorlage, Geol. Landesamt Baden-Württemberg v. 19.2.1980.